Naturally Occurring Chemicals against Alzheimer's Disease

Naturally Occurring Chemicals against Alzheimer's Disease

Edited by

Tarun Belwal
College of Biosystems Engineering and Food Science,
Zhejiang University, China

Seyed Mohammad Nabavi
Baqiyatallah University of Medical Sciences, Iran

Seyed Fazel Nabavi
Baqiyatallah University of Medical Sciences, Iran

Ahmad Reza Dehpour
Tehran University of Medical Sciences, Iran

Samira Shirooie
Kermanshah University of Medical Sciences, Iran

ELSEVIER

ACADEMIC PRESS
An imprint of Elsevier

Academic Press is an imprint of Elsevier
125 London Wall, London EC2Y 5AS, United Kingdom
525 B Street, Suite 1650, San Diego, CA 92101, United States
50 Hampshire Street, 5th Floor, Cambridge, MA 02139, United States
The Boulevard, Langford Lane, Kidlington, Oxford OX5 1GB, United Kingdom

Library of Congress Cataloging-in-Publication Data
A catalog record for this book is available from the Library of Congress

British Library Cataloguing-in-Publication Data
A catalogue record for this book is available from the British Library

ISBN: 978-0-12-819212-2

For information on all Academic Press publications visit our website at https://www.elsevier.com/books-and-journals

Publisher: Andre Gerhard Wolff
Acquisitions Editor: Erin Hill-Parks
Editorial Project Manager: Billie Jean Fernandez
Production Project Manager: Stalin Viswanathan
Cover Designer: Christian J. Bilbow

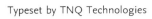

Typeset by TNQ Technologies

Contents

Contributors xix
Editors' biographies xxv

1. Introduction

Fatma Tugce Guragac Dereli and Tarun Belwal

Introduction 1
References 6

2. Alzheimer's disease: ethanobotanical studies

Swati Sharma, Sangita Sharma, Rounak Chourasia, Aseesh Pandey, Amit Kumar Rai and Dinabandhu Sahoo

Introduction 11
Medicinal plants used against AD 12
Gingko biloba 12
Curcuma longa 14
Salvia officinalis 16
Tinospora cordifolia 16
Melissa officinalis 17
Glycyrrhiza glabra 18
Centella asiatica 18
Convolvulus pluricaulis 19
Withania somnifera 20
Bacopa monnieri 20
Gaps and future challenges 21
Proper documentation of traditional medicinal system 21
Lack of information of scientific evidences on dose and ways of administration 22
Globalization effects and species lost 22
Challenges in mass cultivation 23
Conclusions 23
Acknowledgment 23
References 24

Chapter 3
Phytochemicals/plant extract against Alzheimer's

Section 3.1
Phytochemicals (pure compounds)

3.1.1. Resveratrol

Ashwani K. Dhingra, Vaibhav Rathi and Bhawna Chopra

Introduction 33
Sources 33
Natural sources 34
Pharmacokinetics 34
Chemistry 36
Resveratrol in Alzheimer's disease 37
Toxicology and adverse drug reactions 41
Clinical trials 41
References 42

3.1.2. Curcumin

Ashutosh Paliwal, Ashwini Kumar Nigam, Jalaj Kumar Gour, Deepak Singh, Pooja Pandey and Manoj Kumar Singh

Introduction 49
Alzheimer's disease, symptoms, and pathophysiology 50
Phytomedicinal properties of curcumin in relation to Alzheimer's disease 52
Role of curcumin in clinical studies/trials 53
Impact of curcumin on Alzheimer disease 54
Treatment of Alzheimer's disease 57
Conclusion 58
References 59

3.1.3. Omega 3 PUFA

Vipul Chaudhary, Ashwini Kumar Nigam, Ashutosh Paliwal, Manoj Kumar Singh, Jalaj Kumar Gour and Vimlendu Bhushan Sinha

Introduction 65
Alzheimer's disease (AD) 68
Why omega 3? 69
Importance of omega 3 71
Clinical trials 71
Conclusion 78
References 78

3.1.4. Galantamine

Vaibhav Rathi

Introduction 83
Biological and geographical distribution 84
Biological sources 84
In vitro production of galantamine 84
Chemistry 84
Pharmacology 84
Galantamine in Alzheimer's disease 86
Toxicology and adverse drug reactions 88
Conclusion 89
References 89

3.1.5. Rivastigmine

Shahira M. Ezzat, Mohamed A. Salem, Nihal M. El Mahdy and Mai F. Ragab

Introduction 93
Rivastigmine 95
Sources 96
Chemistry of rivastigmine 96
Methods of synthesis 97
Mechanism of action 98
Pharmacokinetics and pharmacodynamics 99
Drug interactions 99
Different hybrids and their actions 100
The available dosage forms of rivastigmine 100
Intranasal route 101
Injections 102
Oral 103
Inhalation 103
Transdermal 103
Tolerability 103
Conclusion 104
References 104
108

3.1.6. Quercetin

Fatma Tugce Guragac Dereli and Tarun Belwal

Introduction 109
Quercetin and its pharmacological properties 110
Preclinical studies on the anti-AD activity of que 110
Clinical studies on the anti-AD activity of que 111
Side effects and toxicological profile of que 111
Conclusion 112
References 112

3.1.7. Valerenic and acetoxyvalerenic acid

Sarita Khatkar, Amit Lather and Anurag Khatkar

Introduction 117
Occurrence 118
Chemical structure 118
Pharmacological potential of valerenic acid and acetoxyvalerenic acid 119
Role in brain-derived neurotrophic factor (BDNF) and Alzheimer's disease 119
Dementia 120
Anesthetic action 120
Sedative and anxiolytic potential 121
Insomnia 121
Physical and psychological stress 121
Antiinflammatory 121
Gastrointestinal activity 122
Inhibitory effects on CYP3A4-mediated metabolism 122
Anticonvulsant 122
Clinical studies 122
Conclusion 122
References 123

3.1.8. Huperzine A

Weaam Ebrahim, Ferhat Can Özkaya and Galal T. Maatooq

Introduction 127
Chemistry of huperzine A 128
Conclusions 133
References 135

3.1.9. Caprylic/capric triglyceride

Manjul Mungali, Navneet Sharma and Gauri

Caprylic/capric triglyceride 139
Synthesis and chemistry of caprylic triglyceride 139
Properties of capric triglyceride 140
Uses of capric triglyceride 140
Alzheimer's disease and capric triglyceride-based medical food 141
Mechanism of action and biochemistry 142
Case studies related to caprylic triglyceride-based treatment in Alzheimer's disease 143
Precautionary advice 145
Conclusion 145
References 145

3.1.10. Berberine

Merve Keskin, Gülsen Kaya, Fatma Tugce Guragac Dereli and Tarun Belwal

Introduction 147
Berberine and its pharmacological properties 148
Anti-oxidant activity of BBR 148
Anti-apoptotic activity of BBR 149
Anti-inflammatory activity of BBR 149
Anti-cholinesterase activity of BBR 149
Anti-amyloidogenic activity of BBR 150
Side effects and toxicological profile of BBR 151
Conclusion 151
References 151

3.1.11. Hypericin and pseudohypericin

Koula Doukani, Ammar Sidi Mohammed Selles and Hasna Bouhenni

Introduction 155
Description of *Hypericum perforatum* L. 156
Bioactive constituents 157
Hypericin and pseudohypericin 157
Pharmacological activities of HP 158
Effect on AChE inhibitor 159
Effects on β-amyloid peptides 160
Antiinflammatory 161
Antioxidant activity 161
Safety 161
Conclusion 162
References 162

3.1.12. Protopine

Bijo Mathew, Della G.T. Parambi, Manjinder Singh, Omnia M. Hendawy, Mohammad M Al-Sanea and Rania B. Bakr

Introduction 167
Synthesis of protopine 169
Crystal structure of protopine hydrochloride (salt of protopine) 169
Biological activities of protopine 170
Acetylcholinesterase inhibition/anti-Alzheimer's activity 170
Other pharmacological activity 171
Hepatoprotective activity 171
Antioxidant and anticancer activity 171
Concluding remarks 172
References 172

3.1.13. Spinosin

Jessica Pandohee and Mohamad Fawzi Mahomoodally

Introduction 175
Causes and mechanism of Alzheimer's disease 176
Diagnosis of Alzheimer's disease 177
Current therapies for Alzheimer's disease 178
Spinosin 178
Potential of spinosin against Alzheimer's disease 179
Future directions and conclusion 181
References 181

3.1.14. Nobiletin

Hari Prasad Devkota, Anjana Adhikari-Devkota, Amina Ibrahim Dirar and Tarun Belwal

Introduction 185
Chemistry and sources of nobiletin 187
Metabolism and distribution in the human body 187
Pharmacological effects of nobiletin in the prevention and treatment of AD 188
Antineuroinflammatory and neuroprotective activities 188
Neuroprotective effect against hydrogen peroxide-induced oxidative stress 189
Cholinesterase inhibitory activities 189
Effects in cognitive impairment 190
Effects on amyloid-β protein (Aβ) 190
Studies in humans 191
Conclusions 192
References 193

Section 3.2
Plants and their extracts

3.2.1. *Ginkgo biloba*

Ashutosh Paliwal, Pooja Pandey, Kushagra Pant, Manoj Kumar Singh, Vipul Chaudhary, Jalaj Kumar Gour, Ashwini Kumar Nigam and Vimlendu Bhushan Sinha

Introduction 199
Ginkgo biloba 200
Alzheimer's disease 203
Major constituents of *Ginkgo biloba* and their modes of action 204
Antioxidant activity 204
Protective effects on mitochondrial function 205
Antiapoptotic effect 205
Antiinflammatory effect 207
Preventive effects on amyloidogenesis and Aβ aggregation 207

Some other mechanisms 208
Possible mechanism of actions 209
Conclusion 209
References 210

3.2.2. *Panax ginseng* c.a. Meyer

Amit Bahukhandi, Shashi Upadhyay and Kapil Bisht

Introduction 217
Taxonomy and distribution 217
Bioactive constituents 217
Mechanism of action 218
Pharmacological studies 219
Dosage 220
Side effects 221
Conclusion and recommendations 221
References 221

3.2.3. *Melissa officinalis* (lemon balm)

Koula Doukani, Ammar Sidi Mohammed Selles and Hasna Bouhenni

Introduction 225
Taxonomy 226
Cultivation 227
Phytochemical profile 227
Traditional uses and pharmacology 227
Treatment 228
Acetylcholinesterase inhibitors (AChEI) 228
N-methyl-D-aspartate (NMDA) receptor antagonists 229
Alternative treatment by *Melissa officinalis* 231
Mechanism of action 231
Effect of *M. officinalis* on AchE 231
Effects of *M. officinalis* on β-amyloid (Aβ) 233
Antioxidant activity of *M. officinalis* 234
Anti-inflammatory activity 235
Safety 235
Conclusion 236
References 236

3.2.4. *Bacopa monnieri* (Brahmi)

Tanuj Joshi, Abhishek Gupta, Prashant Kumar, Anita Singh and Aadesh Kumar

Introduction 243
Taxonomical classification of *B. monnieri* 244
Mechanism of action of *B. monnieri* in Alzheimer's disease (AD) 246

Experiments on *B. monnieri* related to Alzheimer's disease (AD) using *in vitro* and *in vivo* methods 247
Clinical studies on *B. monnieri* with respect to Alzheimer's disease (AD) 250
Adverse effects associated with *B. monnieri* 252
Conclusion 253
References 253

3.2.5 *Centella asiatica*

Arvind Jantwal, Sumit Durgapal, Jyoti Upadhyay, Mahendra Rana, Mohd Tariq, Aadesh Dhariwal and Tanuj Joshi

Introduction 257
Distribution 259
Chemical constituents 259
Pharmacological activity 260
Alzheimer's disease 260
In vitro and in vivo studies 261
Clinical studies 263
Toxicity and interactions 263
Conclusion 264
References 265

3.2.6. *Rosmarinus officinalis* L.

Shashi Upadhyay, Kapil Bisht, Amit Bahukhandi, Monika Bisht, Poonam Mehta and Arti Bisht

Introduction, botanical description, and distribution 271
Bioactive compounds 273
Rosemary—drug interactions 273
Impact of rosemary on diseases of central nervous system 276
Side effects 276
Conclusion 276
References 277

3.2.7 *Valeriana officinalis* (valerian)

Manjul Mungali, Gauri, Alok Tripathi and Surabhi Singhal

Valeriana officinalis 283
Natural chemicals in *V. officinalis* 283
Medicinal uses of *V. officinalis* 285
Mechanism of action of key constituent of *V. officinalis* plant 286
V. officinalis and Alzheimer disease 287
Clinical studies 288
Safety issues 289
Conclusion 289
References 290
Further reading 291

3.2.8. *Matricaria recutita*

Fatma Tugce Guragac Dereli and Tarun Belwal

Introduction 293
***M. recutita* L** 294
Preclinical studies 295
Clinical studies 296
Side effects of *M. recutita* L. 296
Conclusion 296
References 297

3.2.9. *Galanthus nivalis* L. (snowdrop)

Devesh Tewari, Tanuj Joshi and Archana N. Sah

Background 301
Distribution and habitat associated with *Galanthus nivalis* 301
Botany of *Galanthus nivalis* 302
Phytochemistry and structure–activity relationship 303
Importance of snowdrops in neurodegenerative disorders 304
Galanthus nivalis against AD 304
Galanthus nivalis actions against other neurodegenerative diseases 308
Clinical studies on *Galanthus nivalis* with respect to Alzheimer's disease 309
Conclusion 312
References 312

3.2.10. Guggulu [*Commiphora wightii* (Arn.) Bhandari.]

Jyoti Upadhyay, Sumit Durgapal, Arvind Jantwal, Aadesh Kumar, Mahendra Rana and Nidhi Tiwari

Introduction 317
Pathophysiology of Alzheimer's disease (AD) 318
Antioxidants and Alzheimer's disease 320
Guggulu (*Commiphora wightii* (Arn.) Bhandari) 321
Chemistry 321
Antioxidant and antiinflammatory activity (preclinical and clinical studies) 322
Preclinical studies 322
Clinical studies 324
Mechanism of action 324
Safety issues 325
Drug interactions 325
Conclusion 326
References 326

3.2.11. *Lepidium meyenii*

Amit Bahukhandi, Tanuj Joshi and Aadesh Kumar

Introduction 329
Chemical composition of maca 330
Mechanism of action and pharmacological actions of *Lepidium meyenii* 331
Clinical studies of *Lepidium meyenii* 333
Toxicity profile of *Lepidium meyenii* 333
Conclusion 333
References 334

3.2.12. *Acorus calamus*

Ajay Singh Bisht, Amit Bahukhandi, Mahendra Rana, Amita Joshi Rana and Aadesh Kumar

Introduction 337
Phytochemical investigation of *Acorus calamus* L. 339
Pharmacology of *Acorus calamus* L. 342
Cognitive impairment 343
Cholinergic system and cognitive impairment: mechanism of action 343
Clinical trials and safety issues related to *A. calamus* in AD 344
Role of *Acorus calamus* in AD 345
Conclusion 345
References 346
Further reading 349

3.2.13. Tinospora cordifolia

Osama M. Ahmed

Plants and their extracts 351
Tinospora cordifolia 351
References 356

3.2.14. *Magnolia officinalis* Rehder & E.H.Wilson

Ipek Süntar and Gülsüm Bosdancı

Alzheimer disease and treatment approaches 359
Magnolia officinalis Rehder & E.H.Wilson (Magnoliaceae) 360
Neuroprotective effects of *M. officinalis* and its secondary metabolites 363
Safety of *M. officinalis* extract and its principle components 367
Conclusion and future prospects 367
References 368

3.2.15. *Collinsonia canadensis* L.

Mohamad Fawzi Mahomoodally, Preethisha Devi Dursun and Katharigatta N. Venugopala

Introduction 373
Botanical description 373
Traditional uses of *Collinsonia canadensis* 374
Collinsonia canadensis and Alzheimer's disease 374
References 376

3.2.16. *Bertholletia excelsa*

Arti Bisht, Sushil Kumar Singh and Rahul Kaldate

Introduction 379
Distribution 380
Chemical constituents 380
Mode of action of Brazilian nuts (*B. excelsa*) against Alzheimer disease 380
Clinical studies 382
Adverse effect 384
Conclusion 384
References 384

3.2.17. *Urtica diocia*

Sumit Durgapal, Arvind Jantwal, Jyoti Upadhyay, Mahendra Rana, Aadesh Kumar, Tanuj Joshi and Amita Joshi Rana

Introduction 389
Plant description 392
Chemical constituents and mechanism of action of *U. dioica* 392
Oxidative stress and Alzheimer disease 393
Inflammatory cascades and AD 395
Toxicologic studies 396
Conclusion 397
Summary 398
References 398

3.2.18. *Withania somnifera*

Vaibhav Rathi, Ashwani K. Dhingra and Bhawna Chopra

Introduction 401
Pharmacokinetics 401
Ashwagandha in Alzheimer's disease 402
Toxicological studies 404
References 404

3.2.19. *Convolvulus prostratus*

Deepak Kumar Semwal, Ankit Kumar, Ruchi Badoni Semwal and Harish Chandra Andola

Introduction 409
Pharmacological activities of *C. prostratus* 412
Anti-Alzheimer's activity 412
Activity against other CNS-related disorders 416
Effect on oxidation and oxidative stress 420
Clinical evidence to support CNS-related activities of *C. prostrates* 420
Toxicity studies 421
Conclusion 421
References 422

3.2.20. *Celastrus paniculatus*

Harikesh Maurya, Rajewshwar K.K. Arya, Tarun Belwal, Mahendra Rana and Aadesh Kumar

Introduction 425
Botanical description 426
Plant taxonomy 426
Medicinal parts 427
Physical properties 427
Chemical constituents 428
Therapeutic effects 429
Pharmacological activities 430
Benefits and uses 431
Dosage and administration 432
Safety issue 432
Conclusion 432
References 433

3.2.21. *Uncaria rhynchophylla* (Miq.) Jacks

Devina Lobine and Mohamad Fawzi Mahomoodally

Introduction 437
Phytochemistry 438
Overview of the pharmacological activity of *U. rhynchophylla* 441
Antioxidant properties 441
Antiinflammatory properties 441
Hypotensive and cardioprotective effects 442
Antiangiogenic properties 443
Anticancer properties 443
Antiviral activity 444
***Uncaria rhynchophylla* and central nervous system (CNS)-related activity** 444

Anticonvulsion 444
Anti-Parkinson's disease 445
Alzheimer's disease 445
Conclusion 447
Acknowledgment 447
References 447

3.2.22. *Alpinia officinarum*

Arpan Mukherjee, Gowardhan Kumar Chouhan, Saurabh Singh, Koustav Chatterjee, Akhilesh Kumar, Anand Kumar Gaurav, Durgesh Kumar Jaiswal and Jay Prakash Verma

Introduction 453
***A. officinarum* used for medicinal purpose** 454
Varieties around the world of *Alpinia* 454
A. officinarum application in other disease 455
A. officinarum application in Alzheimer disease 456
Working mechanism of *A. officinarum* on Alzheimer disease 457
Clinical studies and safety issues 458
Conclusion 459
Acknowledgment 459
References 459

3.2.23. *Himatanthus lancifolius* (Müll.Arg.) Woodson

Devina Lobine and Mohamad Fawzi Mahomoodally

Introduction 463
Phytochemistry 464
Overview of pharmacological properties 464
Conclusion 465
Acknowledgment 466
References 466

3.2.24. *Nelumbo nucifera*

Firoz Akhter, Asma Akhter, Victor W. Day, Erika D. Nolte, Suman Bhattacharya and Mohd Saeed

Nutrition and cognitive function 467
Glycative and oxidative stress and Alzheimer's disease 468
Constituents of *Nelumbo nucifera* 470
Role of *Nelumbo nucifera* as a potent antiglycation treatment and antioxidant 470
Preventive effect of *Nelumbo nucifera* in amyloid accumulation and cognitive impairment 473
***Nelumbo nucifera* and cognitive impairment: in vitro evidence** 474

Nelumbo nucifera and cognitive impairment: in vivo evidence 474
Nelumbo nucifera, behavioral studies, and neuropathology 475
Future direction 476
References 476

3.2.25. *Zingiber officinale*

Tanuj Joshi, Laxman Singh, Arvind Jantwal, Sumit Durgapal, Jyoti Upadhyay, Aadesh Kumar and Mahendra Rana

Introduction 481
Distribution and cultivation 481
Chemical composition 482
Role of *Zingiber officinale* in Alzheimer's disease 482
In vitro and *in vivo* studies of *Zingiber officinale* with respect to Alzheimer's disease 485
Clinical studies on *Zingiber officinale* with respect to Alzheimer's disease 487
Toxicological studies 492
Conclusion 492
References 492

Index 495

Contributors

Anjana Adhikari-Devkota, Graduate School of Pharmaceutical Sciences, Kumamoto University, Kumamoto, Japan

Osama M. Ahmed, Physiology Division, Zoology Department, Faculty of Science, Beni-Suef University, Salah Salem St., Beni-Suef, Egypt

Asma Akhter, Department of Surgery, Columbia University Irving Medical Center, New York, NY, United States

Firoz Akhter, Department of Surgery, Columbia University Irving Medical Center, New York, NY, United States

Mohammad M Al-Sanea, College of Pharmacy, Department of Pharmaceutical Chemistry, Jouf University, Sakaka, Saudi Arabia

Harish Chandra Andola, School of Environment & Natural Resources, Doon University, Dehradun, Uttarakhand, India

Rajewshwar K.K. Arya, Department of Pharmaceutical Sciences, Sir. J.C. Bose Technical Campus, Bhimtal, Nainital, Uttarakhand, India

Amit Bahukhandi, G.B. Pant National Institute of Himalayan Environment, Almora, Uttarakhand, India

Rania B. Bakr, College of Pharmacy, Department of Pharmaceutical Chemistry, Jouf University, Sakaka, Saudi Arabia; Department of Pharmaceutical Organic Chemistry, Faculty of Pharmacy, Beni Suef University, Beni Suef, Egypt

Tarun Belwal, College of Biosystems Engineering and Food Science, Zhejiang University, China

Suman Bhattacharya, Department of Surgery, Columbia University Irving Medical Center, New York, NY, United States

Ajay Singh Bisht, Himalayan Institute of Pharmacy and Research, Dehradun, Uttarakhand, India

Arti Bisht, G.B. Pant National Institute of Himalayan Environment, Almora, Uttarakhand, India

Kapil Bisht, G.B. Pant National Institute of Himalayan Environment, Almora, Uttarakhand, India

Monika Bisht, G.B. Pant National Institute of Himalayan Environment, Almora, Uttarakhand, India

Gülsüm Bosdancı, Department of Pharmacognosy, Faculty of Pharmacy, Gazi University, Etiler, Ankara, Turkey; Department of Pharmacognosy, Faculty of Pharmacy, Selcuk University, Campus-Konya, Turkey

Hasna Bouhenni, Faculty of Nature and Life Sciences, University of Ibn Khaldoun, Tiaret, Algeria

Koustav Chatterjee, Department of Biotechnology, Visva-Bharati, Santiniketan, West Bengal, India

Vipul Chaudhary, Department of Biotechnology, Deenbandhu Chhotu Ram University of Science and Technology, Sonepat, Haryana, India

Bhawna Chopra, Guru Gobind Singh College of Pharmacy, Yamuna Nagar, Haryana, India

Gowardhan Kumar Chouhan, Institute of Environment and Sustainable Development, Banaras Hindu University, Varanasi, Uttar Pradesh, India

Rounak Chourasia, Institute of Bioresources and Sustainable Development, Sikkim Centre, Tadong, Sikkim, India

Victor W. Day, X-ray Crystallography Laboratory, University of Kansas, Lawrence, KS, United States

Hari Prasad Devkota, Graduate School of Pharmaceutical Sciences, Kumamoto University, Kumamoto, Japan

Preethisha Devi Dursun, Department of Health Sciences, Faculty of Medicine and Health Sciences, University of Mauritius, Réduit, Mauritius

Aadesh Dhariwal, Department of Pharmaceutical Sciences Bhimtal Campus, Kumaun University, Nainital, Uttarakhand, India

Ashwani K. Dhingra, Guru Gobind Singh College of Pharmacy, Yamuna Nagar, Haryana, India

Amina Ibrahim Dirar, Graduate School of Pharmaceutical Sciences, Kumamoto University, Kumamoto, Japan

Koula Doukani, Faculty of Nature and Life Sciences, University of Ibn Khaldoun, Tiaret, Algeria

Sumit Durgapal, Department of Pharmaceutical Sciences, Kumaun University, Nainital, Uttarakhand, India

Weaam Ebrahim, Department of Pharmacognosy, Faculty of Pharmacy, Mansoura University, Mansoura, Egypt

Nihal M. El Mahdy, Department of Pharmaceutics and Industrial Pharmacy, Faculty of Pharmacy, October University for Modern Sciences and Arts (MSA), Giza, Egypt

Shahira M. Ezzat, Department of Pharmacognosy, Faculty of Pharmacy, Cairo University, Cairo, Egypt; Department of Pharmacognosy, Faculty of Pharmacy, October University for Modern Sciences and Arts (MSA), Giza, Egypt

Anand Kumar Gaurav, Institute of Environment and Sustainable Development, Banaras Hindu University, Varanasi, Uttar Pradesh, India

Gauri, S.M.P. Government Girls PG College, Meerut, Uttar Pradesh, India

Jalaj Kumar Gour, Department of Biochemistry, Faculty of Science, University of Allahabad, Prayagraj, Uttar Pradesh, India

Abhishek Gupta, Hygia Institute of Pharmaceutical Education and Research, Lucknow, Uttar Pradesh, India

Fatma Tugce Guragac Dereli, Department of Pharmacognosy, Faculty of Pharmacy, Suleyman Demirel University, Isparta, Turkey

Omnia M. Hendawy, College of Pharmacy, Department of Pharmacology, Jouf University, Sakaka, Saudi Arabia

Durgesh Kumar Jaiswal, Institute of Environment and Sustainable Development, Banaras Hindu University, Varanasi, Uttar Pradesh, India

Arvind Jantwal, Department of Pharmaceutical Sciences, Kumaun University, Nainital, Uttarakhand, India

Tanuj Joshi, Department of Pharmaceutical Sciences, Faculty of Technology, Kumaun University, Nainital, Uttarakhand, India

Rahul Kaldate, Department of Agriculture Biotechnology, Assam Agricultural University, Jorhat, Assam, India

Gülsen Kaya, Scientific and Technological Research Center, Inonu University, Malatya, Turkey

Merve Keskin, Vocational School of Health Services, Bilecik Seyh Edebali University, Bilecik, Turkey

Anurag Khatkar, Laboratory for Preservation Technology and Enzyme Inhibition Studies, Department of Pharmaceutical Sciences, Maharshi Dayanand University, Rohtak, Haryana, India

Sarita Khatkar, Vaish Institute of Pharmaceutical Education & Research, Rohtak, Haryana, India

Aadesh Kumar, Department of Pharmaceutical Sciences, Kumaun University, Nainital, Uttarakhand, India

Ankit Kumar, Research and Development Centre, Faculty of Biomedical Sciences, Uttarakhand Ayurved University, Dehradun, Uttarakhand, India

Akhilesh Kumar, Institute of Environment and Sustainable Development, Banaras Hindu University, Varanasi, Uttar Pradesh, India

Prashant Kumar, School of Pharmacy, Bharat Institute of Technology, Meerut, Uttar Pradesh, India

Amit Lather, Vaish Institute of Pharmaceutical Education & Research, Rohtak, Haryana, India

Devina Lobine, Department of Health Sciences, Faculty of Medicine and Health Sciences, University of Mauritius, Réduit, Mauritius

Galal T. Maatooq, Department of Pharmacognosy, Faculty of Pharmacy, Mansoura University, Mansoura, Egypt; Department of Pharmaconosy, College of Pharmacy, The Islamic University in Najaf, Iraq

Mohamad Fawzi Mahomoodally, Department of Health Sciences, Faculty of Medicine and Health Sciences, University of Mauritius, Réduit, Mauritius

Bijo Mathew, Department of Pharmaceutical Chemistry, Amrita School of Pharmacy, Amrita Vishwa Vidyapeetham, Amrita Health Science Campus, Kochi, India

Harikesh Maurya, M.G.B. Rajat College of Pharmacy, Ambedkar Nagar, Uttar Pradesh, India

Poonam Mehta, G.B. Pant National Institute of Himalayan Environment, Almora, Uttarakhand, India

Arpan Mukherjee, Institute of Environment and Sustainable Development, Banaras Hindu University, Varanasi, Uttar Pradesh, India

Manjul Mungali, Department of Biotechnology, School of Life Sciences & Technology IIMT University, Meerut, Uttar Pradesh, India

Ashwini Kumar Nigam, Department of Zoology, Udai Pratap College, Varanasi, Uttar Pradesh, India

Erika D. Nolte, Department of Pharmacology and Toxicology, Higuchi Bioscience Center, University of Kansas, Lawrence, KS, United States

Ferhat Can Özkaya, Faculty of Fisheries, İzmir Katip Çelebi University, İzmir, Turkey

Ashutosh Paliwal, Department of Biotechnology, Kumaun University Nainital, Bhimtal Campus, Bhimtal, Uttarakhand, India

Aseesh Pandey, GB Pant National Institute of Himalayan Environment and Sustainable Development, Sikkim Regional Centre, Gangtok, Sikkim, India

Pooja Pandey, Department of Biotechnology, Kumaun University Nainital, Bhimtal Campus, Bhimtal, Uttarakhand, India

Jessica Pandohee, Department of Health Sciences, Faculty of Medicine and Health Sciences, University of Mauritius, Réduit, Mauritius

Kushagra Pant, Directorate of Cold Water Fisheries Research, ICAR, Bhimtal, Uttarakhand, India

Della G.T. Parambi, College of Pharmacy, Department of Pharmaceutical Chemistry, Jouf University, Sakaka, Saudi Arabia

Mai F. Ragab, Department of Pharmacology and Toxicology, Faculty of Pharmacy, October University for Modern Sciences and Arts (MSA), Giza, Egypt

Amit Kumar Rai, Institute of Bioresources and Sustainable Development, Sikkim Centre, Tadong, Sikkim, India

Amita Joshi Rana, Department of Pharmaceutical Sciences, Kumaun University, Nainital, Uttarakhand, India

Mahendra Rana, Department of Pharmaceutical Sciences, Center for Excellence in Medicinal Plants and Nanotechnology, Kumaun University, Nainital, Uttarakhand, India

Vaibhav Rathi, School of Health Sciences, Quantum University, Roorkee, Uttarakhand, India

Mohd Saeed, Department of Biology, College of Sciences, University of Hail, Hail, Saudi Arabia

Archana N. Sah, Department of Pharmaceutical Sciences, Faculty of Technology, Bhimtal Campus, Kumaun University, Nainital, Uttarakhand, India

Dinabandhu Sahoo, Institute of Bioresources and Sustainable Development, Sikkim Centre, Tadong, Sikkim, India

Mohamed A. Salem, Department of Pharmacognosy, Faculty of Pharmacy, Menoufia University, Shibin Elkom, Menoufia, Egypt

Ammar Sidi Mohammed Selles, Institute of Veterinary Sciences, University of Ibn Khaldoun, Tiaret, Algeria

Deepak Kumar Semwal, Department of Phytochemistry, Faculty of Biomedical Sciences, Uttarakhand Ayurved University, Dehradun, Uttarakhand, India

Ruchi Badoni Semwal, Department of Chemistry, Pt. Lalit Mohan Sharma Government Postgraduate College, Rishikesh, Uttarakhand, India

Navneet Sharma, Department of Biotechnology, School of Life Sciences & Technology IIMT University, Meerut, Uttar Pradesh, India

Swati Sharma, Institute of Bioresources and Sustainable Development, Sikkim Centre, Tadong, Sikkim, India

Sangita Sharma, Institute of Bioresources and Sustainable Development, Sikkim Centre, Tadong, Sikkim, India

Anita Singh, Department of Pharmaceutical Sciences, Bhimtal, Kumaun University (Nainital), Uttarakhand, India

Deepak Singh, Department of Biotechnology, Kumaun University Nainital, Bhimtal Campus, Bhimtal, Uttarakhand, India

Laxman Singh, G.B. Pant National Institute of Himalayan Environment & Sustainable Development, Kosi-katarmal, Almora, Uttarakhand, India

Manjinder Singh, Chitkara College of Pharmacy, Chitkara University, Chandigarh, Punjab, India

Manoj Kumar Singh, Center for Noncommunicable Diseases (NCD), National Centre for Disease Control, Delhi, India

Saurabh Singh, Institute of Environment and Sustainable Development, Banaras Hindu University, Varanasi, Uttar Pradesh, India

Surabhi Singhal, Department of Biotechnology, School of Life Sciences & Technology, IIMT University, Meerut, Uttar Pradesh, India

Sushil Kumar Singh, Department of Agriculture Biotechnology, Assam Agricultural University, Jorhat, Assam, India

Vimlendu Bhushan Sinha, Department of Biotechnology, School of Engineering and Technology, Sharda University, Greater Noida, Uttar Pradesh, India

Ipek Süntar, Department of Pharmacognosy, Faculty of Pharmacy, Gazi University, Etiler, Ankara, Turkey

Mohd Tariq, G.B. Pant National Institute of Himalayan Environment & Sustainable Development, Kosi-Katarmal, Almora, Uttarakhand, India; Department of Biotechnology and Microbiology, Meerut Institute of Engineering and Technology, Meerut, Uttar Pradesh, India

Devesh Tewari, Department of Pharmacognosy, School of Pharmaceutical Sciences, Lovely Professional University, Phagwara, Punjab, India

Nidhi Tiwari, Department of Pharmaceutical Sciences, Bhimtal Campus, Kumaun University, Nainital, Uttarakhand, India

Alok Tripathi, Department of Biotechnology, School of Life Sciences & Technology, IIMT University, Meerut, Uttar Pradesh, India

Jyoti Upadhyay, School of Health Sciences, University of Petroleum and Energy Studies, Dehradun, Uttarakhand, India

Shashi Upadhyay, G.B. Pant National Institute of Himalayan Environment, Almora, Uttarakhand, India

Katharigatta N. Venugopala, Department of Biotechnology and Food Technology, Durban University of Technology, Durban, South Africa

Jay Prakash Verma, Institute of Environment and Sustainable Development, Banaras Hindu University, Varanasi, Uttar Pradesh, India

Editors' biographies

Dr. Tarun Belwal, Ph.D.

Dr. Belwal is currently working at the College of Biosystems Engineering and Food Science, Zhejiang University, China. He is actively engaged in research pertinent to food science, the potential of plant nutraceuticals, and human health, including the effect of food bioprocess techniques on its quality and other functional attributes.

He obtained his undergraduate degree in pharmaceutical science and a master's degree in biotechnology before receiving a Ph.D. from India. Before his move to Zhejiang University, he worked on using natural products as nutraceutical agents, examining their functional activities (animal model), and developing in vitro cell culture and advanced extraction techniques. For his exemplary work, the Indian government awarded him the prestigious Governor Award for Best Research for 2 consecutive years (2016 and 2017). He has successfully completed a project as principle investigator (2018–20), funded by the China Postdoctoral Science Foundation. His excellent research credentials are reflected by more than 60 peer-reviewed scientific publications and eight book chapters with over 800 citations. He is currently engaged in editing six scientific books and three special issues as guest editor. Moreover, he is serving as an associate editorial board member of *Mini-Reviews in Medicinal Chemistry* and *Current Pharmaceutical Biotechnology*. He is an active member of the European Society of Sonochemistry and was also an invited speaker at the Fourth International Symposium on Phytochemicals in Medicine and Food (Xi'an, China, Nov. 30 to Dec. 4, 2020). In addition, he is an active reviewer of several prestigious journals.

Seyed Mohammad Nabavi

Seyed Mohammad Nabavi is a biotechnologist and senior scientist at the Applied Biotechnology Research Centre, Baqiyatallah University of Medical Sciences, Tehran, Iran. His research interests are focused on natural products. To date, he has authored or coauthored over 250 publications in international journals indexed in ISI and has more than 7000 citations in published articles and more than 10 books.

Seyed Fazel Nabavi

Seyed Fazel Nabavi is a biotechnologist and senior scientist at the Applied Biotechnology Research Center, Baqiyatallah University of Medical Science and a member of Iran's National Elites Foundation. His research focuses on the health-promoting effects of natural products. He is the author or coauthor of 170 publications in international journals, 45 communications at national and international congresses, and four chapters in book series. He is a referee of several international journals. He is also in the top 1% of scientists in the world in the fields of agricultural science and pharmacology and toxicology, according to the Essential Science Indicator from Thompson Reuters ISI.

Ahmad-Reza Dehpour

Ahmad-Reza Dehpour is an Iranian pharmacologist and biomedical scientist, among the top 1% in the world, and one of the world's highly cited researchers announced by Thomson Reuters ISI. He is currently a distinguished professor of pharmacology at the School of Medicine, Tehran University of Medical Sciences.

He was born in Siahkal, Gilan province, north of Iran, in 1948. He earned his doctorate in pharmacy from Tehran University in 1972. He earned a Ph.D. in pharmacology from the Medical School of Tehran University under the supervision of Professor Frank Michal, who was a guest professor from Cambridge University living in Iran at that time. His thesis title was *Some Biophysical and Biochemical Aspects of Blood Cell Membranes*.

He began his academic career by joining Tehran University of Medical Sciences as an assistant professor of pharmacology in 1978. He was promoted to associate professor of pharmacology in 1983 and consequently to professor of pharmacology in 1993 and was awarded a distinguished professorship in 2011. He has devoted himself to teaching fundamental concepts in pharmacology to undergraduate and graduate students, also serving as an undergraduate and graduate research mentor and advisor. In addition, he was a visiting professor at the Swiss Federal Institute of Technology, Zurich from 1998 to 1999.

In addition to his academic contributions, he was general secretary of the Iranian Society of Physiology and Pharmacology from 1981 to 1986. He became a member of the Iranian Board of Pharmacology in 1999 and has been a member of the Iranian Academy of Medical Sciences since 2007. He was president of the Iranian Society of Physiology and Pharmacology from 2005 to 2007. He is also an adjunct professor at the Institute of Biochemistry and Biophysics, University of Tehran.

After his executive roles, he became the founder and director of the Experimental Medicine Research Center in Tehran University of Medical Sciences. He has supervised more than 50 Ph.D. theses and many undergraduate student theses; also, he has published more than 700 original research papers in prestigious international journals.

Beyond his scientific research, he is editor in chief of *Acta Medica Iranica* journal, published by Tehran University of Medical Sciences. He is also a member of the editorial board of *Liver International*, *World Journal of Gastrointestinal Pharmacology and Therapeutics*, *World Journal of Clinical Urology*, *World Journal of Pharmacology*, and *Journal of Family and Reproductive Health*.

Samira Shirooie

Samira Shirooie received a Pharm.D. in 2011 from Kermanshah University of Medical Sciences. After graduation, she worked as a pharmacist in a pharmacy until 2014; then, she earned a Ph.D. in pharmacology from the Medical School of Tehran University under the supervision of Professor Ahmad Reza Dehpour in 2018. She began her academic career by joining Kermanshah University of Medical Sciences as an assistant professor of pharmacology, from 2018 to date. Her current research interests are in neurodegenerative disorders, depression, addiction, hepatotoxicity, and cell culture.

Chapter 1

Introduction

Fatma Tugce Guragac Dereli[1], **Tarun Belwal**[2]
[1]*Department of Pharmacognosy, Faculty of Pharmacy, Suleyman Demirel University, Isparta, Turkey;* [2]*College of Biosystems Engineering and Food Science, Zhejiang University, China*

Introduction

Alzheimer's disease (AD) is an untreatable neurodegenerative health problem that was first diagnosed by Dr. Alois Alzheimer in 1906 (Cipriani et al., 2011).

AD is known to be the main cause of dementia and decreases the quality of life of individuals by hindering their social, physiological and psychological functioning (Singh, 2020). The three early warning symptoms of this disease are mood and behavior changes, impaired cognitive performance, and progressive memory deficits. Hallucinations, muscle spasms, restlessness, irritability, incontinence, and swallowing difficulty may occur in the later stages of AD (Singhal et al., 2012).

Multiple environmental and genetic risk factors have been confirmed for AD pathology. Genetic variations of the apolipoprotein E gene, tau protein/amyloid beta (Aβ) aggregation, acetylcholine deficiency, aging, long-term silicon/aluminum/toxin exposures, diet, alcohol and nicotine consumption, oxidative stress, inflammation, hormones, traumatic brain injury and several vascular risk factors—including hypercholesterolemia, diabetes, and hypertension—are believed to feature in the etiology of AD (Letenneur et al., 2004; Rosler et al., 1998; Shin, 1997). Different hypotheses have been put forth to explain this multifactorial disease, including cholinergic, inflammation, insulin, Aβ, tau, vascular, and gene mutation hypotheses (Fig. 1.1) (Aisen, 2008; Francis et al., 1999; Liu et al., 2019; Pansari et al., 2002; Torreilles and Touchon, 2002). The occurrence and progression of the disease have been found to be related to neuronal dysplasia, angiogenic alterations, neuroinflammation, mitochondrial damage, and decreased cholinergic transmission (Geekiyanage et al., 2012).

The early and correct diagnosis of the disease allows for effective treatment as early as possible. Increases in the cerebrospinal fluid and plasma levels of biological markers such as Aβ (1−42), and phosphorylated and total tau are important in the diagnosis (Blennow et al., 2010).

Naturally Occurring Chemicals against Alzheimer's Disease. https://doi.org/10.1016/B978-0-12-819212-2.00009-8

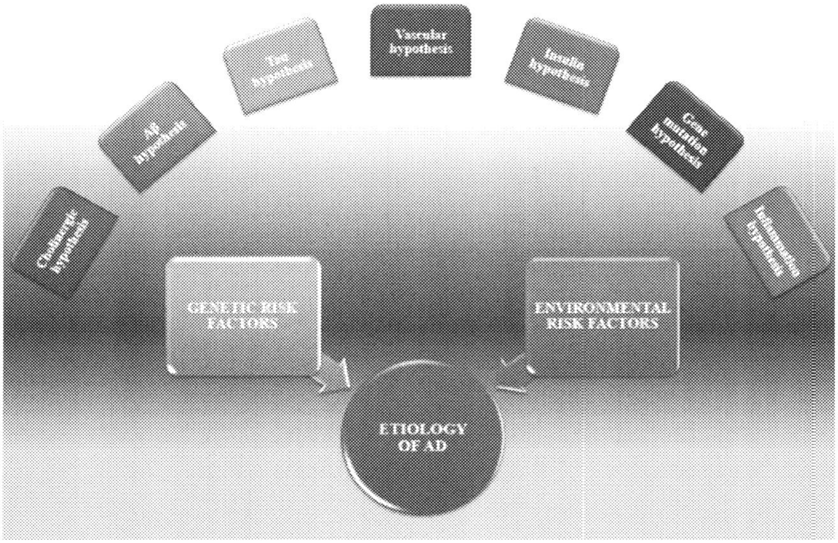

FIGURE 1.1 Etiology of Alzheimer's disease.

The incidence of this global, life-threatening health problem is high, especially for the elderly population (Cummings and Cole, 2002). Today, AD affects about 50 million individuals worldwide, and this number is estimated to triple within 30 years (Alzheimer's Disease International, 2019). Though AD has been recognized for more than a century, and despite its constantly growing global burden, AD is still irremediable (Citron, 2010). US Food and Drug Administration-approved medications are palliative and offer only temporary relief of symptoms by changing the level of neurotransmitters in the brain. These clinically available single-targeted drugs including four acetylcholinesterase inhibitors (donepezil, galantamine, rivastigmine) and one glutamatergic antagonist (memantine) are not capable of stopping or reversing disease progression, and the use of them may result in enormous side effects. Given all of these issues, there is an urgent need to find an effective and safe disease-modifying medication to overcome AD. This necessity has stimulated the search for new medications with potential clinical value (Atri, 2019; Cummings, 2018; Miranda et al., 2015).

Since antiquity, people have been attracted to the search for plant-based medications (Petrovska, 2012). The bioactive secondary metabolites contained in medicinal plants are responsible for diverse medicinal activities and have been used as raw materials in the drug industry. These natural phytochemicals can be categorized in three classes: phenolics, terpenoids, and alkaloids. This classification is based on the synthesis route or chemical structure of the compounds (Harborne, 2000; Li et al., 2020; Satish et al., 2020).

Thus far, thanks to the efforts of scientists, various secondary metabolites have been determined for multiple targets involved in AD pathology. Increasing evidence from experimental investigations has indicated that various medicinal plants could possess anti-AD activity that can counteract the different etiological causes of AD. They exert this activity by virtue of the antioxidant, antiinflammatory, neuroprotective, antiamyloidogenic, anti–tau hyperphosphorylation, anti–neurofibrillary tangle, antiapoptotic, antiamnesic, heavy metal–chelating, and anticholinesterase (anti- ChE) properties of their secondary metabolites (Tundis et al., 2018; Williams et al., 2011; Zhu et al., 2019). A list of medicinal plants/phytochemicals that have been identified to act against AD is presented in Table 1.1.

TABLE 1.1 List of the potential medicinal plants/phytochemicals identified to act against Alzheimer's disease.

Medicinal plant/ Phytochemical	Main mechanism/s of action	References
Acorus calamus L.	Antiacetylcholinesterase (Anti-AChE)	Ahmed et al. (2009)
Allium sativum L.	Antiinflammatory anti–neurofibrillary tangle, antiamyloidogenic	Chauhan (2006)
Alpinia officinarum Hance	Antioxidant, anti-AChE	Kose et al. (2015)
Angelica sinensis (Oliv.) Diels	Anti–tau hyperphosphorylation	Zhang et al. (2011)
Apigenin	Antiamyloidogenic, neurotrophic, neuroprotective	Zhao et al. (2013)
Artemisia asiatica Nakai ex Pamp.	Anti-AChE, antiamyloidogenic	Heo et al. (2000)
Bacopa monnieri (L.) Wettst	Cholinergic, neuroprotective	Uabundit et al. (2010)
Berberine	Antioxidant, antiapoptotic, antiinflammatory, anticholinesterase, antiamyloidogenic	Cai et al. (2016)
Cajanus cajan (L.) Huth	Antioxidant, choline acetyltransferase activation	Ruan et al. (2009)
Collinsonia canadensis L.	Anti-AChE	Singhal et al. (2012)

Continued

TABLE 1.1 List of the potential medicinal plants/phytochemicals identified to act against Alzheimer's disease.—cont'd

Medicinal plant/ Phytochemical	Main mechanism/s of action	References
Camellia sinensis (L.) Kuntze	Aβ-producing secretase modulatory, antiamyloidogenic	Lee et al. (2009a)
Celastrus paniculatus (Jyotishmati)	Anti-AChE, antioxidant	Alama and Haque (2011)
Centella asiatica (L.) Urban	Antiamyloidogenic	Dhanasekaran et al. (2009)
Convolvulus prostrates Forssk.	Anti-AChE	Anupama et al. (2019)
Curcumin	Antioxidant, antiinflammatory, antiamyloidogenic	Lim et al. (2001)
Dendrobium nobile Lindl	Anti—tau hyperphosphorylation, antiapoptotic	Yang et al. (2014)
Galantamine	Anti-AChE, modulator of nicotinic ACh receptors, enhancer of cholinergic nicotinic neurotransmission	Pearson (2001)
Ginkgo biloba L.	Antiamnesic, antioxidant, antiapoptotic, mitochondria-mediated apoptosis signaling inhibitory	Bastianetto et al. (2000)
Ginseng	Antiamyloidogenic, antioxidant, antiapoptotic	Chen et al. (2006), Lee et al. (2009b)
Himatanthus lancifolius (Mqell. Arg) Woodson	Anti-AChE	Seidl et al. (2010)
Hypericin	Antiinflammatory	Zhang et al. (2016)
Huperzine A	Anti-AChE, antioxidant	Little et al. (2008)
Lepidium meyenii Walp.	Antioxidant, anti-AChE	Rubio et al. (2011)
Magnolia officinalis Rehder & E.H. Wilson	Memory enhancing, antioxidant, Aβ-producing Secretase modulatory, antiinflammatory	Lee et al. (2011, 2012)

TABLE 1.1 List of the potential medicinal plants/phytochemicals identified to act against Alzheimer's disease.—cont'd

Medicinal plant/ Phytochemical	Main mechanism/s of action	References
Matricaria recutita L.	Antiinflammatory, antiamnesic, antioxidant, antiamyloidogenic	Alibabaei et al. (2014), Ionita et al. (2018)
Melissa officinalis L.	Antioxidant, antiamyloidogenic, anti-AChE	Mahboubi (2019)
Nelumbo nucifera Gaertn.	Choline acetyltransferase expression inducer, antiapoptotic	Kumaran et al. (2018), Oh et al. (2009)
Nobiletin	Anti–tau hyperphosphorylation, antioxidant, Antiamyloidogenic	Nakajima et al. (2014)
Ptychopetalum olacoides Benth. (Marapuama)	Anti-AChE, antiamyloidogenic	Figueiro et al. (2010, 2011)
Panax ginseng C.A. Mey.	Cognitive performance enhancer	Heo et al. (2011)
Panax notoginseng (Burkill) F.H. Chen ex C.H. Chow	Cholinergic neuron protector	Zhong et al. (2005)
Pinus pinaster Aiton	Antioxidant, antiapoptotic	Peng et al. (2002)
Protopine	Antiamnesic, anti-AChE	Kim et al. (1999)
Quercetin	Plaque formation and mitochondrial dysfunction reduction, anti–tau hyperphosphorylation, antioxidant	Patil et al. (2003), Paula et al. (2019), Wang et al. (2014)
Resveratrol	Antioxidant, antiamyloidogenic, antiinflammatory	Rege et al. (2014)
Rivastigmine	Anti-AChE	Williams et al. (2003)
Rosmarinus officinalis L.	Antioxidant, antiinflammatory, antiamyloidogenic, anti-AChE	Habtemariam (2016)
Spinosin	Antioxidant, antiinflammatory,	Xu et al. (2019)

Continued

TABLE 1.1 List of the potential medicinal plants/phytochemicals identified to act against Alzheimer's disease.—cont'd

Medicinal plant/ Phytochemical	Main mechanism/s of action	References
	antiapoptotic expression of neurotrophic factor	
Uncaria rhynchophylla (Miq.) Jacks	Antiamyloidogenic, antiinflammatory, anti-AChE	Jung et al. (2013), Xian et al. (2012), Yang et al. (2012)
Withania somnifera (L.) dun.	Anti-AChE, antioxidant	Kumar et al. (2011), Mahrous et al. (2017)
Zingiber officinale	Anti-AChE, antiamyloidogenic	Kim et al. (2019)

The book (*Naturally Occurring Chemicals Against Alzheimer's Disease*) addresses, in the following chapters, both AD-protective and AD-treatment effects of natural products (nutrients, bioactive compounds, and plant extracts).

References

Ahmed, F., Chandra, J.N.N.S., Urooj, A., Rangappa, K.S., 2009. *In vitro* antioxidant and anticholinesterase activity of *Acorus calamus* and *Nardostachys jatamansi* rhizomes. J. Pharm. Res. 2 (5), 830–883.

Aisen, P.S., 2008. The inflammatory hypothesis of Alzheimer disease: dead or alive? Alzheimer Dis. Assoc. Disord. 22 (1), 4–5.

Alama, B., Haque, E., 2011. Anti-Alzheimer and antioxidant activity of *Celastrus paniculatus* seed. Iran. J. Pharm. Sci. 7 (1), 49–56.

Alibabaei, Z., Rabiei, Z., Rahnama, S., Mokhtari, S., Rafieian-kopaei, M., 2014. *Matricaria chamomilla* extract demonstrates antioxidant properties against elevated rat brain oxidative status induced by amnestic dose of scopolamine. Biomed. Aging Pathol. 4 (4), 355–360.

Alzheimer's Disease International, 2019. World Alzheimer Report 2019: Attitudes to Dementia. Alzheimer's Disease International, London.

Anupama, K.P., Shilpa, O., Antony, A., Siddanna, T.K., Gurushankara, H.P., 2019. *Convolvulus pluricaulis* (Shankhapushpi) ameliorates human microtubule-associated protein tau (hMAP tau) induced neurotoxicity in Alzheimer's disease Drosophila model. J. Chem. Neuroanat. 95, 115–122.

Atri, A., 2019. Current and future treatments in Alzheimer's disease. Semin. Neurol. 39 (2), 227–240.

Bastianetto, S., Ramassamy, C., Dore, S., Christen, Y., Poirier, J., Quirion, R., 2000. The *Ginkgo biloba* extract (EGb 761) protects hippocampal neurons against cell death induced by beta-amyloid. Eur. J. Neurosci. 12 (6), 1882–1890.

Blennow, K., Hampel, H., Weiner, M., Zetterberg, H., 2010. Cerebrospinal fluid and plasma biomarkers in Alzheimer disease. Nat. Rev. Neurol. 6 (3), 131–144.

Cai, Z.Y., Wang, C.L., Yang, W.M., 2016. Role of berberine in Alzheimer's disease. Neuropsychiatric Dis. Treat. 12, 2509–2520.

Chauhan, N.B., 2006. Effect of aged garlic extract on APP processing and tau phosphorylation in Alzheimer's transgenic model Tg2576. J. Ethnopharmacol. 108 (3), 385–394.

Chen, F., Eckman, E.A., Eckman, C.B., 2006. Reductions in levels of the Alzheimer's amyloid β-peptide after oral administration of ginsenosides. FASEB. J. 20 (8), 1269–1271.

Cipriani, G., Dolciotti, C., Picchi, L., Bonuccelli, U., 2011. Alzheimer and his disease: a brief history. Neurol. Sci. 32 (2), 275–279.

Citron, M., 2010. Alzheimer's disease: strategies for disease modification. Nat. Rev. Drug Discov. 9 (5), 387–398.

Cummings, J., 2018. Lessons learned from Alzheimer disease: clinical trials with negative outcomes. Clin. Transl. Sci. 11 (2), 147–152.

Cummings, J.L., Cole, G., 2002. Alzheimer disease. J. Am. Med. Assoc. 287 (18), 2335–2338.

Dhanasekaran, M., Holcomb, L.A., Hitt, A.R., Tharakan, B., Porter, J.W., Young, K.A., Manyam, B.V., 2009. *Centella asiatica* extract selectively decreases amyloid-β levels in hippocampus of Alzheimer's disease animal model. Phytother. Res. 23 (1), 14–19.

Figueiro, M., Ilha, J., Pochmann, D., Porciuncula, L.O., Xavier, L.L., Achaval, M., Nunes, D.S., Elisabetsky, E., 2010. Acetylcholinesterase inhibition in cognition-relevant brain areas of mice treated with a nootropic Amazonian herbal (Marapuama). Phytomedicine 17 (12), 956–962.

Figueiro, M., Ilha, J., Linck, V.M., Herrmann, A.P., Nardin, P., Menezes, C.B., Achaval, M., Goncalves, C.A., Porciuncula, L.O., Nunes, D.S., Elisabetsky, E., 2011. The Amazonian herbal Marapuama attenuates cognitive impairment and neuroglial degeneration in a mouse Alzheimer model. Phytomedicine 18 (4), 327–333.

Francis, P.T., Palmer, A.M., Snape, M., Wilcock, G.K., 1999. The cholinergic hypothesis of Alzheimer's disease: a review of progress. J. Neurol. Neurosurg. Psychiatry 66 (2), 137–147.

Geekiyanage, H., Jicha, G.A., Nelson, P.T., Chan, C., 2012. Blood serum miRNA: non-invasive biomarkers for Alzheimer's disease. Exp. Neurol. 235 (2), 491–496.

Habtemariam, S., 2016. The therapeutic potential of rosemary (*Rosmarinus officinalis*) diterpenes for Alzheimer's disease. Evid. Based Complement. Alternat. Med. 2680409.

Harborne, J.B., 2000. Arsenal for survival: secondary plant products. Taxon 49 (3), 435–449.

Heo, H.J., Yang, H.C., Cho, H.Y., Hong, B., Lim, S.T., Park, H.J., Kim, K.H., Kim, H.K., Shin, D.H., 2000. Inhibitory effect of *Artemisia asiatica* alkaloids on acetylcholinesterase activity from rat PC12 cells. Mol. Cells 10 (3), 253–262.

Heo, J.H., Lee, S.T., Oh, M.J., Park, H.J., Shim, J.Y., Chu, K., Kim, M., 2011. Improvement of cognitive deficit in Alzheimer's disease patients by long term treatment with Korean red ginseng. J. Ginseng. Res. 35 (4), 457–461.

Ionita, R., Postu, P.A., Mihasan, M., Gorgan, D.L., Hancianu, M., Cioanca, O., Hritcu, L., 2018. Ameliorative effects of *Matricaria chamomilla* L. hydroalcoholic extract on scopolamine-induced memory impairment in rats: a behavioral and molecular study. Phytomedicine 47, 113–120.

Jung, H.Y., Nam, K.N., Woo, B.C., Kim, K.P., Kim, S.O., Lee, E.H., 2013. Hirsutine, an indole alkaloid of *Uncaria rhynchophylla*, inhibits inflammation-mediated neurotoxicity and microglial activation. Mol. Med. Rep. 7 (1), 154–158.

Kim, S.R., Hwang, S.Y., Jang, Y.P., Park, M.J., Markelonis, G.J., Oh, T.H., Kim, Y.C., 1999. Protopine from *Corydalis ternata* has anticholinesterase and antiamnesic activities. Planta Med. 65 (3), 218–221.

Kim, J.E., Shrestha, A.C., Kim, H.S., Ham, H.N., Kim, J.H., Kim, Y.J., Noh, Y.J., Kim, S.J., Kim, D.K., Jo, H.K., Kim, D.S., Moon, K.H., Lee, J.H., Jeong, K.O., Leem, J.Y., 2019. WS-5 extract of *Curcuma longa, Chaenomeles sinensis*, and *Zingiber officinale* contains anti-AChE compounds and improves β-amyloid-induced memory impairment in mice. Evid. Based Complement. Alternat. Med. 2019, 5160293.

Kose, L.P., Gulcin, I., Goren, A.C., Namiesnik, J., Martinez-Ayala, A.L., Gorinstein, S., 2015. LC-MS/MS analysis, antioxidant and anticholinergic properties of galanga (*Alpinia officinarum* Hance) rhizomes. Ind. Crop. Prod. 74, 712–721.

Kumar, S., Seal, C.J., Okello, E.J., 2011. Kinetics of acetylcholinesterase inhibition by an aqueous extract of *Withania somnifera* roots. Int. J. Pharm. Sci. Res. 2 (5), 1188–1192.

Kumaran, A., Ho, C.C., Hwang, L.S., 2018. Protective effect of *Nelumbo nucifera* extracts on beta amyloid protein induced apoptosis in PC12 cells, *in vitro* model of Alzheimer's disease. J. Food Drug Anal. 26 (1), 172–181.

Lee, J.W., Lee, Y.K., Ban, J.O., Ha, T.Y., Yun, Y.P., Han, S.B., Oh, K.W., Hong, J.T., 2009a. Green tea (-)-epigallocatechin-3-gallate inhibits beta-amyloid-induced cognitive dysfunction through modification of secretase activity via inhibition of ERK and NF-kappaB pathways in mice. J. Nutr. 139 (10), 1987–1993.

Lee, M.S., Yang, E.J., Kim, J.I., Ernst, E., 2009b. Ginseng for cognitive function in Alzheimer's disease: a systematic review. J. Alzheimers Dis. 18 (2), 339–344.

Lee, Y.K., Choi, I.S., Ban, J.O., Lee, H.J., Lee, U.S., Han, S.B., Jung, J.K., Kim, Y.H., Kim, K.H., Oh, K.W., Hong, J.T., 2011. 4-O-methylhonokiol attenuated β-amyloid-induced memory impairment through reduction of oxidative damages via inactivation of p38 MAP kinase. J. Nutr. Biochem. 22 (5), 476–486.

Lee, Y.J., Choi, D.Y., Lee, Y.K., Lee, Y.M., Han, S.B., Kim, Y.H., Kim, K.H., Nam, S.Y., Lee, B.J., Kang, J.K., Yun, Y.W., Oh, K.W., Hong, J.T., 2012. 4-O-methylhonokiol prevents memory impairment in the Tg2576 transgenic mice model of Alzheimer's disease via regulation of β-secretase activity. J. Alzheimers Dis. 29 (3), 677–690.

Letenneur, L., Larrieu, S., Barberger-Gateau, P., 2004. Alcohol and tobacco consumption as risk factors of dementia: a review of epidemiological studies. Biomed. Pharmacother. 58 (2), 95–99.

Li, Y.Q., Kong, D.X., Fu, Y., Sussman, M.R., Wu, H., 2020. The effect of developmental and environmental factors on secondary metabolites in medicinal plants. Plant Physiol. Biochem. 148, 80–89.

Lim, G.P., Chu, T., Yang, F.S., Beech, W., Frautschy, S.A., Cole, G.M., 2001. The curry spice curcumin reduces oxidative damage and amyloid pathology in an Alzheimer transgenic mouse. J. Neurosci. 21 (21), 8370–8377.

Little, J.T., Walsh, S., Aisen, P.S., 2008. An update on huperzine A as a treatment for Alzheimer's disease. Expert. Opin. Inv. Drug 17 (2), 209–215.

Liu, P.P., Xie, Y., Meng, X.Y., Kang, J.S., 2019. History and progress of hypotheses and clinical trials for Alzheimer's disease. Signal Transduct. Target. Ther. 4 (29).

Mahboubi, M., 2019. *Melissa officinalis* and rosmarinic acid in management of memory functions and Alzheimer disease. Asian Pac. J. Trop. Biomed. 9 (2), 47–52.

Mahrous, R., Ghareeb, D.A., Fathy, H.M., EL-Khair, R.M.A., Omar, A.A., 2017. The protective effect of Egyptian *Withania somnifera* against Alzheimer's. Med. Aromatic Plants 6 (285), 2167-0412.1000285.

Miranda, L.F., Gomes, K.B., Silveira, J.N., Pianetti, G.A., Byrro, R.M., Peles, P.R., Pereira, F.H., Santos, T.R., Assini, A.G., Ribeiro, V.V., Tito, P.A., Matoso, R.O., Lima, T.O., Moraes, E.N., Caramelli, P., 2015. Predictive factors of clinical response to cholinesterase inhibitors in mild and moderate Alzheimer's disease and mixed dementia: a one-year naturalistic study. J. Alzheimers Dis. 45 (2), 609–620.

Nakajima, A., Ohizumi, Y., Yamada, K., 2014. Anti-dementia activity of nobiletin, a Citrus flavonoid: a review of animal studies. Clin. Psychopharm. Neurosci. 12 (2), 75–82.

Oh, J.H., Choi, B.J., Chang, M.S., Park, S.K., 2009. *Nelumbo nucifera* semen extract improves memory in rats with scopolamine-induced amnesia through the induction of choline acetyltransferase expression. Neurosci. Lett. 461 (1), 41–44.

Pansari, K., Gupta, A., Thomas, P., 2002. Alzheimer's disease and vascular factors: facts and theories. Int. J. Clin. Pract. 56 (3), 197–203.

Patil, C.S., Singh, V.P., Satyanarayan, P.S.V., Jain, N.K., Singh, A., Kulkarni, S.K., 2003. Protective effect of flavonoids against aging- and lipopolysaccharide-induced cognitive impairment in mice. Pharmacology 69 (2), 59–67.

Paula, P.C., Maria, S.G.A., Luis, C.H., Patricia, C.G.G., 2019. Preventive effect of quercetin in a triple transgenic Alzheimer's disease mice model. Molecules 24 (12), 2287.

Pearson, V.E., 2001. Galantamine: a new Alzheimer drug with a past life. Ann. Pharmacother. 35 (11), 1406–1413.

Peng, Q.L., Buz'Zard, A.R., Lau, B.H.S., 2002. Pycnogenol protects neurons from amyloid-beta peptide-induced apoptosis. Mol. Brain Res. 104 (1), 55–65.

Petrovska, B.B., 2012. Historical review of medicinal plants' usage. Pharmacogn. Rev. 6 (11), 1–5.

Rege, S.D., Geetha, T., Griffin, G.D., Broderick, T.L., Babu, J.R., 2014. Neuroprotective effects of resveratrol in Alzheimer disease pathology. Front. Aging Neurosci. 6, 218.

Rosler, M., Retz, W., Thome, J., Riederer, P., 1998. Free radicals in Alzheimer's dementia: currently available therapeutic strategies. J. Neural. Transm. Suppl. 54, 211–219.

Ruan, C.J., Si, J.Y., Zhang, L., Chen, D.H., Du, G.H., Su, L., 2009. Protective effect of stilbenes containing extract-fraction from *Cajanus cajan* L. on Aβ (25-35)-induced cognitive deficits in mice. Neurosci. Lett. 467 (2), 159–163.

Rubio, J.L., Qiong, W., Liu, X.M., Jiang, Z., Dang, H.X., Chen, S.L., Gonzales, G.F., 2011. Aqueous extract of Black Maca (*Lepidium meyenii*) on memory impairment induced by ovariectomy in mice. Evid. Based Complement. Alternat. Med. 2011, 253958.

Satish, L., Shamili, S., Yolcu, S., Lavanya, G., Alavilli, H., Swamy, M.K., 2020. Biosynthesis of secondary metabolites in plants as influenced by different factors. In: Swamy, M. (Ed.), Plant-derived Bioactives. Springer, Singapore, pp. 61–100.

Seidl, C., Correia, B.L., Stinghen, A.E.M., Santos, C.A.M., 2010. Acetylcholinesterase inhibitory activity of uleine from *Himatanthus lancifolius*. Z. Naturforsch. C Biosci. 65 (7–8), 440–444.

Shin, R.W., 1997. Interaction of aluminum with paired helical filament tau is involved in neurofibrillary pathology of Alzheimer's disease. Gerontology 43, 16–23.

Singh, R.K., 2020. Antagonism of cysteinyl leukotrienes and their receptors as a neuroinflammatory target in Alzheimer's disease. Neurol. Sci. (published online ahead of print).

Singhal, A.K., Naithani, V., Bangar, O.P., 2012. Medicinal plants with a potential to treat Alzheimer and associated symptoms. Int. J. Nutr. Pharmacol. Neurol. Dis. 2 (2), 84–91.

Torreilles, F., Touchon, J., 2002. Pathogenic theories and intrathecal analysis of the sporadic form of Alzheimer's disease. Prog. Neurobiol. 66 (3), 191–203.

Tundis, R., Loizzo, M.R., Nabavi, S.M., Erdogan Orhan, I., Skalicka-Woźniak, K., D'Onofrio, G., Aiello, F., 2018. Natural compounds and their derivatives as multifunctional agents for the treatment of Alzheimer disease. In: Brahmachari, G. (Ed.), Discovery and Development of Neuroprotective Agents from Natural Products. Elsevier, pp. 63−102.

Uabundit, N., Wattanathorn, J., Mucimapura, S., Ingkaninan, K., 2010. Cognitive enhancement and neuroprotective effects of Bacopa monnieri in Alzheimer's disease model. J. Ethnopharmacol. 127 (1), 26−31.

Wang, D.M., Li, S.Q., Wu, W.L., Zhu, X.Y., Wang, Y., Yuan, H.Y., 2014. Effects of long-term treatment with quercetin on cognition and mitochondrial function in a mouse model of Alzheimer's disease. Neurochem. Res. 39 (8), 1533−1543.

Williams, B.R., Nazarians, A., Gill, M.A., 2003. A review of rivastigmine: a reversible cholinesterase inhibitor. Clin. Ther. 25 (6), 1634−1653.

Williams, P., Sorribas, A., Howes, M.J.R., 2011. Natural products as a source of Alzheimer's drug leads. Nat. Prod. Rep. 28 (1), 48−77.

Xian, Y.F., Lin, Z.X., Mao, Q.Q., Hu, Z., Zhao, M., Che, C.T., Ip, S.P., 2012. Bioassay-guided isolation of neuroprotective compounds from *Uncaria rhynchophylla* against beta-amyloid-induced neurotoxicity. Evid. Based Complement. Alternat. Med, 802625. https://doi.org/10.1155/2012/802625.

Xu, F.X., He, B.S., Xiao, F., Yan, T.X., Bi, K.S., Jia, Y., Wang, Z.Z., 2019. Neuroprotective effects of spinosin on recovery of learning and memory in a mouse model of Alzheimer's disease. Biomol. Ther. 27 (1), 71−77.

Yang, Z.D., Duan, D.Z., Du, J., Yang, M.J., Li, S., Yao, X.J., 2012. Geissoschizine methyl ether, a corynanthean-type indole alkaloid from *Uncaria rhynchophylla* as a potential acetylcholinesterase inhibitor. Nat. Prod. Res. 26 (1), 22−28.

Yang, S., Gong, Q.H., Wu, Q., Li, F., Lu, Y.F., Shi, J.S., 2014. Alkaloids enriched extract from *Dendrobium nobile* Lindl. attenuates tau protein hyperphosphorylation and apoptosis induced by lipopolysaccharide in rat brain. Phytomedicine 21 (5), 712−716.

Zhang, Z.X., Zhao, R.P., Qi, J.P., Wen, S.R., Tang, Y., Wang, D.S., 2011. Inhibition of glycogen synthase kinase-3β by *Angelica sinensis* extract decreases β -amyloid-induced neurotoxicity and tau phosphorylation in cultured cortical neurons. J. Neurosci. Res. 89 (3), 437−447.

Zhang, M., Wang, Y.Y., Qian, F., Li, P., Xu, X.J., 2016. Hypericin inhibits oligomeric amyloid β42-induced inflammation response in microglia and ameliorates cognitive deficits in an amyloid beta injection mouse model of Alzheimer's disease by suppressing MKL1. Biochem. Biophys. Res. Commun. 481 (1−2), 71−76.

Zhao, L., Wang, J.L., Liu, R., Li, X.X., Li, J.F., Zhang, L., 2013. Neuroprotective, anti-amyloidogenic and neurotrophic effects of apigenin in an Alzheimer's disease mouse model. Molecules 18 (8), 9949−9965.

Zhong, Z., Qu, Z., Wang, N., Wang, J., Xie, Z., Zhang, F., Zhang, W., Lu, Z., 2005. Protective effects of *Panax notoginseng* saponins against pathological lesion of cholinergic neuron in rat model with Alzheimer' s disease. Zhong Yao Cai 28 (2), 119−122.

Zhu, Y., Le Peng, J.H., Chen, Y., Chen, F., 2019. Current anti-Alzheimer's disease effect of natural products and their principal targets. J. Integr. Neurosci. 18 (3), 327−339.

Chapter 2

Alzheimer's disease: ethanobotanical studies

Swati Sharma[1], Sangita Sharma[1], Rounak Chourasia[1], Aseesh Pandey[2], Amit Kumar Rai[1], Dinabandhu Sahoo[1]
[1]*Institute of Bioresources and Sustainable Development, Sikkim Centre, Tadong, Sikkim, India;* [2]*GB Pant National Institute of Himalayan Environment and Sustainable Development, Sikkim Regional Centre, Gangtok, Sikkim, India*

Introduction

Plant are one of the best gift of nature to human kind, not just as a source of food but also as a resource of medicine for prevention and treatment of several diseases. Ayurveda, Siddha, and Unani medicines are best examples of medical treatment where medicinal plants are applied for the cure of diseases. Since many generations people have use medicinal plants as a part of their local culture for treating disease by applying their traditional knowledge (Basha and Sudarshanam, 2010). AD is a very widespread neurodegenrative diseases associated with sudden or gradual memory loss and many other cognitive disorders (Rao et al., 2012; Mahboubi et al., 2019). According to World Alzheimer's report a case of AD have been increased globally to 41 million numbers and is predicted to reach 131 million people by 2050 if not controlled (Martin et al., 2016). Among elderly patients AD is one of the major reasons for occurrence of dementia. AD could be seen as the result of Amyloid β peptide (Aβ), which are produced from the amyloid precursor protein (APP). Abnormalities in metabolism of APP results in higher deposition of Aβ, which further leads to neuronal cell death in AD (Sepand et al., 2013). Oxidative stress also plays an significant role in AD as it is presumed that toxicity of Aβ is partially induced by free radicals (Sepand et al., 2013). Along with the memory loss, AD patients also suffer from vocabulary loss, poor judgments, and behavioral issues.

In the recent decade's research on medicinal plants and their bioactive compounds have attracted researchers due its reliability and authenticity apart from very least or no side effects. The study on medicinal plants mainly acquired on the basis of traditional knowledge is called as ethnobotanical studies or ethnopharmacology. Many medicinal plants have been applied in traditional

Naturally Occurring Chemicals against Alzheimer's Disease. https://doi.org/10.1016/B978-0-12-819212-2.00044-X

knowledge having a significant effect on treatment of AD since many generations. In ayurvedic medicine system, Bhrahmi scientifically known as *Baccopa monneri*, Haldi (*Curcuma longa*), Ashwagandha (*Whitania somnifera*) have been already used to enhance memory and have proven for their therapeutic values against the AD (Rao et al., 2012; Shakeri et al., 2016; Farooqui et al., 2018). Similarly, yokukansan a blend of six herbs is used in Chinese and Japanese traditional medicine system to treat dementia (Matsunaga et al., 2016). Many medicinal plants have also been documented for the treatment of AD in traditional Chinese medicine system (TCM), Unani and Siddha. Hence, ethnobotanical studies are the potential source for therapeutic active chemical compound from medicinal plants to treat memory loss, dementia and AD. Documenting the traditional knowledge by recognizing the authentic traditional healers and finding the chemical compound could result in inhibiting the association of amyloid β peptide and by decreasing the amyloid precursor protein (APP) expression. In this chapter we will discuss the medicinal plants used in the treatment of AD along with the gaps and challenges for future research.

Medicinal plants used against AD

Medicinal plants have been used since many generations as a part of traditional knowledge against AD. In this section, we summarize the information of research being carried out on selected medicinal plants, which are being used in traditional medicine system that has potential in drug discovery for the treatment of AD. The selected medicinal plants and their ways of applications are displayed in Tables 2.1 and 2.2.

Gingko biloba

Gingko biloba belongs to Gingkgoaceae family, commonly known as Maidenhair tree and is found in China, Japan, and Korea, and cultivated in many parts of Europe, America and New Zealand, Argentina and India. The tree is deciduous with height of 4 m, with a reddish bark. In Ayurveda, TCM system and Japan, *Gingko biloba* extract is used for the treatment of blood insufficiency, especially in the brain resulting in memory loss, headache and loss of consciousness (Lou, 2006; Sun et al., 2013). In China, the leaves and seeds have been traditionally used for 100 years for age related declining of memory and cognitive functions (Howes et al., 2003). In Iran, *G. biloba* is traditionally used for the improvement of memory loss, which is associated with abnormalities in blood circulation (Ross, 2001).

The extract of plant contains approximately 24% flavonoids and 6% tarpene lactones. There are convincing evidences, which shows that *Gingko biloba* extracts have impact on several cellular and molecular neuro-protective mechanisms, including the shrinking of apoptosis, anti-inflammatory effects,

TABLE 2.1 Medicinal plants used against Alzheimer's disease in traditional medicine system.

Plant name (scientific name)	Local name	Ethnic community/ Region	Country/ Traditional medicine system	References
Gingko biloba	Ginko, Madeinhair	Indian, Chinese, Korean, Japanese	Ayurveda, Chinese traditional medicines, Korean traditional medicines, Japense traditional medicines	Lou et al. (2006)
Curcuma longa	Haldi, turmeric	Indian, South Asian sub continents, Chinese	Ayurveda, Unani, Chinese traditional medicines	Chattopadhyay et al. (2004), Kikusawa et al. (2007)
Salvia officinalis	Sage	Across the globe	Across the globe	Akhondzadeh et al. (2003)
Tinospora cordifolia	Guduchi, Giloy	Indian, Pakistan, Bangladesh, Sri Lankan, Malaysian & Indonesian	Ayurveda, Unani	Burton (2009), Tillostson et al. (2001)
Melissa officinalis	Lemon balm, BalmMint, common balm	European, central Asian, Iranian	Iranian traditional medicine	Houghton et al. (2005)
Glycyrrhiza glabra	Liquorice	Indian, European	Ayurveda, Unani	(Omar et al., 2012)
Centella asiatica	Gotu Kola	Indian, Malaysian, Indonesian	Ayurveda, Unani	Tiwari et al. (2008)
Convolvulus pluricaulis	Shankhpushpi	India	Ayurveda	Asolkar (1992)
Withania somnifera	Ashwagandha, Indian Gingsen	Nepal, Yemen, China, India	Ayurveda, Chinese traditional	Pandit et al. (2013)

Continued

TABLE 2.1 Medicinal plants used against Alzheimer's disease in traditional medicine system.—cont'd

Plant name (scientific name)	Local name	Ethnic community/ Region	Country/ Traditional medicine system	References
			medicines, Unani	
Bacopa monnieri	Brahmi	Australia, India, Europe, Africa, Asia	Uniani, Ayurveda	Correll (2019)

an inhibition of membrane lipid peroxidation and inhibition of amyloid-β-aggregation (Luo et al., 2002). The flavonoid fraction of *G. biloba* extracts have shown to possess antioxidant properties including free-radical scavenging properties. The main chemical constituent of *G. biloba* is gingkolides, which is a prominent antioxidant along with cholinergic and neuroprotective activities that helps in managing of AD. Many studies have shown that the dose used for the treatment of Alzheimer ranges from 80 to 720 mg/dL in the durations ranging from 2 weeks to 2 years (Mary and Bailey, 2012).

Curcuma longa

Curcuma longa commonly known as turmeric in English is a flowering plant that belongs to family Zingiberaceae and is native to Southeast Asia and Indian Subcontinent (Priyadarshini, 2014). The plant requires temperatures ranging from 20 to 30°C and a significant amount of rainfall to flourish. The plants are cultivated and collected each year for their rhizomes, which are consumed in dried powdered form and some kept for propagation in the following season. Rhizomes of *C. longa* are used fresh or in dried powdered form having deep orange-yellow color as a dye and flavoring agent in curries and spicy dishes, which is known as "Haldi" in India. Turmeric powder has bitter, earthy and mustards like aroma (Swapna et al., 2016), which is being used by Asians since decades as an integral part of Ayurveda, Siddha medicine, TCM and Unani medicine for the prevention and treatment of wide range of ailments (Chattopadhyay et al., 2004; Kikusawa and Lawrence, 2007). Cucurmin is traditionally taken up orally or tropically depending on the type of ailments. In Ayurveda system, it is used for neurodegenerative diseases and as a rejuvenator for slowing down the aging process (Auddy et al., 2003).

The most important chemical constituent in turmeric is called *curcuminoids*. *Curcuminoids* has significant anti-inflammatory and antioxidant properties

TABLE 2.2 Different ways of using medicinal plants against Alzheimer's disease.

Plant name	Formulation/ Preparation	Dose regime/ Administration ways	References
Gingko biloba	Capsules	80–720 mg/dL (2 weeks–2 years)	Mary and Bailey (2012)
Curcuma longa	Powder	125 mg/per day	https://alzheimer.neurology.ucla.edu/Curcumin.html
Salvia officinalis	Oil, tincture	60 drops/day	Akhondzadeh et al. (2003)
Tinospora cordifolia	Guduchi Imminity Wellness (Himalaya) Guduchi tablets (Alka ayurvedic pharmacy)	500–1000 mg/per day	Singh (2015)
Melissa officinalis	Traditional medicines	Dry leaves (40 gms) Fresh leaves (80 gms) Dry seeds (9 gms)	Mahboubi (2019)
Glycyrrhiza glabra	Ayurveda, Herbalism traditional medicine	150 mg/kg	Akram and Nawaz, 2017
Centella asiatica	Powder	500 mg twice a day or 1000 mg per day	Tiwari et al. (2008), Dhingra et al. (2004)
Withania somnifera	Powder extract of roots	500 mg	Changappa et al. (2013)
Bacopa monnieri	Extract of leaves	300–600 mg	Asthana et al. (1996), Phase et al. (2012)

which are being combating for AD Patients (Mishra and Palanivelu, 2008). *C. longa* is used as a source curcumin (diferulomethane), an orange-yellow constituent of turmeric. Curcumin plays many important roles by following different mechanism against AD. It plays a protective role against Aβ protein and cure inflammation (Wang et al., 2010) and has shown to reduce oxidative stress by exhibiting strong antioxidant properties (Lim et al., 2001). Apart from curcuminoid, many other chemical components such as tetrahydro-curcumin, demethoxy-curcumin, and bismethoxy-curcumin, helps in striking to key step responsible for AD by inhibiting the formation of Aβ protein (Yang et al., 2005; Zhang et al., 2006). Curcuminoids emerge to protect long-term potentiation, which is an indication of functional memory and may result in reversing physiological damage by disruption of existing plaques and restoration of distorted neurites (Giri et al., 2004; Akhondzadeh et al., 2003). Curcumin has a greater binding affinity for copper and iron, which may add to its protective effect against AD, which is an iron-mediated damage (Mandel et al., 2007).

Salvia officinalis

Salvia officianlis is also known as common sage, garden sage or culinary sage. It is a perennial evergreen sub shrub with grayish leaves, woody stems and blue to purple color flowers. *S. officianlis* is a member of the Lamiaceae family and native to the Mediterranean region. It is being used as medicinal and culinary since ages and is enlisted in old herbals. *S. officinalis* is used by ancient people for waving off witch crafts, evil, snakebites, treating women infertility and improving memory loss (Lorraine, 2012; Houghton and Howes, 2005). *S. officinalis* is 2 feet tall and wide plant, with grayish green leaves on the upper side, and white color leaves underneath. The plant blooms in spring or summer and the color of the flower is generally white, pink, or purple. *S. officinalis* has significant anti-inflammatory, antioxidant properties and acetylcholine inhibitory properties against AD (Eidi et al., 2006; Jivad and Zahra, 2014). The main chemical constituent of *S. officinalis* which reduce plaques with the alteration in Aβ along with lipid peroxidation, DNA fragmentation is Rosmarinie (Jivad and Zahra, 2014).

In Europe, *S. officinalis* has been traditionally used to treat age related cognitive disorders along with inflammation and mild dyspepsia (Ghorbani and Esmaeilizadeh, 2017). Since decades *S. officinalis* is being used as memory enhancing agents due to its cholinergic properties for AD by local healers (Jivad and Zahra, 2014; Tildesley et al., 2005). Based on many research, *S. officinalis* has proven the efficiency in wandering the symptoms of AD. Rosmarinic acid, an important component of *S. officinalis* has neuroprotective property against Aβ (Iuvone et al., 2006), as it reduces Aβ accumulation and entanglements in vitro and modulates other pathways to relieve the Alzheimer, including DNA fragmentation and hyperphosphorylation (Iqbal and Grundke-Iqbal, 2005).

Tinospora cordifolia

Tinospora cordifolia also known as "Guduchi" is a large, climber shrub with many elongated branches. The leaves of this plant are simple, alternate, petioles (15 cm long), round and pulvinate. Flowers of this plant are small in size, greenish yellow in color and apprears when the plant is leafless on terminal and axillary racemes (Sinha et al., 2004). The species are widely distributed in India extending from the Himalaya region to the Southern part India. *T. cordifolia* one of the most important herbs in Ayurveda has remarkable affect in memory impairment and learning enhancement (Mutalik and Mutalik, 2011). It stimulates the immune system by synthesizing acetylcholine, which has significant role in the enhancement of cognitive functions. Role of aqueous extract of *T. cordifolia* roots for improving memory has been mentioned in TCM as well as in western and Ayurvedic herbal medicine (Tillotson et al., 2001). *T. cordifolia* has been studied and shown to have significant effect in treatments of deficit hyperactivity disorder (Burtron, 2009). It is also used in the form of a tablet of a polyherbal formulation for general improvement of memory function (Mutalik and Mutalik, 2011). *T. cordifolia* is one of the important components in several polyherbal formulations including Chaihu-Shugan-San, Banxia-houpu, EuMil, Catuama, Mentot, Sho-ju-sen, Siotone, and Kami-shoyo-san prescribed for depressive disorders (Mutalik and Mutalik, 2011). The role of *T. cordifolia* in cholinergic mechanisms needs to be clearly understood and the outcome can lead to have positive impact to AD patients.

Melissa officinalis

Melissa officinalis is also well-known as Lemon balm, common balm, or balm mint which is a perennial herbaceous plant belonging to the Lamiaceae family. The plant is native to Mediterranean Basin, south-central Europe, Central Asia and Iran. The height of the plant maximum ranges from 70 to 150 cm and the leaves has a mild lemony fragrance similar to mint. Since many decades *M. officinalis* is being applied as a remedy in traditional medicine for memory impairment and promoting long life (Houghton and Howes, 2005). It has common use in Iranian traditional medicine system as memory enhancing herb and for the treatment of neurological disorders such as anxiety and depression. The leaves of plant have monoterpene which has mild anti-acetylcholine activity and possess phenol carboxylic acids and rosmarinic acid as main chemical constituent, which possess antioxidant, anti-amylodogenic and anti-apoptotic effects (Shakeri et al., 2016). Consumption of canary wine having *M. officinalis* essential oil acts as a brain tonic as it is believed to strengthen the activity of brain (Mahboubi, 2019). *M. officinalis* has acetylcholine receptor activity in central nervous system and including binding properties with muscurinic and nicotinic, which helps in modulating the cognitive performance in AD patients (Wake et al., 2000; Kennedy et al., 2002; Schhultz et al., 1999). As per studies the daily oral dose of dried herb for adult is 1.5–4.5 g, while for the infusion in 150 mL water, 2–4 mL of 45% ethanol extract in the ratio 1:1 (3 times/day), and 2–6 mL in tincture (3 times/day) (Edwards et al.,

2015). In a recent study on Iranian Traditional Medicine, dry leaves (40 g), fresh leaves (80 g) and dry seeds (9 g) showed statistically significant therapeutic results against AD (Mahboubi, 2019).

Glycyrrhiza glabra

Glycyrrhiza glabra also known as Liquorice is a part of traditional medicine and has application as flavoring agent. The plant grows in the form of herb, with the height of approximately 1 m, with pinnate leaves ranging from 7 to 15 cm long. The fruits of plant is oblong, 2–3 cm long, persisting of several seeds (Brown, 1995). Liquorice is native to southern Europe, the Middle East and parts of Asia, particularly in India. It consists of wide range compounds including polysaccharides, triterpene, flavonoids, pectins, simple sugars, saponins, amino acids and mineral salts. *G. glabra* is used in many traditional system of medicine such as TCM, Ayurveda, Unani, SIddha and Homeopathy to treat wide range of diseases. Liquorice extracts have shown to possess several pharmacological properties and is being used by local healer as an important traditional medicine. Liquorice contain glycyrrhizinic and volatile molecules such as antheole acid, which doesn't have any adverse effect such as high blood pressure, hypokalemia and muscle weakness in dose lower than 2 mg/kg/day (Omar et al., 2012, 2019). Sweetness in liquorice is credited to glycyrrhizin, which impart sweet taste, 30–50 times more than sugar (Somjen et al., 2004; Tamir et al., 2001). Licorice extract root have positive impact in the treatment and prevention of brain cell death in AD patients (Singhal et al., 2012; Bilge and Ilkay, 2005). In a dose dependent study, *G. glabra* extracts were given to mice at three different dose (75, 150, 300 mg/kg) for 7 days and found that the dose of 150 mg/kg was found to be effective against memory impairment (Dhingra et al., 2004; Akram and Nawaz, 2017). In their study (Song et al., 2013), have concluded that Glycyrrhizin, a main component of *G. glabra* is a putative candidate for drug discovery against neurogenerative disease such as AD.

Centella asiatica

Centella asiatica is known as Indian pennywort in English and mandukaparni in the Ayurvedic medicine system of India. It is a highly rated herb in traditional medicine system, which is referred as rejuvenating herb and helps in improving memory and intelligence. It is grown as medicinal herb, belonging to family Apiaceae and is also used for its culinary property (Verma and Gurmaita, 2019). The color of the *C. asiatica* flowers ranges from slight pinkish to red. *C. asiatica* consist of pentacyclic triterpenoids, including asiaticoside, brahmoside, asiatic acid, madecassoside, brahmic acid, centellose and centelloside (James and Dubery, 2009). *C. asiatica* is used in traditional medicinal herb for many years in India, China, Sri Lanka, Nepal, and

Madagascar for wound healing, revitalization of nerve and brain cell, thus known as a brain food (Singh et al., 2010). Since 3000 years, *C. asiatica* have been studied and used in Ayurvedic medicine for significant neuroprotective properties and boosting memory. It has been used in Ayurvedic system as revitalizing agent to improve nervous system and restore youthness. Some of the studies have shown that *C. asiatica* possess antioxidant property and helps in reducing in-vitro and in-vivo oxidative stress (Kumari et al., 2016).

Specific parts of the plant are applied in different traditional medicine system for improving memory function. Leaves of *C. asiatica* are used as a revitalizing agent in Ayurveda that improves memory function. In TCM, *C. asiatica* is also applied for fighting mental and physical exhaustion (Sun et al., 2013). Whole extract of the plant have shown a very promising in-vivo effect against many neurological disorders (Lokanathan et al., 2016). Many researchers across the globe are focusing on confirming the traditional use of *C. asiatica* against AD. Recent studies on *C. asiatica* extract are focused on the molecular mechanism of neuroprotection (Lokanathan et al., 2016). According to their study, aqueous extract of *C. asiatica* showed positive impact in a dose of 500 mg twice a day for 24 weeks for mild cognitive impairment (Tiwari et al., 2008). *C. asiatica* has also been proven to improve memory performance in rats induced with memory dysfunction due to oxidative stress (Amjad and Umesalma, 2015; Doknark et al., 2014).

Convolvulus pluricaulis

Convolvulus pluricaulis is an herb cultivated in India and Burma and is commonly used in Ayurveda. It has prostrate branches and small oblong, lanceolate and obtuse leaves. It has light blue color flowers, which most likely appears like a marine shell "Shankh" and therefore called as shankhpushpi (Asolkar, 1992). Pale yellow oil with a green tinge and characteristic odor is extracted with the help of steam distillation of fresh flowers, while dry flower is also stored for application in many formulations (Asolkar, 1992). Traditionally, it is one of the important medications for improving memory function and treating brain related ailments. It is used as tonic to improve memory and cognitive function in addition to other health benefits (Rao et al., 2012). C. *pluricaulis* consists of many phytochemicals such as glycosides, coumarines, flavonoids, alkaloids, sitosterol, hydroxy cinnamic acid, octacosanol, tetracosane along with glucose. Since many decades, C. *pluricaulis* has been used as brain tonic tranquilizer and calms the nerves by regulating stress hormones, adrenaline and cortisol (Sharma et al., 2010). Results of studies on extracts of Shankhpushpi concluded that ethanolic extract enhanced memory by enhancing neurite growth (Farooqui et al., 2018).

Withania somnifera

Withania somnifera is well-known as ashwagandha, the Indian ginseng, winter cherry or poison gooseberry and belongs to Solanaceae family (Schmelzer and Gurib-Fakim, 2012). *W. somnifera* is a short and perennial shrub, height ranges from 35 to 75 cm, flowers are small, green and bell-shaped (Stearn, 1995). The name of the plant "*ashwagandha*" is a combination of two sanskrit words *ashva* (meaning horse) and *gandha* (meaning smell). The plant is cultivated in India at bigger scale and also found in Nepal, China and Yemen (Pandit et al., 2013). Traditionally, it is one of the highly rated herbs used in Ayurveda since more than 4000 years. According to Ayurvedic scholar Charaka in 10 BCE, the herb is recommended to get sharp memory, obtain longevity, get freedom from disease and get strength of horse (Howes et al., 2003). The main phytochemical constituents of *W. somnifera* are 40 withanolides, 12 alkaloids, and many sitoindosides. Among them withanolides share structural similarity to the ginsenosides from *Panax ginseng*, therefore are known as "Indian ginseng" (Stearn, 1995). The roots of *W. somnifera* exhibit antioxidant activity including free radical scavenging activity that helps in enhancing the immune system (Russo et al., 2001). In a recent study, *W. somnifera* showed stress relieving activity and improving the ability to focus and prevent memory impairment in a dose-dependent manner (Auddy et al., 2008). *W. somnifera* contains wide range of bioactive compounds which includes withasomniferols A to C, withanolides A to Y, withasomniferin A, withaferin A, dehydro withanolide R, withasomidienone, withanone, amino acids, phytosterols sitoindosides VII to X, β-sitosterol, alkaloids and iron (Matsuda et al., 2001). Many studies have shown that the phytochemical withanamides A and C binds perfectly to active motifs of Aβ peptide resulting in prevention of fibril formation (Kumar et al., 2012; Rao et al., 2012). A study further proved that *Withania somnifera* taken 500 mg/day helps in improving the vocabulary and cognition (Chengappa et al., 2013). In Ayurvedic medicine system Ashwagandha has been used as memory booster since ages. In a study performed on 50 adults showed that 300 mg extract of *W. somnifera* (taken twice a day) can significantly improved memory (Farooqui et al., 2018). Although the dosage of *W. somnifera* depends on the type of formulation but extracts are seen more effective than the dried root powder.

Bacopa monnieri

Bacopa monnieri belongs to family Plantaginaceae found in India, Australia, Europe, Asia and Africa. *B. monnieri* is a medicinal herb used in Ayurveda for centuries for memory loss, loss of concentration and cognition (Nemetchek et al., 2017; Correll, 2019). In *B. monnieri*, the main photochemical is triterpenoid saponins well-known as bacosides, with aglycone units (Sivaramakrishna et al., 2005). Bacosides consists of 12 known analogs

(Garai et al., 2009; Chakravarty et al., 2003). The alkaloids present in the plant contain brahmine, nicotine, herpestine, apigenin, hersaponin, D-mannitol, plantainoside B monnierasides I—III, and cucurbitacin (Bhandari et al., 2007; Russo et al., 2003). The methanolic extract of the plant possess radical scavenging properties which has defensive effect on cleavage of DNA (Russo et al., 2003). *B. monnieri* is supposed to exhibit antioxidant, anti-inflammatory, anti-stress, cognition fascilatatory, immunomodulatory and anti-aging effects. The anti lipid per-oxidation property of the plant is responsible for memory enhancing actions. *B. monnieri* has been used as a memory booster since many decades and to treat memory dysfunction (Sharma, 1998). *B. monnieri* helps in preventing the breakdown of acetylcholine to help in memory enhancement by acting on its cognitive and related brain function.

Bramhi has been applied in Ayurveda since many generations as a nerve tonic and treatment of neurological disease. In the 16th century *Bacopa* is recommended in Ayurvedic text (*Caraka Samhita*) for the treatment of wide range of mental illness (Russo and Borrelli, 2005). The traditional knowledge has become the base of several studies, which concludes that extracts of *B. monnieri* per day can improve memory and other cognitive functions (Pase et al., 2012; Raghav et al., 2006). Inspite of background of traditional knowledge and research across the globe no commercial drug derived from this plant has been formulated and marketed.

Gaps and future challenges

Gaps in ethanobotanical studies may be defined well, as the unavailability of many traditionally used medicinal plants in the modern medicines or in pharmacy stores. There are several factors due to which medicinal plants are facing trouble to come to pharmacy from the traditional medicine system (Fig. 2.1). The factors include (i) Proper documentation of traditional medicinal system (ii) Lack of scientific evidences in dose and ways of administration, (iii) Globalization and species lost, and (iv) Challenges of mass cultivation.

Proper documentation of traditional medicinal system

Among the factors, documentation of the traditional knowledge in the most important one as it is the baseline for discovery of potential drugs. In ancient times family used to run patrimonial occupation by which the legacy of knowledge used to transfer from one generation to another but in modern times, the traditional knowledge in getting extinct gradually. We are already running on the edge of time and there is an urgent need for documentation of the traditionally available against AD. Secondly, our ancient books like the Ayurveda, TCM, Korean medicines, etc which contain traditional medicine knowledge are written in ancient languages which are difficult for the present

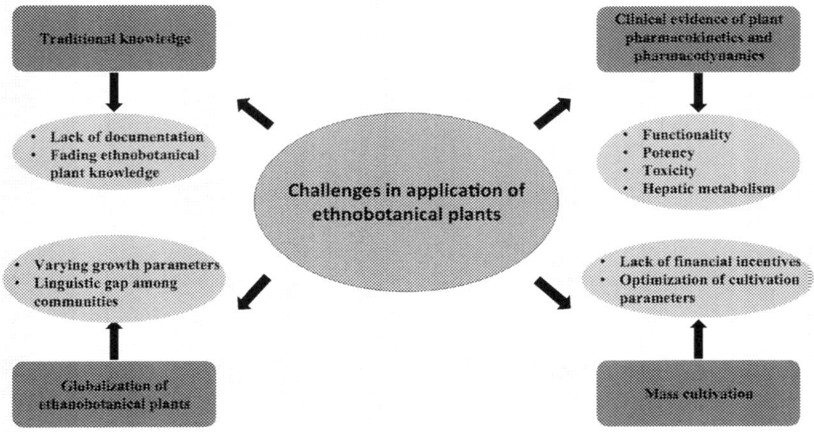

FIGURE 2.1 Gaps and challenges for drug discovery from traditional knowledge on medicinal plants.

generation to understand translate. Therefore systematic programmes must be undertaken to document the available traditional knowledge against AD.

Lack of information of scientific evidences on dose and ways of administration

Although many plants have been reported to have activity against AD, lack the scientific evidences on the dose and administration is a challenge. Identification and characterization of bioactive compounds responsible for positive effect against AD need to be thoroughly studied. Mechanism of action of the identified bioactive molecule against AD needs to be studied at molecular level to validate the traditional knowledge. Pharmacokinetics and phamacodynamics are important parameters for application of any drug, which need to standardized. There are no clinical evidences available in most of the drugs about their therapeutic window, toxicity, side effects, adverse effects, potency, hepatic metabolism, which are a necessary mandate to be studied before giving a status of a drug from traditional medicine. After screening of bioactive compounds, we should run proper clinical trials on the promising plant candidates not just in the case of AD but for associated problems as well. This will help in getting scientific evidences for the traditional claims and information regarding on dose needed for treatment of AD.

Globalization effects and species lost

Globalization is undoubtedly an important topic to be discussed in every aspects and scenario as it refers to the economic system that allows free flow of intellectual property, raw material of plant and manufactured goods across international borders. Globalisation lead to loss of wild species, which might be a potential and natural source of any high value drug. If the decline or massive utilization of any plant species occurs at higher rate, it may results in its extinction. The rapid increase in urbanization and technologies also results in exploitation of plant species in an uncontrolled manner. There is a need to have sustainable utilization strategy for any high demand plant species to maintain the balance of the ecosystem by preserving specific plant species before they become extinct.

Challenges in mass cultivation

Mass cultivation of medicinal plants is an important and challenging factor for drug production. Every plant has its own growth and climatic condition, which doesn't favor their cultivation across the globe. Cultivation of a foreign species also needs same soil condition for proper growth and metabolite profile. Proper irrigation, field management and preparation for specific plan species need to be standardised. In the case of medicinal plants choosing right manure and fertilizer which do not interfere with the quality and efficacy of the plant is necessary. Lack of awareness and monetary benefits regarding some of the ethnobotanical plant is also a challenge before mass cultivation. Farmers training and awareness programmes can result in solving such issues. Apart from these issues urbanization and global warming can also effect the growth of specific medicinal plant with same metabolic profile.

Conclusions

The traditional medicine system is considered as a potential source for discovery of novel drugs against AD. The present chapter has discussed about the medicinal plants having promising effects on AD. Some of the medicinal plants have been already cultivated and applied since decades in traditional medicinal system, which is the base of researchers for drug discovery against AD. Despite of many research on these medicinal plants, which have shown positive effects on cognitive function and promising activity by enhancing memory, the pharmaceutical industries are facing challenges to develop a novel and efficient formulation in the drug discovery process. Combination of modern science and strong base of traditional knowledge against AD can result in discovery of novel drug for the global market.

Acknowledgment

The authors of the chapter acknowledge Institute of Bioresources and Sustainable Development (IBSD), a National Institute of Department of Biotechnology, Govt. of India for all the support, encouragement and providing necessary help to undertake the study.

References

Akhondzadeh, S., Noroozian, M., Mohammadi, M., Ohadinia, S., Jamshidi, A., Khani, M., 2003. Extract in the treatment of patients with mild to moderate Alzheimer's disease: a double blind, randomised, placebo controlled trial. J. Neurol. Neurosurg. 74 (7), 863–866.

Akram, M., Nawaz, A., April 2017. Effects of medicinal plants on Alzheimer's disease and memory deficits. Neural Regen. Res. 12 (4), 660–670.

Amjad, S., Umesalma, S., 2015. Protective effect of Centella *asiatica* against aluminium-induced neurotoxicity in cerebral cortex, striatum, hypothalamus and hippocampus of rat brain-histopathological, and biochemical approach. J. Mol. Biomarkers Diagn. 6 (1), 212. https://doi.org/10.4172/2155-9929.1000212.CrossRef (Google Scholar).

Asolkar, L.V., Kakkar, K.K., Chakre, O.J., 1992. Second supplement of glossary of Indian medicinal plant with active principles: part 1 (A-K), (1965–1981). Publications and information Directorate, New Delhi.

Asthana, O.P., Shrivastava, J.S., Ghatak, A., Gaur, S.P.S., Dhawan, B.N., 1996. Safety and tolerability of Bacosides A and B in the healthy human volunteers. J. Indian Pharmacol. 28, 37.

Auddy, B., Ferreira, M., Blasina, F., Lafon, L., Arredondo, F., Dajas, F., et al., 2003. Screening of antioxidant activity of three Indian medicinal plants, traditionally used for the management of neurodegenerative diseases. J. Ethnopharmacol. 84, 131–138.

Auddy, B., Hazra, J., Mitra, A., et al., 2008. "A standardized *Withania somnifera* extract signifi cantly reduces stress–related parameters in chronically stressed humans: a double–blind randomized placebocontrolled study. JANA 11 (1), 50–56.

Basha, S.K., Sudarshanam, G., 2010. Ethnobotanical studies on medicinal plants used by sugalis of yerramalais in kurnool district, Andhra Pradesh, India. Int. J. Phytomed. 2, 349–353.

Bhandari, P., Kumar, N., Singh, B., Kaul, V.K., 2007. Cucurbitacins from *Bacopa monnieri*. Phytochemistry 68 (9), 1248–1254.

Bilge, S., Ilkay, O., 2005. Discovery of drug candidates from some Turkish plants and conservation of biodiversity. Pure Appl. Chem. 77, 53–64.

Brown, D., 1995. Encyclopedia of Herbrs and Their Uses. Dorling Kindersley, London.

Burton, D., 2009. Attention-deficit Hyperactivity Disorder: Natural and Herbal Treatments. Ohlone Herbal Center. Research Papers. http://www.ohlonecenter.org/research-papers/attention AD deficit-hyperactivity disorder- herbal-and-natural-treatments. (Accessed 7 June 2011).

Chakravarty, A.K., Garai, S., Masuda, K., Nakane, T., Kawahara, N., 2003. Bacopasides III– V: three new triterpenoid glycosides from *Bacopa monniera*. Chem. Pharm. Bull. 51, 215–217. https://doi.org/10.1248/cpb.51.215.

Chattopadhyay, I., Kaushik, B., Uday, B., Ranajit, K.B., 2004. Turmeric and curcumin: biological actions and medicinal applications. Curr. Sci. 87 (1), 44–53.

Chengappa, K.N., Bowie, C.R., Schlicht, P.J., Fleet, D., Brar, J.S., Jindal, R., 2013. Randomized placebo-controlled adjunctive study of an extract of *Withania somnifera* for cognitive, dysfunction, inbipolar, disorder. J. Clin. Phsychiatry 74 (11), 10761083. https://doi.org/10.4088/JCP.13m08413.

Correll Jr., W.A., February 5, 2019. FDA Warning Letter: Peak Nootropics LLC Aka Advanced Nootropics. Office of Compliance, Center for Food Safety and Applied Nutrition, Inspections, Compliance, Enforcement, and Criminal Investigations, US Food and Drug Administration. Retrieved 11 May 2019.

Dhingra, D., Parle, M., Kulkarni, S.K., 2004. Memory-strengthening activity of *Glycyrrhiza glabra* in exteroceptive and interoceptive behavioral models. J. Med. Food 7 (4), 462–466.

Doknark, S., Mingmalairak, S., Vattanajun, A., Tantisira, B., Tantisira, M.H., 2014. Study of ameliorating effects of ethanolic extract of *Centella asiatica* on learning and memory deficit in animal models. J. Med. Assoc. Thail. 97, S68–S76.

Edwards, S.E., Da-Costa-Rocha, I., Williamson, E.M., Heinrich, M., 2015. Phytopharmacy: An Evidence-Based Guide to Herbal Medicinal Products. Willey.

Eidi, M., Eidi, A., Bahar, M., 2006. Effects of *Salvia officinalis L* (sage) leaves on memory retention and its interaction with the cholinergic system in rats. Nutrition 22 (3), 321–326.

Farooqui, A.A., Farooqui, T., Madan, A., Jing Ong, J., Ong, W.Y., 2018. Ayurvedic medicine for the treatment of dementia: mechanistic aspects. Evid. Based Complement. Altern. Med. 2481076.

Garai, M.S., Ohtani, S.B., Yamasaki, K.K., 2009. Dammarane triterpenoid saponins from *Bacopa monnieri*. J. Canad. Chem. 87, 1230–1234.

Ghorbani, A., Esmaeilizadeh, M., 2017. Pharmacological properties of *Salvia officinalis* and its components. J. Tradit. Complement. Med. 7, 433–440.

Giri, R.K., Rajagopal, V., Kalra, V.k., 2004. Curcumin, the active constituent of turmeric, inhibits amyloid peptide-induced cytochemokine gene expression and CCR5-mediated chemotaxis of THP-1 monocytes by modulating early growth response-1 transcription factor. J. Neurochem. 91 (5), 1199–1210.

Houghton, P., Howes, M.J., 2005. Natural products and derivatives affecting neurotransmission relevant to Alzheimer's and Parkinson's disease. Neurosignals 14 (1–2), 6–22.

Howes, M.R., Perry, N.S.L., Houghton, P.J., 2003. Plants with traditional uses and activities, relevant to the management of Alzheimer's disease and other cognitive disorders. Phytother. Res. 17, 1–18.

Iqbal, K., Grundke-Iqbal, I., 2005. Pharmacological approaches of neuro fibrillary degeneration. Curr. Alzheimer Res. 2, 335–341.

Iuvone, T., De Filippis, D., Esposit, G.D., Amico, A., Izzo, A.A., 2006. The spice sage and its active ingredient rosmarinic acid protect PC12 cells from amyloid-β peptide-induced neurotoxicity. J. Pharmacol. Exp. Therapeut. 317, 1143–1149.

James, J.T., Dubery, I.A., 2009. Pentacyclic triterpenoidsfrm the medicinal herb, Centella asiatica(L.)Urban. Molecules 14, 3922–3941.

Jivad, N., Zahra, R., 2014. A review study on medicinal plants used in the treatment of learning and memory impairments. J. Asian Pac. Trop. Biomed. 4 (10), 780–789.

Kennedy, D.O., Scholey, A.B., Tildesley, N.T., Perry, E.K., Wesnes, K.A., 2002. Modulation of mood and cognitive performance following acute administration of *Melissa officinalis* (lemon balm). J. Pharmacol. Biochem. Behav. 72 (4), 953–964.

Kikusawa, R.R., Lawrence, A., 2007. Proto who utilized turmeric, and how? In: Siegel, J., Lynch, J., Eades, D. (Eds.), Language Description, History and Development: Linguistic Indulgence in Memory of Terry Crowley. John Benjamins Publishing Company, pp. 339–352.

Kumar, S., Harris, R.J., Seal, C.J., Okello, E.J., 2012. An aqueous extract of *Withania somnifera* root inhibits amyloid β fibril formation in vitro. J. Phytother. Res. 26 (1), 113–117. https://doi.org/10.1002/ptr.3512.

Kumari, S., Deori, M., Elancheran, R., Kotoky, J., Devi, R., 2016. In vitro and in vivo antioxidant, anti-hyperlipidemic properties and chemical characterization of *Centella asiatica* (L.) extract. Front. Pharmacol. 7, 400.

Lim, G.P., Chu, T., Yang, F., Beech, W., Frautschy, S.A., Cole, G.M., 2001. The curry spice curcumin reduces oxidative damage and amyloid pathology in an Alzheimer transgenic mouse. J. Neurosci. 21, 8370–8377.

Lokanathan, Y., Omar, N., Ahmad Puzi, N.N., Saim, A., Hj Idrus, R., 2016. Recent updates in neuroprotective and neuroregenerative potential of *Centella asiatica*. J. Malaysia Med. Sci. 23 (1), 4–14.

Lorraine, H., 2012. RHS Latin for Gardeners More than 1,500 Essential Plant Names and the Secrets They Contain. Mitchell Beazley, United Kingdom, p. 224. ISBN 9781845337315.

Luo, Y., 2006. "Alzheimer's disease, the nematode *Caenorhabditis elegans*, and *Ginkgo biloba* leaf extract". J. Life Sci. 78 (18), 2066–2072.

Luo, Y., Smith, J.V., Paramasivam, V., Burdick, A., Curry, K.J., Buford, J.P., Khan, I., Netzer, W.J., Xu, H., Butko, P., 2002. Inhibition of amyloid-β aggregation and caspase-3 activation by the *Ginko biloba* extract EGb761. PNAS 99, 12197–12202.

Mahboubi, M., 2019. *Mellissa officinalis* and rosmarinic acid in management of memory functions and Alzheimer disease. J. Asian Pac. Trop. Biomed. 9, 47–52.

Mandel, S., Tamar, A., Bar-Am, O., Youdim, M.B.H., 2007. Iron dysregulation in Alzheimer's disease: multimodal brain permeable iron chelating drugs, possessing neuroprotective-neurorescue and amyloid precursor protein-processing regulatory activities as therapeutic agents. Programme Neurobiol. 82 (6), 348–360. PMID: 17659826.

Martin, P., Adelina, C., Martin, K., Maelenn, G., Maria, K., 2016. World Alzheimer Report 2016: Improving Healthcare for People Living with Dementia: Coverage, Quality and Costs Now and in the Future. Alzheimer's Disease International (ADI), London, UK.

Mary, R., Bailey, M.A., 2012. Neuropsychology, cognitive neuroscience, and clinical outcomes laboratory, department of psychology. Univ. J. Psychiatr. Clin. https://doi.org/10.1016/j.psc..12.006.

Matsuda, H., Murakami, T., Kishi, A., et al., 2001. Structures of withanosides I, II, III, IV, V, VI, and VII, new withanolide glycosides, from the roots of Indian *Withania somnifera* DUNAL and inhibitory activity for tachyphylaxis to clonidine in isolated Guinea–pig ileum. J. Bioorgan. Med. Chem. 9 (6), 1499–1507.

Matsunaga, S., Taro Kishi, T., Iwata, N., 2016. Yokukansan in the treatment of behavioral and psychological symptoms of dementia: an updated meta-analysis of randomized controlled trials. J. Alzheimers Dis. 54635–54643.

Mishra, S., Palanivelu, K., 2008. "The effect of *Curcumin* (turmeric) on Alzheimer's disease". An overview. Annu. Indian Acad. Neurochem. 11 (1), 13–19.

Murray, 2012. In: Joseph, E.P. (Ed.), T.Textbook of Natural Medicine, fourth ed., vol. 650. Churchill Livingstone, Edinburgh. ISBN 9781437723335.

Mutalik, M., Mutalik, M., 2011. *Tinospora cordifolia*: role in depression, cognition, and memory. Aust. J. Med. Herbal. 23 (4), 168–173.

Nemetchek, M.D., Stierle, A.A., Stierle, D.B., Lurie, D., 2017. The Ayurvedic plant *Bacopa monnieri* inhibits inflammatory pathways in the brain. J. Ethnopharmacol. 197, 92–100.

Omar, A., Farida, A., Shimaa, A., Haredya, Reham, M., Niazya, Linhardtb, R.J., Wardac, M., 2019. Chem. Biol. Interact. 308, 279–287.

Omar, H., Komarova, R., El-Ghonemi, I., Ahmed, M., Fathy, R., Rania, A., Hany Yerramadha, D., Reddy, M., Ali, Y., Camporesi, E.,M., 2012. How much is too much?. In: Licorice Abuse: Time to Send a Warning Message from "Therapeutic Advances in Endocrinology and Metabolism", vol. 3, pp. 125–138. https://doi.org/10.1177/2042018812454322.PMC.3498851.PMID.23185686, 4.

Pandit, S., Chang, K.W., Jeon, J.G., 2013. Effects of *Withania somnifera* on the growth and virulence properties of *Streptococcus mutans* and *Streptococcus sobrinus* at sub-MIC levels. J. Anaerobe 19, 1–8.

Pase, M.P., Kean, J., Sarris, J., Neale, C., Scholey, A.B., Stough, C., 2012. The cognitive enhancing effects of *Bacopa monnieri* a systematic review of randomized, controlled human clinical trials. J. Altern. Complement. Med. 18 (7), 647–652. https://doi.org/10.1089/acm.2011.0367.

Priyadarsini, K.I., 2014. The chemistry of curcumin: from extraction to therapeutic agent. J. Molecules 19 (12), 20091–20112. https://doi.org/10.3390/molecules191220091.

Raghav, S., Singh, H., Dalal, P.K., Srivastava, J.S., Asthana, O.P., 2006. Randomized controlled trial of standardized *Bacopa monniera* extract in age-associated memory impairment. J. Indian Psychiatry 48 (4), 238–242. https://doi.org/10.4103/0019-5545.31555.

Rao, R.V., Descamps, O., Varghese, J., Dale, E.B., 2012. Ayurvedic medicinal plants for Alzheimer's disease: a review. Alzheimers Res. Ther. 4 (3), 22.

Ross, I.A., 2001. Medicinal plants of the World. In: Chemical Constituents, Traditional and Modern Medicinal Uses, vol. 2. Humana Press, Totowa, NJ.

Russo, A., Borrelli, F., 2005. Bacopa monniera, a Reputed Nootropic plant: An overview. Phytomedium 4, 305–318.

Russo, A., Izzo, A.A., Borrelli, F., Renis, M., Vanella, A., 2003. Free radical scavenging capacity and protective effect of Bacopamonnieri L. on DNA damage. Phytother. Res. 17 (18), 870–875.

Russo, A., Izzo, A.A., Cardile, V., Borrelli, F., Venella, A., 2001. Indian medicinal plants as antiradicals and DNA cleavage protectors". Phytomedicine 8 (2), 125–132.

Schhultz, V., Hansel, R., Tyler, V., 1999. Rational Phytotherapy: A Physician's Guide to Herbal Medicine. Springer-Verlag, New York.

Schmelzer, G.H., Gurib-Fakim, A., 2012. Plant Resource of Tropical Africa 11 (1) Medicinal Plants 1, p. 791. PROTA Foundation, Wageningen, Netherlands/Backhuys Publishers, Leiden, Netherlands/CTA, Wageningen, Netherlands.

Sepand, M.R., Soodi, M., Hajimehdipoor, H., Soleimani, M., Sahraei, E., 2013. Comparison of neuroprotective effects of Melissa officinalis total extract and its acidic and non-acidic fractions against A β-induced toxicity. Iran J. Pharm. Res. 12 (2), 415–423. Springer.

Shakeri, A., Sahebkar, A., Javadi, B., 2016. Melissa officinalis L. - a review of its traditional uses, phytochemistry and pharmacology. J. Ethnopharmacol. 188, 204–228.

Sharma, P.V., Dravyaguna V., 1998. 2nd vol. Varanasi, India: Chaukhambha Bharati Academy. pp. 6–8.

Sharma, k., Bhatnagar, M., Kulkarni, S.K., 2010. Effect of Convolvulus pluricaulis Choisy and Asparagus racemosus Wild on learning and memory in young and old mice: A comparative evaluation. Ind. J. Exp. Biol. 48, 479–485.

Singh, J., 2015. "Giloy.GhanVati". "AyurTimes.". https://www.google.com/amp/s/www.ayurtimes.com/giloy-ghan-vati/amp.

Singh, S., Gautam, A., Sharma, A., Batra, A., 2010. *Centella asiatica* (L.): a plant with immense medicinal potential but threatened. Int. J. Pharm. Sci. Rev. Res. 4, 9–17.

Singhal, A.k., Naithani, V., Bangar, O.P., 2012. Medicinal plants with a potential to treat Alzheimer and associated symptoms. J. Int. Nutr. Pharmacol. Neurol. Dis. 2, 84–91.

Sinha, K., Mishra, N.P., Singh, J., Khanuja, S.P.S., 2004. *Tinospora cordifolia* (Guduchi), a reservoir plant for therapeutic applications: a review. J. Indian Tradit. Know. 3 (3), 257–270.

Sivaramakrishna, C., Rao, C.V., Trimurtulu, Vanisree, G., Subbaraju, M.G.V., 2005. Triterpenoid, glycosides, from *Baccopamonnerii*. Phytochemistry 66, 2719–2728. https://doi.org/10.1016/j.phytochem.2005.09.016.

Somjen, K.D., Vaya, S., Kaye, J., Hendel, A.M., Posner, D., Tamir, G.H.,S., 2004. Estrogenic activity of glabridin and glabrene from licorice roots on human osteoblasts and prepubertal rat skeletal tissues. J. Steroid Biochem. Mol. Biol. 91 (4–5), 241–246.

Song, J.H., et al., 2013. Glycyrrhizin alleviates neuroinflammation and memory deficit induced by systemic lipopolysaccharide treatment in mice. Molecules 18, 15788−15803.

Stearn, W.T., 1995. Botanical Latin History, Grammar, Syntax, Terminology and Vocabulary. Timber Press. ISBN 978-0-88192-321-6.

Sun, Z.K., Yang, H.Q., Chen, S.D., 2013. Traditional Chinese Medicine: a promising candidate for the treatment of Alzheimer's disease. Transl. Neurodegener. 2, 6.

Swapna, M., Bhaumik, A., K. Devi, N., Neelamma, G., Sreedevi, 2016. Evaluation of in vitro cytotoxic activity of various extracts of turmeric powder (*Curcuma longa*) Agaunst human prostate cancer cell line. DU"-145. World J. Pharm. Med. Res. 2 (3), 127−133. SJIF Research Article ISSN 2455-3301.

Tamir, S., Eizenberg, M., Somjen, D., Izrael, S., Vaya, J., 2001. Estrogen-like activity of glabrene and other constituents isolated from licorice root. J. Steroid Biochem. Mol. Biol. 78 (3), 291−298.

Tildesley, N.T., Kennedy, D.O., Perry, E.K., Ballard, C.G., Wesnes, K.A., Scholey, A.B., 2005. Positive modulation of mood and cognitive performance following administration of acute doses of *Salvia lavandulaefolia* essential oil to healthy young volunteers. J. Physiol. Behav. 83 (5), 699−709.

Tillotson, A.K., Tillotson, N.H., Abel Jr., R., 2001. "The One Earth Herbal Sourcebook" Everything You Need to Know about Chinese, Western, and Ayurvedic Herbal Treatments". Twin Streams, Kesington Publishing Corp, New York, NY.

Tiwari, S., Singh, S., Patwardhan, K., Gehlot, S., Gambhir, I., 2008. Effect of *Centella asiatica* on mild cognitive impairment (MCI) and other common age-related clinical problems. J. Digest Nanomater. Biosci. 3, 215−220.

Verma, R., Gurmaita, A., 2019. A review on anticarcinogenic activity of *Centella asiatica*. World J. Pharm. https://doi.org/10.20959/wjpr20193-14284.

Wake, G., Court, J., Pickering, A., Lewis, R., Wilkins, R., Perry, E., 2000. CNS acetylcholine receptor activity in European medicinal plants traditionally used to improve failing memory". J. Ethnopharmacol. 69 (2), 105−114.

Wang, H.M., Zhao, Y.X., Zhang, S., Liu, G., Kang, W.Y., Tang, H.D., Ding, J., Chen, S., 2010. PPAR gamma agonist curcumin reduces the amyloid-beta-stimulated inflammatory responses in primary astrocytes. J. Alzheimers Dis. 20 (4), 1189−1199.

Yang, F., Lim, G.P., Begum, A.N., Ubeda, O.J., Simmons, M.R., Ambegaokar, S.S, Chen, P.P., Kaead, R., Glabe, C.G., Fratschy, S.A., Cole, G.M., 2005. Curcumin inhibits formation of amyloid beta oligomers and fibrils, binds plaques, and reduces amyloid in vivo. J. Biol. Chem. 280, 5892−5901.

Zhang, C., Andrew, B., Rudolph, D.C., Tanzi, E., 2006. Curcumin decreases amyloid-beta peptide levels by attenuating the maturation of amyloid-beta precursor protein. J. Gastroenterol. (1), 120−126.

Chapter 3

Phytochemicals/plant extract against Alzheimer's

Section 3.1

Phytochemicals (pure compounds)

Chapter 3.1.1

Resveratrol

Ashwani K. Dhingra[1], Vaibhav Rathi[2], Bhawna Chopra[1]
[1]*Guru Gobind Singh College of Pharmacy, Yamuna Nagar, Haryana, India;* [2]*School of Health Sciences, Quantum University, Roorkee, Uttarakhand, India*

Introduction

The bioactive component 3,4′,5-trihydroxystilbene, commonly known as resveratrol, is present in purple grapes, red wine, blueberries, mulberries, rhubarb, groundnuts, cranberries, chocolate, pines, and peanuts. It was first obtained in 1940 and is known to promote antiaging effects and provide protection from cardiovascular diseases and various cancers. Recent studies suggest that it modulates multiple mechanisms in the pathology of Alzheimer's disease (AD) (Thimmappa, 2006). Resveratrol is biosynthesized by the enzyme stilbene synthase (Stervbo et al., 2007; Fornara et al., 2008). This enzyme undergoes condensation reaction between coumaroyl-coenzyme and malonyl CoA; catalyzes the loss of the terminal carboxyl group and thus generates a 14-carbon-containing molecule, resveratrol (Fornara et al., 2008; Wang et al., 2010). Resveratrol has been found to destruct and inhibit the formation of preformed aggregates, reduces microglia activation (NF-kB inhibitor) and activation of the sirtuin1 (Sirt1) pathway to mediate neuroprotection (Sousa et al., 2020). Thus, it mitigates pathophysiological aspects of the disease; still, further studies are needed to prove the safety and efficacy of this compound in humans.

Sources

Resveratrol was first reported in 1939 by a Japanese researcher, Dr. Michio Takaoka, from the poisonous medicinal herb *Veratrum grandiflorum* (Takaoka, 1939) and later from the roots of *Polygonum cuspidatum* (Nonomura et al., 1963). It can also be isolated from more than 70 plant species including grapes and red wine (Rege et al., 2014).

Naturally Occurring Chemicals against Alzheimer's Disease. https://doi.org/10.1016/B978-0-12-819212-2.00037-2

Natural sources

Pediomelum cuspidatum is an important medicine in China, with a resveratrol content of 1.8 mg/g (Liu et al., 2019). Its root extract plays a very important role in Japanese medicine, as it is the main ingredient in ko-jo-kon (Nonomura et al., 1963).

V. grandiflorum leaves have a high resveratrol content (Hanawa et al., 1992; Chung et al., 1992). In addition, the roots and rhizomes of *Veratrum formosanum* are rich in resveratrol (Thimmappa, 2006).

Vitis vinifera, the common grape, contains resveratrol in its skin and seeds (Thimmappa, 2006). In red wine, the resveratrol content ranges between 1 and 14 mg/L; higher and lower levels are frequently found (Goldberg et al., 1996). Table 3.1.1.1 shows the resveratrol content of various biological sources.

Pharmacokinetics

Absorption: Resveratrol has high oral absorption and rapid and extensive metabolism, with adverse effects in both rodents (Bertelli et al., 1996a,b; Bertelli et al., 1998; Marier et al., 2002; Soleas et al., 2001) and humans (Goldberg et al., 2003; Vitaglione et al., 2005; Walle et al., 2004; Boocock et al., 2007b). In humans, 70% of resveratrol administered orally at a dose of 25 mg is rapidly (<30 min) absorbed (Goldberg et al., 2003; Walle et al., 2004; Gescher and Steward, 2003). Drug absorption varies significantly from person to person (Walle et al., 2004; Vitaglione et al., 2005).

Distribution: Resveratrol binds to protein transporters (Khan et al., 2002) and serum proteins (lipoproteins, hemoglobin, and albumin) and forms complexes that are spontaneous and exothermic in nature (Jannin et al., 2004). Orally administered resveratrol accumulates extensively in the liver of rats and mice (Bertelli et al., 1998; Vitrac et al., 2003).

Metabolism: Resveratrol metabolizes with a half-life of 9–10 h with a peak plasma concentration of $\approx 2\ \mu M$ (Walle et al., 2004; Goldberg et al., 2003; Gescher and Steward, 2003). The process of metabolism varies from individual to individual, and its extent depends on hepatic function and the metabolic activity of local intestinal flora. Resveratrol undergoes phase I and phase II metabolism in the liver and intestinal epithelial cells immediately after ingestion (Marier et al., 2002; Vitaglione et al., 2005; Gescher and Steward, 2003; Kaldas et al., 2003; Maier et al., 2006; Wenzel et al., 2005; Wenzel and Somoza, 2005). During metabolism, presystemic and systemic conversion yields glucuronic (resveratrol-3-glucuronide, resveratrol-4-glucuronide) conjugates by glucuronidation in the presence of uridine 5′-diphospho-glucuronosyltransferases in intestine and sulfate (*trans*-resveratrol-3-O-4′-O-disulfate, trans-resveratrol-4′-O-sulfate, and *trans*-resveratrol-3-O-sulfate) conjugates by sulfation in the presence of sulfotransferases in liver as major metabolites; others are dihydro-resveratrol and piceatannol mediated by

TABLE 3.1.1.1 Resveratrol content of various sources (a serving corresponds to a typical consumed portion of each food source).

Natural source	Biological source	Serving a total resveratrol	Serving (μg)	References
Itadori roots	*Reynoutria japonica*	1 g	2200	Burns et al. (2002)
Itadori tea	*R. japonica*	200 mL	2000	Burns et al. (2002)
Grapes	Genus *Vitis*	100 g	150 −780	Burns et al. (2002), Rimando et al. (2004)
Red wine	–	150 mL	80 −2700	Burns et al. (2002), Hurst et al. (2008), Wang et al. (2002b), Sanders et al. (2000), Bolling et al. (2010)
Grape juice	Genus *Vitis*	240 mL	0.12 −0.26	Hurst et al. (2008)
Blueberries	Genus *Vaccinium*	100 g	86–170	Wang et al. (2002b)
Cranberries	Genus *Vaccinium*	100 g	90	Wang et al. (2002b)
Bilberries	Genus *Vaccinium*	100 g	77	Wang et al. (2002b)
Pistachios	Genus *Pistacia*	28 g	2.5–47	Burns et al. (2002)
Raw peanuts	*Arachis hypogaea*	28 g	0.6–50	Burns et al. (2002), Sobolev and Cole (1999), Hurst et al. (2008), Sanders et al. (2000)
Roasted peanuts	*Arachis hypogaea*	28 g	0.5–2.2	Sobolev and Cole (1999), Hurst et al. (2008)
Peanut butter	Genus *Arachis*	32 g	4.7–24	Sobolev and Cole (1999), Rimando et al. (2004), Hurst et al. (2008)
Cocoa powder	Genus *Theobroma*	15 g	19–34	Hurst et al. (2008)
Dark chocolate	*Theobroma cacao*	15 g	3.8–6.5	Counet et al. (2006), Hurst et al. (2008)
Milk chocolate	*T. cacao*	15 g	0.8–2.6	Hurst et al. (2008)

microbial fermentation of transform in GIT. The *cis* forms of metabolites identified in human urine are *cis* resveratrol -4′-sulphate, cis-resveratrol-3-O-glucuronide, and *cis* resveratrol-4′-O-glucuronide (Urpi et al., 2007; Zamora et al., 2006).

Excretion: Resveratrol's metabolites are eliminated in urine and feces. Complete elimination has been observed 72 h after a single dose. Total recovery of glucuronic and sulfate conjugates was 71%−98% after an oral dose and 54%−91% after an intravenous dose; the aglycone form of resveratrol presents zero recovery (Vitaglione et al., 2005; Williams et al., 2009).

Chemistry

Resveratrol, or trans-3, 4′, 5-trihydroxy stilbene, is a monomer stilbene with the molecular formula $C_{14}H_{12}O_3$ and a molecular weight of 228.25 g/mol. It exists as a white powder with a yellow tinge. This molecule consists of two aromatic rings, A and B, linked by an ethylene bridge with a double bond (Fig. 3.1.1.1). Ring A contains two hydroxyl groups at positions 3 and 5; ring B contains one hydroxyl group at position 4′ (Tsai et al., 2017). Chemically, it is also described as a 6-2-6 carbon skeleton with m-hydroquinone and 4′-hydroxystyryl moieties involving rings A and B (Niesen et al., 2013).

Resveratrol exists in the geometrical isomeric forms *cis* and *trans* (Fig. 3.1.1.2). Resveratrol can be consumed in food products in *trans* form rather than in *cis* form (Anisimova et al., 2011). On exposure to UV and visible rays, isomerization occurs in the molecule, i.e., from *trans* to *cis* (Silva et al., 2013). *Cis*-resveratrol is less stable and thus cannot be utilized commercially (Cottart et al., 2010).

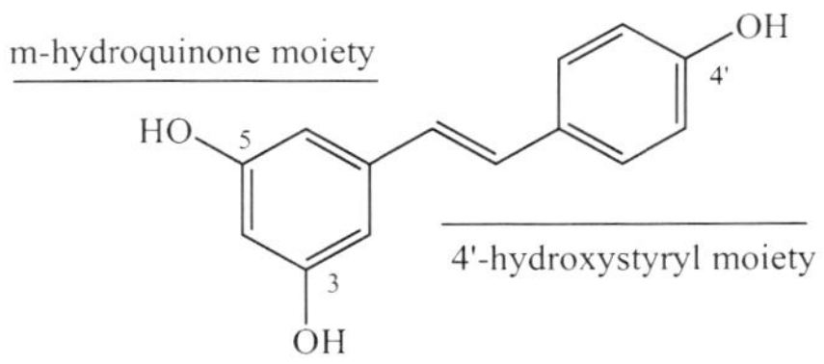

FIGURE 3.1.1.1 The molecular structure of resveratrol.

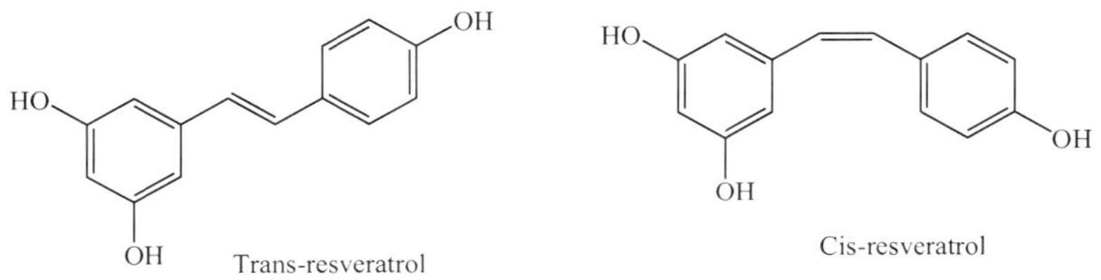

FIGURE 3.1.1.2 *Cis* and *trans* forms of resveratrol.

Resveratrol is nonplanar due to the steric effects of its aromatic rings. Its *cis* isomer is sensitive to low pH (1.0) and converts to *trans* form in 23 h. Both isomeric forms are stable at a pH of 3–8 but can be degraded at a pH of 10 or above (Trela and Waterhouse, 1996).

Resveratrol in Alzheimer's disease

- In vitro and in vivo experimental models have demonstrated the neuroprotective potential of resveratrol. One study evidenced that resveratrol facilitates nonamyloidogenic breakdown of the amyloid precursor protein and promotes the removal of neurotoxic amyloid-β peptides. It reduces damage to neuronal cells via activation of NAD (+) dependent histone deacetylase enzymes known as sirtuins (Braidy et al., 2016).
- Resveratrol modulates via multiple mechanisms; one of them is modulation of Sirt1 protein, a human homologue of the yeast silent information regulator (Sir)-2 and NAD^+ dependent histone deacetylases. It can mediate by the calorie-restriction regimen and calorie-restriction mimetics in a variety of organisms. It was found to protect neurons against the ployQ toxicity tested in a Huntington's disease mouse model and also found to protect neurons from degeneration as proven in an experimental study of a Wallerian degeneration slow mice model (Thimmappa, 2006).
- Resveratrol acts as a calorie-restriction mimetic by affecting sirtuin pathways (Baur and Sinclair, 2006). In addition, the suppression of p53 in neurons was observed in several studies (Anekonda and Reddy, 2006).
- Studies conclude that it increases the eNOS, modulates Sirt1 expression, and affects cellular metabolic function. Amyloid peptides trigger activation by acting with a number of toll-like receptors (TLRs) including TLR4. This study suggests that in brains with AD, resveratrol may increase eNOS, modulate the expression of Sirt1, and affect other cellular metabolic functions. Resveratrol also protects both nerves and blood vessels against Aβ insults. Moreover, amyloid peptides trigger activation by acting with a number of TLRs including TLR4.
- Evidence suggests resveratrol as a neuroprotective agent as tested in PC12 cells. It can protect from Aβ25-35 induced toxicity, attenuates apoptotic cell death, reduces changes in mitochondrial potential, and thus inhibits the accumulation of reactive oxygen species (Jang and Surh, 2003). It interacts with the ubiquitin-proteasome system by affecting the proteasome-mediated degradation of Aβ (Marambaud et al., 2005).
- Resveratrol was found to be safe and well tolerated in mild to moderate AD patients, as no statistically significant difference was observed. Weight loss in a resveratrol-treated group was due to enhancement of mitochondrial biogenesis mediated by Sirt1 activation of PGC1α, with nausea as an adverse effect (Pasinetti et al., 2015; Kulkarni and Canto, 2015; Bastianetto et al., 2015).

- Resveratrol reduces the number of activated microglia in APP/PS1 mice and affects amyloid deposition (Capiralla et al., 2012).
- Resveratrol enhances cognitive function in HFD-fed mice and reduces serum TNF-alpha, macrophage filtration, neuroinflammation and oxidative stress of the hippocampus in male C67BL/6 J mice (Jeon et al., 2012).
- Literature suggests that resveratrol exhibits neuroprotective benefits in animal models of AD. It also promotes nonamyloidogenic cleavage of the amyloid precursor protein, enhancing amyloid-β peptide clearance. In addition, it reduces neuronal damage (Fei et al., 2012).
- Resveratrol acts in both in vivo and in vitro models to determine the amyloid-β hypothesis, a key factor involved in the etiology of AD. A variety of research has now focused on the dynamic equilibrium, target production, and clearance of Aβ. This suggests that resveratrol can be employed for AD treatment (Yongming et al., 2017).
- Epidemiological studies indicate that consuming a moderate amount of wine lowers the risk of AD. The resveratrol present in red wine or grapes lowers the level of secreted and intracellular amyloid-β peptides. Resveratrol acts on the enzyme proteasome, thereby decreasing the intracellular degradation of Aβ. Studies suggest that resveratrol does not inhibit Aβ production. This demonstrates the proteasome-dependent anti-amyloidogenic activity of resveratrol (Marambaud et al., 2005).
- Several epidemiological studies indicate that moderate consumption of red wine is associated with a lower incidence of dementia and AD. Red wine is rich in antioxidant polyphenols with potential neuroprotective activities. Despite skepticism concerning the bioavailability of these polyphenols, in vivo data have clearly demonstrated the neuroprotective properties of the naturally occurring polyphenol resveratrol in rodent models of stress and disease. Furthermore, recent work in cell cultures and animal models has shed light on the molecular mechanisms potentially involved in the beneficial effects of resveratrol intake against the neurodegenerative process in AD (Vingtdeux et al., 2008).
- The effect of oral administration in AβPP/PS1 mice was tested by the object recognition test. It was observed that long-term treatment with resveratrol prevents memory loss. Resveratrol also reduces the amyloid burden and increases the mitochondrial complex IV protein level in mouse brain. Protective effects are mediated due to increased activation of Sirt1 and AMPK pathways in mice. In addition, it promotes changes in inflammatory processes by causing increased levels of IL1β and TNF (Porquet et al., 2014).
- Resveratrol increases memory performance and has a neuroprotective effect due to the depolymerization of amyloid β fibrils. It mainly targets Sirt1, known to be a homeostatic regulator in AD (Sathya et al., 2018).
- A 3.8-fold increase in the protein level of transthyretin and a 2.3-fold increase in drebrin were observed in resveratrol-fed mice and demonstrate

that it was due to the increased level of GSK3-β phosphorylation. Thus, this work provides another insight into neuroprotection (Varamini et al., 2014).

- Resveratrol and other natural components like curcumin, apigenin, docosahexaenoic acid, epigallocatechin gallate and α-lipoic acid exhibit great potential in the prevention and treatment of AD. Their use might be a possible remedy, lead to a safe strategy to delay the onset of AD, and slow the progression of this pervasive disorder (Venigalla et al., 2016).
- By increasing the clearance of β-amyloid and modulating intracellular effectors associated with oxidative stress, neuronal energy homeostasis, program cell death, and longevity, resveratrol helps produce the neuroprotective action (Bastianetto et al., 2015).
- Literature suggests that regulating mRNA may become a potential target to prevent or treat AD. Both in vitro and in vivo AD models demonstrate that resveratrol exerts a neuroprotective role (Kou and Chen, 2017).
- Resveratrol and its derivative pterostilbene protect against dementia syndromes such as AD as evaluated in cellular and mammalian models. Upon comparing the two, it was well documented that pterostilbene was more effective in combating brain changes with aging in comparison with resveratrol. This is because pterostilbene is more lipophilic, having two methoxy groups in its structure, whereas resveratrol contains two hydroxyl groups. To determine the bioavailability of both, clinical trials would need to be explored (Klaus and Shiming, 2018).
- Semisynthetic prenylated derivatives of resveratrol were evaluated as inhibitors of amyloid−β and β-secretase (BACE1). One derivative, named (E)-3,5,4-Trihydroxy-4-prenylstilbene, exhibits good anti-Aβ aggregation with an IC 50 value of 4.78 μM and moderate anti-BACE1 inhibitory activity, i.e., 23.70% at 50 μM. Moreover, this compound shows no neurotoxicity along with a greater ability to inhibit oxidative stress on P19-derived neuronal cells (Puksasook et al., 2017).
- Curcumin, demethoxycurcumin, and resveratrol appear to be beneficial as anti−AD agents due to their ability to prevent the aggregation of amyloid-β peptides. Still, clinical data are limited in evaluating the clinical efficiency and potential toxicity of their use in AD treatment (Villaflores et al., 2012).
- Prolonged use of NSAIDs reduces the risk of developing AD because AD is characterized by Aβ peptides and neurofibrillary tangles that are surrounded by inflammatory cells. NSAIDs also target pathological hallmarks of AD by interacting with COX (inhibition) and peroxisome proliferator-activated receptor gamma (activation). Drugs like ibuprofen, flurbiprofen, indomethacin, and sulindac possess Aβ-lowering properties that have been tested in both AD transgenic mice and cell cultures of peripheral, glial, and neuronal origin. In this study, it was contradictory that COX inhibition occurs at low concentration in vitro, whereas the Aβ-lowering effect was

produced at high concentration. Further study is required to assess whether an adequate concentration reaches the brain to define the dose window and the length of the treatment in future clinical trials (Gasparini et al., 2004).

- Studies confirmed that NDDS, resveratrol-loaded with lipid core nanocapsule treatment, was able to rescue the deleterious effects of Aβ1-42, while treatment with resveratrol alone presents only partial beneficial effects against intracerebroventricular injection of Aβ1-42 in rats. The results also show that Aβ1-42-infused animals show significant impairment on learning memory ability, activated astrocytes and microglial cells, and activation of glycogen synthase kinase-3β. The study concludes that this effect is antiamyloidogenic in nature as an increased resveratrol concentration in the brain (Frozza et al., 2013).
- Multitarget-directed ligands are an innovative approach used to develop tacrine–resveratrol-fused hybrid compounds as anti-AD. A series of compounds was prepared, out of which four molecules—6-chloro-N-(4-(3,5-dimethoxyphenethyl)phenyl)-1,2,3,4-tetrahydroacridin-9-amine, N-(4-(3,5-dimethoxyphenethyl)phenyl)-6-methoxy-1,2,3,4-tetrahydroacridin-9-amine, 5-(4-(6-chloro-1,2,3,4-tetrahydroacridin-9-ylamino)phenethyl) benzene-1,3-diol, and (E)-5-(4-((6-chloro-1,2,3,4-tetrahydroacridin-9-yl)amino)styryl)benzene-1,3-diol—inhibit human acetyl cholinesterase at micromolar concentration and modulate Aβ self-aggregation effectively. These compounds also possess high BBB permeability with a low toxicity profile on neurons (Jerabek et al., 2017).
- Resveratrol modulates the activity of Sirt1, AMPK, and PGC-1α (Pasinetti et al., 2015). It also inhibits the polymerization of β-amyloid peptide and reduces NF-kB signaling via activation of Sirt1 (Berman et al., 2017).
- Resveratrol is also effective in reducing biomarkers associated with AD and helps in bearing adverse effects (Kursvietiene et al., 2016). Resveratrol nanocapsules enhance its effectiveness (Bastianetto et al., 2015).
- Human umbilical cord derived mesenchymal stem cells (hUC-MSCs) and resveratrol produce additive effects on Sirt1 signaling in the hippocampus of mice and contribute neuroprotection in mice (Wang et al., 2016; Perasso et al., 2010).
- Resveratrol extends spatial memory with decreased accumulation of Aβ (25–35) and lipid peroxidation inside the hippocampus. The included biomarkers, HO-1 and iNOS, in the subject (mice) can be normalized by administration of resveratrol (Yang et al., 2011; Chen, 2012; Yin et al., 2013; Yu et al., 2013; Lee et al., 2012).
- Resveratrol is capable of ending the aggregation of amyloid β peptides (Perasso et al., 2010).
- Researchers view resveratrol as a feasible anti-AD compound due to its capability to end the aggregation of amyloid-β peptides. For many of the pharmacologic activities and anti-AD effects noted for resveratrol, clinical information is very limited. Clinical effectiveness and the conceivable

toxicity of these compounds in larger trials requires the distinction between past suggestions associated with resveratrol use and its use today for AD therapy (Villaflores et al., 2012).

- Resveratrol mimics caloric restriction with the useful effect of extending the life span by activating deacetylases from the sirtuin family. Sirt1 decreases the Michaelis constant of theses enzymes. Sirtuin deacetylation controls the activity of a number of transcription factors. Kim and colleagues state that resveratrol intracerebroventricular injection reduces neurodegeneration in the hippocampus and averts reading impairment in p25 transgenic AD mouse model. Thus, it attenuates amyloid deposition, Sirt1 activation, and Aβ-associated neuropathy, thus contributing to anti-amyloidogenic properties (Vingtdeux et al., 2008).
- Thus, it can be concluded that resveratrol acts on a number of AD models both in vitro and in vivo by various molecular mechanisms: Sirt1, Aβ-protein, and senile plaques (Diego et al., 2010; Touqeer et al., 2017). Due to the bioavailability factor, researchers have been trying various methods to improve its efficiency (Teng et al., 2014).

Toxicology and adverse drug reactions

Several studies report that resveratrol doesn't show skin or eye irritation effects, allergy signs (Williams et al., 2009), or carcinogenicity (Schmitt et al., 2002). In addition, it doesn't affect the bone density and reproductive capacity of the individual (Boocock et al., 2007a). *Trans*-resveratrol is well tolerated in humans, and a dose of 450 mg/day represents a safe dose for a 70 kg person. It was also evident from the study that there was a reduction (50%) in breast cancer after the consumption of a high level of resveratrol (Levi et al., 2005). Thus, it can be concluded that resveratrol contains a weak toxicity profile and therefore should prove to be an interesting candidate for chemoprevention in humans (Jang et al., 1997). It also penetrates the BBB, suggesting its potential therapeutic role in brain disorders (Wang et al., 2002a). Resveratrol shows no adverse effects in humans with a dose of up to 5000 mg (Boocock et al., 2007a,b). Animal studies show a safety profile in the consumption of resveratrol (Juan et al., 2002).

Clinical trials

In a 52-week study, 119 patients were screened to assess safety and tolerability. Resveratrol 500 mg, oral dose once a day with dose escalation by a 500 mg increment every 13 weeks until a dose of 1000 mg twice daily was reached. Resveratrol reached cerebrospinal fluid at lower nanomolar concentration even at a high dose (Turner et al., 2015). A clinical trial registered as NCT00678431 conducted a study of 39 subjects with mild to moderate AD.

Results of oral administration of resveratrol at 12 months indicated that it is safe at low doses and well tolerated. Further study is required on a larger scale in other phases (Carolyn et al., 2018).

A phase 2 clinical trial study was completed (NCT01504854) in 2014 and included 119 patients with mild to moderate AD. They were administered capsules containing placebo or resveratrol with a starting dose of 500 mg a day which was increased by up to 1 g twice a day. Another clinical study (NCT01716637), sponsored by Life Extension Foundation, evaluated the effect of a resveratrol-containing dietary supplement. This study included 12 participants with mild to moderate AD. The phase I pilot study was completed in 2016, but results have not been reported. Another phase I study (NCT02502253) was conducted on 48 patients who took a combination of resveratrol and grape seed and who had a prediabetic condition and mild cognitive impairment (who were at risk of the onset of Alzheimer's disorder). This study was cosponsored by Johns Hopkins University and the Icahn School of Medicine at Mount Sinai. In 2017, a 50% reduction in MMP-9 levels in CBS was observed when patients were treated with resveratrol (https://alzheimersnewstoday.com).

Further research conducted in 2019 analyzed 19 patients taking resveratrol in placebo, and the results proved that it restores the integrity of the BBB. In another study, participant blood levels of β-amyloid were measured. No changes in levels were observed in patients, whereas there was a decrease in the level of protein in placebo group studies (https://alzheimersnewstoday.com).

References

Anekonda, T.S., Reddy, P.H., 2006. Neuronal protection by sirtuins in Alzheimer's disease. J. Neurochem. 96, 305–313.

Anisimova, N.Y., Kiselevsky, M.V., Sosnov, A.V., Sadovnikov, S.V., Stankov, I.N., Gakh, A.A., 2011. Trans-, cis- and dihydro-resveratrol: a comparative study. Chem. Cent. J. 5, 88.

Bastianetto, S., Menard, C., Quirion, R., 2015. Neuroprotective action of resveratrol. Biochim. Biophys. Acta 1852 (6), 1195–1201.

Baur, J.A., Sinclair, D.A., 2006. Therapeutic potential of resveratrol: the in vivo evidence. Nat. Rev. Drug Discov. 5, 493–506.

Berman, A.Y., Motechin, R.A., Wiesenfeld, M.Y., Holz, M.K., 2017. The therapeutic potential of resveratrol: a review of clinical trials. NPJ Precis. Oncol. 1, 35.

Bertelli, A., Bertelli, A.A.E., Gozzini, A., Giovannini, L., 1998. Plasma and tissue resveratrol concentrations and pharmacological activity. Drugs Exp. Clin. Res. 24 (3), 133–138.

Bertelli, A.A.E., Giovannini, L., Stradi, R., Bertelli, A., Tillement, J.P., 1996a. Plasma, urine and tissue levels of trans- and cis-resveratrol (3,4',5-trihydroxystilbene) after short term or prolonged administration of red wine to rats. Int. J. Tissue React. 18 (2–3), 67–71.

Bertelli, A.A.E., Giovannini, L., Stradi, R., Urien, S., Tillement, J.P., Bertelli, A., 1996b. Kinetics of trans- and cis-resveratrol (3,4',5-trihydroxystilbene) after red wine oral administration in rats. Int. J. Clin. Pharmacol. Res. 16 (4–5), 77–81.

Bolling, B.W., McKay, D.L., Blumberg, J.B., 2010. The phytochemical composition and antioxidant actions of tree nuts. Asia Pac. J. Clin. Nutr. 19 (1), 117–123.

Boocock, D.J., Faust, G.E.S., Patel, K.R., Schinas, A.M., Brown, V.A., Ducharme, M.P., Booth, T.D., Crowell, J.A., Perloff, M., Gescher, A.J., Steward, W.P., Brenner, D.E., 2007a. Phase I dose escalation pharmacokinetic study in healthy volunteers of resveratrol, a potential cancer chemopreventive agent. Cancer Epidemiol. Biomarkers Prev. 16 (6), 1246–1252.

Boocock, D.J., Patel, K.R., Faust, G.E.S., Normolle, D.P., Marczylo, T.H., Crowell, J.A., Brenner, D.E., Booth, T.D., Gescher, A., Steward, W.P., 2007b. Quantitation of trans-resveratrol and detection of its metabolites in human plasma and urine by high performance liquid chromatography. J. Chromatogr. B 848 (2), 182–187.

Braidy, N., Jugder, B.E., Poljak, A., Jayasena, T., Mansour, H., Nabavi, S.M., Sachdev, P., Grant, R., 2016. Resveratrol as a potential therapeutic candidate for the treatment and management of Alzheimer's disease. Curr. Top. Med. Chem. 16 (17), 1951–1960.

Burns, J., Yokota, T., Ashihara, H., Lean, M.E.J., Crozier, A., 2002. Plant foods and herbal sources of resveratrol. J. Agric. Food Chem. 50 (11), 3337–3340.

Capiralla, H., Vingtdeux, V., Zhao, H., Sankowski, R., Al-Abed, Y., Davies, P., Marambaud, P., 2012. Resveratrol mitigates lipopolysaccharide- and Aβ-mediated microglial inflammation by inhibiting the TLR4/NF-κB/STAT signaling cascade. J. Neurochem. 120 (3), 461–472.

Carolyn, W.Z., Hillel, G.M., Judith, N., Susan, P., Amanda, B., Xiaodong, L., Mary, S., 2018. A randomized, double-blind, placebo-controlled trial of resveratrol with glucose and malate (RGM) to slow the progression of Alzheimer's disease: a pilot study. Alzheimers Dement. 4, 609–616.

Chen, L.F., 2012. Inhibitory effect of resveratrol on tumor growth in Lewis C57BL6J and its antioxidation activity in vivo and in vitro. Zhong Guo Yi Yuan Yao Xue Za Zhi 32 (21), 1696–1699.

Chung, M.I., Teng, C.M., Cheng, K.L., Ko, F.N., Lin, C.N., 1992. An antiplatelet principle of *Veratrum formosanum*. Planta Med. 58 (3), 274–276.

Cottart, C.H., Nivet, A.V., Laguillier, M.C., Beaudeux, J.L., 2010. Resveratrol bioavailability and toxicity in humans. Mol. Nutr. Food Res. 54, 7–16.

Counet, C., Callemien, D., Collin, S., 2006. Chocolate and cocoa: new sources of trans-resveratrol and trans-piceid. Food Chem. 98 (4), 649–657.

Diego, A., Letizia, P., Alessandra, S., Gianluigi, F., 2010. Neuroprotective properties of resveratrol in different neurodegenerative disorders. Biofactors 36 (5), 370–376.

Fei, L., Qihai, G., Hongxin, D., Jingshan, S., 2012. Resveratrol, a neuroprotective supplement for Alzheimer's disease. Curr. Pharmaceut. Des. 18 (1), 27–33.

Fornara, V., Onelli, E., Sparvoli, F., Rossoni, M., Aina, R., Marino, G., Citterio, S., 2008. Localization of stilbene synthase in *Vitis vinifera* L. during berry development. Protoplasma 233 (1–2), 83–93.

Frozza, R.L., Bernardi, A., Hoppe, J.B., Meneghetti, A.B., Matte, A., Battastini, A.M.O., Pohlmann, A.R., Guterres, S.S., Salbego, C., 2013. Neuroprotective effects of resveratrol against Aβ administration in rats are improved by lipid-core nanocapsules. Mol. Neurobiol. 47 (3), 1066–1080.

Gasparini, L., Ongini, E., Wenk, G., 2004. Non-steroidal anti-inflammatory drugs (NSAIDs) in Alzheimer's disease: old and new mechanisms of action. J. Neurochem. 91 (3), 521–536.

Gescher, A.J., Steward, W.P., 2003. Relationship between mechanisms, bioavailibility, and preclinical chemopreventive efficacy of resveratrol: a conundrum. Canc. Epidemiol. Biomarkers Prev. 12 (10), 953–957.

Goldberg, D.A., Yan, J., Soleas, G.J., 2003. Absorption of three wine-related polyphenols in three different matrices by healthy subjects. Clin. Biochem. 36 (1), 79–87.

Goldberg, D.M., Tsang, E., Karumanchiri, A., Diamandis, E.P., Soleas, G., Ng, E., 1996. Method to assay the concentrations of phenolic constituents of biological interest in wines. Anal. Chem. 68 (10), 1688–1694.

Hanawa, F., Tahara, S., Mizutani, J., 1992. Antifungal stress compounds from *Veratrum grandiflorum* leaves treated with cupric chloride. Phytochemistry 31 (9), 3005–3007.

Hurst, W.J., Glinski, J.A., Miller, K.B., Apgar, J., Davey, M.H., Stuart, D.A., 2008. Survey of the trans-resveratrol and trans-piceid content of cocoa-containing and chocolate products. J. Agric. Food Chem. 56 (18), 8374–8378.

Jang, J.H., Surh, Y.J., 2003. Protective effect of resveratrol on beta-amyloid-induced oxidative PC12 cell death. Free Radic. Biol. Med. 34, 1100–1110.

Jang, M.S., Cai, E.N., Udeani, G.O., Slowing, K.V., Thomas, C.F., Beecher, C.W.W., Fong, H.H.S., Farnsworth, N.R., Kinghorn, A.D., Mehta, R.G., Moon, R.C., Pezzuto, J.M., 1997. Cancer chemopreventive activity of resveratrol, a natural product derived from grapes. Science 275 (5297), 218–220.

Jannin, B., Menzel, M., Berlot, J.P., Delmas, D., Lancon, A., Latruffe, N., 2004. Transport of resveratrol, a cancer chemopreventive agent, to cellular targets: plasmatic protein binding and cell uptake. Biochem. Pharmacol. 68 (6), 1113–1118.

Jeon, B.T., Jeong, E.A., Shin, H.J., Lee, Y., Lee, D.H., Kim, H.J., Kang, S.S., Cho, G.J., Choi, W.S., Roh, G.S., 2012. Resveratrol attenuates obesity-associated peripheral and central inflammation and improves memory deficit in mice fed a high-fat diet. Diabetes 61 (6), 1444–1454.

Jerabek, J., Uliassi, E., Guidotti, L., Korabecny, J., Soukup, O., Sepsova, V., Hrabinova, M., Kuca, K., Bartolini, M., Peña-Altamira, L.E., Petralla, S., Monti, B., Roberti, M., Bolognesi, M.L., 2017. Tacrine-resveratrol fused hybrids as multi-target-directed ligands against Alzheimer's disease. Eur. J. Med. Chem. 127, 250–262.

Juan, M.E., Vinardell, M.P., Planas, J.M., 2002. The daily oral administration of high doses of trans-resveratrol to rats for 28 days is not harmful. J. Nutr. 132 (2), 257–260.

Kaldas, M.I., Walle, U.K., Walle, T., 2003. Resveratrol transport and metabolism by human intestinal Caco-2 cells. J. Pharm. Pharmacol. 55 (3), 307–312.

Khan, M.A., Muzammil, S., Musarrat, J., 2002. Differential binding of tetracyclines with serum albumin and induced structural alterations in drug-bound protein. Int. J. Biol. Macromol. 30 (5), 243–249.

Klaus, W.L., Shiming, L., 2018. Resveratrol, pterostilbene, and dementia. Biofactors 44 (1), 83–90.

Kou, X., Chen, N., 2017. Resveratrol as a natural autophagy regulator for prevention and treatment of Alzheimer's disease. Nutrients 9 (9), 927.

Kulkarni, S.S., Canto, C., 2015. The molecular targets of resveratrol. Biochim. Biophys. Acta 1852 (6), 1114–1123.

Kursvietiene, L., Staneviciene, I., Mongirdiene, A., Bernatoniene, J., 2016. Multiplicity of effects and health benefits of resveratrol. Medicinal 52, 148–155.

Lee, K.A., Lee, Y.J., Ban, J.O., Lee, S.H., Cho, M.K., Nam, H.S., Hong, J.T., Shim, J.H., 2012. The flavonoid resveratrol suppresses growth of human malignant pleural mesothelioma cells through direct inhibition of specificity protein 1. Int. J. Mol. Med. 30 (1), 21–27.

Levi, F., Pasche, C., Lucchini, F., Ghidoni, R., Ferraroni, M., La Vecchia, C., 2005. Resveratrol and breast cancer risk. Eur. J. Cancer Prev. 14 (2), 139–142.

Liu, Z., Xu, J., Wu, X., Wang, Y., Lin, Y., Wu, D., Zhang, H., Qin, J., 2019. Molecular analysis of UV-C induced resveratrol accumulation in *Polygonum cuspidatum* leaves. Int. J. Mol. Sci. 20 (24), 6185.

Maier, S.A., Hagenauer, B., Wirth, M., Gabor, F., Szekeres, T., Jager, W., 2006. Increased transport of resveratrol across monolayers of the human intestinal Caco-2 cells is mediated by inhibition and saturation of metabolites. Pharmaceut. Res. 23 (9), 2107–2115.

Marambaud, P., Zhao, H., Davies, P., 2005. Resveratrol promotes clearance of Alzheimer's disease amyloid-beta peptides. J. Biol. Chem. 280, 37377–37382.

Marier, J.F., Vachon, P., Gritsas, A., Zhang, J., Moreau, J.P., Ducharme, M.P., 2002. Metabolism and disposition of resveratrol in rats: extent of absorption, glucuronidation, and enterohepatic recirculation evidenced by a linked-rat model. J. Pharmacol. Exp. Therapeut. 302 (1), 369–373.

Niesen, D.B., Hessler, C., Seeram, N.P., 2013. Beyond resveratrol: a review of natural stilbenoids identified from 2009–2013. J. Berry Res. 3, 181–196.

Nonomura, S., Kanagawa, H., Makimoto, A., 1963. Chemical constituents of polygonaceous plants. I. Studies on the components of Ko-jo-kon (*Polygonum cuspidatum* Sieb. et Zucc.). Yakugaku Zasshi 83, 988–990.

Pasinetti, G.M., Wang, J., Ho, L., Zhao, W., Dubner, L., 2015. Roles of resveratrol and other grape-derived polyphenols in Alzheimer's disease prevention and treatment. Biochim. Biophys. Acta 1852, 1202–1208.

Perasso, L., Cogo, C.E., Giunti, D., Gandolfo, C., Ruggeri, P., Uccelli, A., Balestrino, M., 2010. Systemic administration of mesenchymal stem cells increases neuron survival after global cerebral ischemia in vivo (2VO). Neural Plast. 2010, 534925.

Porquet, D., Grinan, F.C., Isidre, F., Antoni, C., Coral, S., Jaume, D.V., Merce, P., 2014. Neuro-protective role of trans-resveratrol in a murine model of familial Alzheimer's disease. J. Alzheim. Dis. 42 (4), 1209–1220.

Puksasook, T., Kimura, S., Tadtong, S., Jiaranaikulwanitch, J., Pratuangdejkul, J., Kitphati, W., Suwanborirux, K., Saito, N., Nukoolkarn, V., 2017. Semisynthesis and biological evaluation of prenylated resveratrol derivatives as multi-targeted agents for Alzheimer's disease. J. Nat. Med. 71 (4), 665–682.

Rege, S.D., Geetha, T., Griffin, G.D., Broderick, T.L., Babu, J.R., 2014. Neuroprotective effects of resveratrol in Alzheimer disease pathology. Front. Aging Neurosci. 6, 218.

Rimando, A.M., Kalt, W., Magee, J.B., Dewey, J., Ballington, J.R., 2004. Resveratrol, pterostilbene, and piceatannol in *Vaccinium* berries. J. Agric. Food Chem. 52 (15), 4713–4719.

Sanders, T.H., McMichael, R.W., Hendrix, K.W., 2000. Occurrence of resveratrol in edible peanuts. J. Agric. Food Chem. 48 (4), 1243–1246.

Sathya, M., Thilagar, G., Arul, S., Meher, N., Mahesh, K., Musthafa, M.E., Kesavan, S.J., Muthuswamy, A., 2018. Biophysical interaction of resveratrol with sirtuin pathway: significance in Alzheimer's disease. Front. Biosci. (Landmark Ed). 23, 1380–1390.

Schmitt, E., Lehmann, L., Metzler, M., Stopper, H., 2002. Hormonal and genotoxic activity of resveratrol. Toxicol. Lett. 136 (2), 133–142.

Silva, C.G., Monteiro, J., Marques, R.R., Silva, A.M., Martínez, C., Canle, M., Faria, J.L., 2013. Photochemical and photocatalytic degradation of trans-resveratrol. Photochem. Photobiol. Sci. 12, 638–644.

Sobolev, V.S., Cole, R.J., 1999. Trans-resveratrol content in commercial peanuts and peanut products. J. Agric. Food Chem. 47 (4), 1435–1439.

Soleas, G.J., Angelini, M., Grass, L., Diamandis, E.P., Goldberg, D.M., 2001. Absorption of trans-resveratrol in rats. In: Flavonoids and Other Polyphenols, vol. 335. Academic Press Inc, San Diego, pp. 145–154.

Sousa, J.C.E, Santana, A.C.F, MagalhÃes, G.J.P, 2020. Resveratrol in Alzheimer's disease: a review of pathophysiology and therapeutic potential. Arq. Neuropsiquiatr. 78 (8), 501–511.

Stervbo, U., Vang, O., Bonnesen, C., 2007. A review of the content of the putative chemopreventive phytoalexin resveratrol in red wine. Food Chem. 101 (2), 449–457.

Takaoka, M., 1939. Resveratrol, a new phenolic compound from *Veratrum grandiflorum*. Nippon Kagaku Kaishi 60, 1090–1100.

Teng, M., Meng, S.T., Jin, T.Y., Lan, T., 2014. Resveratrol as a therapeutic agent for Alzheimer's disease. Biomed. Res. Int. 2014, 350516.

Thimmappa, S.A., 2006. Resveratrol-a boon for treating Alzheimer's disease? Brain Res. Rev. 52 (2), 316–326.

Touqeer, A., Sehrish, J., Sana, J., Ameema, T., Dunja, S., Silvia, T., Seyed, F.N., Nady, B., Seyed, M.N., 2017. Resveratrol and Alzheimer's disease: mechanistic insights. Mol. Neurobiol. 54 (4), 2622–2635.

Trela, B.C., Waterhouse, A.L., 1996. Resveratrol: isomeric molar absorptivities and stability. J. Agric. Food Chem. 44, 1253–1257.

Tsai, H.Y., Ho, C.T., Chen, Y.K., 2017. Biological actions and molecular effects of resveratrol, pterostilbene and 3'-hydroxypterostilbene. J. Food Drug Anal. 25, 134–147.

Turner, R.S., Thomas, R.G., Craft, S., van Dyck, C.H., Mintzer, J., Reynolds, B.A., Brewer, J.B., Rissman, R.A., Raman, R., Aisen, P.S., 2015. A randomized, double-blind, placebo-controlled trial of resveratrol for Alzheimer disease. Neurology 85 (16), 1383–1391.

Urpi, S.M., Zamora, R.R., Lamuela, R.R., Cherubini, A., Jauregui, O., de la Torre, R., Covas, M.I., Estruch, R., Jaeger, W., Andres, L.C., 2007. HPLC-tandem mass spectrometric method to characterize resveratrol metabolism in humans. Clin. Chem. 53 (2), 292–299.

Varamini, B., Sikalidis, A.K., Bradford, K.L., 2014. Resveratrol increases cerebral glycogen synthase kinase phosphorylation as well as protein levels of drebrin and transthyretin in mice: an exploratory study. Int. J. Food Sci. Nutr. 65 (1), 89–96.

Venigalla, M., Sonego, S., Gyengesi, E., Sharman, M.J., Munch, G., 2016. Novel promising therapeutics against chronic neuroinflammation and neurodegeneration in Alzheimer's disease. Neurochem. Int. 95, 63–74.

Villaflores, O.B., Chen, Y.J., Chen, C.P., Yeh, J.M., Wu, T.Y., 2012. Curcuminoids and resveratrol as anti-Alzheimer agents. Taiwan. J. Obstet. Gynecol. 51 (4), 515–525.

Vingtdeux, V., Dreses, W.U., Zhao, H., Davies, P., Marambaud, P., 2008. Therapeutic potential of resveratrol in Alzheimer's disease. BMC Neurosci. 9 (2), 1–5.

Vitaglione, P., Sforza, S., Galaverna, G., Ghidini, C., Caporaso, N., Vescovi, P.P., Fogliano, V., Marchelli, R., 2005. Bioavailability of trans-resveratrol from red wine in humans. Mol. Nutr. Food Res. 49 (5), 495–504.

Vitrac, X., Desmouliere, A., Brouillaud, B., Krisa, S., Deffieux, G., Barthe, N., Rosenbaum, J., Merillon, J.M., 2003. Distribution of C-14-trans-resveratrol, a cancer chemopreventive polyphenol, in mouse tissues after oral administration. Life Sci. 72 (20), 2219–2233.

Walle, T., Hsieh, F., DeLegge, M.H., Oatis, J.E., Walle, U.K., 2004. High absorption but very low bioavailability of oral resveratrol in humans. Drug Metabol. Dispos. 32 (12), 1377–1382.

Wang, W., Tang, K., Yang, H.R., Wen, P.F., Zhang, P., Wang, H.L., Huang, W.D., 2010. Distribution of resveratrol and stilbene synthase in young grape plants (*Vitis vinifera* L. cv. Cabernet Sauvignon) and the effect of UV-C on its accumulation. Plant Physiol. Biochem. 48 (2–3), 142–152.

Wang, Y., Catana, F., Yang, Y.N., Roderick, R., van Breemen, R.B., 2002b. An LC-MS method for analyzing total resveratrol in grape juice, cranberry juice, and in wine. J. Agric. Food Chem. 50 (3), 431–435.

Wang, X., Ma, S., Meng, N., Yao, N., Zhang, K., Li, Q., Zhang, Y., Xing, Q., Han, K., Song, J., Yang, B., Guan, F., 2016. Resveratrol exerts dosage-dependent effects on the self-renewal and neural differentiation of hUC-MSCs. Mol Cells 39 (5), 418–425.

Wang, Q., Xu, J.F., Rottinghaus, G.E., Simonyi, A., Lubahn, D., Sun, G.Y., Sun, A.Y., 2002a. Resveratrol protects against global cerebral ischemic injury in gerbils. Brain Res. 958 (2), 439–447.

Wenzel, E., Soldo, T., Erbersdobler, H., Somoza, V., 2005. Bioactivity and metabolism of trans-resveratrol orally administered to Wistar rats. Mol. Nutr. Food Res. 49 (5), 482–494.

Wenzel, E., Somoza, V., 2005. Metabolism and bioavailability of trans-resveratrol. Mol. Nutr. Food Res. 49 (5), 472–481.

Williams, L.D., Burdock, G.A., Edwards, J.A., Beck, M., Bausch, J., 2009. Safety studies conducted on high-purity trans-resveratrol in experimental animals. Food Chem. Toxicol. 47 (9), 2170–2182.

Yang, K., He, J., Zhang, P., 2011. Inhibitory effects of resveratrol on growth of lewis lung cancer cell in mice and possible mechanism. Zhongliu Fangzhi Yanjiu 38 (8), 871–874.

Yin, H.T., Tian, Q.Z., Guan, L., Zhou, Y., Huang, X.E., Zhang, H., 2013. In vitro and in vivo evaluation of the antitumor efficiency of resveratrol against lung cancer. Asian Pac. J. Cancer Prev. 14 (3), 1703–1706.

Yongming, J., Wang, N., Liu, X., 2017. Resveratrol and amyloid-beta: mechanistic insights. Nutrients 9 (10), 1122.

Yu, Y.H., Chen, H.A., Chen, P.S., Cheng, Y.J., Hsu, W.H., Chang, Y.W., Chen, Y.H., Jan, Y., Hsiao, M., Chang, T.Y., Liu, Y.H., Jeng, Y.M., Wu, C.H., Huang, M.T., Su, Y.H., Hung, M.C., Chien, M.H., Chen, C.Y., Kuo, M.L., Su, J.L., 2013. MiR-520h-mediated FOXC2 regulation is critical for inhibition of lung cancer progression by resveratrol. Oncogene 32 (4), 431–443.

Zamora, R.R., Urpi, S.M., Lamuela, R.R.M., Estruch, R., Vaázquez, A.M., Serrano, M.M., Jaeger, W., Andres, L.C., 2006. Diagnostic performance of urinary resveratrol metabolites as a biomarker of moderate wine consumption. Clin. Chem. 52 (7), 1373–1380.

Chapter 3.1.2

Curcumin

Ashutosh Paliwal[1], Ashwini Kumar Nigam[2], Jalaj Kumar Gour[3], Deepak Singh[1], Pooja Pandey[1], Manoj Kumar Singh[4]

[1]*Department of Biotechnology, Kumaun University Nainital, Bhimtal Campus, Bhimtal, Uttarakhand, India;* [2]*Department of Zoology, Udai Pratap College, Varanasi, Uttar Pradesh, India;* [3]*Department of Biochemistry, Faculty of Science, University of Allahabad, Prayagraj, Uttar Pradesh, India;* [4]*Center for Noncommunicable Diseases (NCD), National Centre for Disease Control, Delhi, India*

Introduction

Plants are important sources of natural bioactive compounds with tremendous therapeutic potential. Extracts of different parts of plants have been used in traditional medicinal systems since ancient times. The unavoidable side effects of synthetic drugs during disease treatment and management are necessitating the exploration/need/search for safer substitutes. Herbal medication is now emerging as the most promising alternative and is expected to revolutionize the modern methods of disease treatment.

The Indian subcontinent is blessed with diverse but less explored medicinal flora. The traditional Indian health system is based on herbal formulations. There are many medicinal plants available in various parts of India that possess a broad array of phytomedicinal properties including having antimicrobial, antiinflammatory, antioxidant antineoplastic, antianalgesic, and antipyretic potentials. These bioactivities are due to the presence of various chemical compounds called phytochemicals, which are synthesized in plants naturally. The interest of public and scientific communities in the phytochemical domain in which curcumin is very prevalent is very considerable. Curcumin, an important phytochemical with a well-known medicinal spectrum of activities is an active ingredient of the rhizome of the turmeric plant. This plant is an annual herb (*Curcuma longa* Linn.), known as turmeric in English, Haldi in Hindi, and has been used in Asia for its medicinal properties since the second millennium BC. In Indian household daily practices, turmeric is generally considered and recommended as an antiseptic agent. It is used along with many other natural combinations. Curcumin (diferuloylmethane) is a low-molecular-weight polyphenol (Fig. 3.1.2.1) that has antioxidant, antiinflammatory,

Naturally Occurring Chemicals against Alzheimer's Disease. https://doi.org/10.1016/B978-0-12-819212-2.00018-9

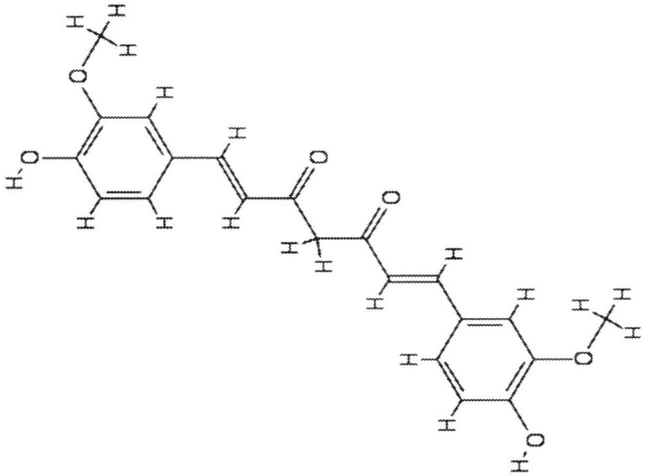

FIGURE 3.1.2.1 Chemical structure of curcumin.

anticancer, chemopreventive, and decisively chemotherapeutic-like characteristics (Fig. 3.1.2.2). Many of these bioactivities endorse the candidature of curcumin in the prevention and halting or slowing of age-related cognitive decline and dementia.

Alzheimer's disease, symptoms, and pathophysiology

The brain is considered as the command center for regulating various types of tasks, metabolism, and day-to-day activities. For proper functioning of all physiological, metabolic-related activities in the human body, the brain needs to function appropriately. In some cases, proper signaling between neurons in nerve cells can be disrupted, leading to accretion of amyloids in the brain, and so resulting in dysfunction of daily activities in AD-affected persons. The disease is pathologically defined by impairments in memory and cognitive tasks, which finally results in loss of self-sufficiency. Alzheimer's disease is a devastating neural condition in which the degeneration of neurons results in loss of memory, cerebral diminution, and a prominent form of dementia (accounting for 50%–70%). There are some common symptoms that can be associated with AD, i.e., short-term memory loss, difficulty in performing daily activities such as dressing, eating with a spoon, difficulty in communicating, becoming lost, etc. It is a collective category of presenile and senile dementia.

Alzheimer's disease progressively worsens over time, leading to an inability to verbally communicate and incorrect responses to surroundings. Alzheimer's Association (2015) estimated that one out of nine patients over the age of 65 is affected with AD, and this rate increases greatly after reaching

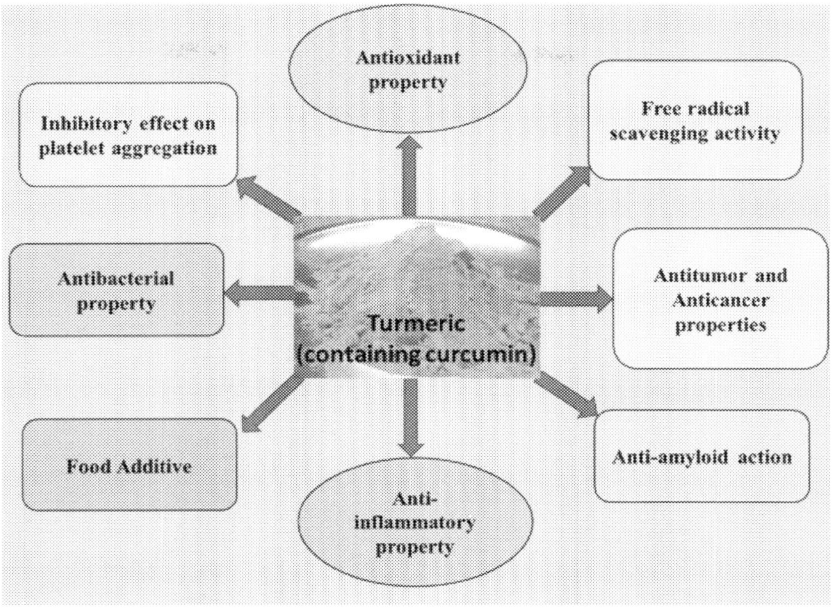

FIGURE 3.1.2.2 Bioactivities of curcumin.

an age of 85 years. As dementia worsens with time, AD patients cut themselves from society, and the final outcome is death. It has been observed that the life span after diagnosis ranges between 3–9 years (Todd et al., 2013). In 2015, approximately 44 million people globally were suffering from AD or related dementia. In India, 33.6% of the population is reported to be affected by or diagnosed with Alzheimer's disease. This number is projected to increase to 115.4 million world wide by 2050, and it is predicted that, every 33 seconds, a new AD patient will be reported somewhere in the world (Alzheimer's Disease International World Alzheimer Report, 2010). AD exhibited tremendous worst effect on human health as well as on economy and society, and there would be nothing surprising about becoming of AD epidemic shortly (Hampel et al., 2011).

Alzheimer's disease, like other chronic diseases, can arise from multiple factors and not only from a single cause. However, the precise reason behind the onset of disease remains unclear (LaFerla and Green, 2012). Clinical symptoms of AD, as documented by Huang and Mucke (2012), include changes in behavior and personality, weakness of memory, and cognitive disorders. Any abnormality observed during neurodegeneration is due to two proteins, i.e., amyloid-beta (Aβ) and tau protein. Accretion of extracellular amyloid plaques and intracellular neurofibrillary tangles (NFTs) in the brain of AD patients are considered as pathological symptoms. Amyloid-β (Aβ) peptides are produced from breakdown of the amyloid precursor proteins (APPs) by the action of enzymes, namely β- and γ-secretases. Amyloid-β breakdown results in native Aβ monomers that have prosurvival effects on neurons and shield mature neurons against excitotoxic demise (Giuffrida et al., 2009). In contrast, under pathological conditions, disproportionate amassing of monomers results in their assembly into soluble, diffusible toxic oligomeric Aβ species which are low-molecular-weight aggregates consisting of 2–30 Aβ peptides. When the oligomers reach the critical/optimum concentration, they form insoluble fibrils/aggregates and plaques. It is important to note that soluble Aβ oligomers are more toxic than insoluble deposits (Verma et al., 2015). Hyperphosphorylation of tau protein prompts the synthesis of NFTs, another hallmark of AD that disrupts the structure of microtubules and also have a role in the impairment in axonal transport (Beharry et al., 2014; Metaxas and Kempf, 2016; Ye et al., 2017). Besides general neuro-inflammation, oxidative damage is also an observable characteristic in the brains of AD-affected victims. Importantly, a vicious cycle develops among Aβ, NFTs, oxidative stress, and inflammation. Aβ and NFTs activate microglia and induce the generation of reactive oxygen species (ROS) and inflammatory factors. ROS and inflammatory cytokines affect the neurons and increase the emergence of Aβ and neurons (Glass et al., 2010; Broussard et al., 2012; Luque-Contreras et al., 2014).

Several reports are available on cloning of the amyloid precursor gene and also on its localization at chromosome 21 that previously was documented while studying trisomy 21 (Olson and Shaw, 1969; Robakis et al., 1987). The occurrence of mutation close to the Aβ region of the coding sequence of the APP gene is also reported (Goate et al., 1991; Hendriks et al., 1992). Based on these studies, it can be concluded that any change in the metabolism of APP by secretases enzyme or any nonsignificant behavior/change in APP metabolism might be crucial for AD pathogenesis. According to Iwatsubo et al. (1994), there are two isoforms of Aβ protein—Aβ40 and Aβ42—differing from each other only in the presence of the carboxy-terminus of the Aβ end at the 40th or 42nd amino acid positions, respectively. Based on scientific reports, it was concluded that Aβ42 could manifest the plaque formation at the early age of 12 years. On the other hand, Aβ40 is involved later, at the age of 20 years (Lemere et al., 1996).

Phytomedicinal properties of curcumin in relation to Alzheimer's disease

Curcumin has been associated with improved cognitive abilities and lowered pervasiveness of dementia. One hypothesis suggests that, in animal studies, curcumin could hinder cognitive diseases as they have the potential to preclude or revert insufficiency in memory and cognition associated with aging, stress, epilepsy, oxidative stress, anxiety, traumatic brain injury, and dementia.

Curcumin has an active role in poor behavioral and neurotransmitter aberrations in depression and anxiety. It has antimicrobial, antisporadic, and antiscleroderma activity that help to treat many diseases such as hyperglycemia, cancer, and high blood pressure. Curcumin downregulates the cyclooxygenase and lipoxygenase pathways that prevent the formation of leukotrienes and prostaglandins, which are mediators of inflammation; this is why it is reported to have antiinflammatory property. Curcumin has the potential to affect endorsement of glucose due to enhanced expression of glucose transporter and reduce blood glucose level and also induce pancreatic cells to secrete more insulin, which illustrates its antidiabetic effect.

Excessive deterioration of cholinergic neurons in AD led to the expansion of treatments targeted toward acceleratory cholinergic activity (Fisher, 2008). In this way, it can be concluded that the most successful drugs for treatment of AD could be cholinesterase inhibitors which would upregulate acetylcholine in synaptic cleft, further enhancing the role of left cholinergic neurons. There are various reports available on the current standard of treatment including donepezil, galantamine, and rivastigmine (Birks, 2006; Atri et al., 2008; Hansen et al., 2008; Osborn and Saunders, 2010). The implications of cholinesterase inhibitor use are linked with a drawback because their potency decreases due to continued deterioration of cholinergic neurons. The studies of Molinuevo et al. (2004), McShane et al. (2006), Atri et al. (2008), and Herrmann et al. (2011) revealed that another approved method for treatment of AD is memantine, an *N*-methyl-D-aspartic acid (NMDA) receptor antagonist. There are several animal and cell culture study outcomes indicating that the use of antiinflammatory drugs could mitigate the risks of AD by reducing the levels of β-amyloid and tau, possibly by the suppression of γ-secretase activity (Tomita, 2009). Unfortunately, however, after many successful clinical trials of antiinflammatory drugs in older AD patients, the recovery rates are unsatisfactory (Martin et al., 2008).

Role of curcumin in clinical studies/trials

Detailed research by various scientific groups has revealed the therapeutic prospects of curcumin against a wide spectrum of terrible human diseases. Before entering clinical trials of curcumin in humans, Prasad et al. (2014) tested the impact of curcumin on animal models; during their study, they tested oral administration to animal models, while some other modes, such as intraperitoneal injection as well as intravenal delivery, were also tested in those models during the study period. To elucidate a more effective role of curcumin, different research groups have discovered various modes of administration. Studies suggest possible mode of administrations including dry emulsions, lauryl sulfated chitosan, flexible liposomes, etc. to achieve better absorption and maximum concentrations (Shelma and Sharma, 2013; Jang et al., 2014). Additionally, various reports have proclaimed the interactive potential of curcumin with numerous cell-signaling molecules such as proinflammatory cytokines, apoptotic proteins, and phosphorylase kinases. These studies set a strong foundation to examine the effectiveness of curcumin in clinical trials (Table 3.1.2.1). A number of studies conducted in animal models with curcumin treatment have given fruitful results which could not be satisfactorily observed during clinical trials (LaFerla and Green, 2012). This highlights the urgent need for a next generation of animal models, which better recapitulates critical aspects of the disease spectrum and facilitates success in presymptomatic studies and human clinical trials. In view of this, more investigations with better bioavailability and targeting strategies, large study size, and longer treatment duration are needed. Baum et al. (2008) carried out a clinical trial of a 6-month treatment with curcumin on AD patients and reported no significant variation between curcumin-treated patients and placebo-treated ones. In addition, a difference in absorption rate was not noticed between the patients receiving a 1 g dose and those receiving 4 g (Baum et al., 2008). A summary of clinical trial studies is given in Table 3.1.2.1.

Impact of curcumin on Alzheimer disease

Indian natives are long familiar with curcumin, as it is a very common household spice with many medicinal qualities. The scientific community has focused its attention toward curcumin, as some epidemiological studies have claimed that cases of AD were reduced 4.4-fold in Indian populations through the use of curcumin (Ganguli et al., 2000). Curcumin is a well-known and widely explored compound for its medicinal potential and also its clinical and therapeutic potential against various diseases. There are a number of in vivo studies providing detailed information about the curative potency of curcumin against AD. Several studies have claimed that curcumin could overcome the hurdle of crossing the blood—brain barrier and exhibited a preventive role by decreasing Aβ fibril synthesis and eliminating Aβ fibrils along with their extensions (Garcia-Alloza et al., 2007; Yang et al., 2005). An experiment performed on an AD mouse model exhibited a reductive effect on serum Aβ levels as well as on Aβ burden in the brain after curcumin treatment, and curative results were effectively shown in the neocortex and hippocampus (Wang et al., 2009). Treatment with curcumin decreased the amounts of deformed neuritic structures in immediate vicinity to plaques, and this phenomenon strongly supports the therapeutic potential of curcumin in AD treatment (Garcia-Alloza et al., 2007). Alzheimer's disease interferes in the synthetic pathway and final maturation of amyloid precursor protein (APP) (Suh and Checler, 2002). Curcumin reduces the synthesis of Aβ with a synchronal lessening of presenilin 1 gene expression, which intimates its role in γ-secretase inflection (Xiong et al., 2011). Furthermore, curcumin has also been reported to disrupt the synthetic pathway of APP, resulting in diminished APP production (Zhang et al., 2010). To support the role of curcumin, a study

TABLE 3.1.2.1 Clinical trial reports of curcumin to improve cognitive functions in Alzheimer's disease (AD) or other conditions.

Participants	Mean age group/sample size	Treatment duration	Dose of compound	Cognitive measurement	Study/ references
A. Studies resulting in no significant cognitive improvement in AD patients/healthy adults					
Elderly AD patients	78 y; N = 27	6 months	1000 mg/4000 mg curcumin	MMSE score	Baum et al. (2008)
Elderly AD patients	74 y; N = 30	6 months	2000 mg/4000 mg curcumin	MMSE score; ADAS-cog	Ringman et al. (2012)
Healthy adults	66 y; N = 96	12 months	1500 mg curcumin	RAVLT, COWAT, WDSS, and CogState battery of assessments	Rainey-Smith et al. (2016)
B. Studies resulting in significant improvement in cognitive and memory in healthy or non-AD patients					
Untreated prediabetic adults	73 y; N = 48	Acute effect for 6 h	Single dose of 1000 mg curcumin alone or with 2000 mg cinnamon	–	Lee et al. (2014)
Healthy adults	68 y; N = 60	Acute effect for 1–3 h; Chronic effects for 4 weeks	400 mg of curcumin	COMPASS, DASS21	Cox et al. (2015)
Nondemented adults to elderly persons	51–84 y; N = 40	18 months	90 mg curcumin	SRT, BVMT-R, Trail Making Test	Small et al. (2018)

ADAS-cog, Alzheimer's Disease Assessment Scale, cognitive subportion; *BVMT-R*, Brief Visual Memory Test-Revised; *COMPASS*, Computerized Mental Performance Assessment System; *COWAT*, Controlled Oral Word Association Test; *DASS21*, 21-item version of the Depression, Anxiety and Stress Scales; *MMSE*, Mini-Mental State Examination; *RAVLT*, Rey Auditory Verbal Learning Test; *SRT*, Buschke Selective Reminding Test; *WDSS*, Wechsler Digit Symbol Scale.

conducted on two AD populations has demonstrated an improved cognitive function in the population consuming curcumin in their diet as compared to those deprived of curcumin (Ng et al., 2006). Other investigations have revealed that the free radical scavenging ability of curcumin is higher than that of vitamin E (α-tocopherol) and tetrahydro-curcumin (Ak and Gülçin, 2008; Begum et al., 2008). Application of curcumin in a mouse model demonstrated a reduction in the levels of Aβ40 and Aβ42 (Zhang et al., 2010). It has been reported that curcumin exerts its effect by delaying the exit of immature APP from the endoplasmic reticulum (ER), thereby stabilizing an immature form of APP at ER site, and at the same time curcumin reduced the process of endocytosis of APP (Zhang et al., 2010). Additionally, some reports also have suggested the advantageous outcomes of curcumin on brain cells by enhancing the level of heme oxygenase-1, which protects brain cells from stress (Scapagnini et al., 2006; Yang et al., 2009).

Recently, Teter et al. (2019) presented an interesting insight into the immunomodulatory potential of curcumin through regulating the expression of the TREM2-CD33-TyroBP hub to promote favorable inflammatory conditions in the brain and thereby improve Alzheimer pathophysiology and symptoms. Curative potential of curcumin on "sporadic dementia of the Alzheimer's type" (SDAT) was studied using a comparable model (Agrawal et al., 2010). During this study, curcumin was given through the oral route for 14 days prior to artificial AD induction or up to next 6 days after induction. The outcomes of this curcumin administration were exhibited by increasing memory execution in a Morris Water Maze (MWM) with time under both treatment situations. In addition, the application of curcumin helped to put back the level of oxidative stress, acetylcholine, and insulin. In another study, Awasthi et al. (2010) examined the therapeutic potential of curcumin at different doses, i.e., at 10, 20, and 50 mg/kg, through the oral route beginning from the first day of SDAT induction. The medicinal efficacy of curcumin was also evaluated by administering doses at 25 and 50 mg/kg for 7 days post induction. During the study, it was observed that at 20 and 50 mg/kg doses, curcumin exhibited a preventive ability against memory difficulties; on the other hand, at doses of 25 and 50 mg/kg, it improved memory-related damage. At higher doses, additional beneficial results were achieved. Moreover, curcumin also improved blood circulation in the cerebrum, oxidative damage, and the level of acetylcholinesterase. The study of Isik et al. (2009) claimed therapeutic and curative potentials of curcumin at a comparatively higher dose of 300 mg/kg. Zhang et al. (2015) demonstrated a familial form of AD (FAD) and assessed preventive outcomes of curcumin at doses of 50, 100, and 200 mg/kg, on intraventricularly injected Aβ1-42 animal models. Likewise, Wang et al. (2013) and Yin et al. (2014) investigated the impact of curcumin (300 mg/kg) on an Aβ1−40-induced rat model and reported beneficial effects of curcumin in the improvement of spatial learning and memory damage associated with enhanced hippocampal rejuvenation. The impact of different doses of curcumin on cognitive impairment was investigated by Wang et al. (2014). They used an APP/PS1 double transgenic AD model to evaluate the efficacy of two different concentrations of curcumin doses at 160 and 1000 ppm. Different investigators working in the field ascertained a remarkable advancement in cognitive impairment in both doses in comparison to an untreated group. Moreover, a significant response with a high dose of curcumin was recorded with time, as high amount of curcumin exhibited greater cognitive improvement. The conclusive outcome of this study was that curcumin decreased Aβ accumulation by promoting autophagy.

Several experiments have also been performed to investigate the effects of curcumin on neurofibrillary tangles (NFT) in AD pathogenesis. Longvida, a dietary solid-lipid nanoparticle at a dose of 500 ppm, was applied on a tau mouse model (Ma et al., 2013). The outcome of this study advocated that subjection of curcumin may modify cognitive function and also resulted in various biochemical changes which lead to suppressed aggregation of tau protein. Sundaram et al. (2017) investigated the effect of dietary curcumin on abnormal accumulation of Aβ and NFT in a p25 transgenic mice model. Both characteristics of AD were found to be decreased, probably due to repressed levels of neuroinflammatory cytokines, e.g., MIP-1α, TNF-α, and IL-1β. Recently, McClure et al. (2017) investigated the effectiveness of an inhaled formulation of curcumin against AD, which was attributed to enhanced permeability across the blood–brain barrier. In their study, young 5XFAD mice, treated with nasal curcumin formulation exhibited improved impact on memory impairments and also on Aβ plaque burden in adulthood when compared with control mice. Fortunately, no reports of any noxious reactions in the respiratory or circulatory systems of tested models were received, because curcumin was applied in a nebulized form. Taken together, these reports on AD animal models confirm the protective and therapeutic impact of curcumin on cognitive traits. Advantageous and promising outcomes have been reported at both molecular and behavioral levels.

Treatment of Alzheimer's disease

To date, limited synthetic drugs are available to treat and manage AD. Nowadays, research is focusing mainly on herbal remedies which not only treat AD but also exert a less toxic effect during medication (Ansari and Khodagholi, 2013). Also, there are no proper diagnosis or management recommendations available for AD. Early diagnosis of AD patients based on clinical symptoms is not an easy task, and most patients are diagnosed in the final stage of the disease (Dubois et al., 2007). Diagnosis is only possible through observable changes in behavior and mental performance of individuals. As no permanent treatment is available for AD, only an ideal diet, physical exercise, and mental exercise can stimulate and improve mental health (Vivar, 2015). Additionally, some studies have revealed that a phenolic-rich diet could attenuate the threat of AD (Yamada et al., 2015). Safouris et al. (2015) documented the consequence of a Mediterranean-type diet on the prevalence of AD. In contrast, other reports have advised that an Asian diet, which is generally rich in soy and/or turmeric, containing substantial amounts of isoflavones and curcumin, respectively, also reduces the risk/proliferation of AD (Ng et al., 2006). There are many synthetic commercialized drugs for many chronic diseases, but these drugs could alleviate cytotoxicity so attenuating its effect; therefore, herbal remedies like curcumin are recommended to treat AD. Since ancient times curcumin has been used as a medicinal alternative in the traditional Indian health system. There are various scientific reports available that suggest the beneficial role of curcumin on chronic diseases. Curcumin is a phenolic-based natural compound that shows positive effects on obesity, diabetes, and depression (Arun and Nalini, 2002; Kim and Kim, 2010; Rinwa et al., 2013). Reitz et al. (2011) reported that uptake/oral inhalation of curcumin may prevent the progression of AD. Moreover, in healthy rodents, curcumin helps in promoting memory function by enhancing synaptic plasticity and neurogenesis (Kim et al., 2008; Dong et al., 2012; Belviranlı et al., 2013).

Curcumin may also increase the synthesis of docosahexaenoic acid (DHA), an important omega-3 fatty acid that results in better plasma membrane integrity, which further maintains normal mitochondrial and synaptic function (Pinkaew et al., 2015; Wu et al., 2015). Many reports have explored the role of curcumin in old people. DiSilvestro et al. (2012) revealed that a small quantity of lipidated curcumin produced several potential health benefits in healthy middle-aged people by enhancing their nitric oxide levels and lowering levels of soluble intercellular adhesion molecule. Both molecules also have relevance for cardiovascular disease risk. Further, curcumin suppressed alanine aminotransferase activity, a general marker of liver injury, and elevated plasma myeloperoxidase, a consequence related to inflammation (DiSilvestro et al., 2012). Cox et al. (2015) demonstrated that supplementation with solid lipid curcumin preparation (80 mg as Longvida) improved cognitive function, reduced fatigue, and lessened the injurious effect of psychological stress on mood, which possibly could increase the quality of life of old patients. Therefore, dietary intake of curcumin may reduce AD risk, enhance cognitive function, and delay and counteract the effects of aging and neurodegenerative disease.

Conclusion

Alzheimer's is a chronic diseases which is fatal if not treated/diagnosed in time. Symptoms of AD are not seen at an early stage, and so the diagnosis can be quite difficult. Alzheimer's is most prevalent in aged/old people and leads to memory loss, cognitive disorders, and behavioral changes. There is various scientific research supporting the potent role of curcumin against AD. Also, the abundance of curcumin (as turmeric) in the Asian subcontinent makes it a strong candidate against AD as a treatment option. Evidence also suggests that curcumin consumption has diverse potential health benefits in the aged population. Curcumin is one of the most studied phytochemicals in turmeric, displaying complex and multifaceted activities. There have been many reports on curcumin and its roles in AD. Curcumin would be an alternative medication to manage conditions of AD through destabilization and increased phagocytosis of Aβ plaques, and reduced overall inflammation. Additionally, curcumin exhibits a more comprehensive range of treatment without or with insignificant cytotoxic effects in animal models and clinical trials. Various preclinical trials have indicated promising outcomes of curcumin treatment of the cognitive impairment of AD and undue aging. However, the number of human trials to explore the efficacy of curcumin on AD remains limited. Primary data of human-based studies were in corroboration with preclinical outcomes that curcumin may have the potential to prevent cognitive deterioration rather than enhance it in a healthy community. However, the results of some clinical trials have been found to be less consistent than those of preclinical investigations. As curcumin is a natural and widely procurable compound, it would be a very cost-effective treatment for AD patients. In addition, curcumin exhibits strong cognitive impairment capability with a safer profile, which puts it in the category of safer treatments. Curcumin also shows antioxidative, antiinflammatory, and antiamyloid potential in vitro and in earlier reports in animal models. Also, curcumin lowers the COX2 and NFL levels which lead to attenuation in all types of inflammation. As studies reported a higher concentration of NFL in AD cases, this supports the strong candidature of curcumin in the treatment of AD. Other studies on curcumin have shown a reduction in pathology of amyloid in a concentration-dependent manner. Additionally, various preclinical studies have advocated the potential candidature of curcumin against AD pathophysiological features. Curcumin could become a promising alternative for dementia treatment over the presently available treatments that are accompanied by critical reactions and side effects. Nevertheless, in the present scenario of increasing AD patients, extensive clinical trials are needed to better assess the efficacy of curcumin.

References

Agrawal, R., Mishra, B., Tyagi, E., Nath, C., Shukla, R., 2010. Effect of curcumin on brain insulin receptors and memory functions in STZ (ICV) induced dementia model of rat. Pharmacol. Res. 61 (3), 247–252.

Ak, T., Gülçin, İ., 2008. Antioxidant and radical scavenging properties of curcumin. Chem. Biol. Interact. 174 (1), 27–37.

Alzheimer's, A., 2015. 2015 Alzheimer's disease facts and figures. Alzheimer's Dement. 11 (3), 332.

Alzheimer's Disease International World Alzheimer Report 2010: The Global Economic Impact of Dementia. Alzheimer's Disease International 2010.

Ansari, N., Khodagholi, F., 2013. Natural products as promising drug candidates for the treatment of Alzheimer's disease: molecular mechanism aspect. Curr. Neuropharmacol. 11 (4), 414–429.

Arun, N., Nalini, N., 2002. Efficacy of turmeric on blood sugar and polyol pathway in diabetic albino rats. Plant Foods Hum. Nutr. 57 (1), 41–52.

Atri, A., Shaughnessy, L.W., Locascio, J.J., Growdon, J.H., 2008. Long-term course and effectiveness of combination therapy in Alzheimer disease. Alzheimer Dis. Assoc. Disord. 22 (3), 209–221.

Awasthi, H., Tota, S., Hanif, K., Nath, C., Shukla, R., 2010. Protective effect of curcumin against intracerebral streptozotocin induced impairment in memory and cerebral blood flow. Life Sci. 86 (3–4), 87–94.

Baum, L., Lam, C.W.K., Cheung, S.K.K., Kwok, T., Lui, V., Tsoh, J., 2008. Six-month randomized, placebo-controlled, double-blind, pilot clinical trial of curcumin in patients with Alzheimer disease. J. Clin. Psychopharmacol. 28, 110–113.

Begum, A.N., Jones, M.R., Lim, G.P., Morihara, T., Kim, P., Heath, D.D., Hu, S., 2008. Curcumin structure-function, bioavailability, and efficacy in models of neuroinflammation and Alzheimer's disease. J. Pharmacol. Exp. Therapeut. 326 (1), 196–208.

Beharry, C., Cohen, L.S., Di, J., Ibrahim, K., Briffa-Mirabella, S., Alonso, A.D.C., 2014. Tau-induced neurodegeneration: mechanisms and targets. Neurosci. Bull. 30 (2), 346–358.

Belviranlı, M., Okudan, N., Atalık, K.E.N., Öz, M., 2013. Curcumin improves spatial memory and decreases oxidative damage in aged female rats. Biogerontology 14 (2), 187–196.

Birks, J., 2006. Cholinesterase inhibitors for Alzheimer's disease. Cochrane Database Syst. Rev. 1, CD005593.

Broussard, G.J., Mytar, J., Li, R.C., Klapstein, G.J., 2012. The role of inflammatory processes in Alzheimer's disease. Inflammopharmacology 20, 109–126.

Cox, K.H., Pipingas, A., Scholey, A.B., 2015. Investigation of the effects of solid lipid curcumin on cognition and mood in a healthy older population. J. Psychopharmacol. 29 (5), 642–651.

DiSilvestro, R.A., Joseph, E., Zhao, S., Bomser, J., 2012. Diverse effects of a low dose supplement of lipidated curcumin in healthy middle aged people. Nutr. J. 11 (1), 79.

Dong, S., Zeng, Q., Mitchell, E.S., Xiu, J., Duan, Y., Li, C., Zhao, Z., 2012. Curcumin enhances neurogenesis and cognition in aged rats: implications for transcriptional interactions related to growth and synaptic plasticity. PLoS One 7 (2), e31211.

Dubois, B., Feldman, H.H., Jacova, C., DeKosky, S.T., Barberger-Gateau, P., Cummings, J., Meguro, K., 2007. Research criteria for the diagnosis of Alzheimer's disease: revising the NINCDS–ADRDA criteria. Lancet Neurol. 6 (8), 734–746.

Fisher, A., 2008. Cholinergic treatments with emphasis on M1 muscarinic agonists as potential disease-modifying agents for Alzheimer's disease. Neurotherapeutics 5 (3), 433–442.

Ganguli, M., Chandra, V., Kamboh, M.I., Johnston, J.M., Dodge, H.H., Thelma, B.K., DeKosky, S.T., 2000. Apolipoprotein E polymorphism and Alzheimer disease: the Indo–US cross-national dementia study. Arch. Neurol. 57 (6), 824–830.

Garcia-Alloza, M., Borrelli, L.A., Rozkalne, A., Hyman, B.T., Bacskai, B.J., 2007. Curcumin labels amyloid pathology in vivo, disrupts existing plaques, and partially restores distorted neurites in an Alzheimer mouse model. J. Neurochem. 102 (4), 1095–1104.

Giuffrida, M.L., Caraci, F., Pignataro, B., Cataldo, S., De Bona, P., Bruno, V., Garozzo, D., 2009. β-amyloid monomers are neuroprotective. J. Neurosci. 29 (34), 10582–10587.

Glass, C.K., Saijo, K., Winner, B., Marchetto, M.C., Gage, F.H., 2010. Mechanisms underlying inflammation in neurodegeneration. Cell 140 (6), 918–934.

Goate, A., Chartier-Harlin, M.C., Mullan, M., Brown, J., Crawford, F., Fidani, L., Mant, R., 1991. Segregation of a missense mutation in the amyloid precursor protein gene with familial Alzheimer's disease. Nature 349 (6311), 704.

Hampel, H., Prvulovic, D., Teipel, S., Jessen, F., Luckhaus, C., Frölich, L., Hoffmann, W., 2011. The future of Alzheimer's disease: the next 10 years. Prog. Neurobiol. 95 (4), 718–728.

Hansen, R.A., Gartlehner, G., Webb, A.P., Morgan, L.C., Moore, C.G., Jonas, D.E., 2008. Efficacy and safety of donepezil, galantamine, and rivastigmine for the treatment of Alzheimer's disease: a systematic review and meta-analysis. Clin. Interv. Aging 3 (2), 211–225.

Hendriks, L., Van Duijn, C.M., Cras, P., Cruts, M., Van Hul, W., Van Harskamp, F., Hofman, A., 1992. Presenile dementia and cerebral haemorrhage linked to a mutation at codon 692 of the β-amyloid precursor protein gene. Nat. Genet. 1 (3), 218.

Herrmann, N., Li, A., Lanctot, K., 2011. Memantine in dementia: a review of the current evidence. Expert Opin. Pharmacother. 12 (5), 787–800.

Huang, Y., Mucke, L., 2012. Alzheimer mechanisms and therapeutic strategies. Cell 148 (6), 1204–1222.

Isik, A.T., Celik, T., Ulusoy, G., Ongoru, O., Elibol, B., Doruk, H., Akman, S., 2009. Curcumin ameliorates impaired insulin/IGF signalling and memory deficit in a streptozotocin-treated rat model. Age 31 (1), 39–49.

Iwatsubo, T., Odaka, A., Suzuki, N., Mizusawa, H., Nukina, N., Ihara, Y., 1994. Visualization of Aβ42 (43) and Aβ40 in senile plaques with end-specific Aβ monoclonals: evidence that an initially deposited species is Aβ42 (43). Neuron 13 (1), 45–53.

Jang, D.J., Kim, S.T., Oh, E., Lee, K., 2014. Enhanced oral bioavailability and antiasthmatic efficacy of curcumin using redispersible dry emulsion. Biomed. Mater. Eng. 24, 917–930.

Kim, M., Kim, Y., 2010. Hypocholesterolemic effects of curcumin via up-regulation of cholesterol 7a-hydroxylase in rats fed a high fat diet. Nutr. Res. Pract. 4 (3), 191–195.

Kim, S.J., Son, T.G., Park, H.R., Park, M., Kim, M.S., Kim, H.S., Lee, J., 2008. Curcumin stimulates proliferation of embryonic neural progenitor cells and neurogenesis in the adult hippocampus. J. Biol. Chem. 283 (21), 14497–14505.

LaFerla, F.M., Green, K.N., 2012. Animal models of Alzheimer disease. Cold Spring Harb. Perspect. Med. 2 (11), a006320.

Lee, M.S., Wahlqvist, M.L., Chou, Y.C., Fang, W.H., Lee, J.T., Kuan, J.C., Andrews, Z.B., 2014. Turmeric improves post-prandial working memory in pre-diabetes independent of insulin. Asia Pac. J. Clin. Nutr. 23 (4), 581.

Lemere, C.A., Blusztajn, J.K., Yamaguchi, H., Wisniewski, T., Saido, T.C., Selkoe, D.J., 1996. Sequence of deposition of heterogeneous amyloid β-peptides and APO E in Down syndrome: implications for initial events in amyloid plaque formation. Neurobiol. Dis. 3 (1), 16–32.

Luque-Contreras, D., Carvajal, K., Toral-Rios, D., Franco-Bocanegra, D., Campos-Peña, V., 2014. Oxidative stress and metabolic syndrome: cause or consequence of Alzheimer's disease? Oxid. Med. Cell. Longev. 1–11. Article ID 497802.

Ma, Q.L., Zuo, X., Yang, F., Ubeda, O.J., Gant, D.J., Alaverdyan, M., Teter, B., 2013. Curcumin suppresses soluble tau dimers and corrects molecular chaperone, synaptic, and behavioral deficits in aged human tau transgenic mice. J. Biol. Chem. 288 (6), 4056–4065.

Martin, B.K., Szekely, C., Brandt, J., 2008. ADAPT Research Group. Cognitive function over time in the Alzheimer's Disease Anti-inflammatory Prevention Trial (ADAPT): results of a randomized, controlled trial of naproxen and celecoxib. Arch. Neurol. 65 (7), 896–905.

McClure, R., Ong, H., Janve, V., Barton, S., Zhu, M., Li, B., Gore, J.C., 2017. Aerosol delivery of curcumin reduced amyloid-β deposition and improved cognitive performance in a transgenic model of Alzheimer's disease. J. Alzheimers Dis. 55 (2), 797–811.

McShane, R., Areosa, Sastre, A., Minakaran, N., 2006. Memantine for dementia. Cochrane Database Syst. Rev. 2, CD003154.

Metaxas, A., Kempf, S.J., 2016. Neurofibrillary tangles in Alzheimer's disease: elucidation of the molecular mechanism by immunohistochemistry and tau protein phospho-proteomics. Neural Regen. Res. 11 (10), 1579.

Molinuevo, J.L., Garcia-Gil, V., Villar, A., 2004. Memantine: an antiglutamatergic option for dementia. Am. J. Alzheimer's Dis. Other Dementias 19 (1), 10–18.

Ng, T.P., Chiam, P.C., Lee, T., Chua, H.C., Lim, L., Kua, E.H., 2006. Curry consumption and cognitive function in the elderly. Am. J. Epidemiol. 164 (9), 898–906.

Olson, M.I., Shaw, C.M., 1969. Presenile dementia and Alzheimer's disease in mongolism. Brain 92 (1), 147–156.

Osborn, G.G., Saunders, A.V., 2010. Current treatments for patients with Alzheimer disease. J. Am. Osteopath. Assoc. 110 (9 Suppl. 8), S16–S26.

Pinkaew, D., Changtam, C., Tocharus, C., Thummayot, S., Suksamrarn, A., Tocharus, J., 2015. Di-O-demethylcurcumin protects SK-N-SH cells against mitochondrial and endoplasmic reticulum-mediated apoptotic cell death induced by Aβ25-35. Neurochem. Int. 80, 110–119.

Prasad, S., Tyagi, A.K., Aggarwal, B.B., 2014. Recent developments in delivery, bioavailability, absorption and metabolism of curcumin: the golden pigment from golden spice. Cancer Res. Treat. 46, 2–18.

Rainey-Smith, S.R., Brown, B.M., Sohrabi, H.R., Shah, T., Goozee, K.G., Gupta, V.B., Martins, R.N., 2016. Curcumin and cognition: a randomised, placebo-controlled, double-blind study of community-dwelling older adults. Br. J. Nutr. 115 (12), 2106–2113.

Reitz, C., Brayne, C., Mayeux, R., 2011. Epidemiology of Alzheimer disease. Nat. Rev. Neurol. 7 (3), 137.

Ringman, J.M., Frautschy, S.A., Teng, E., Begum, A.N., Bardens, J., Beigi, M., Porter, V., 2012. Oral curcumin for Alzheimer's disease: tolerability and efficacy in a 24-week randomized, double blind, placebo-controlled study. Alzheimer's Res. Ther. 4 (5), 43.

Rinwa, P., Kumar, A., Garg, S., 2013. Suppression of neuroinflammatory and apoptotic signaling cascade by curcumin alone and in combination with piperine in rat model of olfactory bulbectomy induced depression. PLoS One 8 (4), e61052.

Robakis, N.K., Ramakrishna, N., Wolfe, G., Wisniewski, H.M., 1987. Molecular cloning and characterization of a cDNA encoding the cerebrovascular and the neuritic plaque amyloid peptides. Proc. Natl. Acad. Sci. U.S.A. 84 (12), 4190–4194.

Safouris, A., Tsivgoulis, G., N Sergentanis, T., Psaltopoulou, T., 2015. Mediterranean diet and risk of dementia. Curr. Alzheimer Res. 12 (8), 736–744.

Scapagnini, G., Colombrita, C., Amadio, M., D'Agata, V., Arcelli, E., Sapienza, M., Calabrese, V., 2006. Curcumin activates defensive genes and protects neurons against oxidative stress. Antioxid. Redox Signal. 8 (3–4), 395–403.

Shelma, R., Sharma, C.P., 2013. *In vitro* and *in vivo* evaluation of curcumin loaded lauroyl sulphated chitosan for enhancing oral bioavailability. Carbohydr. Polym. 95, 441–448.

Small, G.W., Siddarth, P., Li, Z., Miller, K.J., Ercoli, L., Emerson, N.D., et al., 2018. Memory and brain amyloid and tau effects of a bioavailable form of curcumin in non-demented adults: a double-blind, placebo-controlled 18-month trial. Am. J. Geriatr. Psychiatry 26 (3), 266–277.

Suh, Y.H., Checler, F., 2002. Amyloid precursor protein, presenilins, and α-synuclein: molecular pathogenesis and pharmacological applications in Alzheimer's disease. Pharmacol. Rev. 54 (3), 469–525.

Sundaram, J.R., Poore, C.P., Sulaimee, N.H.B., Pareek, T., Cheong, W.F., Wenk, M.R., Kesavapany, S., 2017. Curcumin ameliorates neuroinflammation, neurodegeneration, and memory deficits in p25 transgenic mouse model that bears hallmarks of Alzheimer's disease. J. Alzheimers Dis. 60 (4), 1429–1442.

Teter, B., Morihara, T., Lim, G.P., Chu, T., Jones, M.R., Zuo, X., Cole, G.M., 2019. Curcumin restores innate immune Alzheimer's disease risk gene expression to ameliorate Alzheimer pathogenesis. Neurobiol. Dis. 127, 432–448.

Todd, S., Barr, S., Roberts, M., Passmore, A.P., 2013. Survival in dementia and predictors of mortality: a review. Int. J. Geriatr. Psychiatry 28 (11), 1109–1124.

Tomita, T., 2009. Secretase inhibitors and modulators for Alzheimer's disease treatment. Expert Rev. Neurother. 9 (5), 661–679.

Verma, M., Vats, A., Taneja, V., 2015. Toxic species in amyloid disorders: oligomers or mature fibrils. Ann. Indian Acad. Neurol. 18 (2), 138.

Vivar, C., 2015. Adult hippocampal neurogenesis, aging and neurodegenerative diseases: possible strategies to prevent cognitive impairment. Curr. Top. Med. Chem. 15 (21), 2175–2192.

Wang, C., Zhang, X., Teng, Z., Zhang, T., Li, Y., 2014. Down regulation of PI3K/Akt/mTOR signaling pathway in curcumin-induced autophagy in APP/PS1 double transgenic mice. Eur. J. Pharmacol. 740, 312–320.

Wang, Y.J., Thomas, P., Zhong, J.H., Bi, F.F., Kosaraju, S., Pollard, A., Zhou, X.F., 2009. Consumption of grape seed extract prevents amyloid-β deposition and attenuates inflammation in brain of an Alzheimer's disease mouse. Neurotox. Res. 15 (1), 3–14.

Wang, Y., Yin, H., Li, J., Zhang, Y., Han, B., Zeng, Z., Li, J., 2013. Amelioration of β-amyloid-induced cognitive dysfunction and hippocampal axon degeneration by curcumin is associated with suppression of CRMP-2 hyperphosphorylation. Neurosci. Lett. 557, 112–117.

Wu, A., Noble, E.E., Tyagi, E., Ying, Z., Zhuang, Y., Gomez-Pinilla, F., 2015. Curcumin boosts DHA in the brain: implications for the prevention of anxiety disorders. Biochim. Biophys. Acta Mol. Basis Dis. 1852 (5), 951–961.

Xiong, Z., Hongmei, Z., Lu, S., Yu, L., 2011. Curcumin mediates presenilin-1 activity to reduce beta-amyloid production in a model of Alzheimer's disease. Pharmacol. Rep. (63), 1101–1108.

Yamada, M., Ono, K., Hamaguchi, T., Noguchi-Shinohara, M., 2015. Natural phenolic compounds as therapeutic and preventive agents for cerebral amyloidosis. Adv. Exp. Med. Biol. 863, 79–94.

Yang, C., Zhang, X., Fan, H., Liu, Y., 2009. Curcumin up regulates transcription factor Nrf2, HO-1 expression and protects rat brains against focal ischemia. Brain Res. 1282, 133–141.

Yang, F., Lim, G.P., Begum, A.N., Ubeda, O.J., Simmons, M.R., Ambegaokar, S.S., Cole, G.M., 2005. Curcumin inhibits formation of amyloid β oligomers and fibrils, binds plaques, and reduces amyloid in vivo. J. Biol. Chem. 280 (7), 5892–5901.

Ye, S., Wang, T.T., Cai, B., Wang, Y., Li, J., Zhan, J.X., Shen, G.M., 2017. Genistein protects hippocampal neurons against injury by regulating calcium/calmodulin dependent protein kinase IV protein levels in Alzheimer's disease model rats. Neural Regen. Res. 12 (9), 1479.

Yin, H.L., Wang, Y.L., Li, J.F., Han, B., Zhang, X.X., Wang, Y.T., Geng, S., 2014. Effects of curcumin on hippocampal expression of NgR and axonal regeneration in Aβ-induced cognitive disorder rats. Genet. Mol. Res. 13 (1), 2039–2047.

Zhang, C., Browne, A., Child, D., Tanzi, R.E., 2010. Curcumin decreases amyloid-β peptide levels by attenuating the maturation of amyloid-β precursor protein. J. Biol. Chem. 285 (37), 28472–28480.

Zhang, L., Fang, Y., Xu, Y., Lian, Y., Xie, N., Wu, T., Wang, Z., 2015. Curcumin improves amyloid β-peptide (1-42) induced spatial memory deficits through BDNF-ERK signaling pathway. PLoS One 10 (6), e0131525.

Chapter 3.1.3

Omega 3 PUFA

Vipul Chaudhary[1], Ashwini Kumar Nigam[2], Ashutosh Paliwal[3], Manoj Kumar Singh[4], Jalaj Kumar Gour[5], Vimlendu Bhushan Sinha[6]
[1]*Department of Biotechnology, Deenbandhu Chhotu Ram University of Science and Technology, Sonepat, Haryana, India;* [2]*Department of Zoology, Udai Pratap College, Varanasi, Uttar Pradesh, India;* [3]*Department of Biotechnology, Kumaun University Nainital, Bhimtal Campus, Bhimtal, Uttarakhand, India;* [4]*Center for Noncommunicable Diseases (NCD), National Centre for Disease Control, Delhi, India;* [5]*Department of Biochemistry, Faculty of Science, University of Allahabad, Prayagraj, Uttar Pradesh, India;* [6]*Department of Biotechnology, School of Engineering and Technology, Sharda University, Greater Noida, Uttar Pradesh, India*

Introduction

Fat, or more appropriately lipid, is an essential macronutrient which not only provides nutrition but also is essential for the delivery of energy and cell growth. Lipids are mainly comprised of carbon and hydrogen atoms which puts them in the hydrophobic category. Chemically, lipids are the esters of fatty acids with an alcohol. Alcoholic moiety in lipids may be glycerol as in triglycerides and phospholipids, sphingosine as in many glycolipids, or triacontanol as in waxes. Fatty acids are long-chain carboxylic acids containing more than four carbon atoms in their long hydrocarbon tail. Structurally, fatty acids are categorized into two classes—saturated and unsaturated fatty acids. Saturated fatty acids do not contain a double bond in the chain, e.g., myristic acid (C14:0), palmitic acid (C16:0), stearic acid (C18:0), etc. Lipids or triglycerides containing only saturated fatty acids are solid at room temperature, and are known as fats with high melting points. Unsaturated fatty acids contain one or more double bonds in their long hydrocarbon chain. Lipids or triglycerides having abundant unsaturated fatty acids are called oils, and remain in a liquid state at room temperature, i.e., they have a low melting point. Further, unsaturated fatty acids are of two types—monounsaturated fatty acids (MUFAs) with only one double bond and polyunsaturated fatty acids (PUFAs) having two or more double bonds. MUFAs include palmitolic acid (C16:1, *cis*-Δ^9) and oleic acid (C18:1, *cis*-Δ^9). There is a long list of PUFAs, which have more than one double bond.

On the basis of nutritional requirements, fatty acids are classified as nonessential and essential fatty acids. Most organisms are capable of

Naturally Occurring Chemicals against Alzheimer's Disease. https://doi.org/10.1016/B978-0-12-819212-2.00005-0

synthesizing nonessential fatty acids, including all saturated fatty acids and MUFAs. PUFAs are essential fatty acids that are not synthesized by mammals, including humans, due to the absence of specific desaturase enzymes (Δ^{12} and Δ^{15} desaturases), and therefore they must come from food directly. Plants are capable of synthesizing these essential fatty acids, and are good sources to be consumed by human. Saturated and unsaturated fatty acids, described as good or bad fats, exist in nature. *Trans*-fats, which are synthesized artificially or produced from good fats during prolonged high-temperature cooking, are categorized under bad fats.

Essential PUFAs are grouped into two families—omega 6 fatty acids and omega 3 fatty acids. In the omega (ω) convention of fatty acid nomenclature, the carbon atom of methyl group, i.e., the most distant carbon or the end opposite to the carboxyl terminal, is called the omega (ω) carbon, and is given the number 1 (Fig. 3.1.3.1). Linoleic and α-linolenic acids are parent molecules of omega 6 and omega 3 fatty acid groups, respectively.

The recommended ratio of omega 6 to omega 3 fatty acids is 5−10:1 in our diet. Of these essential fatty acids, omega 3 (ω-3) fatty acids play important roles in our body. Consumption of long-chain omega 3 PUFAs may hamper progression of this mental disorder and increase cognitive ability. This is why omega 3 fatty acids are also regarded as brain food. As they are not

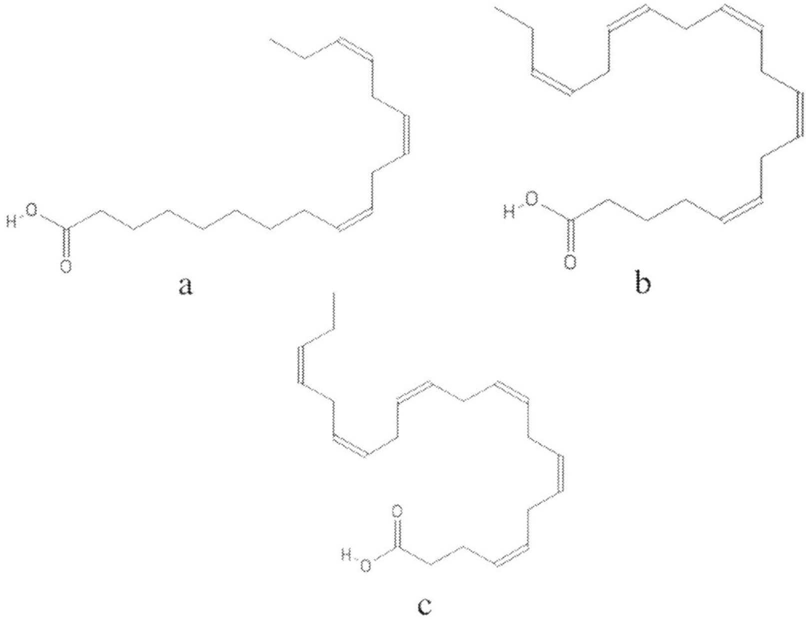

FIGURE 3.1.3.1 Chemical structure of α-linolenic acid (A), eicosapentaenoic acid (B), and docosahexaenoic acid (C).

synthesized in the body, they must be included in the diet. Important foods rich in ω-3 fatty acids include walnuts, almonds, chia seeds, flaxseeds, olive oil, green leafy vegetables, sea foods, and fish (sardine, salmon, hilsa, etc.). In nature, there are various types of omega 3 fatty acids, as listed in Table 3.1.3.1.

There are three major types of PUFA reported, i.e., ALA (alpha-linolenic acid), DHA (docosahexaenoic acid), and EPA (eicosapentaenoic acids)—all differ at their long carbon chains. ALA is converted into EPA and DHA in the liver. However, this conversion process is very insufficient and only accounts for 0.5% of the DHA and EPA requirements of the body. Therefore, these three omega 3 PUFAs must be supplied in the diet. DHA is the major structural lipid portion of brain, comprising 40% of PUFA and is mostly distributed in the cerebral cortex and synapses (Singh et al., 2005). Various scientific reports have illustrated that DHA is the most common type of omega 3.

There are a number of reports which strongly demonstrate a correlation of imbalance of intake of omega 6 and omega 3 fatty acids with various chronic diseases including heart-related problems, hypertension, diabetes, obesity, premature aging, and some forms of cancer. It was suggested that higher uptake of omega 3 fatty acids and their food sources reduces the risks of AD and dementia (Kalmijn et al., 1997a,b; Morris et al., 2003; Huang et al., 2005;

TABLE 3.1.3.1 Most common omega-3 fatty acids and their dietary sources.

S.no.	Common name	Carbon skeleton	Common dietary sources
1.	α-Linolenic acid (ALA)	C18:3;*cis*-$\Delta^{9,12,15}$	All listed PUFA: Plant sources: walnuts, almonds, chiaseed, flaxseeds, olive oil, green leafy vegetables, sea weeds etc. Animal sources: marine animals specially fishes—sardine, salmon, hilsa etc.
2.	Stearidonic acid (SDA)	C18:4;*cis*-$\Delta^{6,9,12,15}$	
3.	Eicosatrienoic acid (ETE)	C20:3;*cis*-$\Delta^{11,14,17}$	
4.	Eicosapentaenoic acid (EPA)	C20:5;*cis*-$\Delta^{5,8,11,14,17}$	
5.	Heneicosapentaenoic acid (HPA)	C21:5;*cis*-$\Delta^{6,9,12,15,18}$	
6.	Docosapentaenoic acid (DPA)	C22:5;*cis*-$\Delta^{7,10,13,16,19}$	
7.	Clupanodonic acid	C22:6;*cis*-$\Delta^{4,7,10,13,16,19}$	
8.	Docosahexaenoic acid (DHA)	C22:6;*cis*-$\Delta^{4,7,10,13,16,19}$	

Barberger-Gateau et al., 2007). As the prevalence and evidence of psychological and neurological disorders has increased daily, there is a need for therapeutic interventions to prevent illness and ensure the safety of pharmacological treatments. Apart from that our diet is considered as an important factor for the prevention of this neurological disorder (Thomas et al., 2015). There is a great deal of evidence and studies available that show that omega 3 PUFAs play a critical role in neuronal functions and cognition. Both the beneficial and protective effects attributed to omega 3 fatty acid consumption were seen in Alzheimer's disease (AD) patients (Ajith, 2018). Thus nutrition has been recognized as an important factor for the prevention and treatment of AD. Both DHA and EPA are blessed with antiinflammatory, antiapoptotic, and neurotrophic properties, and are considered as agents with the ability to increase the nerve growth factor level, and also help improve cognition and depressive disorders.

Alzheimer's disease (AD)

Alzheimer's disorder is a disorder related to neurons that decrease the thinking ability of individuals, including cognition and progression, leading to destruction of memory. The disease was named after its discovery by the German psychiatrist Alios Alzheimer. He noticed changes in the brain where some clumps were abnormal (amyloid plaques), along with tangled protein bundles (neurofibrillary, tau, or tangles) in patients suffering from cognition dysfunction, memory loss, language difficulties, and abnormal behavior. Positive and negative neuropathological changes are associated with AD. Excess accumulation of amyloid plaques, neurofibrillary tangles, and neuropil threads leads to hyperphosphorylation of tau protein, causing AD (Mandelkow and Mandelkow, 1998; Trojanowski and Lee, 2000; Iqbal and Grundke-Iqbal, 2002; Crews et al., 2010).

Congophilic amyloid angiopathy is a frequent concurrent feature. Unique lesions, found primarily in the hippocampal formation, include Hirano bodies and granulovacuolar degeneration. In addition to these positive lesions, characteristic losses of neurons, neuropils, and synaptic elements are core negative features of AD (Terry et al., 1991; Scheff et al., 2006, 2007). Neuropathological characteristics such as degeneration of the limbic system (Arnold et al., 1991; Klucken et al., 2003), neocortical regions (Terry et al., 1981), and basal forebrain (Teipel et al., 2005) in AD patients are closely associated with cognitive impairment. This neurodegenerative process is characterized by early damage to the synapses (Crews et al., 2010) with retrograde degeneration of the axons and eventual atrophy of the dendritic tree (Coleman and Perry, 2002; Grutzendler et al., 2007; Perlson et al., 2010) and perikaryon (Lippa et al., 1992). A report by DeKosky and Scheff (1990), Terry et al. (1991), and DeKosky et al. (1996) explained the correlation between loss of synapses in the neocortex and limbic system with cognitive impairment in AD patients.

Tangles and amyloid are the hallmarks of Alzheimer's disease and AD is the primary cause of dementia in elderly people due to which it has now became a public health burden and epidemic. The precise etiology/risk factors of AD are unknown but a number of factors work cumulatively in promoting the disease, including genetic, environmental factors such as advancing age, family history, diabetes, and poor diet. During aging, the normal physiology of the brain is changed, including shortening of the long chain of omega 3 fatty acids and reduced DHA levels. In 2019, over 5.8 million Americans of all ages has Alzheimer's disease, with aging brains about twice as likely to have AD. Alzheimer's is not only memory loss; it is the fifth leading cause of death in the United States among those aged 65 and over.

Why omega 3?

As stated earlier in this chapter, PUFA comprises omega 6 and omega 3 fatty acids. They are one of the most important sources of energy in the diet. At least a 10:1 proportion of ω-6 and ω-3 should be present in the diet to ensure proper structural maintenance and functioning of the nervous and cardiovascular systems. Omega 6 fatty acids promote the aggregation of platelets, and are thought to be proinflammatory because the end product of ω-6 fatty acid (linoleic acid) is arachidonic acid, which is a precursor for synthesis of inflammatory mediators (eicosanoids), e.g., thromboxanes, leukotrienes, and prostaglandins. Therefore, a greater proportion of ω-6 fatty acids is often associated with greater risk of inflammation in blood vessels causing atherosclerosis (Ghafoorunissa, 1998) and enhanced glutamate-mediated neuronal destruction through oxidative damage (Singh et al., 2005). On the other hand, end metabolism of ω-3 fatty acid produces EPA and DHA, which are crucial for brain development and neuronal functioning. DHA is supposed to reduce the level of thromboxane A_2 and to increase prostacyclins. Hence, the effect of ω-3 fatty acids is antiinflammatory, inhibiting platelet aggregation and reducing cellular and vascular inflammation (Singh et al., 2005). Since omega 3 PUFA is an essential fatty acid as it is very slowly formed in the cell membranes, it has to be taken from foods like fish oils (mackerel, tuna, herring, and salmon) which are considered to be its richest sources. It plays a vital role in the development and maintenance of brain cells. Membranes of brain cells have higher concentrations and help in the fetus to adult phase with roles in growth, communication with cells, and other cognitive functions including memory and learning.

The biological properties of omega 3 PUFA (Fig. 3.1.3.2) and structural changes to the brain should provide a better way to understand the possible pathway involved in preventing or treating mental disorders. Brain cells with a high level of omega 3 PUFA are better for communication with other cells, which is important for brain function. Cellular pathways, cell signaling, and structure of membrane are involved in achieving homeostasis along with brain-regulating fatty acids influenced by omega 3 fatty acids and this works as a model for considering it as a therapeutic intervention in AD. A study by He et al. (2009) showed the omega 3 fatty acid, which was generated by the conversion of omega 6 in a transgenic mouse, increased the quantity of DHA

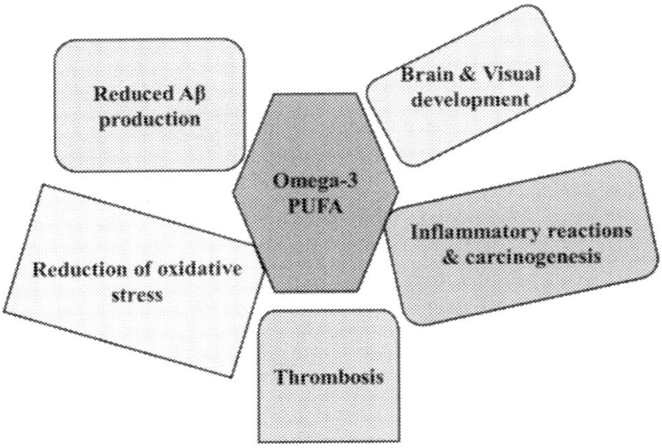

FIGURE 3.1.3.2 Biological effects of omega-3 PUFA on Alzheimer's disease.

in the brain (Wani et al., 2015). In low blood glucose states or in hypoglycemic states, triglyceride is the substrate for energy metabolism, while PUFAs are integral lipids which help in maintaining the structure and functions of the neuronal membrane (Jicha and Markesbery, 2010). The incorporation of both PUFAs with neuronal membrane increases the membrane fluidity and regulates signal transduction by enhancing G-protein-mediated communication (Lee and Hamm, 1989; Ahmad et al., 1989).

Dopaminergic neurodevelopmental functioning is affected with low levels of ω-3 PUFAs and, in association with other genetic factors, leads to development of some metabolic disorders and its role in mature brain may help in further research for the development of new interventions for the prevention of metabolic disorders. The alterations in serotonergic and noradrenergic neurotransmission are the main pathophysiological changes observed in mentally or psychologically ill patients. However, reports are available suggesting that a higher level of DHA is important to increase serotonergic neurotransmission (Hibbeln et al., 1998). An omega 3 intake showed the positive effect on depressive status as it has a potential interaction with serotonergic and dopaminergic transmission and receptor function. Overall, omega 3 PUFAs are beneficial fatty acids with the ability to improve cognitive function (Hooijmans et al., 2012). Omega 3 PUFA not only regulates transmembrane Na^+/K^+ATPase but also enhances G-protein-mediated signal transduction. Omega 3 is inversely related to dopamine receptor (DR-2) and serotonin (5-HT), which are regulatory neurotransmitters for cognition (Chiu et al., 2008).

Omega 3 PUFA-induced decreased levels of the Bax/Bcl activities illustrated the antiapoptotic activity of these essential fatty acids (Sanchez-Villegas and Martínez-González, 2013). Omega-3 PUFA acts as a stimulant for the brain, endothelial cell proliferation inhibitor, glucose uptake influencer, reduces glucose transporter isoforms, i.e., GLUT 1 in brain, and elevates plasma homocysteine levels. An animal preclinical study has shown that erythrocyte

DHA can be inversely correlated, and AA (arachidonic acid) and the AA/DHA and AA/EPA ratios can be positively correlated with plasma, C-reactive protein levels, interleukin, and TNFα (Grosso et al., 2014).

The field of pharmacogenomics is an emerging trend in which currently designed drugs for AD do not amend the neurodegeneration progression and are effective for only mild symptomatic improvements (Reitz, 2012). Most of these drugs diminish Aβ generation from amyloid precursor protein (APP) and hit pathways in the amyloid cascade, helping removal of toxic Aβ aggregates. The drugs currently approved for AD are not capable of addressing progressive neurodegeneration in AD and yield only superficial symptomatic improvements. The famous target pathways for the approved drugs lie in the amyloid cascade, Aβ generation from APP, formation and facilitating the removal of a few toxic aggregates, hyperphosphorylation prevention, and tau protein aggregation (Gilman et al., 2005). The immunization rationale which involves the use of anti-Aβ antibodies has witnessed success in clearing Aβ and finally decreasing the plaque load but has largely failed to improve cognitive abilities (Gilman et al., 2005; Vellas et al., 2009; Salloway et al., 2009) and also caused side effects like brain atrophy and meningoencephalitis (Gilman et al., 2005). Compared with new medications, omega 3 PUFA fortification could be a promising intervention with long-term safety. The administration of omega 3 showed protection against AD, proved efficacious in reducing mortality from cardiovascular disease, and particularly DHA has proven its role in the pathogenic cascade as biological expression of fulminate AD.

Importance of omega 3

The brain is considered to be the master organ of the human body because of its regulatory functions. The human brain is one of the largest "consumers" of DHA. Low DHA levels have been linked to low brain serotonin levels, which again are connected to an increased tendency to depression, suicide, and violence. Memory-related problems which worsen with age can be aided by adding omega 3-rich food to the diet. This would also reduce the risk of developing AD. In a scientific report, use of an omega 3-rich diet/supplement caused a significant improvement in Alzheimer's patients (Yehuda et al., 1996; Levine, 1997; Kalmijn et al., 1997a,b).

Clinical trials

As the elderly population is most vulnerable to Alzheimer's disease and it becomes fatal with age, it is becoming a worldwide social and economic burden (Wimo et al., 2010). Although there is no specific drug available for the treatment of AD, various scientific groups are working on several compounds with different modes of action, e.g., affecting neurotransmission, accumulation of amyloid β and tau protein, restoration of balanced growth factors, etc. Some of the compounds under trial or experimentation are listed in Table 3.1.3.2.

TABLE 3.1.3.2 List of compounds tested for treatment of Alzheimer's disease.

S. no.	Name of compound	Nature of compound	Mode of action	References
Acetylcholinesterase inhibitors				
1.	(−)Phenserine	Derivative of physostigmine	Reduce Aβ precursor protein (APP) and Aβ concentrations by decreasing the translation of APP mRNA	Klein (2007)
2.	(+) Phenserine	Enantiomer, posiphen	Substantially decrease APP production by reducing APP mRNA translation	Klein (2007)
3.	Talsaclidine	M1 muscarinic receptor agonists	Decreased CSF Aβ concentrations in patients with Alzheimer's disease	Fisher (2007)
4.	AF-102B	M1 muscarinic receptor agonists	Decreased CSF Aβ concentrations in patients with AD	Hock et al. (2003)
5.	AF-267B (NGX-267)	M1 muscarinic receptor agonists		Nitsch et al. (2000)
6.	Ispronicline (AZD-3480)	A selective agonist of the nicotinic receptor α4β2	Positive effects on cognition in healthy individuals and in people with age associated memory impairment	Dunbar et al. (2007)
7.	ABT-089	A partial agonist of nicotinic α4β2 and α6β2 receptors	Reverse cognitive deficits induced by hyoscine (scopolamine) in healthy volunteers	Baker et al. (2009)
8.	EVP-6124	An agonist of the nicotinic α7 receptor	Cognitive improvement	Hilt et al. (2009)
β-Secretase inhibitors				
9.	Thiazolidinediones rosiglitazone and pioglitazone	β-secretase inhibitor	Stimulating the nuclear peroxisome proliferator-activated receptor γ (PPARγ)	Landreth et al. (2008)

γ-Secretase inhibitors				
10.	Semagacestat (LY-450139), MK-0752, E −2012, BMS-708163, PF-3084014, begacestat (GSI-953), and NIC5-15	Gamma secretase inhibitors have been clinically tested	Reduces Aβ concentrations in the plasma and Aβ production in the CNS	Henley et al. (2009), Bateman et al. (2009)
11.	Begacestat	Notch-sparing γ-secretase inhibitors (second-generation inhibitors)	Under clinical trials	Jacobsen et al. (2009)
12.	BMS-708163		Dose-dependent decrease of Aβ1−40 in the CSF	Imbimbo (2008), Ereshefsky et al. (2009)
13.	PF-3084014	High selectivity for APP	Showed decreases in Aβ in the plasma, CSF, and brain, without a rebound effect on plasma Aβ	Wood et al. (2008)
14.	NIC5-15	Naturally occurring monosaccharide	Can act as a Notch-sparing γ-secretase inhibitor and insulin sensitizer	Wang et al. (2005), Humanetics Pharmaceuticals (2009)
15.	Ibuprofen, indometacin, and sulindac sulfide	Bind to APP, gamma secretase modulator	Decreasing Aβ1−40 and Aβ1−42 production, with increased generation of Aβ1−38 fragments	Tomita (2009)
16.	CHF-5074	γ-Secretase modulator	Reduced Aβ brain load and improved behavioral deficits in animals	Imbimbo et al. (2009)
α-Secretase activators				
17.	Etazolate (EHT-0202)	A selective GABAA receptor modulator	Stimulates neuronal α-secretase and increases sAPPα production	Marcade et al. (2008)

Continued

TABLE 3.1.3.2 List of compounds tested for treatment of Alzheimer's disease.—cont'd

S. no.	Name of compound	Nature of compound	Mode of action	References
18.	Bryostatin-1	A macrocyclic lactone	Reduces brain Aβ1–40 and Aβ1–42 and improves behavioral outcomes in mouse models of AD	Etcheberrigaray et al. (2004)
19.	Exebryl-1	Modulates β-secretase and α-secretase activity	Causes substantial reduction of Aβ formation and accumulation in the mouse brain, with memory improvements	Snow et al. (2009)
Drugs to prevent Aβ aggregation				
20.	Tramiprosate	Commercialized as Vivimind	Nutraceutical that protects memory functions	Alzheimer Research Forum and McCaffrey (2008))
21.	Clioquinol (PBT1)	An inhibitor of Aβ aggregation	Works by interfering with interactions between Aβ, copper, and zinc, showed positive results in phase II RCTs	Adlard et al. (2008)
22.	PBT2	A second-generation inhibitor of metal-induced Aβ aggregation	Binds to the Aβ–copper or Aβ–zinc complex, and studies in animals showed that PBT2 prevents Aβ oligomerization, promotes Aβ oligomer clearance, reduces soluble and insoluble brain Aβ, and decreases plaque burden, with positive effects on cognition	Adlard et al. (2008)

23.	Scyllo-inositol (scyllo-cyclohexanehexol, AZD-103, ELND-005)	An orally administered stereoisomer of inositol	Thought to bind to Aβ, modulate its misfolding, inhibit its aggregation, and promote dissociation of aggregates	McLaurin et al. (2006)
24.	Epigallocatechin-3-gallate (EGCg)	A polyphenol from green tea	Induced α-secretase and prevented Aβ aggregation in animals by directly binding to the unfolded peptide	Mandel et al. (2008)
Drugs that promote Aβ clearance				
25.	CAD-106	A vaccine	Can induce Aβ-specific antibodies and reduce amyloid accumulation in animals. In patients with mild-to-moderate AD, CAD-106 induced a substantial anti-Aβ IgG response, was well tolerated, and did not induce meningoencephalitis	Winblad et al. (2009)
26.	ACC-001 and V-950	Vaccines which contains the B-cell epitope	Based on the N-terminal Aβ fragment	
27.	ACI-24	A vaccine that contains Aβ1−15 embedded within the aliposomal surface	Reduces brain amyloid load and restores memory deficits in mice	Muhs et al. (2007)
28.	UB-311	A vaccine	Immunogen Aβ1−14 is associated with the UBITh peptide and a mineral salt suspension adjuvant, is being tested in patients with mild-to-moderate Alzheimer's disease in a phase I RCT (NCT00965588).	Wang et al. (2007)

Continued

TABLE 3.1.3.2 List of compounds tested for treatment of Alzheimer's disease.—cont'd

S. no.	Name of compound	Nature of compound	Mode of action	References
29.	Affitopes AD-01 and AD-02	Short peptides	Target the N-terminal Aβ fragment and both had disease-modifying properties in animal models of AD	Schneeberger et al. (2009)
30.	Bapineuzumab	A humanized anti-Aβ monoclonal antibody	A phase II RCT in patients with mild-to-moderate AD	Salloway et al. (2009)
31.	Solanezumab	A monoclonal antibody	Binds specifically to soluble Aβ, promotes Aβ clearance from the brain through the blood	Siemers et al. (2008)
32.	GSK-933776, R-1450 (RO-4909832), and MABT-5102A	Monoclonal antibodies	Target Aβ and have been tested in patients with Alzheimer's disease in phase I studies	Dodel et al. (2004)
Drugs that target tau protein				
33.	Valproate; valproic acid	Tau-directed compound	No effects on cognition and functional status	Tariot and Aisen (2009), Tariot et al. (2009)
34.	Lithium and valproate	Well known for the treatment of psychiatric disorders	Inhibit GSK3, to reduce tau phosphorylation and prevent or reverse aspects of tauopathy in animal models	Tariot and Aisen (2009)
35.	NP-031112 (NP-12)	GSK3 inhibitors, under development, a thiadiazolidinone-derived compound	Reduce brain concentrations of phosphorylated tau and amyloid deposition, and prevent neuronal death and cognitive deficits in animals.	Sereno et al. (2009)

36.	Methylthioninium chloride	Methylene blue, Rember	A widely used histology dye, acts as a tau antiaggregant	Wischik et al. (1996)
37.	Leuco-methyl thioninium	New formulation		Wischik (2009)
38.	Davunetide (AL-108, NAP)	An intranasally administered, 8 amino acid peptide fragment derived from the activity-dependent neuroprotective protein		
39.	Latrepirdine	Weakly inhibits acetylcholinesterase and butyrylcholinesterase	This drug also inhibits NMDA receptors and voltage-gated calcium channels	Wu et al. (2008)
Other potential therapeutic strategies				
40.	Docosahexaenoic acid		Supplementation in elderly people with cognitive decline or Alzheimer's disease	Yurko-Mauro et al. (2009), Chiu et al. (2008)
41.	Statins		Reduce cholesterol concentrations and have several pleiotropic effects (i.e., can lower Aβ production and reduce Aβ-mediated neurotoxicity, as well as having antioxidant and antiinflammatory properties)	Solomon and Kivipelto (2009)
42.	Pitavastatin		Treatment for group of mild to moderate Alzheimer's disease (PIT-ROAD)	
43.	Simvastatin		Used in amnestic mild cognitive impairment (SIMaMCI) studies, to test the effectiveness of statins in prevention and therapy of Alzheimer's disease	

Conclusion

The aging brain is more prone to the development of neurodegenerative diseases, which also depends on sex, genetic constitution, diet, and life style. Omega 3 PUFAs contribute to lipid metabolism of the brain, influencing enzymatic production of bioactive mediators that regulate microglial signaling and function. Omega 3 PUFAs exhibit neuroprotective properties and thus could be promising agents for the development of novel neurotherapeutics.

References

Adlard, P.A., Cherny, R.A., Finkelstein, D.I., Gautier, E., Robb, E., Cortes, M., Laughton, K., 2008. Rapid restoration of cognition in Alzheimer's transgenic mice with 8-hydroxy quinoline analogs is associated with decreased interstitial Aβ. Neuron 59 (1), 43–55.

Ahmad, S.N., Alma, B.S., Alam, S.Q., 1989. Dietary omega-3 fatty acids increase guanine nucleotide binding proteins and adenylatecyclase activity in rat salivary glands. FASEB J. 3.

Ajith, T.A., 2018. A recent update on the effects of omega-3 fatty acids in Alzheimer's disease. Curr. Clin. Pharmacol. 13 (4), 252–260.

Alzheimer Research Forum, McCaffrey, P., 2008. Experts Slam Marketing of Tramiprosate (Alzhemed) as Nutraceutical.

Arnold, S.E., Hyman, B.T., Flory, J., Damasio, A.R., Van Hoesen, G.W., 1991. The topographical and neuroanatomical distribution of neurofibrillary tangles and neuritic plaques in the cerebral cortex of patients with Alzheimer's disease. Cereb. Cortex 1 (1), 103–116.

Baker, J.D., Lenz, R.A., Locke, C., Wesnes, K., Maruff, P., Abi-Saab, W.M., Saltarelli, M.D., 2009. ABT-089, a neuronal nicotinic receptor partial agonist, reverses scopolamine-induced cognitive deficits in healthy normal subjects. Alzheimers Dement. 5 (4), P325.

Barberger-Gateau, P., Raffaitin, C., Letenneur, L., Berr, C., Tzourio, C., Dartigues, J.F., Alpérovitch, A., 2007. Dietary patterns and risk of dementia: the Three-City cohort study. Neurology 69 (20), 1921–1930.

Bateman, R.J., Siemers, E.R., Mawuenyega, K.G., Wen, G., Browning, K.R., Sigurdson, W.C., et al., 2009. A γ-secretase inhibitor decreases amyloid-β production in the central nervous system. Ann. Neurol. 66 (1), 48–54.

Chiu, C.C., Su, K.P., Cheng, T.C., Liu, H.C., Chang, C.J., Dewey, M.E., Huang, S.Y., 2008. The effects of omega-3 fatty acids monotherapy in Alzheimer's disease and mild cognitive impairment: a preliminary randomized double-blind placebo-controlled study. Prog. Neuro Psychopharmacol. Biol. Psychiatry 32 (6), 1538–1544.

Coleman, M.P., Perry, V.H., 2002. Axon pathology in neurological disease: a neglected therapeutic target. Trends Neurosci. 25 (10), 532–537.

Crews, L., Rockenstein, E., Masliah, E., 2010. APP transgenic modeling of Alzheimer's disease: mechanisms of neurodegeneration and aberrant neurogenesis. Brain Struct. Funct. 214 (2–3), 111–126.

DeKosky, S.T., Scheff, S.W., 1990. Synapse loss in frontal cortex biopsies in Alzheimer's disease: correlation with cognitive severity. Ann. Neurol. 27 (5), 457–464.

DeKosky, S.T., Scheff, S.W., Styren, S.D., 1996. Structural correlates of cognition in dementia: quantification and assessment of synapse change. Neurodegeneration 5 (4), 417–421.

Dodel, R.C., Du, Y., Depboylu, C., Hampel, H., Frölich, L., Haag, A., Spottke, A., 2004. Intravenous immunoglobulins containing antibodies against β-amyloid for the treatment of Alzheimer's disease. J. Neurol. Neurosurg. Psychiatr. 75 (10), 1472–1474.

Dunbar, G.C., Inglis, F., Kuchibhatla, R., Sharma, T., Tomlinson, M., Wamsley, J., 2007. Effect of ispronicline, a neuronal nicotinic acetylcholine receptor partial agonist, in subjects with age associated memory impairment (AAMI). J. Psychopharmacol. 21 (2), 171–178.

Ereshefsky, L., Jhee, S.S., Yen, M., Moran, S.V., 2009. The role for CSF dynabridging studies in developing new therapies for Alzheimer's disease. Alzheimers Dement. 5 (4), P414–P415.

Etcheberrigaray, R., Tan, M., Dewachter, I., Kuipéri, C., Van der Auwera, I., Wera, S., Van Leuven, F., 2004. Therapeutic effects of PKC activators in Alzheimer's disease transgenic mice. Proc. Natl. Acad. Sci. U.S.A. 101 (30), 11141–11146.

Fisher, A., 2007. M1 muscarinic agonists target major hallmarks of Alzheimer's disease-an update. Curr. Alzheimer Res. 4 (5), 577–580.

Gilman, S., Koller, M., Black, R.S., Jenkins, L., Griffith, S.G., Fox, N.C., Orgogozo, J.M., 2005. Clinical effects of Aβ immunization (AN1792) in patients with AD in an interrupted trial. Neurology 64 (9), 1553–1562.

Grosso, G., Galvano, F., Marventano, S., Malaguarnera, M., Bucolo, C., Drago, F., Caraci, F., 2014. Omega-3 fatty acids and depression: scientific evidence and biological mechanisms. Oxid. Med. Cell. Longev. 2014.

Grutzendler, J., Helmin, K., Tsai, J., Gan, W.B., 2007. Various dendritic abnormalities are associated with fibrillar amyloid deposits in Alzheimer's disease. Ann. N. Y. Acad. Sci. 1097 (1), 30–39.

Ghafoorunissa, November 1, 1998. Requirements of dietary fats to meet nutritional needs & prevent the risk of atherosclerosis-An Indian perspective. Indian J. Med. Res. 108, 191–202.

He, C., Qu, X., Cui, L., Wang, J., Kang, J.X., 2009. Improved spatial learning performance of fat-1 mice is associated with enhanced neurogenesis and neuritogenesis by docosahexaenoic acid. Proc. Natl. Acad. Sci. U.S.A. 106 (27), 11370–11375.

Henley, D.B., May, P.C., Dean, R.A., Siemers, E.R., 2009. Development of semagacestat (LY450139), a functional γ-secretase inhibitor, for the treatment of Alzheimer's disease. Expert Opin. Pharmacother. 10 (10), 1657–1664.

Hibbeln, J.R., Linnoila, M., Umhau, J.C., Rawlings, R., George, D.T., Salem Jr., N., 1998. Essential fatty acids predict metabolites of serotonin and dopamine in cerebrospinal fluid among healthy control subjects, and early-and late-onset alcoholics. Biol. Psychiatry 44 (4), 235–242.

Hilt, D., Gawryl, M., Koenig, G., July 11–16, 2009. EVP-6124_StudyGroup. EVP-6124: safety, tolerability and cognitive effects of a novel A7 nicotinic receptor agonist in Alzheimer's disease patients on stable donepezil or rivastigmine therapy. In: International Conference on Alzheimer's Disease; Vienna, Austria.

Hock, C., Maddalena, A., Raschig, A., Müller-Spahn, F., Eschweiler, G., Huger, K., Wienrich, M., 2003. Treatment with the selective muscarinic ml agonist talsaclidine decreases cerebrospinal fluid levels of Aβ42 in patients with Alzheimer's disease. Amyloid 10 (1), 1–6.

Hooijmans, C.R., Pasker-de Jong, P., de Vries, R., Ritskes-Hoitinga, M., 2012. The effects of long-term omega-3 fatty acid supplementation on cognition and Alzheimer's pathology in animal models of Alzheimer's disease: a systematic review and meta-analysis. J. Alzheim. Dis. 28 (1), 191–209.

Huang, T.L., Zandi, P.P., Tucker, K.L., Fitzpatrick, A.L., Kuller, L.H., Fried, L.P., Carlson, M.C., 2005. Benefits of fatty fish on dementia risk are stronger for those without APOE ε4. Neurology 65 (9), 1409–1414.

Humanetics Pharmaceuticals, 2009. A Disease Modifying Drug Candidate for Alzheimer's Disease. NIC5-15.

Imbimbo, B.P., 2008. Alzheimer's disease: γ-secretase inhibitors. Drug Discov. Today Ther. Strat. 5 (3), 169–175.

Imbimbo, B.P., Hutter-Paier, B., Villetti, G., Facchinetti, F., Cenacchi, V., Volta, R., Windisch, M., 2009. CHF5074, a novel γ-secretase modulator, attenuates brain β-amyloid pathology and learning deficit in a mouse model of Alzheimer's disease. Br. J. Pharmacol. 156 (6), 982–993.

Iqbal, K., Grundke-Iqbal, I., 2002. Neurofibrillary pathology leads to synaptic loss and not the other way around in Alzheimer disease. J. Alzheim. Dis. 4 (3), 235–238.

Jacobsen, S., Comery, T., Kreft, A., Mayer, S., Zaleska, M., Riddell, D., Forlow, S., 2009. GSI-953 is a potent APP-selective gamma-secretase inhibitor for the treatment of Alzheimer's disease. Alzheimers Dement. 5 (4), P139.

Jicha, G.A., Markesbery, W.R., 2010. Omega-3 fatty acids: potential role in the management of early Alzheimer's disease. Clin. Interv. Aging 5, 45.

Kalmijn, S., Feskens, E.J.M., Launer, L.J., Kromhout, D., 1997. Polyunsaturated fatty acids, antioxidants, and cognitive function in very old men. Am. J. Epidemiol. 145 (1), 33–41.

Kalmijn, S., Launer, L.J., Ott, A., Witteman, J.C., Hofman, A., Breteler, M.M., 1997. Dietary fat intake and the risk of incident dementia in the Rotterdam Study. Ann. Neurol. 42 (5), 776–782.

Klein, J., 2007. Phenserine. Expert Opin. Invest. Drugs 16 (7), 1087–1097.

Klucken, J., McLean, P.J., Gomez-Tortosa, E., Ingelsson, M., Hyman, B.T., 2003. Neuritic alterations and neural system dysfunction in Alzheimer's disease and dementia with Lewy bodies. Neurochem. Res. 28 (11), 1683–1691.

Landreth, G., Jiang, Q., Mandrekar, S., Heneka, M., 2008. PPARγ agonists as therapeutics for the treatment of Alzheimer's disease. Neurotherapeutics 5 (3), 481–489.

Lee, C.R., Hamm, M.W., 1989. Effect of dietary fat and cholesterol supplements on glucagon receptor binding and adenylatecyclase activity of rat liver plasma membrane. J. Nutr. 119 (4), 539–546.

Levine, B.S., November/December 1997. Most frequently asked questions about DHA. Nutr. Today 32, 248–249.

Lippa, C.F., Hamos, J.E., Pulaski-Salo, D., DeGennaro, L.J., Drachman, D.A., 1992. Alzheimer's disease and aging: effects on perforant pathway perikarya and synapses. Neurobiol. Aging 13 (3), 405–411.

Mandel, S.A., Amit, T., Kalfon, L., Reznichenko, L., Weinreb, O., Youdim, M.B., 2008. Cell signaling pathways and iron chelation in the neurorestorative activity of green tea polyphenols: special reference to epigallocatechingallate (EGCG). J. Alzheim. Dis. 15 (2), 211–222.

Mandelkow, E.M., Mandelkow, E., 1998. Tau in Alzheimer's disease. Trends Cell Biol. 8 (11), 425–427.

Marcade, M., Bourdin, J., Loiseau, N., Peillon, H., Rayer, A., Drouin, D., Désiré, L., 2008. Etazolate, a neuroprotective drug linking GABAA receptor pharmacology to amyloid precursor protein processing. J. Neurochem. 106 (1), 392–404.

McLaurin, J., Kierstead, M.E., Brown, M.E., Hawkes, C.A., Lambermon, M.H., Phinney, A.L., Chen, F., 2006. Cyclohexanehexol inhibitors of Aβ aggregation prevent and reverse Alzheimer phenotype in a mouse model. Nat. Med. 12 (7), 801.

Morris, M.C., Evans, D.A., Bienias, J.L., Tangney, C.C., Bennett, D.A., Wilson, R.S., Schneider, J., 2003. Consumption of fish and n-3 fatty acids and risk of incident Alzheimer disease. Arch. Neurol. 60 (7), 940–946.

Muhs, A., Hickman, D.T., Pihlgren, M., Chuard, N., Giriens, V., Meerschman, C., Bechinger, B., 2007. Liposomal vaccines with conformation-specific amyloid peptide antigens define immune response and efficacy in APP transgenic mice. Proc. Natl. Acad. Sci. U.S.A. 104 (23), 9810–9815.

Nitsch, R.M., Deng, M., Tennis, M., Schoenfeld, D., Growdon, J.H., 2000. The selective muscarinic M1 agonist AF102B decreases levels of total Aβ in cerebrospinal fluid of patients with Alzheimer's disease. Ann. Neurol. 48 (6), 913–918.

Perlson, E., Maday, S., Fu, M.M., Moughamian, A.J., Holzbaur, E.L., 2010. Retrograde axonal transport: pathways to cell death? Trends Neurosci. 33 (7), 335–344.

Reitz, C., 2012. Alzheimer's disease and the amyloid cascade hypothesis: a critical review. Int. J. Alzheimers Dis. 2012.

Salloway, S., Sperling, R., Gilman, S., Fox, N.C., Blennow, K., Raskind, M., Mulnard, R., 2009. A phase 2 multiple ascending dose trial of bapineuzumab in mild to moderate Alzheimer disease. Neurology 73 (24), 2061–2070.

Sanchez-Villegas, A., Martínez-González, M.A., 2013. Diet, a new target to prevent depression? BMC Med. 11 (1), 3.

Scheff, S.W., Price, D.A., Schmitt, F.A., Mufson, E.J., 2006. Hippocampal synaptic loss in early Alzheimer's disease and mild cognitive impairment. Neurobiol. Aging 27 (10), 1372–1384.

Scheff, S.W., Price, D.A., Schmitt, F.A., DeKosky, S.T., Mufson, E.J., 2007. Synaptic alterations in CA1 in mild Alzheimer disease and mild cognitive impairment. Neurology 68 (18), 1501–1508.

Schneeberger, A., Mandler, M., Otava, O., Zauner, W., Mattner, F., Schmidt, W., 2009. Development of AFFITOPE vaccines for Alzheimer's disease (AD)—from concept to clinical testing. J. Nutr. Health Aging 13 (3), 264–267.

Sereno, L., Coma, M., Rodriguez, M., Sanchez-Ferrer, P., Sanchez, M.B., Gich, I., Clarimon, J., 2009. A novel GSK-3β inhibitor reduces Alzheimer's pathology and rescues neuronal loss in vivo. Neurobiol. Dis. 35 (3), 359–367.

Siemers, E.R., Friedrich, S., Dean, R.A., Sethuraman, G., DeMattos, R., Jennings, D., Seibyl, J., 2008. P4-346: safety, tolerability and biomarker effects of an Abeta monoclonal antibody administered to patients with Alzheimer's disease. Alzheimers Dement. 4 (4), T774.

Singh, U., Devaraj, S., Jialal, I., 2005. Vitamin E, oxidative stress, and inflammation. Annu. Rev. Nutr. 25 (1), 151–174.

Snow, A.D., Cummings, J., Lake, T., Hu, Q., Esposito, L., Cam, J., Runnels, S., 2009. Exebryl-1: a novel small molecule currently in human clinical trials as a disease-modifying drug for the treatment of Alzheimer's disease. Alzheimers Dement. 5 (4), P418.

Solomon, A., Kivipelto, M., 2009. Cholesterol-modifying strategies for Alzheimer's disease. Expert Rev. Neurother. 9 (5), 695–709.

Tariot, P.N., Aisen, P.S., 2009. Can lithium or valproate untie tangles in Alzheimer's disease? J. Clin. Psychiatry 70 (6), 919–921.

Tariot, P.N., Aisen, P., Cummings, J., Jakimovich, L., Schneider, L., Thomas, R., Loy, R., 2009. The ADCS valproate neuroprotection trial: primary efficacy and safety results. Alzheimers Dement. 5 (4), P84–P85.

Teipel, S.J., Flatz, W.H., Heinsen, H., Bokde, A.L., Schoenberg, S.O., Stöckel, S., Hampel, H., 2005. Measurement of basal forebrain atrophy in Alzheimer's disease using MRI. Brain 128 (11), 2626–2644.

Terry, R.D., Masliah, E., Salmon, D.P., Butters, N., DeTeresa, R., Hill, R., Katzman, R., 1991. Physical basis of cognitive alterations in Alzheimer's disease: synapse loss is the major correlate of cognitive impairment. Ann. Neurol. 30 (4), 572–580.

Terry, R.D., Peck, A., DeTeresa, R., Schechter, R., Horoupian, D.S., 1981. Some morphometric aspects of the brain in senile dementia of the Alzheimer type. Ann. Neurol. 10 (2), 184–192.

Thomas, J., Thomas, C.J., Radcliffe, J., Itsiopoulos, C., 2015. Omega-3 fatty acids in early prevention of inflammatory neurodegenerative disease: a focus on Alzheimer's disease. BioMed. Res. Int. 2015.

Tomita, T., 2009. Secretase inhibitors and modulators for Alzheimer's disease treatment. Expert Rev. Neurother. 9 (5), 661–679.

Trojanowski, J.Q., LEE, V.M.Y., 2000. "Fatal attractions" of proteins: a comprehensive hypothetical mechanism underlying Alzheimer's disease and other neurodegenerative disorders. Ann. N. Y. Acad. Sci. 924 (1), 62–67.

Vellas, B., Black, R., Thal, L.J., Fox, N.C., Daniels, M., McLennan, G., Grundman, M., 2009. Long-term follow-up of patients immunized with AN1792: reduced functional decline in antibody responders. Curr. Alzheimer Res. 6 (2), 144–151.

Wang, C.Y., Finstad, C.L., Walfield, A.M., Sia, C., Sokoll, K.K., Chang, T.Y., Windisch, M., 2007. Site-specific UBITh® amyloid-β vaccine for immunotherapy of Alzheimer's disease. Vaccine 25 (16), 3041–3052.

Wang, J., Ho, L., Pasinetti, G.M., 2005. The Development of NIC5-15, a natural anti-diabetic agent, in the treatment of Alzheimer's disease. Alzheimers Dement. 1 (1), S62.

Wani, A.L., Bhat, S.A., Ara, A., 2015. Omega-3 fatty acids and the treatment of depression: a review of scientific evidence. Integr. Med. Res. 4 (3), 132–141.

Wimo, A., Winblad, B., Jönsson, L., 2010. The worldwide societal costs of dementia: estimates for 2009. Alzheimers Dement. 6 (2), 98–103.

Winblad, B.G., Minthon, L., Floesser, A., Imbert, G., Dumortier, T., He, Y., Orgogozo, J.M., 2009. Results of the first-in-man study with the active Aβ immunotherapy CAD106 in Alzheimer patients. Alzheimers Dement. 5 (4), P113–P114.

Wischik, C., 2009. Rember: issues in design of a phase 3 disease modifying clinical trial of tau aggregation inhibitor therapy in Alzheimer's disease. Alzheimers Dement. 5 (4), P74.

Wischik, C.M., Edwards, P.C., Lai, R.Y., Roth, M., Harrington, C.R., 1996. Selective inhibition of Alzheimer disease-like tau aggregation by phenothiazines. Proc. Natl. Acad. Sci. U.S.A. 93 (20), 11213–11218.

Wood, K.M., Lanz, T.A., Coffman, K.J., Becker, S.L., van Deusen, J., Nolan, C.E., et al., 2008. P2-375: efficacy of the novel γ-secretase inhibitor, PF-3084014, in reducing Aβ in brain, CSF, and plasma in Guinea pigs and Tg2576 mice. Alzheimers Dement. 4 (4), T482–T483.

Wu, J., Li, Q., Bezprozvanny, I., 2008. Evaluation of Dimebon in cellular model of Huntington's disease. Mol. Neurodegener. 3 (1), 15.

Yehuda, S., Rabinovtz, S., Carasso, R.L., Mostofsky, D.I., 1996. Essential fatty acids preparation (SR-3) improves Alzheimer's patients quality of life. Int. J. Neurosci. 87 (3–4), 141–149.

Yurko-Mauro, K., McCarthy, D., Bailey-Hall, E., Nelson, E.B., Blackwell, A., 2009. Results of the MIDAS trial: effects of docosahexaenoic acid on physiological and safety parameters in age-related cognitive decline. Alzheimers Dement. 5 (4), P84.

Chapter 3.1.4

Galantamine

Vaibhav Rathi
School of Health Sciences, Quantum University, Roorkee, Uttarakhand, India

Introduction

Galantamine is a selective competitive and reversible Ach inhibitor alkaloid derived from various plant sources, while in vitro production is also only carried out via plant parts. It has been used and studied extensively for many years for various uses, previously being a drug of choice for myasthenia gravis, but as there are many other Ach inhibitors it also has shown excellent potential in the treatment of Alzheimer's disease with minimal side effects—the side effects are usually mild and the discontinuation rate is also low as compared to other Ach inhibitors. This chapter compiles the data from various research and peer reviews regarding the use of galantamine in Alzheimer disease.

Galantamine is found in various plant sources, mainly from the genera *Amaryllis*, *Lycoris*, *Hippeastrum*, *Ungernia*, *Leucojum*, *Zephyranthes*, *Narcissus*, *Galanthus*, *Hymenocallis*, and *Haemanthus* and is a naturally occurring alkaloid of the family Amaryllidaceae, which is a selective competitive and reversible acetylcholinesterase (Ach) inhibitor, which can also be prepared synthetically by various methods. Galantamine is used in various neurodegenerative diseases, especially dementia and Alzheimer's disease which is the main cause of dementia (around 75%) worldwide. Galantamine is prescribed to treat mild-to-moderate Alzheimer's disease. It has relatively few adverse effects and is considered relatively safer to use as compared to tacrine, which was withdrawn due to its severe side effects. Galantamine is sold as galantamine hydrobromide salt as Razadine or Reminyl, and galantamine hydrobromide is also used for the treatment of other neurological diseases such as poliomyelitis (Emilien et al., 2004; Marco-Contelles et al., 2006; Coelho dos Santos et al., 2018; Viegas et al., 2005; Heinrich and Teoh, 2004; Cherkasov, 1978).

This chapter discusses the biological and geographical distribution, various sources from which galantamine is obtained, and the chemistry, pharmacology, and Alzheimer disease clinical study profiles of galantamine.

Naturally Occurring Chemicals against Alzheimer's Disease. https://doi.org/10.1016/B978-0-12-819212-2.00006-2

Biological and geographical distribution

Galantamine can be extracted from various genera of plants, namely *Amaryllis*, *Lycoris*, *Hymenocallis*, *Ungernia*, *Hippeastrum*, *Leucojum*, *Narcissus*, *Galanthus*, *Zephyranthes*, and *Haemanthus* and it can also be prepared synthetically using various methods (Cherkasov, 1978; Cherkasov and Tokachev, 2002).

Biological sources

1. *Leucojum aestivum* L. (snowflake)—This is found in the Mediterranean region and eastern Europe with a yield range from 0.1% to 0.3% (Stefanov, 1990)
2. *Narcissus* species—the bulbs of *Narcissus* plant are utilized for the extraction of galantamine, around 0.1% yield is found in the plant, some good plants of this genus include *N. pseudonarcissus* which gives a relatively high yield (0.13%) (Cherkasov and Tokachev, 2002; Kreh, 2002)
3. *Ungernia victoris*—*Ungernia victoris* Vved is found in Tajikistan and Uzbekistan is perennial species. Total alkaloid in leaves and bulbs is about 0.27%−0.0.71% and 1.18%−1.65% respectively. (Sadykov and Khodzimatov, 1988)
4. *Lycoris radiata*—*Lycoris radiata* Grey is a plant which is widespread in China, Korea and Japan. The *L. radiata* exact yield was not reported by the authors (Hayashi et al., 2005).

In vitro production of galantamine

Narcissus confusus and *Narcissus aestivum* are the only two species used for in vitro production of galantamine, the in vitro production was done using callus culture and micropropagation. Twelve μg/g from *L. aestivum* and 0.03 μg/g dry weight yield from *L. confusus* was obtained from the callus culture (Pavlov et al., 2007; Proskurina; Yakovleva, 1955).

Chemistry

Galantamine, also known as galanthamaine, is (1S,12S,14R)-9-methoxy-4methyl-11-oxa-4-azatetracycloheptadeca-6(17),7,9,15-tetreb-14-ol, and is an alkaloid found in different plants of the family Amaryllidaceae, some examples of which include *Narcissus pseudonarcissus* (daffodil) and *Leucojum aestivum* (snowflake).The chemical structure of galantamine is given in Fig. 3.1.4.1 (Attar-ur-Rahman, 2019; Thomas et al., 2008; Zohra and Uriel, 2005):

Pharmacology

Galantamine is reported to have a distinctive dual mode of action, which includes reversible and competitive inhibition of AChE, along with positive modulation of nAChRs (Albuquerque et al., 2001).

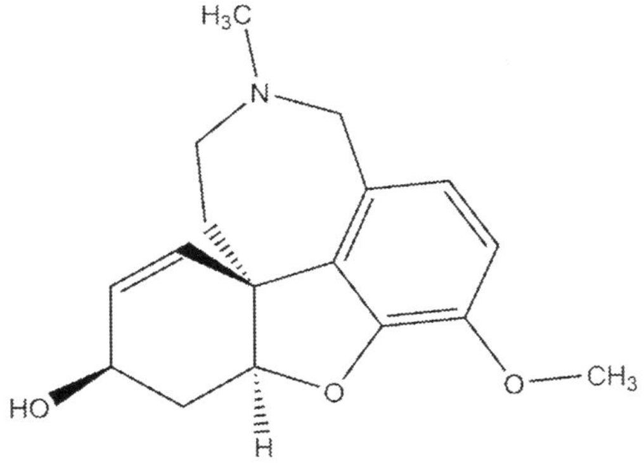

FIGURE 3.1.4.1 Chemical structure of galantamine.

AChE inhibition: Cholinergic neuron reduction has been correlated with cognitive deficits in AD, which affects learning and memory, in turn reducing the breakdown of Ach. Galantamine inhibits the interruption and breaking of Ach by binding to the active site of AChe competitively and reversibly (Albuquerque et al., 2001; Thomsen et al., 1990).

The two areas most affected in AD patients are the hippocampus and frontal cortex of the brain, where galantamine binds, and inhibit transmission by binding to AChE receptors (Thomsen et al., 1991).

nACHR modulation: the drop in the numbers of nACHR receptors plays a major role in AD. Many types of nACHR protein subunits are known, nine types of α-subunits are found in the brain, on which Ach binds, and three types of β-subunits are known, which are basically structural subunits. Loss of nACHRs is seen in the hippocampus and neocortex area of the brain in cases of AD. Galantamine binds to nACHRs at a secondary binding site on the α7 subunit, although galantamine cannot activate the site itself, rather it behaves like an allosteric modulator; when ACh and galantamine bind together on their respective site on nACHRs, galantamine starts a cellular response which is induced by ACh. Galantamine actually makes nACHRs more sensitive to ACh (Albuquerque et al., 2001; Martin-Ruiz et al., 1999; Paterson and Nordberg, 2000; Perry et al., 1995).

Absorption: galantamine has shown linear pharmacokinetics with a recommended dosage of 8–24 mg/day. It is freely absorbed, and has high bioavailability and a half-life in plasma of about 7 h, but plasma binding is low (only about 18%) (Bickel et al., 1991; Mihailova et al., 1989).

Distribution: repeated dosage of galantamine at about 12–16 mg twice a day gave a mean plasma concentration of about 42–137 ng/mL and the plasma concentration varied from 29 to 97 ng/mL. Galantamine does not accumulate, even after continuous usage of 6 months. When ingested with food, the absorption of the drug is delayed but it does not affect overall absorption, and peak plasma concentration falls to about 25% (Jones et al., 1996).

Metabolism: Galantamine is mostly metabolized by cytochrome P-450 isozyme in the liver. Metabolites that are formed include O-desmethyl-galantamine, norgalantamine, O-desmethyl-norgalantamine, galantaminone, and epigalantamine. None of these are present in any notable AChE activities, and activity is largely on unmetabolized galantamine (Janssen, 2000).

Excretion: Renal excretion in a healthy person is about 20%−25% of total plasma clearance, elimination is decreased in patients with renal impairment, and a dose reduction has been recommended for such patients (Paterson and Nordberg, 2000; Janssen, July 11, 2000).

Galantamine in Alzheimer's disease

1. Considerable help from galantamine was shown on the ADAS-Cog ($P < .0001$) and CIBIC (74% of the galantamine and 59% of the placebo patients were found to be stable and improved afterward). The galantamine also presented a significant improvement in patients suffering from Alzheimer disease and cerebrovascular disease (Erkinjuntti et al., 2002).
2. A total of 345 subjects were studied, of which 229 patients completed the 3- and 6-month visits. Patients withdrew because of adverse events and insufficient response to the drug treatment. Forty-five percent of patients were male. The patients were taking a variety of drugs including antiepileptic, antipsychotic, and herbal medications. The regular starting dosage of galantamine was 9.7 ± 6.7 mg/day, and after continuing the treatment for 3 months the average dosage was increased to 14.6 ± 5.0 mg/day and after being on the treatment for 6 months the average dosage given was 15.2 ± 2.7 mg/day. After 3 months of assessment 65% of patients had an increased MMSE (mini mental state examination score) with a response rate up to 92%. Most of the patients also showed improvement in MMSE score after the assessment for 6 months, providing a response rate of up to 91%. Overall, after 3 months, 38% of subjects were marginally improved, 22% were well improved, and 4% improved greatly, whereas 28% were unchanged. After 6 months, 20% were unchanged, 26% were marginally improved, 32% were well improved, and 7% responded very well to galantamine. A follow-up study was conducted for 12 months on 194 patients, in which the long-term effects of galantamine were found to be positive on the progression of AD (Henry et al., 2006, 2007).
3. Several ad hoc analyses of patients treated with galantamine with an MMSE score ≤ 14 were reported, suggesting improved functional and cognitive behavior in mild to moderate levels of AD (Lanctot et al., 2003).
4. A statistically significant difference between the active treatment and placebo groups showed an improved NPI score (Tariot et al., 2000).

5. The authors conducted a mass study on patients suffering from AD with CVD, they used a four-point study for over 6 months. Twice as many patients were found to be responders as compared to placebo (33.6% vs. 17.2% on placebo, $P = 0.003$). The authors also recommended a starting dose of 4 mg/day as compared to the recommended dosage of 8 mg/day, and gradually increased it to 24 mg/day for a 6-month follow-up program. This outcome was also found to be consistent with some previous mass studies on AD patients (Erkinjuntti et al., 2003, 2008).
6. A 12-week, open-label study was conducted to check the tolerability of a 1-week titration schedule that ranged from 8 to 16 mg/day of galantamine. The study showed 1-week titration to extended-release galantamine 16 mg/day was found to be well tolerated. There were some gastrointestinal (GI)-related adverse reactions (Scharre et al., 2008).
7. Various research on cholinesterase inhibitors like galantamine, donepezil, and rivastigmine have shown that this class of drugs only offers primary symptomatic relief and temporary cognitive improvement, but does not slow down disease progression (Hardy and Selkoe, 2002; Li et al., 2004).
8. Galantamine has shown considerable protection against Ab and okadaic acid-induced toxicity in neuroblastoma cells, which in turn shows an improved stress response (Calciano et al., 2010).
9. Initiating early treatment with cholinesterase inhibitors in mild to moderate AD can help in maintaining higher functions, independence, and improved quality of life. In patients who were studied using a double-blind phased trial, the placebo group did not catch up with the treatment group (Raskind et al., 2000).
10. In an extended 12-month study conducted on 26 AD and 11 HC (Healthy Control) subjects using galantamine, the posterior cingulate and corpus callosum cross-sectional area of the brain were put in focus for DTI studies, with all images from previous studies and newer studies being aligned to minimize errors. The cross-sectional finding at the baseline showed the reduction in genu and splenium of corpus callosum and FA in the posterior body of the corpus callosum in the patients treated with galantamine as compared to the placebo group. This result was also found to be consistent with previous studies, but the effect of galantamine was not preserved after 6 months open-label treatment. FA changes over the time seemed more important as it was found to be declining in both neurodegenerative and healthy aging individuals (Head et al., 2004; Likitjaroen et al., 2012).
11. Galantamine and other second-generation cholinesterase inhibitors, such as donepezil and rivastigmine are FDA-approved drugs for AD. Despite having different pharmacological properties these drugs have been shown to improve efficacy. The medications belonging to cholinesterase inhibitors were also found to be most effective when taken four times per

week. The MMSE score was also seen to be improving (Herrmann and Lanctot, 2007; Mielke et al., 2012).

12. Many cholinesterase inhibitors, especially galantamine, produce undesirable side effects in patients, sometimes even resulting in discontinuation of therapy. To counter this, the authors have continued to enhance the development of intranasal sprays of galantamine and chitosan. They found that GH−chitosan complex did not negatively alter the pharmacological efficiency of the drug, instead they recorded a decreased AChE protein level as compared to oral and nasal galantamine. No signs of toxicity or histopathological manifestations were found, indicating the biocompatibility of galantamine and chitosan (Lilienfeld, 2002; Hanafy et al., 2016).
13. A study conducted in China indicated that galantamine improves cognition and also has antineuroinflammatory effects in mice. Galantamine inhibits gliosis, proinflammatory signaling molecules (NF-κB p65), cytokines, and increased proteins associated with synapse in the hippocampus. Galantamine also shows reduced inflammation in microglia (Yi et al., 2018)
14. Ten trails with a total of 6805 subjects was analyzed, in which treatment with galantamine showed a greater proportion of subjects with improved rating. All dosages were found to cause improvements except for 8 mg/day, and the range of 16−36 mg/day was most favorable. A greater effect was achieved in 6 months than 3 months, which showed an ADAS-cog score improvement. Prolonged-release formulations also had the same efficacy and adverse reactions as the twice-daily regimen (Loy and Schneider, 2006).
15. In seven industry-oriented multicenter phase II and III trails, one of 5 months, one of 13 weeks, one of 29 weeks, two of 6 months, and two of 12 weeks, galantamine showed significant effects at doses of 16−32 mg/day which were statistically significant. Only the dose of 6 mg/kg failed to show any significance. Two 3-months trial with ADAS-cog scale showed significant improvement, using 24−32 mg/day dosages (Olin and Schneider, 2002)
16. In a 6-month follow-up assessment study of 33 AD patients, galantamine was found to benefit about 66.7% patients in different areas (visual construction, concentration, orientation, short- and long-term memory, attention, language ability, judgment, fluency, CASI total, and CDR-SB) except for fluency, which was on a par with donepezil, whereas rivastigmine showed improvements in all aspects. This study also did not find the 8 mg/day dosage to be significant and did not reach an optimum plasma concentration (Lin et al., 2019).

Toxicology and adverse drug reactions

Galantamine is a relatively safe drug, and shows good tolerability with no particular adverse drug reactions to vital signs or changes in laboratory reports.

Adverse reactions occurred in less than 5% of patients when studied on a 5-month randomized trial. The maintenance dose was 16 mg/day, which was escalated to 24 mg/day after 4 weeks at least. The main side effects that caused patients to quit the therapy were nausea, vomiting, anorexia, and diarrhea (Tariot et al., 2000).

Galantamine's adverse effects are similar to those of other cholinesterase inhibitors; it causes gastrointestinal stress with increasing dosage, those on 8 and 16 mg/day are less likely to forfeit therapy as compared to increased dosages of galantamine (Olin and Schneider, 2002; Physicians Desk Reference, 2005).

Conclusion

As discussed above, galantamine is used in neurodegenerative diseases, it is easier to obtain from various plant sources, can be made in vitro, and also can be synthesized in labs. Galantamine has also been studied extensively for its pharmacology and mechanism of action, and has been described as having a dual mode of action. Galantamine has been used for treatment of Alzheimer's disease for a long time and various studies have been discussed ranging from short-term to long-term follow-up studies, and it has been observed that galantamine can be used for mild to moderate Alzheimer's disease and has produced very little to no side effects in patients. However, the drug is not very promising in advanced stages of the disease and cannot actually halt the progression of the disease. Future studies of this drug could open new ways to understand the pathway of Alzheimer's disease progression and aid in the production of alternative modified drugs.

References

Albuquerque, E.X., Santos, M.D., Alkondon, M., Pereira, E.F.R., Maelicke, A., 2001. Modulation of receptor activity in the central nervous system — a novel approach to the treatment of Alzheimer's disease. Alzheimer Dis. Assoc. Disord. 15 (Suppl. 1), S19–S25.

Attar-ur-Rahman, 2019. Studies in Natural Product Chemistry, 1st ed. vol. 62. Elsevier Publication, p. 13. Bioactive Natural Products.

Bickel, U., Thomsen, T., Weber, W., et al., 1991. Pharmacokinetics of galantamine in humans and corresponding cholinesterase inhibition. Clin. Pharmacol. Ther. 50, 420–428.

Calciano, M.A., Zhou, W., Snyder, P.J., Einstein, R., 2010. Drug treatment of Alzheimer's disease patients leads to expression changes in peripheral blood cells. Alzheimer's Dement. 6, 386–393.

Cherkasov, O., Tokachev, O., 2002. In: Hanks, G. (Ed.), Medicinal and Aromatic Plants – Industrial Profiles: Narcissus and Daffodil, the Genus Narcissus. Taylor and Francis, London and New York, pp. 242–255.

Cherkasov, O., 1978. Khim-Farm. Zhur. 11, 84–87.

dos Santos, T.C., Mota Gomes, T., Serra Pinto, B.A., Camara, A.L., Andrade Paes, A.M.de, 2018. Naturally occurring acetylcholinesterase inhibitors and their potential use for Alzheimer's, disease therapy. Front. Pharmacol. 1192.

Emilien, G., Durlach, C., Minaker, K.L., Winblad, B., Gauthier, S., Maloteaux, J.-M., 2004. Alzheimer Disease: Neuropsychology and Pharmacology, p. 176 (Birkhauser).

Erkinjuntti, T., Kurz, A., Gautheir, S., Bullock, R., Lilienfeld, S., 2002. Efficacy of galantamine in probable vascular dementia and Alzheimer's disease combined with cerebrovascular disease: a randomized trial. Lancet 359, 1283–1290.

Erkinjuntti, T., Kurz, A., Small, G.W., Bullock, R., Lilienfeld, S., Damaraju, C.V., 2003. An open-label extension trial of galantamine in patients with probable vascular dementia and mixed dementia. Clin. Ther. 25, 1765–1782.

Erkinjuntti, T., Small, G.W., Bullock, R., Kurz, A., et al., 2008. Galantamine treatment in Alzheimer's disease with cerebrovascular disease: responder analyses from a randomized, controlled trial (GAL-INT-6). J. Psychopharmacol. 22 (7), 761–768.

Hanafy, A.S., Farid, R.M., Helmy, M.W., ElGamal, S.S., 2016. Pharmacological, toxicological and neuronal localization assessment of galantamine/chitosan complex nanoparticles in rats: future potential contribution in Alzheimer's disease management. Drug Deliv. 23 (8), 3111–3122.

Hardy, J., Selkoe, D.J., 2002. The amyloid hypothesis of Alzheimer's disease: progress and problems on the road to therapeutics. Science 297, 353–356.

Hayashi, A., Saito, T., Mukai, Y., Kurita, S., hori, T., 2005. Genes Genet. Syst. 80, 199–212.

Head, D., Buckner, R.L., Shimony, J.S., Williams, L.E., Akbudak, E., Conturo, T.E., McAvoy, M., Morris, J.C., Snyder, A.Z., 2004. Differential vulnerability of anterior white matter in nondemented aging with minimal acceleration in dementia of the Alzheimer type: evidence from diffusion tensor imaging. Cereb. Cortex 14, 410–423.

Heinrich, M., Teoh, H.J., 2004. Ethnopharmacology 92, 147–162.

Henry, B., Woodward, M., Boundy, K., Barnes, N., Allen, G., 2006. A naturalistic study of galantamine for Alzheimer's disease. CNS Drugs 20 (11), 935–943.

Henry, B., Woodward, M., Boundy, K., Barnes, N., Allen, G., 2007. A naturalistic study of galantamine for Alzheimer's disease 12 month follow-up from the nature study. CNS Drugs 21 (4), 335–336.

Herrmann, N., Lanctot, K.L., 2007. Pharmacologic management of neuropsychiatric symptoms of Alzheimer disease. Can. J. Psychiatry 52, 630–646.

Janssen, P., July 11, 2000. Reminyl (Galantamine) Tablets Prescribing Information. Janssen Pharmaceutica NV.

Jones, R.W., Cooper, D.M., Haworth, J., et al., 1996. The effect of food on the absorption of galanthamine in healthy elderly volunteers. Br. J. Clin. Pharmacol. 42, 671.

Kreh, M., 2002. In: Hanks, G. (Ed.), Medicinal and Aromatic Plants – Industrial Profiles: Narcissus and Daffodil, the Genus Narcissus. Taylor and Francis, London and New York, pp. 256–271.

Lanctot, K.L., Herrmann, N., Yau, K.K., Khan, L.R., Liu, B.A., LouLou, M.M., et al., 2003. Effcacy and safety of cholinesterase inhibitors in Alzheimer's disease: a meta-analysis. CMAJ 169, 557–564.

Li, G., Higdon, R., Kukull, W.A., Peskind, E., Van Valen Moore, K., Tsuang, D., et al., 2004. Statin therapy and risk of dementia in the elderly: a community-based prospective cohort study. Neurology 63, 1624–1628.

Likitjaroen, Y., Meindl, T., Friese, U., Wagner, M., et al., 2012. Longitudinal changes of fractional anisotropy in Alzheimer's disease patients treated with galantamine: a 12-month randomized, placebo-controlled, double-blinded study. Eur. Arch. Psychiatry Clin. Neurosci. 262, 341–350.

Lilienfeld, S., 2002. Galantamine-a novel cholinergic drug with a unique dual mode of action for the treatment of patients with Alzheimer's disease. CNS Drug Rev. 8, 159–176.

Lin, Y.T., Chou, M.C., Wu, S.J., Yang, Y.H., 2019. Galantamine plasma concentration and cognitive response in Alzheimer's disease. Peer J. 6887.

Loy, C., Schneider, L., 2006. Galantamine for Alzheimer's disease and mild cognitive impairment. Cochrane Database Syst. Rev. Issue 1.

Marco-Contelles, J.', Carmo Carreiras, M.do, Rodríguez, C., Villarroya, M., 2006. Synthesis and pharmacology of galantamine. Chem. Rev. 106, 116–133.

Martin-Ruiz, C.M., Court, J.A., Molnar, E., et al., 1999. α4 but not α3 and α7 nicotinic acetylcholine receptor subunits are lost from the temporal cortex in Alzheimer's disease. J. Neurochem. 73, 1635–1640.

Mielke, M.M., Leoutsakos, J.-M., Chris, D., Corcoran, Robert, C., 2012. Green et al. Effects of Food and Drug Administration-approved medications for Alzheimer's disease on clinical progression. Alzheimer's Dement. 8, 180–187.

Mihailova, D., Yamboliev, I., Zhivkova, Z., Tencheva, J., Jovovich, V., 1989. Pharmacokinetics of galantamine hydrobromide after single subcutaneous and oral dosage in humans. Pharmacology 39, 50–58.

Olin, J.T., Schneider, L., 2002. Galantamine for Alzheimer's disease. Cochrane Database Syst. Rev. Issue 3.

Paterson, D., Nordberg, A., 2000. Neuronal nicotinic receptors in the human brain. Prog. Neurobiol. 61, 75–111.

Pavlov, A., Berkov, S., Courot, E., Gocheva, T., et al., 2007. Process Biochem. 42, 734–739.

Perry, E., Morris, C.M., Court, J.A., et al., 1995. Alteration in nicotine binding sites in Parkinson's disease, Lewy body dementia and Alzheimer's disease: possible index of early neuropathology. Neuroscience 64, 385–395.

Physicians Desk Reference, 59th ed., 2005. Thomson PDR, Montvale, NJ, p. 1740.

Proskurina, N.F., Yakovleva, A.P., 1955. Zurnal Obshchei Khimii 25, 1035–1039.

Raskind, M.A., Peskind, E.R., Wessel, T., Yuan, W., 2000. Galantamine in AD: a 6 month randomized, placebo-controlled trial with a 6-month extension. The Galantamine USA-1 Study Group. Neurology 54, 2261–2268.

Sadykov, Y., Khodzimatov, M., 1988. Rastitelnie Resursy 24, 410–414.

Scharre, D.W., Shiovitz, T., Zhu, Y., Amatniek, J., 2008. One-week dose titration of extended release galantamine in patients with Alzheimer's disease. Alzheimer's Dement. 4, 30–37.

Stefanov, J., 1990. Ecological, Biological and Phytochemical Studies on Natural Populations and Introduced Origins of Snow Flake (*Leucojum aestivum* l.) in Bulgaria (D.Sc. thesis). Sofia.

Tariot, P.N., Solomon, P.R., Morris, J.C., Kershaw, P., Lilienfeld, S., Ding, C., 2000. A 5-month, randomized, placebo-controlled trial of galantamine in AD: the Galantamine USA-10 Study Group. Neurology 54, 2269–2276.

Thomas, L.L., David, W.A., Roche, V.F., WilliamFoye's, Z.S., 2008. Principles of Medicinal Chemistry, sixth ed. Lippincott Williams and Wilkins, p. 378.

Thomsen, T., Bickel, U., Fischer, J.P., et al., 1990. Stereoselectivity of cholinesterase inhibition by galantamine and tolerance in humans. Eur. J. Clin. Pharmacol. 39, 603–605.

Thomsen, T., Kaden, B., Fischer, J.P., et al., 1991. Inhibition of acetylcholinesterase activity in human brain tissue and erythrocytes by galantamine, physostigmine and tacrine. Eur. J. Clin. Chem. Clin. Biochem. 29, 487–492.

Viegas, C., Bolzani, V., Barriero, E., Fraga, C., 2005. Mini Rev. Med. Chem. 5, 915–926.

Yi, L., Yuyun, Z., Xian, Z., Tongyong, F., et al., 2018. Galantamine improves cognition, hippocampal inflammation, and synaptic plasticity impairments induced by lipopolysaccharide in mice. J. Neuroinflamm. 15, 112.

Zohra, Y., Uriel, B., 2005. Handbook of Medicinal Plants. The Howarth Medical Press, p. 360.

Chapter 3.1.5

Rivastigmine

Shahira M. Ezzat[1,2], Mohamed A. Salem[3], Nihal M. El Mahdy[4], Mai F. Ragab[5]
[1]*Department of Pharmacognosy, Faculty of Pharmacy, Cairo University, Cairo, Egypt;* [2]*Department of Pharmacognosy, Faculty of Pharmacy, October University for Modern Sciences and Arts (MSA), Giza, Egypt;* [3]*Department of Pharmacognosy, Faculty of Pharmacy, Menoufia University, Shibin Elkom, Menoufia, Egypt;* [4]*Department of Pharmaceutics and Industrial Pharmacy, Faculty of Pharmacy, October University for Modern Sciences and Arts (MSA), Giza, Egypt;* [5]*Department of Pharmacology and Toxicology, Faculty of Pharmacy, October University for Modern Sciences and Arts (MSA), Giza, Egypt*

Introduction

Dementia is a neurodegenerative condition characterized by a progressive decline in cognition and intellectual abilities. It leads to short- and long-term memory impairment, learning difficulties, impaired spatial orientation, and behavioral changes that impede social and occupational functioning (Williams et al., 2003; Jones, 2003). It is a rising global problem that imposes an enormous economic burden on society through direct medical and nonmedical costs as well as indirect costs (unpaid caregivers). The worldwide cost of dementia in 2010 was estimated to be US $604 billion (Thomas et al., 2015). The 2018 World Alzheimer Report revealed that this estimated cost has risen to US $1 trillion and that this figure is expected to double by 2030 and there will be a new case of dementia every 3 s.

Alzheimer's disease (AD) is a neurodegenerative dementing disorder characterized by slowly increasing impairment in memory, cognition, speech, and behavioral functions, such as the recognition of objects and people, as a result of neuronal cell death (Ross and Poirier, 2004). Moreover, individuals with AD face difficulties in performing daily life activities and may experience mood swings and depression, which greatly reduce the quality of life (Garcez et al., 2015).

AD is considered to be the most common neurodegenerative disease that accounts for 60%–80% of the diagnosed chronic dementia cases in elderly individuals above age 60 (Thomas et al., 2015; Singh et al., 2013; Chang et al., 2019). It is the fourth leading cause of death in developed countries. In 2013, it was estimated that nearly 44 million individuals had AD, and this figure is

Naturally Occurring Chemicals against Alzheimer's Disease. https://doi.org/10.1016/B978-0-12-819212-2.00007-4

predicted to reach 135 million by 2050 (Kiskis et al., 2015; Amat-Ur-Rasool and Ahmed, 2015).

The exact etiology of AD is not clear, but it is proposed to be initiated by synaptic impairment, neuronal loss, and microglial cell proliferation followed by inflammation as a result of an interplay among several genetic as well as environmental factors (Esfandiary et al., 2015). AD has two types: late-onset sporadic AD (in patients greater than 65 years) and early-onset familial AD (in persons less than 65 years, between 30 and 50 years) (Lane et al., 2018).

Sporadic AD accounts for 95%–99% of all AD cases. Epigenetic factors, such as aging, family history, and environmental factors, are triggers for sporadic AD. Disease conditions, such as high blood pressure, high cholesterol, atherosclerosis, obesity, and hyperglycemia, have all been associated with the development of AD (Thomas et al., 2015; Kim et al., 2014). However, familial AD is a rare autosomal dominant disorder that accounts for about 1%–5% of all AD cases and is caused by a mutation in the genes involved in the metabolism of amyloid beta (A), such as amyloid precursor protein (APP), presenilin-1, or presenilin-2 (Kim et al., 2014; Blennow et al., 2006).

The development of AD involves numerous mechanisms, such as deposition of aggregated proteins, mitochondrial dysfunction, oxidative stress, reduced neurotransmitters synthesis, and neuroinflammation (Hardy and Higgins, 1992). Macroscopic and microscopic alterations are observed in the brain of AD patients. Macroscopically, there is shrinkage and atrophy, especially in the hippocampus and cortex (Kim et al., 2012). At the microscopic level, two hallmark lesions are characteristic of AD: extracellular A plaques and intracellular neurofibrillary tangles (NFTs). The A plaques are composed of small toxic cleavage products (A40 and A42) of the APP, whereas the NFTs are made up of hyperphosphorylated tau proteins that lead to gradual neuronal loss and significantly reduced cholinergic activity, particularly in the cortex and hippocampus (Amat-Ur-Rasool and Ahmed, 2015).

Acetylcholine (ACh) is a neurotransmitter implicated in several processes, including learning and regulation of cognitive functions (Langguth et al., 2007; Roberson and Harrell, 1997). A significant correlation has been established between acetylcholinesterase (AChE) inhibition and cognition (Darreh-Shori et al., 2004). Cortical cholinergic neurons are lost owing to NFTs, which extensively diminish cholinergic neurotransmission in AD (Williams et al., 2003). This has led to the evolution of the cholinergic hypothesis, which attributes the decline in cognition, memory, and behavioral functions associated with dementia to the failure of neurons in the brain to transmit signals (Davies and Maloney, 1976).

To date, there is no cure for AD, and all current treatments aim merely to alleviate the symptoms of this condition. This can be achieved by improving neurotransmission by inhibiting the breakdown of ACh, thereby improving attention, memory, learning, and cognitive functions (Fig. 3.1.5.1) (Klimova et al., 2015). Acetylcholinesterase inhibitors (AChEIs), such as donepezil,

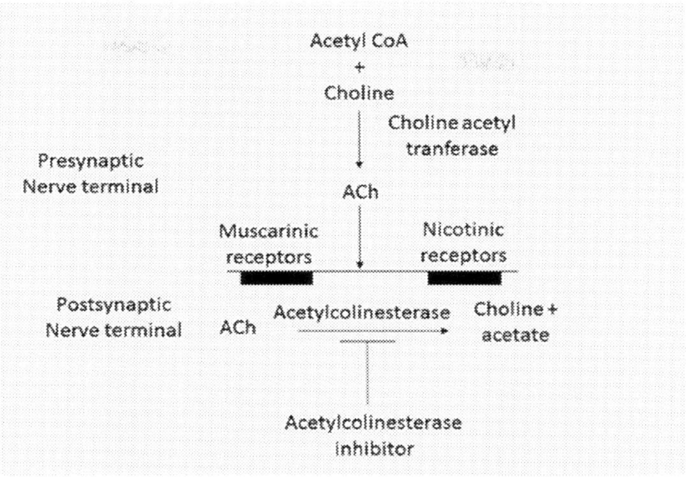

FIGURE 3.1.5.1 Mechanism of action of acetylcholinesterase inhibitors. *ACH*, acetylcholine; *CoA*, coenzyme.

galantamine, and rivastigmine, have been approved by the Food and Drug Administration for the management of mild to moderate stages of AD (Williams et al., 2003).

Rivastigmine

It is one of the new generation of AChE and butyrylcholinesterase (BuChE) inhibitors that were approved in 2000 under the name of Exelon, when it proved great efficacy against various stages of AD (Farlow et al., 2008; Castro and Martinez, 2001).

It is a carbamate-derived AChE reversible inhibitor that acts selectively on the central nervous system (CNS) and proved efficacy against AD as well as Parkinson's disease (PD). Rivastigmine is the generic name of the drug, whereas its commercial name is Exelon. It is a semisynthetic pharmaceutical product of eserine (physostigmine), the parasympathomimetic alkaloid isolated from the seeds of *Physostigma venenosum*, and it acts as a reversible cholinesterase inhibitor. Rivastigmine has a specific inhibitory mechanism through its carbamate structure that reacts covalently slowly with the active site of the AChE and BuChE, so it is a relatively weak inhibitor (Bolognesi et al., 2004).

The pharmacophore of rivastigmine is its carbamate structure, which is responsible for its in vitro and in vivo AChE and BuChE inhibitory activity. In this chapter, we will provide an overview on rivastigmine chemistry and sources, as well as the in vitro, in vivo activity, and clinical trials, bioavailability, pharmacokinetics, toxicity, side effects, and drug interactions.

Sources

Rivastigmine (Fig. 3.1.5.2) is one of the most potent drugs with cholinesterase inhibitory activity, which makes it a promising candidate for treating AD and PD (Polinsky, 1998; Nesi et al., 2017). Rivastigmine is a dual inhibitor for both AChE and BuChE (Reading et al., 2001; Polinsky, 1998). It is a semisynthetic derivative of physostigmine (Kumar, 2006) (Fig. 3.1.5.3).

Chemistry of rivastigmine

The semisynthetic compound rivastigmine ($C_{14}H_{22}N_2O_2$) is an organonitrogen compound (3-[(1S)-1-(dimethylamino)ethyl]phenyl N-ethyl-N-methylcarbamate) that belongs to the aralkylamine (Fig. 3.1.5.1). An aromatic hydrocarbyl group in the structure of rivastigmine replaces the alkyl group at one carbon atom. Rivastigmine has a molecular weight of 3367 Da (melting point of 158°C), and because it is a strong base, it has a pKa value of 8.89. It is only soluble in H_2O_2 at pH 1, but it is insoluble in neutral pH (Rivastigmine, 2019).

It is available as a fine crystalline powder of rivastigmine tartrate (the active form) ($C_{14}H_{22}N_2O_2 \cdot C_4H_6O_6$) (Fig. 3.1.5.2), which is white to off-white and has a molecular weight of 400.43 Da. Its chemical name is 2,6-dioxo-4-phenyl-piperidine-3-carbonitrile. Rivastigmine hydrogen tartrate has positive specific optical rotation. It does not exhibit polymorphism (Rivastigmine, 2019).

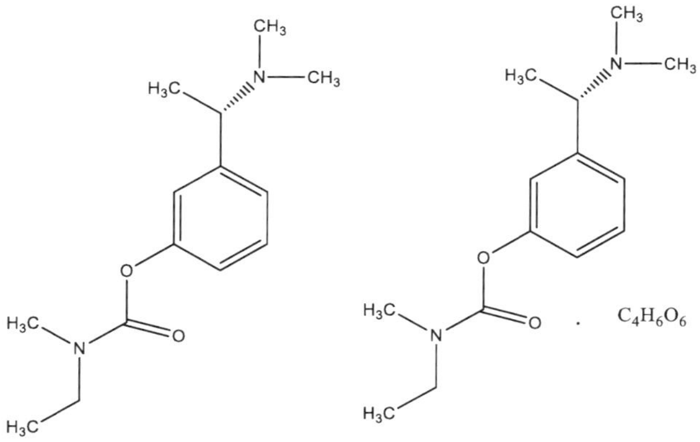

Rivastigmine **Rivastigmine tartrate**

FIGURE 3.1.5.2 Chemical structure of rivastigmine ((S)-3-(1-(dimethylamino)ethyl)phenyl ethylmethylcarbamate) and rivastigmine tartrate.

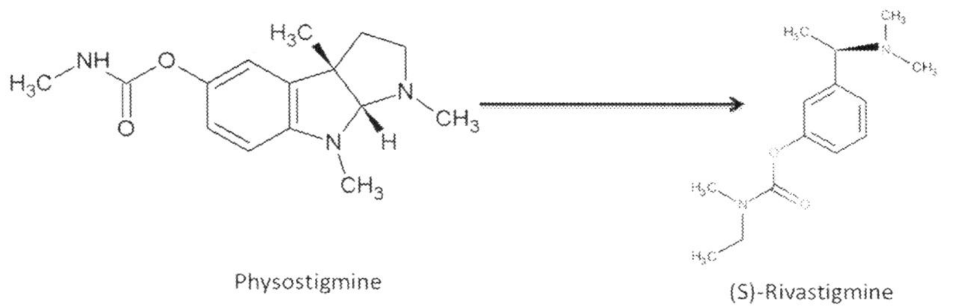

FIGURE 3.1.5.3 Semisynthesis of (S)-rivastigmine.

Methods of synthesis

Only the (S)-enantiomer of rivastigmine is the biologically active form. (S)-Rivastigmine has been synthesized in four steps using an asymmetric enzymatic transamination (Fuchs et al., 2012). The key substrate in this reaction was 3-acetylphenyl ethyl(methyl)carbamate (Fuchs et al., 2012) (Fig. 3.1.5.4).

In addition, the chemoenzymatic synthesis of (S)-rivastigmine has been successfully achieved via ketoreductases with the proton donor NADH/NADPH (Sethi et al., 2013) (Fig. 3.1.5.4).

Among the key chiral intermediates for the synthesis of (S)-rivastigmine is N-ethyl-methyl-carbamic acid-3-[(1S)-hydroxy-ethyl]-phenyl ester. This intermediate was reported to be stereoselectively synthesized by aldo/ketoreductases–glucose dehydrogenase recombinant whole cells (Xie et al., 2019) (Fig. 3.1.5.4).

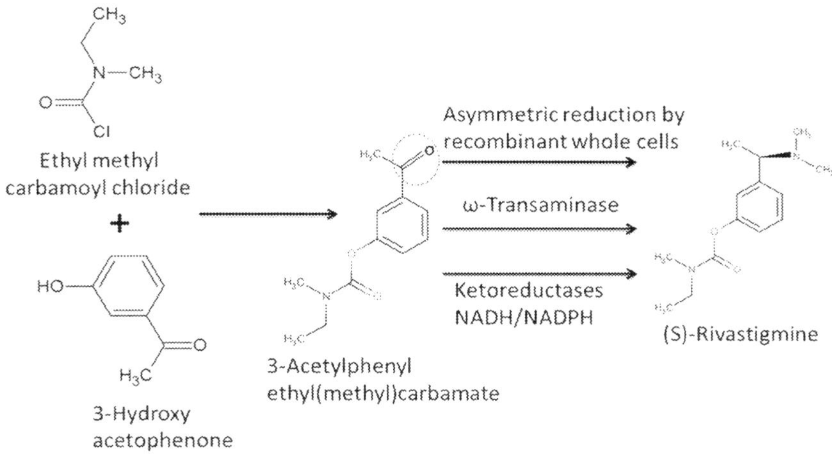

FIGURE 3.1.5.4 Synthesis of (S)-rivastigmine.

Mechanism of action

Rivastigmine is a pseudoirreversible brain-selective cholinesterase inhibitor that regulates the levels of ACh by inhibiting both AChE and BuChE, thus ameliorating the symptoms of dementia observed in mild to moderate AD (Williams et al., 2003; Onor et al., 2007). Studies conducted on human volunteers revealed that rivastigmine is selective to central cholinesterases over the peripheral cholinesterases (Kennedy et al., 1999) and that after 12 months of treating AD patients with rivastigmine, there was a decrease in the activity of AChE and BuChE in the cerebrospinal fluid (CSF) and plasma by 46% and 65%, respectively, in patients with AD (Darreh-Shori et al., 2002). Rivastigmine primarily targets the cerebral cortex and hippocampus, which are the areas that show the most cholinergic deficits in AD patients (Darvesh et al., 1998). AChE is mostly present in the synapses and the cerebral cortical region of the brain, whereas BuChE is mainly present in the glial cells.

Rivastigmine is a carbamate derivative that binds to the ionic and esteratic sites of AChE, inhibiting ACh degradation and improving cholinergic neurotransmission. In AD, the normal tetrameric form, G4 isoform, of AChE is diminished, and the monomeric G1 isoform becomes more abundant. Rivastigmine preferentially inhibits the predominate G1 isoform (Kandiah et al., 2017; Polinsky, 1998; Mesulam and Geula, 1991). As rivastigmine binds AChE, it undergoes cleavage, releasing an inert product that is eliminated through the kidney while the carbamate moiety remains bound to the enzyme for longer as it undergoes hydrolysis slowly (Onor et al., 2007; Polinsky, 1998). The dual inhibition demonstrated by rivastigmine is immensely beneficial because it leads to a significant increase in the levels of ACh, which may result in greater clinical benefits (Ballard, 2002; Poirier, 2002). As AD progresses, AChE activity declines, whereas that of BuChE rises. This depicts the importance of inhibiting AChE in the early stages of AD and inhibiting BuChE activity in the later stages of the disease (Lane et al., 2006). Also, AChE and BuChE are involved in the formation of A plaques and NFTs, where they accumulate within the plaques and in the tangles; thus, through inhibiting both enzymes, rivastigmine decreases the formation of AD pathological lesions (Mesulam and Geula, 1994; Onor et al., 2007). The dual inhibitory action exhibited by rivastigmine was found to produce less cortical atrophy in a study conducted on 26 AD patients, and this sheds light on the implications of rivastigmine as a potential neuroprotective agent (Venneri et al., 2005).

It was also reported that rivastigmine caused a mild upregulation of ACh receptors in AD patients (Almkvist et al., 2004; Darreh-Shori et al., 2004). Studies conducted in animal models revealed that rivastigmine might have an effect of glutaminergic neurotransmission by regulating the activity of some genes, increasing the levels of glutamate (Trabace et al., 2000; Andin et al., 2005). Several randomized, double-blind clinical trials as well as open-label and real-world evidence studies have demonstrated significant improvement or preservation of cognitive and behavioral functions among AD patients using AChE inhibitors such as rivastigmine compared with placebo (Lopez-Pousa et al., 2005; Aguglia et al., 2004; Rosler et al., 1999; Farlow et al., 2000).

Pharmacokinetics and pharmacodynamics

Administered orally, rivastigmine is rapidly and completely (>90%) absorbed with a bioavailability of about 36%. It reaches its peak plasma concentration after about 1 h, but its oral absorption is greatly affected by the concomitant administration of food, which leads to a 30% decrease in rivastigmine concentration and a 90-min delay in reaching its maximum concentration (Polinsky, 1998). Administration of rivastigmine with food improves its tolerability and decreases its gastrointestinal side effects. Rivastigmine has low protein binding (about 40%) and is extensively hydrolyzed by cholinesterases to produce a decarbamylated product; hence, its metabolism is independent of the hepatic microsomal cytochrome P450 system. Its pharmacokinetic half-life is around 1.5 h; about 90% is completely eliminated after 24 h of administration (Onor et al., 2007; Williams et al., 2003). Rivastigmine metabolites are mainly excreted though the kidney; only 1% is eliminated in the feces (Polinsky, 1998; Williams et al., 2003). The pharmacodynamic half-life of rivastigmine is about 10 h (Cutler et al., 1998; Birks and Grimley Evans, 2015); therefore, it should be administered twice daily (Onor et al., 2007). The difference observed between the pharmacokinetic and pharmacodynamic half-lives is because rivastigmine dissociates slowly from its AChE binding site (Cutler et al., 1998; Polinsky, 1998; Williams et al., 2003). The relatively low plasma half-life is considered to be both an advantage and a disadvantage. Rapid elimination of the compound decreases the likelihood of major side effects, but this means more frequent daily dosing that might pose a problem in AD patients as a result of poor compliance (Burns, 2003). After absorption, rivastigmine quickly penetrates the CNS, with a CSF time to maximum plasma concentration of 1.4–3.8 h. Its concentration then drops rapidly with a CSF half-life of 0.31–0.95 h. Rivastigmine acts in a dose-dependent manner, in which the higher concentration of the drug leads to greater enzyme inhibition in the CSF. Doses of 2, 6, 10, and 12 mg/day of rivastigmine lead to 20%, 46%, 55.6%, and 61.7% AChE inhibition and 23.9%, 76.6%, 54.9%, and 61% BuChE inhibition (Onor et al., 2007).

Drug interactions

Rivastigmine has minimal drug–drug interactions, and this is a crucial feature, because this medication is intended for use in elderly people, who typically have comorbidities and are taking different medications. It was reported that there were no interactions between rivastigmine and warfarin, digoxin, diazepam, calcium channel blockers, antihistamines, and fluoxetine (Polinsky,

1998; Spencer and Noble, 1998; Grossberg et al., 2000). This is most probably because rivastigmine undergoes hydrolysis by cholinesterases and hence is independent of the cytochrome P450–metabolizing machinery.

Different hybrids and their actions

The association of rivastigmine with the cholinergic precursors choline or choline alphoscerate on acetylcholine levels was studied in rat frontal cortex, hippocampus, and striatum, in which [^{3}H]hemicholinium-3 binding was used as a high-affinity cholinergic transporter marker (Amenta et al., 2006). The treatments were administered daily for 7 days for comparison. Choline (400 mg/kg) and choline alphoscerate (150 mg/kg) were administered intraperitoneally. Rivastigmine suspension in polyethylene glycol (diluted distilled water at a ratio of 1:3) was tested through daily administration using gastric gavage. The levels of acetylcholine and [^{3}H]hemicholinium-3 binding as well as immunoreactivity of acetylcholine in nerve fibers were enhanced by choline alphoscerate or rivastigmine, whereas choline had no effect. The mixture of choline alphoscerate with rivastigmine showed a dose-dependent increase in acetylcholine levels and [^{3}H]hemicholinium-3 binding. AChE activity was inhibited by rivastigmine alone or its association with choline or choline alphoscerate, whereas choline or choline alphoscerate had no effect on the activity of AChE (Amenta et al., 2006).

Another novel association was made by Wang et al. (2017) of chalcone with rivastigmine. These associations were designed, synthesized, and tested for the in vitro inhibition of AChE and BuChE. Most of the target compounds in the micro- and submicromolar concentrations selectively inhibited BuChE. Compound 3 ([E]-3-(3-[4-Hydroxyphenyl]acryloyl)phenyl dimethylcarbamate) (Fig. 3.1.5.5) showed the highest activity because it inhibited both AChE and BuChE with IC_{50} values (0.87 and 0.36 M). This activity is close to that of commercial rivastigmine. This hybrid also inhibits the production of reactive oxygen species in the neuroblast from neural tissue with low cytotoxic activity, so this association acts as an anti-Alzheimer's factor by different mechanisms.

Chen et al. (2017) also synthesized a group of rivastigmine (caffeic acid and rivastigmine) ferulic acid associations and tested their in vitro anti-Alzheimer's activity. Compound 5 was the most active hybrid with high neuroprotective effect, because it inhibited AChE and -amyloid self-aggregation and showed copper chelation activities. Thus, this hybrid (Fig. 3.1.5.5) could be considered a lead drug in the anti-Alzheimer's drug discovery process.

Available dosage forms of rivastigmine

Currently, rivastigmine is available on the market as 1.5-, 3-, 4.5-, and 6-mg hard capsules with a therapeutic dose of 6–12 mg daily (Khoury et al., 2018). In 2007, rivastigmine became available as a transdermal patch applied to the

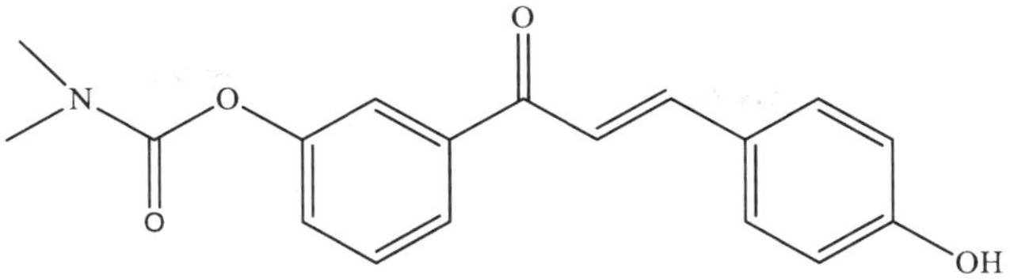

Compound 3

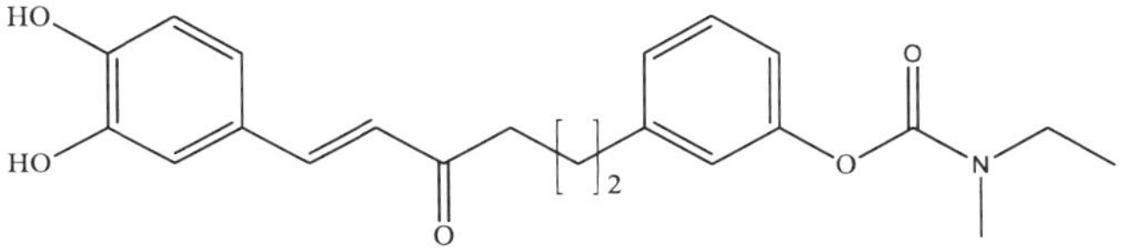

Compound 5

FIGURE 3.1.5.5 Structures of the prepared hybrids.

skin once daily at a dose of 4.6 mg/day for the first 4 weeks and then increased gradually to the recommended maintenance dose of 9.5 mg/day (Adler et al., 2014). The developed patches are safe and well-tolerated and simultaneously allow good adherence because they provide a stable sustained release over time and effective drug penetration via the skin (Blesa González et al., 2011). Research demonstrated that patients and caregivers of AD prefer patches over the oral dosage form, such as capsules for drug delivery, owing to the ease of use, especially with the declining cognitive ability of AD patients over time. It is easier to apply the patch and possibly improves patient compliance and clinical benefits (Winblad et al., 2007; Blesa González et al., 2011). Recent research aims to formulate rivastigmine into other dosage forms to increase its bioavailability and targeted delivery to the brain using different routes of administration.

Intranasal route

The intranasal route remains one of the promising strategies to attain direct nose-to-brain delivery through the olfactory region and the trigeminal nerve pathway. This delivery route avoids the blood-brain barrier (BBB) whereas it achieves better brain distribution and avoids the hepatic first-pass effect (Ganger and Schindowski, 2018). To achieve this aim, researchers worked to incorporate hydrophilic rivastigmine into more lipophilic dosage forms capable of passing the BBB to achieve the targeted drug delivery of rivastigmine, thus making it the most extensively studied route of administration.

One study formulated rivastigmine into a mucoadhesive thermosensitive in situ gel using pluronic F127 as a thermogelling agent and 1% Carbopol 934 as a mucoadhesive agent. The gel showed significant permeation transnasally (84%) compared with the intranasal and intravenous solutions, achieving higher brain delivery and distribution (Abouhussein et al., 2018). Salatin et al. (2017) also developed rivastigmine-loaded poly(lactic-co-glycolic acid) nanoparticles into erodible in situ gel capable of prolonging drug release; it showed adequate cytocompatibility in an A549 cell line with an increase in cellular uptake that was time-dependant in addition to enhanced permeation in nasal sheep mucosa.

Lipid-based carriers have also been used to enhance brain targeting of rivastigmine owing to their hydrophobic nature, which enhances drug permeation through the BBB. One study employed nanostructured lipid carriers that were 266 ± 0.94 nm in diameter and had an entrapment efficiency of 61.82% ± 2.52% and 81.23% ± 2.14% drug diffusion in 12 h through goat nasal mucosa as well as a downregulation in AChE-expressing enzymes (Anand et al., 2019). Liposomes and modified cell-penetrating peptide showed that significantly higher concentrations of rivastigmine could cross the BBB compared with the free rivastigmine.

Haider et al. (2018) developed rivastigmine-loaded nanoemulsion, which showed a higher release, enhanced permeation of rivastigmine across nasal mucosa, and significantly higher levels of rivastigmine achieved in the brain. Shah et al. (2015a) used quality by design to optimize rivastigmine-loaded solid lipid nanoparticles, which showed enhanced in vitro and ex vivo release as well as a high safety profile. The same research group developed microemulsions and mucoadhesive microemulsions, which not exhibit any nasal ciliotoxicity and showed adequate stability for 3 months (Shah et al., 2015b).

Rivastigmine-loaded chitosan nanoparticles coated with Tween 80 were developed by Nagpal et al. (2013), which showed a significant reversal of scopolamine-induced amnesia compared with both free drug and uncoated nanoparticles as well as enhanced rivastigmine bioavailability. Fazil et al. (2012) also prepared chitosan nanoparticles, and rivastigmine concentration was significantly higher compared with the rivastigmine solution in the brain through both the intravenous and intranasal routes.

Injections

Bastiat et al. (2010) developed an injectable subcutaneous organogel–based implant system composed of safflower oil and a modified tyrosine organogelator and concluded that the pharmacokinetics of rivastigmine in the formulation was affected by the composition of the gel as well as the dose and volume of the implant. The formulation also resulted in prolonged inhibition of AChE in the hippocampus.

Oral

Oral bioavailability of rivastigmine was improved by Penhasi and Gomberg (2018) through a once-daily tablet prepared as a tablet-in-tablet using a compression-coating method. This technique offered a double-pulse release profile to enhance patient compliance and avoid the development of tolerance, because the outer tablet released the drug immediately and lasted up to 4 h whereas the film-coated inner tablet gave a time-controlled delivery after 3 h followed by a burst release of the drug.

Inhalation

Pulmonary delivery of rivastigmine was studied by Simon et al. (2016) through loading on microparticles with the addition of L-leucine to alter the smooth surface of the microparticles. The results showed adequate aerodynamic properties, high stability, and better bioavailability through absorption from the lung tissue to the bloodstream.

Transdermal

Rivastigmine-loaded transdermal patches were prepared by Malaiya et al. (2018) using a solvent casting method, which showed adequate drug content uniformity and sustained delivery of the drug into the bloodstream.

Tolerability

The starting dose of rivastigmine is 1.5 mg administered twice daily, and it can be increased to 6 mg twice daily. Increasing the dose is done gradually with 1–2 weeks between every increase in dose depending on the patients's tolerance (Onor et al., 2007). Side effects observed with the use of rivastigmine are cholinergic in nature; they result in gastrointestinal problems such as diarrhea, nausea, vomiting, and anorexia. The use of 6–12 mg/day rivastigmine caused a significant increase in the incidence of dizziness, drowsiness, headache, agitation, cough, myalgia, malaise, fatigue, sweating, and sinusitis compared with placebo. Most of these side effects were dose-dependent and mild to moderate in nature, and had a limited duration. Laboratory investigations and vital signs were normal in patients receiving rivastigmine therapy (6–12 mg/day), except for some minor weight loss observed in 20% of patients. It was observed that food improves drug tolerability; therefore, it is advised to administer rivastigmine with meals (Rosler et al., 1999; Onor et al., 2007; Williams et al., 2003).

Conclusion

AChEIs such as donepezil, galantamine, and rivastigmine are considered promising drug leads for the management of mild to moderate stages of AD. Rivastigmine is a semisynthetic AChEI with proved efficacy and good tolerability in various dosage forms, with mild side effects. It can also be a drug lead for the preparation of more effective derivatives and hybrids. There are still unsolved scientific and clinical inquiries about AChE inhibitors, such as their low efficacy in late stages of AD, their peripheral adverse effects, and the fast development of drug resistance.

References

Abouhussein, D.M.N., Khattab, A., Bayoumi, N.A., Mahmoud, A.F., Sakr, T.M., 2018. Brain targeted rivastigmine mucoadhesive thermosensitive in situ gel: optimization, in vitro evaluation, radiolabeling, in vivo pharmacokinetics and biodistribution. J. Drug Deliv. Sci. Technol. 43, 129–140.

Adler, G., Mueller, B., Articus, K., 2014. The transdermal formulation of rivastigmine improves caregiver burden and treatment adherence of patients with Alzheimer's disease under daily practice conditions. Int. J. Clin. Pract. 68, 465–470.

Aguglia, E., Onor, M.L., Saina, M., Maso, E., 2004. An open-label, comparative study of rivastigmine, donepezil and galantamine in a real-world setting. Curr. Med. Res. Opin. 20, 1747–1752.

Almkvist, O., Darreh-Shori, T., Stefanova, E., Spiegel, R., Nordberg, A., 2004. Preserved cognitive function after 12 months of treatment with rivastigmine in mild Alzheimer's disease in comparison with untreated AD and MCI patients. Eur. J. Neurol. 11, 253–261.

Amat-Ur-Rasool, H., Ahmed, M., 2015. Designing second generation anti-Alzheimer compounds as inhibitors of human acetylcholinesterase: computational screening of synthetic molecules and dietary phytochemicals. PLoS One 10 e0136509.

Amenta, F., Tayebati, S.K., Vitali, D., DI Tullio, M.A., 2006. Association with the cholinergic precursor choline alphoscerate and the cholinesterase inhibitor rivastigmine: an approach for enhancing cholinergic neurotransmission. Mech. Ageing Dev. 127, 173–179.

Anand, A., Arya, M., Kaithwas, G., Singh, G., Saraf, S.A., 2019. Sucrose stearate as a biosurfactant for development of rivastigmine containing nanostructured lipid carriers and assessment of its activity against dementia in *C. elegans* model. J. Drug Deliv. Sci. Technol. 49, 219–226.

Andin, J., Enz, A., Gentsch, C., Marcusson, J., 2005. Rivastigmine as a modulator of the neuronal glutamate transporter rEAAC1 mRNA expression. Dement. Geriatr. Cogn. Disord. 19, 18–23.

Ballard, C.G., 2002. Advances in the treatment of Alzheimer's disease: benefits of dual cholinesterase inhibition. Eur. Neurol. 47, 64–70.

Bastiat, G., Plourde, F., Motulsky, A., Furtos, A., Dumont, Y., Quirion, R., Fuhrmann, G., Leroux, J.C., 2010. Tyrosine-based rivastigmine-loaded organogels in the treatment of Alzheimer's disease. Biomaterials 31, 6031–6038.

Birks, J.S., Grimley Evans, J., 2015. Rivastigmine for Alzheimer's disease. Cochrane Database Syst. Rev. 4 1–153. https://doi.org/10.1002/14651858.CD001191.pub3. CD001191.

Blennow, K., DE Leon, M.J., Zetterberg, H., 2006. Alzheimer's disease. Lancet 368, 387–403.

Blesa González, R., Boada Rovira, M., Martínez Parra, C., Gil-Saladie, D., Almagro, C.A., Gobartt Vázquez, A.L., 2011. Evaluation of the convenience of changing the rivastigmine administration route in patients with Alzheimer disease. Neurologia 26, 262–271.

Bolognesi, M.L., Bartolini, M., Cavalli, A., Andrisano, V., Rosini, M., Minarini, A., Melchiorre, C., 2004. Design, synthesis, and biological evaluation of conformationally restricted rivastigmine analogues. J. Med. Chem. 47, 5945–5952.

Burns, A., 2003. Treatment of cognitive impairment in Alzheimer's disease. Dialogues Clin. Neurosci. 5, 35–43.

Castro, A., Martinez, A., 2001. Peripheral and dual binding site acetylcholinesterase inhibitors: implications in treatment of Alzheimer's disease. Mini Rev. Med. Chem. 1, 267–272.

Chang, C.J., Chou, T.C., Chang, C.C., Chen, T.F., Hu, C.J., Fuh, J.L., Wang, W., Chen, C.M., Hsu, W., Huang, C.C., 2019. Persistence and adherence to rivastigmine in patients with dementia: results from a noninterventional, retrospective study using the National Health Insurance research database of Taiwan. Alzheimers Dement. 5, 46–51.

Chen, Z., Digiacomo, M., Tu, Y., Gu, Q., Wang, S., Yang, X., Chu, J., Chen, Q., Han, Y., Chen, J., Nesi, G., Sestito, S., Macchia, M., Rapposelli, S., Pi, R., 2017. Discovery of novel rivastigmine-hydroxycinnamic acid hybrids as multi-targeted agents for Alzheimer's disease. Eur. J. Med. Chem. 125, 784–792.

Cutler, N.R., Polinsky, R.J., Sramek, J.J., Enz, A., Jhee, S.S., Mancione, L., Hourani, J., Zolnouni, P., 1998. Dose-dependent CSF acetylcholinesterase inhibition by SDZ ENA 713 in Alzheimer's disease. Acta Neurol. Scand. 97, 244–250.

Darreh-Shori, T., Almkvist, O., Guan, Z.Z., Garlind, A., Strandberg, B., Svensson, A.L., Soreq, H., Hellstrom-Lindahl, E., Nordberg, A., 2002. Sustained cholinesterase inhibition in AD patients receiving rivastigmine for 12 months. Neurology 59, 563–572.

Darreh-Shori, T., Hellstrom-Lindahl, E., Flores-Flores, C., Guan, Z.Z., Soreq, H., Nordberg, A., 2004. Long-lasting acetylcholinesterase splice variations in anticholinesterase-treated Alzheimer's disease patients. J. Neurochem. 88, 1102–1113.

Darvesh, S., Grantham, D.L., Hopkins, D.A., 1998. Distribution of butyrylcholinesterase in the human amygdala and hippocampal formation. J. Comp. Neurol. 393, 374–390.

Davies, P., Maloney, A.J., 1976. Selective loss of central cholinergic neurons in Alzheimer's disease. Lancet 2, 1403.

Esfandiary, E., Karimipour, M., Mardani, M., Ghanadian, M., Alaei, H.A., Mohammadnejad, D., Esmaeili, A., 2015. Neuroprotective effects of Rosa damascena extract on learning and memory in a rat model of amyloid-beta-induced Alzheimer's disease. Adv. Biomed. Res. 4, 131.

Farlow, M., Anand, R., Messina Jr., J., Hartman, R., Veach, J., 2000. A 52-week study of the efficacy of rivastigmine in patients with mild to moderately severe Alzheimer's disease. Eur. Neurol. 44, 236–241.

Farlow, M.R., Miller, M.L., Pejovic, V., 2008. Treatment options in Alzheimer's disease: maximizing benefit, managing expectations. Dement. Geriatr. Cogn. Disord. 25, 408–422.

Fazil, M., Md, S., Haque, S., Kumar, M., Baboota, S., Sahni, J.K., Ali, J., 2012. Development and evaluation of rivastigmine loaded chitosan nanoparticles for brain targeting. Eur. J. Pharm. Sci. 47, 6–15.

Fuchs, M., Koszelewski, D., Tauber, K., Sattler, J., Banko, W., Holzer, A.K., Pickl, M., Kroutil, W., Faber, K., 2012. Improved chemoenzymatic asymmetric synthesis of (S)-Rivastigmine. Tetrahedron 68, 7691–7694.

Ganger, S., Schindowski, K., 2018. Tailoring formulations for intranasal nose-to-brain delivery: a review on architecture, physico-chemical characteristics and mucociliary clearance of the nasal olfactory mucosa. Pharmaceutics 10.

Garcez, M.L., Falchetti, A.C., Mina, F., Budni, J., 2015. Alzheimer's disease associated with psychiatric comorbidities. An. Acad. Bras. Cienc. 87, 1461–1473.

Grossberg, G.T., Stahelin, H.B., Messina, J.C., Anand, R., Veach, J., 2000. Lack of adverse pharmacodynamic drug interactions with rivastigmine and twenty-two classes of medications. Int. J. Geriatr. Psychiatry 15, 242–247.

Haider, M.F., Khan, S., Gaba, B., Alam, T., Baboota, S., Ali, J., Ali, A., 2018. Optimization of rivastigmine nanoemulsion for enhanced brain delivery: in-vivo and toxicity evaluation. J. Mol. Liq. 255, 384–396.

Hardy, J.A., Higgins, G.A., 1992. Alzheimer's disease: the amyloid cascade hypothesis. Science 256, 184–185.

Jones, R.W., 2003. The dementias. Clin. Med. 3, 404–408.

Kandiah, N., Pai, M.C., Senanarong, V., Looi, I., Ampil, E., Park, K.W., Karanam, A.K., Christopher, S., 2017. Rivastigmine: the advantages of dual inhibition of acetylcholinesterase and butyrylcholinesterase and its role in subcortical vascular dementia and Parkinson's disease dementia. Clin. Interv. Aging 12, 697–707.

Kennedy, J.S., Polinsky, R.J., Johnson, B., Loosen, P., Enz, A., Laplanche, R., Schmidt, D., Mancione, L.C., Parris, W.C., Ebert, M.H., 1999. Preferential cerebrospinal fluid acetylcholinesterase inhibition by rivastigmine in humans. J. Clin. Psychopharmacol. 19, 513–521.

Khoury, R., Rajamanickam, J., Grossberg, G.T., 2018. An update on the safety of current therapies for Alzheimer's disease: focus on rivastigmine. Ther. Adv. Drug Saf. 9, 171–178.

Kim, D.H., Yeo, S.H., Park, J.M., Choi, J.Y., Lee, T.H., Park, S.Y., Ock, M.S., Eo, J., Kim, H.S., Cha, H.J., 2014. Genetic markers for diagnosis and pathogenesis of Alzheimer's disease. Gene 545, 185–193.

Kim, G.H., Jeon, S., Seo, S.W., Kim, M.J., Kim, J.H., Roh, J.H., Shin, J.S., Kim, C.H., Im, K., Lee, J.M., Qiu, A., Kim, S.T., Na, D.L., 2012. Topography of cortical thinning areas associated with hippocampal atrophy (HA) in patients with Alzheimer's disease (AD). Arch. Gerontol. Geriatr. 54, e122–e129.

Kiskis, J., Fink, H., Nyberg, L., Thyr, J., Li, J.Y., Enejder, A., 2015. Plaque-associated lipids in Alzheimer's diseased brain tissue visualized by nonlinear microscopy. Sci. Rep. 5, 13489.

Klimova, B., Maresova, P., Valis, M., Hort, J., Kuca, K., 2015. Alzheimer's disease and language impairments: social intervention and medical treatment. Clin. Interv. Aging 10, 1401–1407.

Kumar, V., 2006. Potential medicinal plants for CNS disorders: an overview. Phytother. Res. 20, 1023–1035.

Lane, C.A., Hardy, J., Schott, J.M., 2018. Alzheimer's disease. Eur. J. Neurol. 25, 59–70.

Lane, R.M., Potkin, S.G., Enz, A., 2006. Targeting acetylcholinesterase and butyrylcholinesterase in dementia. Int. J. Neuropsychopharmacol. 9, 101–124.

Langguth, B., Bauer, E., Feix, S., Landgrebe, M., Binder, H., Sand, P., Hajak, G., Eichhammer, P., 2007. Modulation of human motor cortex excitability by the cholinesterase inhibitor rivastigmine. Neurosci. Lett. 415, 40–44.

Lopez-Pousa, S., Turon-Estrada, A., Garre-Olmo, J., Pericot-Nierga, I., Lozano-Gallego, M., Vilalta-Franch, M., Hernandez-Ferrandiz, M., Morante-Munoz, V., Isern-Vila, A., Gelada-Batlle, E., Majo-Llopart, J., 2005. Differential efficacy of treatment with acetylcholinesterase inhibitors in patients with mild and moderate Alzheimer's disease over a 6-month period. Dement. Geriatr. Cogn. Disord. 19, 189–195.

Malaiya, M.K., Jain, A., Pooja, H., Jain, A., Jain, D., 2018. Controlled delivery of rivastigmine using transdermal patch for effective management of alzheimer's disease. J. Drug Deliv. Sci. Technol. 45, 408–414.

Mesulam, M.M., Geula, C., 1991. Acetylcholinesterase-rich neurons of the human cerebral cortex: cytoarchitectonic and ontogenetic patterns of distribution. J. Comp. Neurol. 306, 193–220.

Mesulam, M.M., Geula, C., 1994. Butyrylcholinesterase reactivity differentiates the amyloid plaques of aging from those of dementia. Ann. Neurol. 36, 722–727.

Nagpal, K., Singh, S.K., Mishra, D.N., 2013. Optimization of brain targeted chitosan nanoparticles of Rivastigmine for improved efficacy and safety. Int. J. Biol. Macromol. 59, 72–83.

Nesi, G., Chen, Q., Sestito, S., Digiacomo, M., Yang, X., Wang, S., Pi, R., Rapposelli, S., 2017. Nature-based molecules combined with rivastigmine: a symbiotic approach for the synthesis of new agents against Alzheimer's disease. Eur. J. Med. Chem. 141, 232–239.

Onor, M.L., Trevisiol, M., Aguglia, E., 2007. Rivastigmine in the treatment of Alzheimer's disease: an update. Clin. Interv. Aging 2, 17–32.

Penhasi, A., Gomberg, M., 2018. A specific two-pulse release of rivastigmine using a modified time-controlled delivery system: a proof of concept case study. J. Drug Deliv. Sci. Technol. 47, 404–410.

Poirier, J., 2002. Evidence that the clinical effects of cholinesterase inhibitors are related to potency and targeting of action. Int. J. Clin. Pract. (Suppl.), 6–19.

Polinsky, R.J., 1998. Clinical pharmacology of rivastigmine: a new-generation acetylcholinesterase inhibitor for the treatment of Alzheimer's disease. Clin. Ther. 20, 634–647.

Reading, P.J., Luce, A.K., Mckeith, I.G., 2001. Rivastigmine in the treatment of parkinsonian psychosis and cognitive impairment: preliminary findings from an open trial. Mov. Disord. 16, 1171–1174.

Rivastigmine, 2019. CID=77991. PubChem Database [Online].

Roberson, M.R., Harrell, L.E., 1997. Cholinergic activity and amyloid precursor protein metabolism. Brain Res. Brain Res. Rev. 25, 50–69.

Rosler, M., Anand, R., Cicin-Sain, A., Gauthier, S., Agid, Y., Dal-Bianco, P., Stahelin, H.B., Hartman, R., Gharabawi, M., 1999. Efficacy and safety of rivastigmine in patients with Alzheimer's disease: international randomised controlled trial. BMJ 318, 633–638.

Ross, C.A., Poirier, M.A., 2004. Protein aggregation and neurodegenerative disease. Nat. Med. 10 (Suppl.), S10–S17.

Salatin, S., Barar, J., Barzegar-Jalali, M., Adibkia, K., Jelvehgari, M., 2017. Thermosensitive in situ nanocomposite of rivastigmine hydrogen tartrate as an intranasal delivery system: development, characterization, ex vivo permeation and cellular studies. Colloids Surf. B Biointerfaces 159, 629–638.

Sethi, M.K., Bhandya, S.R., Maddur, N., Shukla, R., Kumar, A., Jayalakshmi Mittapalli, V.S.N., 2013. Asymmetric synthesis of an enantiomerically pure rivastigmine intermediate using ketoreductase. Tetrahedron: Asymmetry 24, 374–379.

Shah, B., Khunt, D., Bhatt, H., Misra, M., Padh, H., 2015. Application of quality by design approach for intranasal delivery of rivastigmine loaded solid lipid nanoparticles: effect on formulation and characterization parameters. Eur. J. Pharm. Sci. 78, 54–66.

Shah, B.M., Misra, M., Shishoo, C.J., Padh, H., 2015. Nose to brain microemulsion-based drug delivery system of rivastigmine: formulation and ex-vivo characterization. Drug Deliv. 22, 918–930.

Simon, A., Amaro, M.I., Cabral, L.M., Healy, A.M., De Sousa, V.P., 2016. Development of a novel dry powder inhalation formulation for the delivery of rivastigmine hydrogen tartrate. Int. J. Pharm. 501, 124–138.

Singh, B., Sharma, B., Jaggi, A.S., Singh, N., 2013. Attenuating effect of lisinopril and telmisartan in intracerebroventricular streptozotocin induced experimental dementia of Alzheimer's disease type: possible involvement of PPAR-gamma agonistic property. J. Renin-Angiotensin-Aldosterone Syst. 14, 124–136.

Spencer, C.M., Noble, S., 1998. Rivastigmine. A review of its use in Alzheimer's disease. Drugs Aging 13, 391–411.

Thomas, J., Thomas, C.J., Radcliffe, J., Itsiopoulos, C., 2015. Omega-3 fatty acids in early prevention of inflammatory neurodegenerative disease: a focus on alzheimer's disease. Biomed. Res. Int. 2015, 172801.

Trabace, L., Coluccia, A., Gaetani, S., Tattoli, M., Cagiano, R., Pietra, C., Kendrick, K.M., Cuomo, V., 2000. In vivo neurochemical effects of the acetylcholinesterase inhibitor ENA713 in rat hippocampus. Brain Res. 865, 268–271.

Venneri, A., Mcgeown, W.J., Shanks, M.F., 2005. Empirical evidence of neuroprotection by dual cholinesterase inhibition in Alzheimer's disease. Neuroreport 16, 107–110.

Wang, L., Wang, Y., Tian, Y., Shang, J., Sun, X., Chen, H., Wang, H., Tan, W., 2017. Design, synthesis, biological evaluation, and molecular modeling studies of chalcone-rivastigmine hybrids as cholinesterase inhibitors. Bioorg. Med. Chem. 25, 360–371.

Williams, B.R., Nazarians, A., Gill, M.A., 2003. A review of rivastigmine: a reversible cholinesterase inhibitor. Clin. Ther. 25, 1634–1653.

Winblad, B., Kawata, A.K., Beusterien, K.M., Thomas, S.K., Wimo, A., Lane, R., Fillit, H., Blesa, R., 2007. Caregiver preference for rivastigmine patch relative to capsules for treatment of probable Alzheimer's disease. Int. J. Geriatr. Psychiatry 22, 485–491.

Xie, P., Zhou, X., Zheng, L., 2019. Stereoselective synthesis of a key chiral intermediate of (S)-Rivastigmine by AKR-GDH recombinant whole cells. J. Biotechnol. 289, 64–70.

Chapter 3.1.6

Quercetin

Fatma Tugce Guragac Dereli[1], **Tarun Belwal**[2]
[1]*Department of Pharmacognosy, Faculty of Pharmacy, Suleyman Demirel University, Isparta, Turkey;* [2]*College of Biosystems Engineering and Food Science, Zhejiang University, China*

Introduction

Alzheimer's disease (AD) is known as a progressive, chronic, and age-related neurological disorder of unknown cause (Zverova, 2018). Neuropathologically, it is manifested by the extracellular accumulation of amyloid plaques [consisting of the amyloid-beta (Aβ) peptide] and intracellular deposition of neurofibrillary tangles [(NFTs), composed of tau protein]. The other pathological hallmarks are hyperactivation of acetylcholinesterase (AChE), brain atrophy, neuroinflammation, and a reduction in neuronal activity. This disease is also connected with neuronal cell death and synapse loss. All of these changes lead to a deterioration in cognitive functions (Cummings et al., 2016; De Leon et al., 1997; Ferreira-Vieira et al., 2016; Morris et al., 2014; Zuniga Santamaria et al., 2020).

As the main cause of more than half of dementia cases, AD currently affects about 50 million persons worldwide and it is estimated that the patient number will double every 20 years (Watt et al., 2019; Winblad et al., 2016; Zuniga Santamaria et al., 2020). Presently, pharmacological treatment of AD is limited to symptomatic relief and does not treat the underlying medical cause. Clinically available therapeutics for the treatment include three AChE inhibitors (donepezil, galantamine, and rivastigmine) and one glutamatergic antagonist (memantine). Unfortunately, these agents are not capable of stopping or reversing the disease progression and usage of them is related to several adverse reactions, such as stomach ulcers and seizures. Given all of these factors, there is an urgent necessity to develop disease-modifying medications (Atri, 2019; Cummings, 2018; Miranda et al., 2015).

Medicinal plants are crucial sources for modern drug discoveries. The bioactive secondary metabolites present in them are of enormous interest in designing new therapeutic agents (Balunas and Kinghorn, 2005). Until now, several phytochemicals have been determined to have anti-Alzheimer activity and quercetin is one of them (Bui and Nguyen, 2017).

Naturally Occurring Chemicals against Alzheimer's Disease. https://doi.org/10.1016/B978-0-12-819212-2.00008-6

Quercetin and its pharmacological properties

Quercetin (Que) (Fig. 3.1.6.1) is a phytotherapeutic flavonol derived from diverse vascular plants (Lee and Park, 2020). This promising secondary metabolite has drawn attention due to its multiple pharmacological functions on rodent/human models such as antioxidant, antiviral, antiallergic, antiproliferative, antiulcer, antidiabetic, antihypertensive, immunomodulatory, anti-inflammatory, anticancer, antihypercholesterolemic, antiobesity, anticataract, antiatherosclerotic, cardioprotective, hepatoprotective, neuroprotective, and anti-Alzheimer activities (Bischoff, 2008; Chatterjee et al., 2019; Lee et al., 2017; Li et al., 2016; Mlcek et al., 2016; Rezaei-Sadabady et al., 2016; Russo et al., 2012; Suganthy et al., 2016; Yang et al., 2014; Yao et al., 2018; Zhang et al., 2020).

Que has been shown to be a potential natural option for the prevention and management of AD by reversing primary histological hallmarks and improving cognitive and memory impairments (Khan et al., 2018; Li et al., 2019; Lu et al., 2018; Nakagawa and Ohta, 2019; Patil et al., 2003; Paula et al., 2019; Sabogal-Guaqueta et al., 2015; Vargas-Restrepo et al., 2018; Wang et al., 2014; Zhang et al., 2016). The literature findings of the prophylactic and therapeutic activity mechanisms of Que against AD are presented below.

Preclinical studies on the anti-AD activity of que

To date, numerous preclinical trials have been conducted to research the protective and therapeutic effects of Que against AD.

Nakagawa and colleagues reported that Que prevented cognitive impairments and neurodegeneration by suppressing Aβ production in rodent models of AD (Nakagawa and Ohta, 2019). Wang et al. noted that cognitive functions increased in Que-treated APPswe/PS1dE9 transgenic mice and this increase was found to be related to a reduction in plaque formation and mitochondrial dysfunction (Wang et al., 2014). In a study by Zhang et al. oral administration of Que remarkably reduced plaque formation in the brain cortex of 5xFAD mice. They determined that Que inhibits the degradation of brain apolipoprotein E (ApoE) in astrocytes and ApoE has a pivotal role in the clearance of Aβ (Zhang et al., 2016). In another study, a Que-rich diet during the early

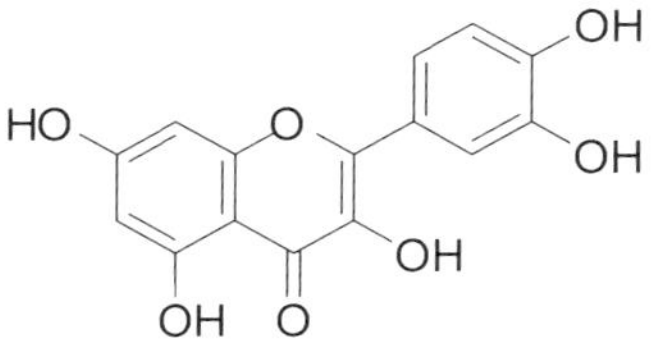

FIGURE 3.1.6.1 The chemical structure of quercetin.

stage of AD was found to be responsible for ameliorating the cognitive deficits by increasing Aβ clearance and reducing astrogliosis in APP/PS1 mice (Lu et al., 2018). In a triple transgenic AD model, the levels of aggregation of Aβ and hyperphosphorylation of tau were determined to decrease in mice orally treated with Que at a dose of 100 mg/kg for 1 year (Paula et al., 2019). Li and colleagues proved that Que administration together with sitagliptin (a dipeptidyl peptidase 4 inhibitor) caused a reduction in Aβ accumulation in rat brain (Li et al., 2019). Patil and colleagues showed that chronic administration of Que exhibited preventive activity against oxidative stress-related neuronal damage and cognitive impairments by inhibiting free radicals in lipopolysaccharide-treated mice (Patil et al., 2003). In various studies, Que has been found to suppress the activation of microgliosis and astrocytosis. The suppression of these neuroinflammation modulators prevents neuronal degeneration related to neuroinflammation (Khan et al., 2018; Lu et al., 2018; Sabogal-Guaqueta et al., 2015; Vargas-Restrepo et al., 2018). Recent findings have indicated that Que also shows protective activity against neurodegeneration mediated by mitochondrial dysfunction (Khan et al., 2018; Wang et al., 2014). Besides these, Que also has been found to exhibit AchE inhibitory activity and thereby improved symptoms of AD regarding cognitive functions (Abdalla et al., 2014; Jung and Park, 2007).

Clinical studies on the anti-AD activity of que

Despite having anti-AD potential, the clinical use of Que has been limited due to its low blood-brain barrier penetration. In addition, the metabolization of Que into metabolites with long half-life may cause accumulation in repeated dosing (Keddy et al., 2012; Ossola et al., 2009).

In total two clinical trials have been registered in clinicaltrials.gov for anti-AD activity of Que as of July 22, 2020. One is related to anti-AD activity of the combination of dasatinib and Que in older adults with early AD. The estimated completion date for this study is specified as August 2023 (clinicaltrials.gov, 2020a). The other is about the anti-AD potential of the coadministration of etanercept and a nutritional supplement. The supplement consists of a mixture of curcumin, omega-3, resveratrol, and Que. The results of this study are not available (clinicaltrials.gov, 2020b).

Side effects and toxicological profile of que

Research concerning the toxicity profile of Que has revealed that it has no adverse effects on the human body up to 4 g per day (Gugler et al., 1975). The administration of high-dose intravenous Que in patients has been found to be associated with nephrotoxicity (Russo et al., 2012). However, the long-term safety of high-dose Que has not yet been investigated (Harwood et al., 2007). CYP3A4 inhibitory activity of Que may result in an interaction with other drugs which are metabolized by this enzyme (Choi and Li, 2005).

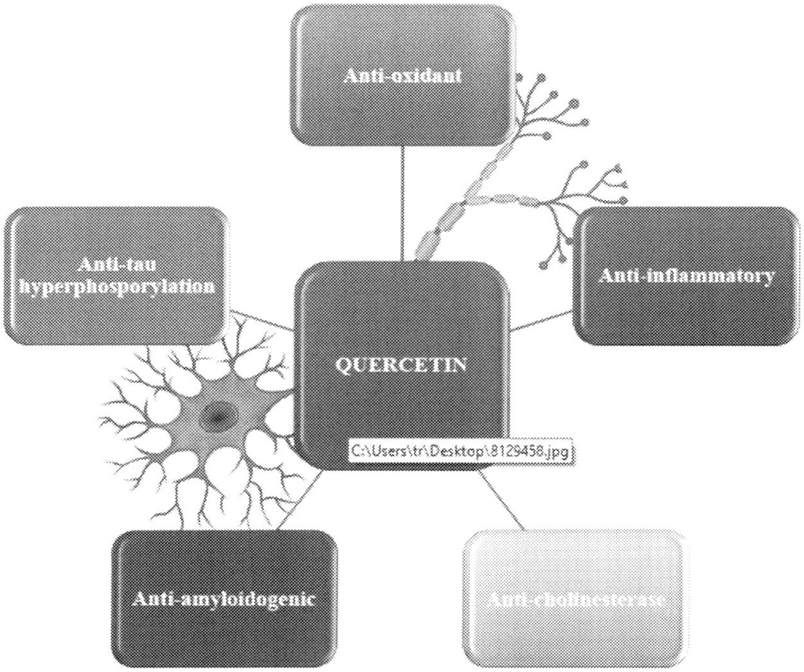

FIGURE 3.1.6.2 Role of Que in preventing and modulating AD.

Conclusion

Que has been documented to show anti-AD activity by several mechanisms including antioxidant, antiinflammatory, anticholinesterase, anti-amyloidogenic, and antitau hyperphosphorylation effects. A summary of the role of Que has been presented in Fig. 3.1.6.2. Future trials should focus on improving the bioavailability and clinical effectiveness of this promising phytochemical and its analogues.

References

Abdalla, F.H., Schmatz, R., Cardoso, A.M., Carvalho, F.B., Baldissarelli, J., de Oliveira, J.S., Rosa, M.M., Nunes, M.A.G., Rubin, M.A., da Cruz, I.B.M., Barbisan, F., Dressler, V.L., Pereira, L.B., Schetinger, M.R.C., Morsch, V.M., Goncalves, J.F., Mazzanti, C.M., 2014. Quercetin protects the impairment of memory and anxiogenic-like behavior in rats exposed to cadmium: possible involvement of the acetylcholinesterase and Na^+,K^+-ATPase activities. Physiol. Behav. 135, 152–167.

Atri, A., 2019. Current and future treatments in Alzheimer's disease. Semin. Neurol. 39 (2), 227–240.

Balunas, M.J., Kinghorn, A.D., 2005. Drug discovery from medicinal plants. Life Sci. 78 (5), 431–441.

Bischoff, S.C., 2008. Quercetin: potentials in the prevention and therapy of disease. Curr Opin Clin Nutr 11 (6), 733–740.

Bui, T.T., Nguyen, T.H., 2017. Natural product for the treatment of Alzheimer's disease. J. Basic Clin. Physiol. Pharmacol. 28 (5), 413–423.

Chatterjee, J., Langhnoja, J., Pillai, P.P., Mustak, M.S., 2019. Neuroprotective effect of quercetin against radiation-induced endoplasmic reticulum stress in neurons. J. Biochem. Mol. Toxicol. 33 (2), 1–8.

Choi, J.S., Li, X.G., 2005. Enhanced diltiazem bioavailability after oral administration of diltiazem with quercetin to rabbits. Int. J. Pharm. 297 (1–2), 1–8.

clinicaltrialsgov, 2020a. Senolytic Therapy to Modulate Progression of Alzheimer's Disease (SToMP-AD). Accessed 27.07.2020. https://clinicaltrials.gov/ct2/show/NCT04063124?term=quercetin&cond=Alzheimer+Disease&draw=2&rank=1.

clinicaltrialsgov, 2020b. Short Term Efficacy and Safety of Perispinal Administration of Etanercept in Mild to Moderate Alzheimer's Disease.

Cummings, J., 2018. Lessons learned from Alzheimer disease: clinical trials with negative outcomes. Cts-Clin. Transl. Sci. 11 (2), 147–152.

Cummings, J., Aisen, P.S., DuBois, B., Frolich, L., Jack, C.R., Jones, R.W., Morris, J.C., Raskin, J., Dowsett, S.A., Scheltens, P., 2016. Drug development in Alzheimer's disease: the path to 2025. Alzheimer's Res. Ther. 8, 39.

De Leon, M.J., George, A.E., Golomb, J., Tarshish, C., Convit, A., Kluger, A., De Santi, S., McRae, T., Ferris, S.H., Reisberg, B., Ince, C., Rusinek, H., Bobinski, M., Quinn, B., Miller, D.C., Wisniewski, H.M., 1997. Frequency of hippocampal formation atrophy in normal aging and Alzheimer's disease. Neurobiol. Aging 18 (1), 1–11.

Ferreira-Vieira, T.H., Guimaraes, I.M., Silva, F.R., Ribeiro, F.M., 2016. Alzheimer's disease: targeting the cholinergic system. Curr. Neuropharmacol. 14 (1), 101–115.

Gugler, R., Leschik, M., Dengler, H.J., 1975. Disposition of quercetin in man after single oral and intravenous doses. Eur. J. Clin. Pharmacol. 9 (2–3), 229–234.

Harwood, M., Danielewska-Nikiel, B., Borzelleca, J.F., Flamm, G.W., Williams, G.M., Lines, T.C., 2007. A critical review of the data related to the safety of quercetin and lack of evidence of *in vivo* toxicity, including lack of genotoxic/carcinogenic properties. Food Chem. Toxicol. 45 (11), 2179–2205.

Jung, M., Park, M., 2007. Acetylcholinesterase inhibition by flavonoids from *Agrimonia pilosa*. Molecules 12 (9), 2130–2139.

Keddy, P.G.W., Dunlop, K., Warford, J., Samson, M.L., Jones, Q.R.D., Rupasinghe, H.P.V., Robertson, G.S., 2012. Neuroprotective and anti-inflammatory effects of the flavonoid-enriched fraction AF4 in a mouse model of hypoxic-ischemic brain injury. PloS One 7 (12).

Khan, A., Ali, T., Rehman, S.U., Khan, M.S., Alam, S.I., Ikram, M., Muhammad, T., Saeed, K., Badshah, H., Kim, M.O., 2018. Neuroprotective effect of quercetin against the detrimental effects of LPS in the adult mouse brain. Front. Pharmacol. 9, 1383.

Lee, S., Lee, H.H., Shin, Y.S., Kang, H., Cho, H., 2017. The anti-HSV-1 effect of quercetin is dependent on the suppression of TLR-3 in Raw 264.7 cells. Arch Pharm. Res. (Seoul) 40 (5), 623–630.

Lee, Y.J., Park, Y., 2020. Green synthetic nanoarchitectonics of gold and silver nanoparticles prepared using quercetin and their cytotoxicity and catalytic applications. J. Nanosci. Nanotechnol. 20 (5), 2781–2790.

Li, Y., Yao, J.Y., Han, C.Y., Yang, J.X., Chaudhry, M.T., Wang, S.N., Liu, H.N., Yin, Y.L., 2016. Quercetin, inflammation and immunity. Nutrients 8 (3), 167.

Li, Y.P., Tian, Q.Y., Li, Z., Dang, M.Y., Lin, Y., Hou, X.Y., 2019. Activation of Nrf2 signaling by sitagliptin and quercetin combination against b-amyloid induced Alzheimer's disease in rats. Drug Dev. Res. 80 (6), 837–845.

Lu, Y.Q., Liu, Q., Yu, Q., 2018. Quercetin enrich diet during the early-middle not middle-late stage of Alzheimer's disease ameliorates cognitive dysfunction. Am. J. Transl. Res. 10 (4), 1237–1246.

Miranda, L.F., Gomes, K.B., Silveira, J.N., Pianetti, G.A., Byrro, R.M., Peles, P.R., Pereira, F.H., Santos, T.R., Assini, A.G., Ribeiro, V.V., Tito, P.A., Matoso, R.O., Lima, T.O., Moraes, E.N., Caramelli, P., 2015. Predictive factors of clinical response to cholinesterase inhibitors in mild and moderate Alzheimer's disease and mixed dementia: a one-year naturalistic study. J. Alzheimers Dis. 45 (2), 609–620.

Mlcek, J., Jurikova, T., Skrovankova, S., Sochor, J., 2016. Quercetin and its anti-allergic immune response. Molecules 21 (5), 622–636.

Morris, J.C., Blennow, K., Froelich, L., Nordberg, A., Soininen, H., Waldemar, G., Wahlund, L.O., Dubois, B., 2014. Harmonized diagnostic criteria for Alzheimer's disease: Recommendations. J. Intern. Med. 275 (3), 204–213.

Nakagawa, T., Ohta, K., 2019. Quercetin regulates the integrated stress response to improve memory. Int. J. Mol. Sci. 20 (11), 2761.

Ossola, B., Kaariainen, T.M., Mannisto, P.T., 2009. The multiple faces of quercetin in neuroprotection. Expet. Opin. Drug Saf. 8 (4), 397–409.

Patil, C.S., Singh, V.P., Satyanarayan, P.S.V., Jain, N.K., Singh, A., Kulkarni, S.K., 2003. Protective effect of flavonoids against aging- and lipopolysaccharide-induced cognitive impairment in mice. Pharmacology 69 (2), 59–67.

Paula, P.C., Maria, S.G.A., Luis, C.H., Patricia, C.G.G., 2019. Preventive effect of quercetin in a triple transgenic Alzheimer's disease mice model. Molecules 24 (12), 2287.

Rezaei-Sadabady, R., Eidi, A., Zarghami, N., Barzegar, A., 2016. Intracellular ROS protection efficiency and free radical-scavenging activity of quercetin and quercetin-encapsulated liposomes. Artif. Cell Nanomed. B 44 (1), 128–134.

Russo, M., Spagnuolo, C., Tedesco, I., Bilotto, S., Russo, G.L., 2012. The flavonoid quercetin in disease prevention and therapy: facts and fancies. Biochem. Pharmacol. 83 (1), 6–15.

Sabogal-Guaqueta, A.M., Munoz-Manco, J.I., Ramirez-Pineda, J.R., Lamprea-Rodriguez, M., Osorio, E., Cardona-Gomez, G.P., 2015. The flavonoid quercetin ameliorates Alzheimer's disease pathology and protects cognitive and emotional function in aged triple transgenic Alzheimer's disease model mice. Neuropharmacology 93, 134–145.

Suganthy, N., Devi, K.P., Nabavi, S.F., Braidy, N., Nabavi, S.M., 2016. Bioactive effects of quercetin in the central nervous system: focusing on the mechanisms of actions. Biomed. Pharmacother. 84, 892–908.

Vargas-Restrepo, F., Sabogal-Guaqueta, A.M., Cardona-Gomez, G.P., 2018. Quercetin ameliorates inflammation in CA1 hippocampal region in aged triple transgenic Alzheimer's disease mice model. Biomedica 38, 62–69.

Wang, D.M., Li, S.Q., Wu, W.L., Zhu, X.Y., Wang, Y., Yuan, H.Y., 2014. Effects of long-term treatment with quercetin on cognition and mitochondrial function in a mouse model of Alzheimer's disease. Neurochem. Res. 39 (8), 1533–1543.

Watt, A.D., Jenkins, N.L., McColl, G., Collins, S., Desmond, P.M., 2019. "To treat or not to treat": informing the decision for disease-modifying therapy in late-stage Alzheimer's disease. J. Alzheimers Dis. 68 (4), 1321–1323.

Winblad, B., Amouyel, P., Andrieu, S., Ballard, C., Brayne, C., Brodaty, H., Cedazo-Minguez, A., Dubois, B., Edvardsson, D., Feldman, H., Fratiglioni, L., Frisoni, G.B., Gauthier, S.,

Georges, J., Graff, C., Iqbal, K., Jessen, F., Johansson, G., Jonsson, L., Kivipelto, M., Knapp, M., Mangialasche, F., Melis, R., Nordberg, A., Rikkert, M.O., Qiu, C., Sakmar, T.P., Scheltens, P., Schneider, L.S., Sperling, R., Tjernberg, L.O., Waldemar, G., Wimo, A., Zetterberg, H., 2016. Defeating Alzheimer's disease and other dementias: a priority for European science and society. Lancet Neurol. 15 (5), 455–532.

Yang, T., Kong, B., Gu, J.W., Kuang, Y.Q., Cheng, L., Yang, W.T., Xia, X., Shu, H.F., 2014. Anti-apoptotic and anti-oxidative roles of quercetin after traumatic brain injury. Cell. Mol. Neurobiol. 34 (6), 797–804.

Yao, C.G., Xi, C.L., Hu, K.H., Gao, W., Cai, X.F., Qin, J.L., Lv, S.Y., Du, C.H., Wei, Y.H., 2018. Inhibition of enterovirus 71 replication and viral 3C protease by quercetin. Virol. J. 15, 116–128.

Zhang, X., Hu, J., Zhong, L., Wang, N., Yang, L.Y., Liu, C.C., Li, H.F., Wang, X., Zhou, Y., Zhang, Y.W., Xu, H.X., Bu, G.J., Zhuang, J.X., 2016. Quercetin stabilizes apolipoprotein E and reduces brain Ab levels in amyloid model mice. Neuropharmacology 108, 179–192.

Zhang, X.W., Chen, J.Y., Ouyang, D., Lu, J.H., 2020. Quercetin in animal models of Alzheimer's disease: a systematic review of preclinical studies. Int. J. Mol. Sci. 21 (2).

Zuniga Santamaria, T., Yescas Gomez, P., Fricke Galindo, I., Gonzalez Gonzalez, M., Ortega Vazquez, A., Lopez Lopez, M., 2020. Pharmacogenetic Studies in Alzheimer Disease. Neurologia (in press).

Zverova, M., 2018. Alzheimer's disease and blood-based biomarkers - potential contexts of use. Neuropsychiatric Dis. Treat. 14, 1877–1882.

Chapter 3.1.7

Valerenic and acetoxyvalerenic acid

Sarita Khatkar[1], Amit Lather[1], Anurag Khatkar[2]
[1]*Vaish Institute of Pharmaceutical Education & Research, Rohtak, Haryana, India;* [2]*Laboratory for Preservation Technology and Enzyme Inhibition Studies, Department of Pharmaceutical Sciences, Maharshi Dayanand University, Rohtak, Haryana, India*

Introduction

The term "herbalism" has been used as a traditional medical practice for the utilization of plants and plant derivatives for the treatment of various body ailments. Plant-based drugs have been used for the treatment of several diseases since ancient time. The use of plants and related derivative parts was not exclusive to any society, tradition, or chronological era. There is some proof from the analysis of pollen from a Neanderthal funeral (in modern-day Iraq) which indicates the use of medicinal plants in ancient times, i.e., to 50,000 BC (Solecki, 1975). The plant *Mandragora officinarum* was utilized as a pain reliever and narcotic, while *Digitalis purpurea* was used in patients suffering from dropsy, CHF (Congestive heart failure) in the late 18th century, and hemlock (*Conium maculatum*), known for its potent central nervous system depressive action. The United States witnessed a radical development in the use of herbal-based drugs and nutraceuticals products in the 20th century (Venkataramanan et al., 2006).

The name valerian has been derived from the Latin word "valere," meaning having the sense to be strong or healthy. *Valeriana* has been obtained from more than 100 plant species with the one most commonly used as a herb being the root of *Valeriana officinalis* (Hobbs, 1989; Brown, 1996; Flynn and Roest, 1995). *Valeriana officinalis* was found to contain more than 100 chemical constituents (Patocka and Jakl, 2010; Gao et al., 2000). Valerenic acid and acetoxyvalerenic acid are considered to be the principal constituents of *Valeriana* plants (Benke et al., 2009; Khom et al., 2007). According to the American Herbal Pharmacopeia, for the extraction of valerenic acids at least 30% alcohol is required (Upton et al., 1999).

Naturally Occurring Chemicals against Alzheimer's Disease. https://doi.org/10.1016/B978-0-12-819212-2.00045-1

Valerenic acid and acetoxyvalerenic acid have wonderful pharmacological potential and a significant role in BDNF, Alzheimer's disease, dementia, anesthetic action, sedative, anxiolytic potential, insomnia, physical as well psychological stress, antiinflammatory, gastrointestinal activity, inhibitory effects on CYP3A4-mediated metabolism, anticonvulsant, etc.

Occurrence

Valerenic acid and acetoxyvalerenic acid were found to be present in various species of *Valeriana* such as *V. officinalis*, *V. wallichii*, *V. jatamansji*, *V. edulis*, *V. faurie*, *V. hardwickii*, *V. dioica*, *V. pyrenaica*, and *V. pyrolaefolia* (Jugran et al., 2019; Srivastava et al., 2018; Tousi et al., 2012).

Chemical structure

Valerenic acid is a monocarboxylic acid, bicyclic sesquiterpenoid chemical constituent of the essential oil of different varieties of herbal drugs. Valerenic acid has 2-methylprop-2-enoic acid, substituted at position 3 by a 3,7-dimethyl-2,4,5,6,7,7 a-hexahydro-1H-inden-4-yl group. It also has been known as a conjugated acid of avalerenate.

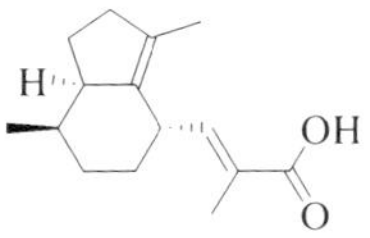

Valerenic acid

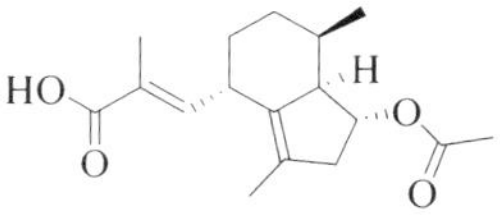

Acetoxyvalerenic acid

IUPAC name:

Valerenic acid: (*E*)-3-[(4*S*,7*R*,7*aR*)-3,7-dimethyl-2,4,5,6,7,7*a*-hexahydro-1*H*-inden-4-yl]-2-methylprop-2-enoic acid

Acetoxyvalereni acid: (*E*)-3-[(1*R*,4*S*,7*R*,7*aR*)-1-acetyloxy-3,7-dimethyl-2,4,5,6,7,7*a*-hexahydro-1*H*-inden-4-yl]-2-methylprop-2-enoic acid

Molecular formula:

Valerenic acid: $C_{15}H_{22}O_2$

Acetoxyvalerenic acid: $C_{17}H_{24}O_4$

Molecular weight:

Valerenic acid: 234.33

Acetoxyvalerenic acid: 292.37

Pharmacological potential of valerenic acid and acetoxyvalerenic acid

The most prominent plant source for valerenic acid is *Valerian officinalis*, belonging to the family Valerianaceae, and commonly used in conventional remedy systems due to of its hypnotic, anticonvulsant, sedative, etc. effects on the central nervous system (CNS). Use of valerenic acid also leads to some adverse reactions after use such as jamming of the chest, nephrotoxicity, headaches, mydriasis, tremors, abdominal pain, etc. This valerian species also has been found to increase the bleeding tendency and cause blood coagulation-related disorders like spontaneous hemorrhage. The valepotriates such as dihydrovaltrate, isovaltrate, and valtrate, have been examined as the chief active tranquilizing constituents and act as good alkylating agents against 4-(p-nitrobenzyl)-pyridine. One more constituent, valiracyl, shows its inhibition of GABA mediators and lowers the intensity of bioenergetics activities in the brain.

Another important source of valerenic acid, i.e., *Valeriana jatamansi* from the plant family Caprifoliaceae, has been cultivated in various Asian countries. The chemical constituents of this species include lignans, flavone glycosides, valepotriates, flavone, phenolic compounds, terpinoids, sesquiterpenoids, etc. *V. jatamansi* has various therapeutic properties such as sedative, antidepressant, neurotoxic, cytotoxic, antioxidant, insect repelling, sedative, anti-helmethic, antimicrobial, and it is utilized for the management of various body disorders in local medicine systems.

Role in brain-derived neurotrophic factor (BDNF) and Alzheimer's disease

The pharmacological effect of valerenic acid in BDNF has been studied. The $GABA_A$ receptor is regulated by BDNF and has a significant function in the CNS. This BDNF has an important role in the modulation of depressive behavior and other neurogenic disorders. This study was conducted to conclude the effects of *V. officinalis* root extract (VO), acetoxy valerenic acid (AVA), and valerenic acid (VA) in the expression of BDNF using SH-SY5Y cell lines. Methanolic extracts of VO, VA, and AVA were found to have a positive impact and increase human BDNF in ELISA kit. The VO and VA extracts have a positive impact and boost BDNF expression as compared to a control. The extracts with increasing amounts of VA lead to a concentration-dependent effect on BDNF expression. The VO extract produced an antidepressant effect primarily due to the occurrence of VA. Low levels of BDNF may lead to depression, Alzheimer's, obesity, and schizophrenia (Bjorkholm and Monteggia, 2016). Treatment of Alzheimer's disease (AD) has not yet been completely exploited due to its multifactorial and multistage nature (Gonulalan et al., 2018).

Dementia

The valeric acid from *V. wallichii* was found to produce a neuroprotective effect and its possible mechanism in the treatment of streptozotocin-induced neurodegeneration was studied in Wistar rats. A dose of 2 mg/kg of picrotoxin was used as a $GABA_A$ antagonist. Streptozotocin leads to a significant increase in retention transfer latency and escape latencies of rats on the 17th and 19th days in a Morris water maze and elevated plus maze, respectively. Treatment with *V. wallichii* extract and valeric acid significantly decreased the escape latency and retention transfer latency in comparison to the streptozotocin-treated group. Extract as well as valeric acid were also found to decrease the lipid per oxidation and increase the glutathione level in the brain of rats. Picrotoxin considerably reversed the effects of plant extract and valeric acid. The research concluded that valeric acid has an important GABAergic effect in experimentally induced dementia (Vishwakarma et al., 2016).

Anesthetic action

Valerenic acid and acetoxyvalerenic acid were found to be pharmacologically active monoterpenes acting through binding to the GABA receptor (Patocka and Jakl, 2010). It was evaluated as an anesthetic agent by potentiating its sedative effect and an effect has been produced via regulation of $GABA_A$ receptor action. Valerenic acid is now established as a $GABA_A$ regulator that may penetrate the CNS transcellularly via inactive diffusion (Yuan et al., 2004). Dietz et al. in 2005 illustrated that valerenic acid could be act as a partial agonist for the 5-HT receptor with strong binding potential (Dietz et al., 2005). Khom et al. (2007) also reported the mechanism of valerenic acid in chloride current stimulation via regulation of $GABA_A$ receptor. The molecular-level mechanism for the action of VA on $GABA_A$ receptors was studied with 13 various *Xenopus oocytes*. It was therefore concluded that VA may have a possible open-channel block mechanism and be a specific allosteric regulator of $GABA_A$ receptors.

The mechanism of action for valerian acid on the GABA receptor was identified in neonatal rat brainstem. Muscimol, a $GABA_A$ agonist, was found to decrease the firing rate in most of the brainstem neurons in a dose-dependent manner. This $GABA_A$ agonistic effect was opposed by bicuculline in both the valerian extract and valerenic acid-treated groups. Pretreatment with valerian extract or valerenic acid also decreased the brainstem inhibitory effects produced by muscimol. These facts also suggested that these compounds play an important role in the regulation of GABA-ergic activity. Therefore, valerian may be found to increase the sedative effects of anesthetics and other medications (Yuan et al., 2004).

Sedative and anxiolytic potential

The valerian root extract containing flavonones, alkaloids, and terpenes (valerenic acid and acetoxyvalerenic acid) was found to produce a sedative and anxiolytic effect. The valerian species was found to trigger the GABA-ergic system by increased GABA release, inhibition of reuptake, and degradation through inhibition of GABA-T activity in male Sprague-Dawley rats. Valerenic acid was found to act on β2/3 subunits of $GABA_A$ receptors and work as an agonist in male rat cortical tissues (Mustafa et al., 2019).

Insomnia

The extracts of *Valeriana officinalis* L. have been used for the management of restlessness and insomnia and studied with the help of radio ligand assay at $GABA_A$ and serotonin receptors. Both dichloromethane (DCM) and petroleum ether (PE) extracts of *Valeriana officinalis* L. have well-built binding affinity for 5-HT and serotonin receptors. Further binding assay experiments paid attention to the 5-HT receptor as it is mainly located in the suprachiasmatic nucleus of the brain and has an important role in the sleep cycle. The PE and DCM extract inhibited lysergic acid diethylamide (LSD) binding to the human 5-HT receptor at up to 51%. The results of this study indicate that valerian and valerenic acid are found to be partial agonists of the 5-HT receptor (Dietz et al., 2005).

Physical and psychological stress

Some studies have confirmed that the use of *Valeriana officinalis* extract containing VA enhances the levels of serum corticosterone in a mouse model. Physical and psychological stress was produced in a mouse model. The *Valeriana officinalis* extract could change 5-HT and norepinephrine (NE) production in the amygdala and hippocampus region of the mouse brain. The mice showed a decline in the period of immobility and serum corticosterone levels at a dose of 0.5 and 1.0 mg/kg. VA administration diminishes the psychological and physical stress due to decreased yield of 5-HT and NE in the hippocampus and amygdala regions of rat brain (Jung et al., 2015).

Antiinflammatory

VA derivatives have the capability to restrain the release of IL-8 and TNF-α. Six synthesized active analogues have moderate activity in the release of IL-8 in an assay with IC_{50} values of 2.8−8.3 μM, but none of these tested compounds have a significant effect on the inhibition of TNF-α release (Egbewande et al., 2017).

Gastrointestinal activity

Valerian is conventionally utilized for its antispasmodic effect in the treatment of intestinal colic and spasms. Due to their bitterness, valerians also have been used as an appetizer and digestive agent. VA also exerts spasmolytic effects in guinea pig ileum by its direct effect on smooth muscle (Hazelhoff et al., 1982; Busanny-Caspari, 1986).

Inhibitory effects on CYP3A4-mediated metabolism

Valerian root containing valerenic acid has an inhibitory effect on CYP3A4 on administration with other therapeutic products. The effect of VA on CYP3A4 inhibition revealed that caution should be warranted during the use of formulations which contain valerian root products (Lefebvre et al., 2004).

Anticonvulsant

Anticonvulsant activity of VA and extracts of valerian were determined using zebrafish as an animal model, and the results of the study were compared with phenytoin as well as clonazepam. Zebrafish were pretreated with antiepileptic drugs, valerenic acid, and valerian extracts throughout the experiment. Pentylenetetrazole (PTZ) was used to induce seizures and a behavioral scale model after some modification was used for scoring of PTZ-induced seizures in control and untreated fish. VA, aqueous, and ethanolic extracts of valerian root were also evaluated for their ability in the improvement of survival time after the administration of PTZ. Both valerenic acid and valerian extracts significantly increased the latency period to onset of the convulsions in zebrafish. Of these extracts, ethanolic valerian extract was found to be a more potent anticonvulsant than the aqueous one. Valerian extracts noticeably improved the anticonvulsant results of both clonazepam and phenytoin (Torres-Hernandez et al., 2015).

Clinical studies

Both valerenic and acetoxyvalerenic acid have been evaluated for their pharmacological potential in clinical studies, as reported in Table 3.1.7.1.

Conclusion

Valerian root was mainly found to contain valerenic acid as the major constituent. VA has a potent modulatory effect on the $GABA_A$ receptor. As the GABA receptor involved in various types of CNS-related disorders, VA could be utilized as a possible novel lead molecule for the discovery of antianxiety drugs.

TABLE 3.1.7.1 Some clinical study outcomes for valerenic acid and acetoxyvalerenic acid.

S. no.	Dose	Experimental design	Outcome measurement	Results	References
1.	100 mg	Double-blinded, randomized placebo-controlled, 4-weeks, 64 nonclinical volunteers	Anxiety (will increase or decrease)	Decreases in theta coherence across another four electrode pairs	Roh et al. (2019)
2.	530 mg	Double-blind RCT, 4 weeks, 51 HIV-positive patients	Antiretroviral therapy	Reduced anxiety	Ahmadi et al. (2017)
3.	765 mg	Placebo-controlled, 56 days in 13 adults having an obsessive-compulsive type of disorder	Obsessive-compulsive disorder symptoms will increase or decrease	Reduced the symptoms of disorder	Pakseresht et al. (2011)

References

Ahmadi, M., Khalili, H., Abbasian, L., Ghaeli, P., 2017. Effect of valerian in preventing neuropsychiatric adverse effects of efavirenz in HIV-positive patients: a pilot randomized, placebo-controlled clinical trial. Ann. Pharmacother. 51 (6), 457−464.

Benke, D., Barberis, A., Kopp, S., Altmann, K.H., Schubiger, M., Vogt, K.E., 2009. GABA A receptors as *in vivo* substrate for the anxiolytic action of valerenic acid, a major constituent of valerian root extracts. Neuropharmacology 56, 174−181.

Björkholm, C., Monteggia, L.M., 2016. BDNF−a key transducer of antidepressant effects. Neuropharmacology 102, 72−79.

Brown, D.J., 1996. Herbal Prescriptions for Better Health: Your Everyday Guide to Prevention, Treatment, and Care. Prima Publishing, Rocklin, CA.

Busanny-Caspari, E., 1986. Indikationen: Funktionelle Herzbeschwerden, Hypotonie und Wetterfuhligkeit. Therapiewoche 36, 2545−2550.

Dietz, B.M., Mahady, G.B., Pauli, G.F., Farnsworth, N.R., 2005. Valerian extract and valerenic acids are partial agonists of the 5-HT_{5a} receptor in vitro. Brain Res. Mol. Brain Res. 138 (2), 191−197.

Egbewande, F.A., Nilsson, N., White, J.M., Coster, M.J., Davis, R.A., 2017. The design, synthesis, and anti-inflammatory evaluation of a drug-like library based on the natural product valerenic acid. Bioorg. Med. Chem. Lett 17, 3185−3189.

Flynn, R., Roest, M., 1995. Your Guide to Standardized Herbal Products. One World Press.

Gao, X.Q., Bjork, L., 2000. Valerenic acid derivatives and valepotriates among individuals, varieties and species of Valeriana. Fitoterapia 71, 19–24.

Gonulalan, E.M., Bayazeid, O., Yalcin, F.N., Demirezer, L.O., 2018. The roles of valerenic acid on BDNF expression in the SH-SY5Y cell. Saudi Pharm. J. 26 (7), 960–964.

Hazelhoff, B., Malingre, T.M., Meijer, D.K., 1982. Antispasmodic effects of valeriana compounds: an in vivo and in vitro study on the Guinea-pig ileum. Arch. Int. Pharmacodyn. Ther. 257, 274–287.

Hobbs, C., 1989. Valerian monograph. HerbalGram 21, 19–34.

Jugran, A.K., Rawat, S., Bhatt, I.D., Valeriana jatamansi, R.R.S., 2019. An herbaceous plant with multiple medicinal uses. Phytother. Res. 33 (3), 482–503.

Jung, H.Y., Yoo, D.Y., Nam, S.M., Kim, J.W., Choi, J.H., Yoo, M., Lee, S., Yoon, Y.S., Hwang, I.K., 2015. Valerenic acid protects against physical and psychological stress by reducing the turnover of serotonin and norepinephrine in mouse hippocampus-amygdala region. J. Med. Food 18 (12), 1333–1339.

Khom, S., Baburin, I., Timin, E., Hohaus, A., Trauner, G., Kopp, B., Hering, S., 2007. Valerenic acid potentiates and inhibits GABAA receptors: molecular mechanism and subunit specificity. Neuropharmacology 53 (1), 178–187.

Lefebvre, T., Foster, B.C., Drouin, C.E., Krantis, A., Arnason, J.T., Livesey, J.F., Jordan, S.A., 2004. *In vitro* activity of commercial valerian root extracts against human cytochrome P450 3A4. J. Pharm. Pharmaceut. Sci. 7 (2), 265–273.

Mustafa, G., Ansari, S.H., Bhat, Z.A., Abdulkareim, A.S., 2019. Antianxiety activities associated with herbal drugs: a review. In: Plant and Human Health, vol. 3. Springer, Cham, pp. 87–100.

Pakseresht, S., Boostani, H., Sayyah, M., 2011. Extract of valerian root (Valeriana officinalis L.) vs. placebo in treatment of obsessive-compulsive disorder: a randomized double-blind study. J. Complement. Integr. Med. 8 (1).

Patocka, J., Jakl, J., 2010. Biomedically relevant chemical constituents of Valeriana officinalis. J. Appl. Biomed. 8 (1), 11–18.

Roh, D., Jung, J.H., Yoon, K.H., Lee, C.H., Kang, L.Y., Lee, S.K., Kim, D.H., 2019. Valerian extract alters functional brain connectivity: a randomized double-blind placebo-controlled trial. Phytother. Res. 33 (4), 939–948.

Solecki, R.S., 1975. Shanidar IV, a Neanderthal flower burial in northern Iraq. Science 190 (4217), 880–881.

Srivastava, R.P., Dixit, P., Singh, L., Verma, P.C., Saxena, G., 2018. Status of Selinum spp. L. a references Himalayan medicinal plant in India: a review of its pharmacology, phytochemistry and traditional uses. Curr. Pharm. Biotechnol. 19 (14), 1122–1134.

Torres-Hernandez, B.A., Del Valle-Mojica, L.M., Ortiz, J.G., 2015. Valerenic acid and Valeriana officinalis extracts delay onset of Pentylenetetrazole (PTZ)-Induced seizures in adult *Danio rerio* (Zebrafish). BMC Complement. Altern. Med. 15 (1), 228.

Tousi, E.S., Radjabian, T., Ebrahimzadeh, H.A.S.A.N., Niknam, V., 2012. Enhanced production of valerenic acids and valepotriates by in vitro cultures of *Valeriana officinalis* L. Int. J. Plant Prod. 4 (3), 209–222.

Upton, R., Graff, A., Williamson, E., Bevill, A., Ertl, F., Reich, E., 1999. Valerian root, *Valeriana officinalis*: analytical quality control, and therapeutic monograph. In: Upton, R. (Ed.), American Herbal Pharmacopoeia and Therapeutic Compendium. American Herbal Pharmacopoeia, Santa Cruz, CA.

Venkataramanan, R., Komoroski, B., Strom, S., 2006. In vitro and in vivo assessment of herb drug interactions. Life Sci. 78 (18), 2105–2115.

Vishwakarma, S., Goyal, R., Gupta, V., Dhar, K.L., 2016. GABAergic effect of valeric acid from *Valeriana wallichii* in amelioration of ICV STZ induced dementia in rats. Rev. Bras. Farmacogn. 26 (4), 1–10.

Yuan, C.S., Mehendale, S., Xiao, Y., Aung, H.H., Xie, J.T., Ang-Lee, M.K., 2004. The gamma-aminobutyric acidergic effects of valerian and valerenic acid on rat brainstem neuronal activity. Anesth. Analg. 98, 353–358.

Chapter 3.1.8

Huperzine A

Weaam Ebrahim[1], Ferhat Can Özkaya[2], Galal T. Maatooq[1,3]
[1]*Department of Pharmacognosy, Faculty of Pharmacy, Mansoura University, Mansoura, Egypt;* [2]*Faculty of Fisheries, İzmir Katip Çelebi University, İzmir, Turkey;* [3]*Department of Pharmaconosy, College of Pharmacy, The Islamic University in Najaf, Iraq*

Introduction

Alzheimer's disease (AD) is a genetically, complex, progressive, and unrepairable neurodegenerative disorder that affects social, work, and mental behaviors. Cerebral atrophy, cerebral senile plaques caused by the deposition of amyloid beta (Aβ) peptide, neurofibrillary tangles, and Aβ plaques responsible for hyperphosphorylated τ-protein, leading to neuronal cell loss and death, are the biochemical characteristic symptoms of AD (Amatsubo et al., 2010). These defects cause aphasia, apraxia, agnosia, and general cognitive symptoms, such as impaired judgment, orientation, and decision-making capacity (Amatsubo et al., 2010; Blennow et al., 2006). People over age 65 years are characteristically attacked by AD, and the prevalence of this disease has been increasing. Moreover, several cases of young adults who have AD are reported. The annual World Alzheimer Report 2018 declared that 50 million people globally developed AD, and this number is expected to increase to 152 million by 2050 (Mishra et al., 2019).

Huperzia serrata (Thunb. ex Murray) Trev. (synonym *Lycopodium serratum* Thunb. ex Murray, a member of Huperziaceae)−derived extracts or pure constituents have traditionally been used in China for the treatment of a number of health problems, such as contusions, strains, swelling, schizophrenia, myasthenia gravis, and organophosphate poisoning (Ma et al., 2006). At the beginning of the 1980s, Chinese researchers proved many therapeutic effects of *Lycopodium* alkaloids extracted from *Lycopodium* sp. (s.l.) on the cardiovascular and neuromuscular systems via in vitro and in vivo experiments. In particular, several positive effects on memory and learning were established (Tang et al., 1986; Zhu and Tang, 1987).

One of the purified constituents isolated from *H. serrata* (Thunb. ex Murray) is huperzine A **(1)** (Fig. 3.1.8.1). This active constituent is considered to be the most famous and patent compound owing to its reverse and selective inhibition of acetylcholinesterase (AChE). Interestingly, neuroscientists have proved that

Naturally Occurring Chemicals against Alzheimer's Disease. https://doi.org/10.1016/B978-0-12-819212-2.00010-4

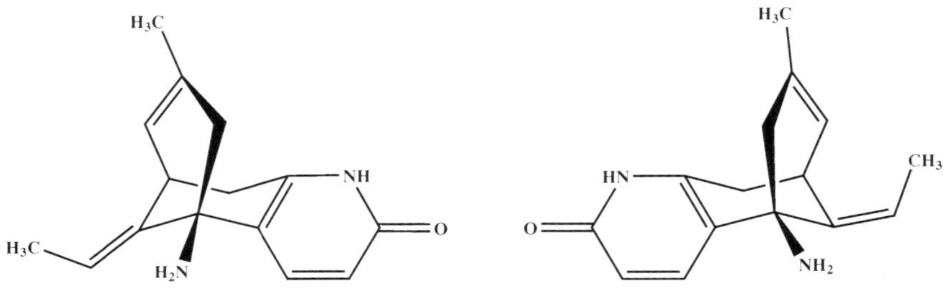

FIGURE 3.1.8.1 The chemical structures of the two enantiomers: (−)-huperzine A (**1**) and (+)-huperzine A (**2**).

the huperzine A inhibition activity of AChE is more effective than donepezil (trade name, Aricept), tacrine (trade name, Cognex), and galanthamine (trade name, Reminyl), which have been approved by the US Food and Drug Administration (Heinrich and Lee Teoh, 2004). Concerning the adverse effects of this drug, the efficiency of cognition and memory treatment in AD and other forms of dementia has been improved without side effects, and this has been proved via clinical experiments in China. Huperzine A has been approved for use as a drug in the treatment of AD in China. Moreover, it is now officially used as a memory-enhancing dietary supplement in the United States (Ma et al., 2006; Ma and Gang, 2008; Wang et al., 2006).

Chemistry of huperzine A

In fact, the phytochemistry of the family Huperziaceae is amazing, as confirmed in the literature, and several compounds isolated from this plant belong to different chemical classes, such as alkaloids, triterpenes, flavones, and phenolic acids. Among them, *Lycopodium* alkaloids are the most famous and bioactive chemical class. The phytochemical investigation of *Lycopodium* alkaloids revealed that they possess unique heterocyclic ring systems, such as $C_{16}N_1$, $C_{16}N_2$, and $C_{27}N$. Moreover, this class of compounds is attractive owing to biological, biogenetic, and chemical properties. To date, more than 400 *Lycopodium* alkaloids have been identified, but their biosynthetic pathways have not yet been completely identified (Dong et al., 2016; Hartrampf et al., 2017; Tao et al., 2013; Xu et al., 2017).

Huperzine A (**1**) involves a complex ring system that belongs to a bicyclo [3.1.1] skeleton that is fused with an α-pyridone ring, an exocyclic ethylidene moiety, and a three-carbon bridge. This structural feature is vital for the electrostatic field of this compound, which is considered a major reason for AChE inhibition (Desilets et al., 2009; Koshiba et al., 2009). Physiochemically, huperzine A is optically active and exists as (−)-huperzine A in nature. Its absolute configuration is found to be crucial in the inhibition of AChE. This is clarified by the fact that the enantiopure compound, (−)-huperzine A, showed a 38-fold increase in the inhibition of AChE activity in the rat cortex (at a *Ki* value of 8 nM) compared with the synthetic (+)-huperzine A **(2)** (at a *Ki* value of 300 nM) (Ma and Gang, 2004), whereas the chemically synthesized racemic mixture ([±]-huperzine A) showed three times less potent activity than the active compound (−)-huperzine A (Wu and Gu, 2006).

Further studies confirmed that these results, such as in (−)-huperzine A, were found to inhibit AChE threefold more than racemic and synthetic (+)-huperzine A **(2)** in rat brains in both in vitro and in vivo experiments (Hanin et al., 1993; Tang et al., 1994; Wang et al., 2011). All of these biochemical studies concerning structural modification and biological effects confirmed that huperzine A is a potent and safe AChE inhibitor and an effective remedy in the therapeutic strategy of AD.

The mass production of natural huperzine A was found to be a big problem because this compound is scarce less in the plant that produces it producing. The yield from its natural source range between 0.0047% and 0.025% from the weight of plant it also depends on the time of collection, the region, the methodology of collection, and the extraction (Bai, 2007; Ha et al., 2011). Moreover, huperzine A−producing plants are not widely abundant in the world, and these plants grow extremely slowly. A successful strategy for the mass production of this compound has not yet been reported (Ma et al., 2007). However, the total synthesis of huperzine A and its related congeners is considered a continuous untapped source of its production (Ma et al., 2007).

The chemical synthesis was started by Qian and Ji (1989) and Xia and Kozikowski (1989). This was followed by Kozikowski and coworkers, who succeeded in synthesizing a large number of related analogs. Among these synthesized derivatives, few showed the inhibition of AChE. An example of this analog is the prodrug ZT-1 **(3)**, which was found to be the most active huperzine A analog (Fig. 3.1.8.2). Chemically, ZT-1 **(3)** was prepared as a semisynthetic Schiff base, which was made by condensation of huperzine A and 5-chloro-O-vanillin. Interestingly, ZT-1 **(3)** is rapidly absorbed in vivo and converted to huperzine A; therefore, it possesses more activity and fortunately less toxicity in mice than the parent derivative, huperzine A. Moreover, a phase I study was conducted and proved that it has good tolerability in humans (Jia et al., 2013; Zhou and Zhu, 2000), whereas phase II clinical trials were completed in both Europe and China (Ma et al., 2007). All of these studies concluded that the synthesis of huperzine A analogs is much more challenging than the total synthesis of huperzine A itself. All of these studies also conclude that the alteration of any structural feature of huperzine A results in a loss of its bioactivity, which is considered a main obstacle to using huperzine A and its derivatives as drugs. To date, no chemical method has been available for the mass production of huperzine-inspired drugs in industry (Ma et al., 2007; Ferreira et al., 2016).

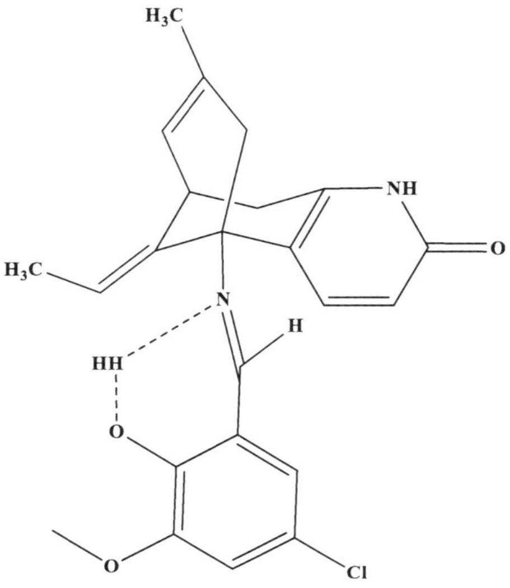

3

FIGURE 3.1.8.2 Chemical structure of ZT-1 (**3**) (Ma and Gang, 2004).

An alternative way to obtain similar biologically related huperzine compounds is from commercially abundant and natural (*R*)-pulegone—based through total synthesis, and this was reported as an efficient industrial process aimed at eliminating the supply problem of (−)-huperzine A. Ding et al. (2014) achieved the synthesis of (−)-huperzine A in 10 steps in which 17% yield was accomplished via palladium-catalyzed Buchwald−Hartwig coupling and Heck cyclization reactions and an Ir-catalyzed olefin isomerization. Moreover, huperzine B (**4**) and huperzine U (**5**) were synthesized successfully from (*R*)-pulegone with a relatively good yield in a 10- to 13-step reaction. Moreover, their absolute configuration was established (Fig. 3.1.8.3).

An in vitro plant tissue culture technique was applied to *Phlegmariurus squarrosus* (a member of the Huperziaceae) to overcome limitations in the production of huperzine A (Fig. 3.1.8.4). The optimum liquid medium formulation and conditions were adjusted for the fast-growing callus and prothallus compared with the intact plant in nature. As interesting results of the experiments, a higher yield was attained and other *Lycopodium* alkaloids were detected in vitro in the extract of propagated *P. squarrosus* tissues, such as huperzine B, des-*N*-methyl-β-obscurine, α-obscurine, and β-obscurine. The in vitro propagated *P. squarrosus* plants are considered a valuable alternative for the mass production of huperzine A in the industry. However, it is still necessary to design detailed development methods, and there is still a major need for sustainable sources for the production of both huperzine A and its analogs (Ma and Gang, 2008).

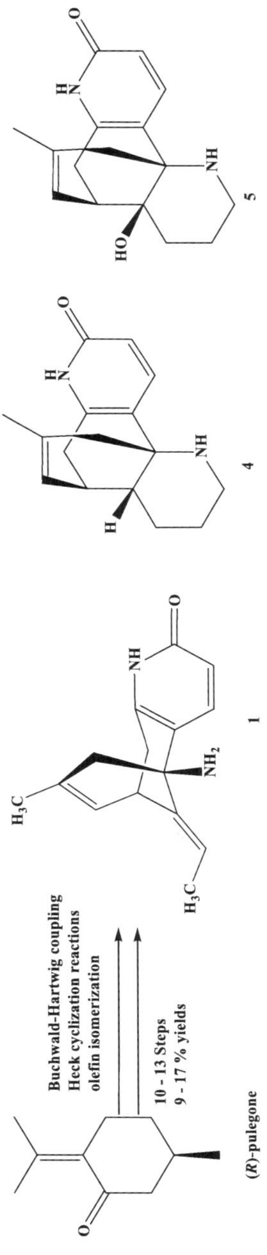

FIGURE 3.1.8.3 Synthesis of (−)-huperzine A (**1**), huperzine B (**4**), and huperzine (U) (**5**) and their absolute configuration (Ding et al., 2014).

FIGURE 3.1.8.4 (A) Over 15-year-old *Huperzia serrata* in the wild, (B) *Phlegmariurus squarrosus* in the green house of the University of Freiburg, (C) *P. squarrosus* in vitro propagated prothalli and sporophyte, (D) *P. squarrosus* in vitro propagated callus, and (E) *P. squarrosus* in vitro rapid callus liquid cultures (Ma and Gang, 2008).

The other biotechnological aspect of meeting the demand of production for huperzine A and its related derivatives is to use plant fungal entophytes. In fact, fungi could be the best sustainable alternative for the production of huperzine A. This technique has the following advantages: fungal fermentation is much more simple, effective, and cheaper than mass production from plants, chemical synthesis, or a cell culture method. Furthermore, microbial fermentation is much safer because it produces no harmful by-products compared with chemical synthesis (Luo et al., 2014; Özkaya et al., 2018). Therefore, entophytic fungi from huperzine A—producing plants were

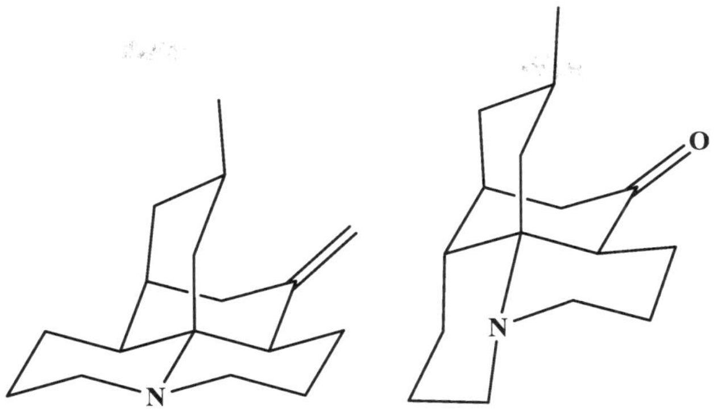

6 7

FIGURE 3.1.8.5 Huperzine derivates 12-*epi*-lycopodine **(6)** and lycopodine **(7)** from UV-irradiated strain of *Lycopodium serratum* var. *longipetiolatum* endophyte *Paraboeremia* (Ishiuchi et al., 2018).

investigated as a source of huperzine A. So far, a few reports have been published about huperzine A–producing entophytic fungi from Huperiaceae plants. *Acremonium* sp. 2F09P03B (Li et al., 2007), *Colletotrichum gloeosporioides* ES026 (Zhang et al., 2015), *Shiraia* sp. Slf14 (Zhu et al., 2010), and *Paecilomyces tenuis* YS-13 (Su and Yang, 2015) were isolated from *H. serrata*. *Blastomyces* sp. and *Botrytis* sp. were isolated from *Phlegmariurus cryptomerianus* (Ju et al., 2009). In addition, plant entophytic fungi were used to produce huperzine A and *Lycopodium* alkaloids as sustainable sources. Ishiuchi et al. (2018) carried out UV irradiation on *L. serratum* Thunb. var. *longipetiolatum* entophyte, *Paraboeremia* sp. Ls 13KI076, and the secondary metabolites; 12-*epi*-lycopodine **(6)**, lycopodine **(7)**, and huperzine A **(1)** were isolated from this irradiated fungal strain (Fig. 3.1.8.5). Interestingly, 12-*epi*-lycopodine was not detected in either the host plant extract or the wild-type fungal extract. It is a unique derivative of huperzine A, and it is known to be scarce in nature. However, 12-*epi*-lycopodine and lycopodine were inactive against AChE.

Conclusions

(−)-Huperzine A is a challenging compound for both chemical synthesis and chemical modification. However, microbial biotransformation might be a valuable alternative to obtain new bioactive huperzine derivatives. This is because of fungal gene clusters and the biochemical advantages of these biotechnological techniques (Ozcinar et al., 2018). In these advanced techniques, hydroxylation and epoxidation are easily obtained compared with chemical routes especially, in a rigid molecule such as (−)-huperzine A. Zhang et al. (2010) succeeded in obtaining hydroxyl and epoxy derivatives with acetyl and formyl derivatives such as 16-hydroxyl huperzine A **(8)**, 14α-hydroxyl huperzine A **(9)**, huperzine A 8α,15α-epoxide **(10)**, 13*N*-formyl huperzine A

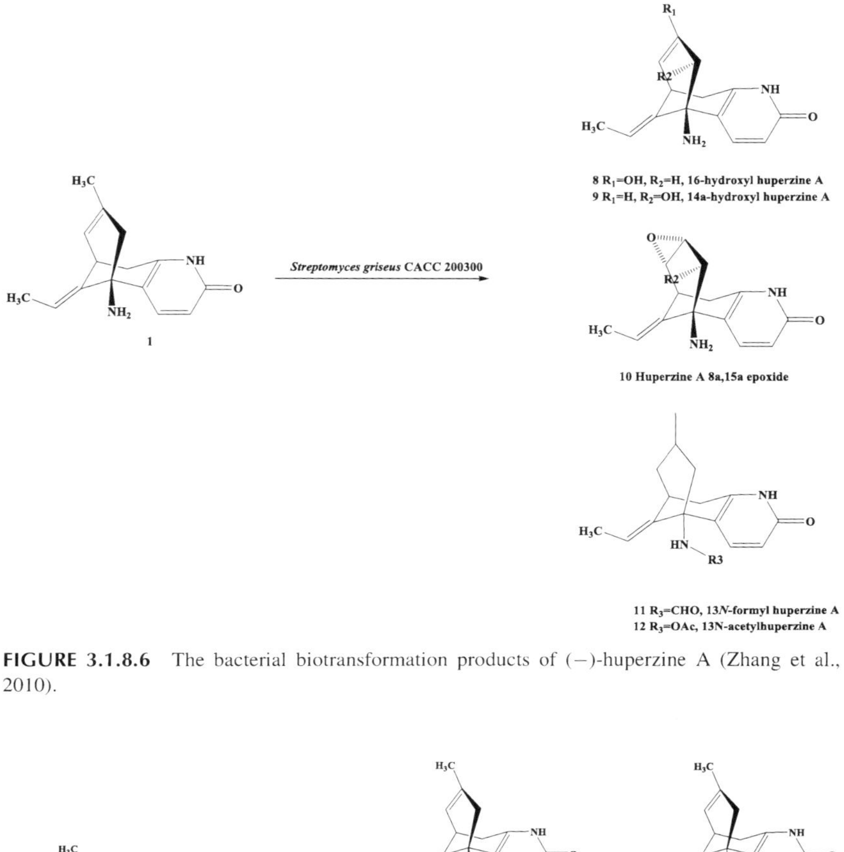

FIGURE 3.1.8.6 The bacterial biotransformation products of (−)-huperzine A (Zhang et al., 2010).

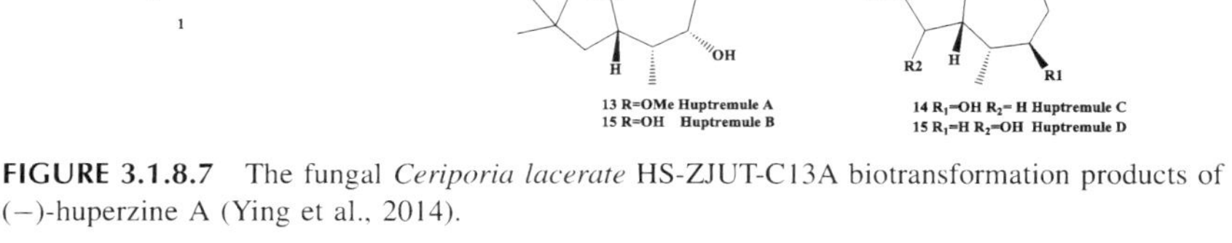

FIGURE 3.1.8.7 The fungal *Ceriporia lacerate* HS-ZJUT-C13A biotransformation products of (−)-huperzine A (Ying et al., 2014).

(11), and 13*N*-acetylhuperzine A **(12)** via cultivation of *Streptomyces griseus* CACC 200300 with (−)-huperzine A (Fig. 3.1.8.6). Moreover, the microbial biotransformation of (−)-huperzine was done in a culture medium of *H. serrata* endophyte *Ceriporia lacerate* HS-ZJUT-C13A. Through this, huptremules A−D **(13−16)** unusual sesquiterpenoid−alkaloid hybrid structures with exogenous substrate huperzine A, were obtained (Fig. 3.1.8.7) (Ying et al., 2014).

References

Amatsubo, T., Yanagisawa, D., Morikawa, S., Taguchi, H., Tooyama, I., 2010. Amyloid imaging using high-field magnetic resonance. Magn. Reson. Med. Sci. 9 (3), 95–99.

Bai, D., 2007. Development of huperzine A and B for treatment of Alzheimer's disease. Pure Appl. Chem. 79, 469.

Blennow, K., de Leon, M.J., Zetterberg, H., 2006. Alzheimer's disease. Lancet 368 (9533), 387–403.

Desilets, A.R., Gickas, J.J., Dunican, K.C., 2009. Role of huperzine a in the treatment of Alzheimer's disease. Ann. Pharmacother. 43 (3), 514–518.

Ding, R., Fu, J.G., Xu, G.Q., Sun, B.F., Lin, G.Q., 2014. Divergent total synthesis of the Lycopodium alkaloids huperzine A, huperzine B, and huperzine U. J. Org. Chem. 79 (1), 240–250.

Dong, L.-B., Wu, X.-D., Shi, X., Zhang, Z.-J., Yang, J., Zhao, Q.-S., Phleghenrines, A.–D., Neophleghenrine, A., 2016. Bioactive and structurally rigid lycopodium alkaloids from *Phlegmariurus henryi*. Org. Lett. 18 (18), 4498–4501.

Ferreira, A., Rodrigues, M., Fortuna, A., Falcão, A., Alves, G., 2016. Huperzine A from Huperzia serrata: a review of its sources, chemistry, pharmacology and toxicology. Phytochem. Rev. 15 (1), 51–85.

Ha, G.T., Wong, R.K., Zhang, Y., 2011. Huperzine A as potential treatment of Alzheimer's disease: an assessment on chemistry, pharmacology, and clinical studies. Chem. Biodivers. 8 (7), 1189–1204.

Hanin, I., Tang, X.C., Kindel, G.L., Kozikowski, A.P., 1993. Natural and synthetic Huperzine A: effect on cholinergic function in vitro and in vivo. Ann. N. Y. Acad. Sci. 695, 304–306.

Hartrampf, F.W., Furukawa, T., Trauner, D., 2017. A Conia-Ene-type cyclization under basic conditions enables an efficient synthesis of (-)-Lycoposerramine R. Angew Chem. Int. Ed. Engl. 56 (3), 893–896.

Heinrich, M., Lee Teoh, H., 2004. Galanthamine from snowdrop—the development of a modern drug against Alzheimer's disease from local Caucasian knowledge. J. Ethnopharmacol. 92 (2), 147–162.

Ishiuchi, K.i., Hirose, D., Suzuki, T., Nakayama, W., Jiang, W.-P., Monthakantirat, O., Wu, J.-B., Kitanaka, S., Makino, T., 2018. Identification of lycopodium alkaloids produced by an ultraviolet-irradiated strain of paraboeremia, an endophytic fungus from *Lycopodium serratum* var. longipetiolatum. J. Nat. Prod. 81 (5), 1143–1147.

Jia, J.-Y., Zhao, Q.-H., Liu, Y., Gui, Y.-Z., Liu, G.-Y., Zhu, D.-Y., Yu, C., Hong, Z., 2013. Phase I study on the pharmacokinetics and tolerance of ZT-1, a prodrug of huperzine A, for the treatment of Alzheimer's disease. Acta Pharmacol. Sin. 34, 976.

Ju, Z., Wang, J., Pan, S., 2009. Isolation and preliminary identification of the endophytic fungi which produce Hupzine A from four species in Hupziaceae and determination of Huperzine A by HPLC. Fudan Univ. J. Med. Sci. 36, 445–449.

Koshiba, T., Yokoshima, S., Fukuyama, T., 2009. Total synthesis of (-)-huperzine A. Org. Lett. 11 (22), 5354–5356.

Li, W., Zhou, J., Lin, Z., Hu, Z., 2007. Study on fermentation for production of huperzine A from endophytic fungus 2F09P03B of *Huperzia serrata*. Chin. Med. Biotechnol. 2, 254–259.

Luo, S.-P., Peng, Q.-L., Xu, C.-P., Wang, A.-E., Huang, P.-Q., 2014. Bio-inspired step-economical, redox-economical and protecting-group-free enantioselective total syntheses of (–)-chaetominine and analogues. Chin. J. Chem. 32 (8), 757–770.

Ma, X., Gang, D.R., 2004. The Lycopodium alkaloids. Nat. Prod. Rep. 21 (6), 752–772.

Ma, X., Gang, D.R., 2008. In vitro production of huperzine A, a promising drug candidate for Alzheimer's disease. Phytochemistry 69 (10), 2022–2028.

Ma, X., Tan, C., Zhu, D., Gang, D.R., 2006. A survey of potential huperzine A natural resources in China: the Huperziaceae. J. Ethnopharmacol. 104 (1), 54–67.

Ma, X., Tan, C., Zhu, D., Gang, D.R., Xiao, P., 2007. Huperzine A from Huperzia species—an ethnopharmacolgical review. J. Ethnopharmacol. 113 (1), 15–34.

Mishra, P., Kumar, A., Panda, G., 2019. Anti-cholinesterase hybrids as multi-target-directed ligands against Alzheimer's disease (1998–2018). Bioorg. Med. Chem. 27 (6), 895–930.

Ozcinar, O., Ozgur, T., Yusufoglu, H., Kivcak, B., Bedir, E., 2018. Biotransformation of neoruscogenin by the endophytic fungus *Alternaria eureka*. J. Nat. Prod. 81 (6), 1357–1367.

Özkaya, F.C., Ebrahim, W., El-Neketi, M., Tansel Tanrıkul, T., Kalscheuer, R., Müller, W.E.G., Guo, Z., Zou, K., Liu, Z., Proksch, P., 2018. Induction of new metabolites from sponge-associated fungus *Aspergillus carneus* by OSMAC approach. Fitoterapia 131, 9–14.

Qian, L., Ji, R., 1989. A total synthesis of (±)-huperzine A. Tetrahedron Lett. 30 (16), 2089–2090.

Su, J., Yang, M., 2015. Huperzine A production by *Paecilomyces tenuis* YS-13, an endophytic fungus isolated from *Huperzia serrata*. Nat. Prod. Res. 29 (11), 1035–1041.

Tang, X.C., Han, Y.F., Chen, X.P., Zhu, X.D., 1986. Effects of huperzine A on learning and the retrieval process of discrimination performance in rats. Zhongguo Yaoli Xuebao 7 (6), 507–511.

Tang, X.C., Kindel, G.H., Kozikowski, A.P., Hanin, I., 1994. Comparison of the effects of natural and synthetic huperzine-A on rat brain cholinergic function in vitro and in vivo. J. Ethnopharmacol. 44 (3), 147–155.

Tao, Y., Fang, L., Yang, Y., Jiang, H., Yang, H., Zhang, H., Zhou, H., 2013. Quantitative proteomic analysis reveals the neuroprotective effects of huperzine A for amyloid beta treated neuroblastoma N2a cells. Proteomics 13 (8), 1314–1324.

Wang, R., Yan, H., Tang, X.C., 2006. Progress in studies of huperzine A, a natural cholinesterase inhibitor from Chinese herbal medicine. Acta Pharmacol. Sin. 27 (1), 1–26.

Wang, Y., Wei, Y., Oguntayo, S., Jensen, N., Doctor, B.P., Nambiar, M.P., 2011. [+]-Huperzine A protects against soman toxicity in Guinea pigs. Neurochem. Res. 36 (12), 2381–2390.

Wu, Q., Gu, Y., 2006. Quantification of huperzine A in Huperzia serrata by HPLC-UV and identification of the major constituents in its alkaloid extracts by HPLC-DAD-MS-MS. J. Pharm. Biomed. Anal. 40 (4), 993–998.

Xia, Y., Kozikowski, A.P., 1989. A practical synthesis of the Chinese "nootropic" agent huperzine A: a possible lead in the treatment of Alzheimer's disease. J. Am. Chem. Soc. 111 (11), 4116–4117.

Xu, B., Lei, L., Zhu, X., Zhou, Y., Xiao, Y., 2017. Identification and characterization of L-lysine decarboxylase from *Huperzia serrata* and its role in the metabolic pathway of lycopodium alkaloid. Phytochemistry 136, 23–30.

Ying, Y.M., Shan, W.G., Zhan, Z.J., 2014. Biotransformation of huperzine A by a fungal endophyte of *Huperzia serrata* furnished sesquiterpenoid-alkaloid hybrids. J. Nat. Prod. 77 (9), 2054–2059.

Zhang, X., Zou, J.-h., Dai, J., 2010. Microbial transformation of (−)-huperzine A. Tetrahedron Lett. 51 (29), 3840–3842.

Zhang, G., Wang, W., Zhang, X., Xia, Q., Zhao, X., Ahn, Y., Ahmed, N., Cosoveanu, A., Wang, M., Wang, J., Shu, S., 2015. De novo RNA sequencing and transcriptome analysis of *Colletotrichum gloeosporioides* ES026 reveal genes related to biosynthesis of huperzine A. PLoS One 10 (3), e0120809.

Zhou, G.-C., Zhu, D.-Y., 2000. Synthesis of 5-substituted analogues of huperzine A. Bioorg. Med. Chem. Lett. 10 (18), 2055–2057.

Zhu, X.D., Tang, X.C., 1987. Facilitatory effects of huperzine A and B on learning amd memory of spatial discrimination in mice. Yao Xue Xue Bao 22 (11), 812–817.

Zhu, D., Wang, J., Zeng, Q., Zhang, Z., Yan, R., 2010. A novel endophytic Huperzine A–producing fungus, *Shiraia* sp. Slf14, isolated from *Huperzia serrata*. J. Appl. Microbiol. 109 (4), 1469–1478.

Chapter 3.1.9

Caprylic/capric triglyceride

Manjul Mungali[1]**, Navneet Sharma**[1]**, Gauri**[2]
[1]*Department of Biotechnology, School of Life Sciences & Technology IIMT University, Meerut, Uttar Pradesh, India;* [2]*SMP Government Girls PG College, Meerut, Uttar Pradesh, India*

Caprylic/capric triglyceride

Caprylic acid is one of many health-promoting nutritional substances like vitamins, trace minerals, antioxidants, flavonoids, lipids, etc. Breast milk, palm kernel oil, coconut oil, and dairy products are some natural sources of caprylic triglyceride. Caprylic acid is an oily liquid sometimes mistakenly known as fractionated coconut oil. Capric triglyceride is a mixed triester that can be formulated by the mixing of coconut oil and glycerin. It falls under the category of medium chain triglyceride (MCT) or medium chain saturated fatty acid (MCFA). It has many nutritional and disease control benefits. Medical formulations are made by using caprylic acid, which slows the progression of some diseases like dementia and Alzheimer's disease (AD). Caprylic acid in the form of medicine is very cost effective, easy to use, and socially acceptable too. The other important property of this acid is that it is safe to use and free of side effects in most cases.

Synthesis and chemistry of caprylic triglyceride

Caprylic acid, with the systematic name of octanoic acid, is a carboxylic acid. The structural formula of caprylic acid is shown in Fig. 3.1.9.1.

Caprylic or capric triglyceride is a blended triester, which can be prepared by blending coconut oil and glycerin. In chemistry, it is also referred to as MFCA. MCT includes caprylic acid (eight carbon), caproic acid (six carbon), and capric acid (10 carbon). MCFA is the common name for the eight-carbon backbone (Stephanie Greenwood, 2015).

According to Anneken et al. (2012), in the production of fatty acids like caprylic, capric, palmitic, steric, and oleic acid, vegetable and animal fats and oils are first hydrolyzed and individual fatty acids are fractioned afterward.

The general method of synthesis of capric/caprylic triglyceride is saponification or steam hydrolysis followed by esterification. In saponification, capric and caprylic acids are originally isolated from the glycerol group of

Naturally Occurring Chemicals against Alzheimer's Disease. https://doi.org/10.1016/B978-0-12-819212-2.00011-6

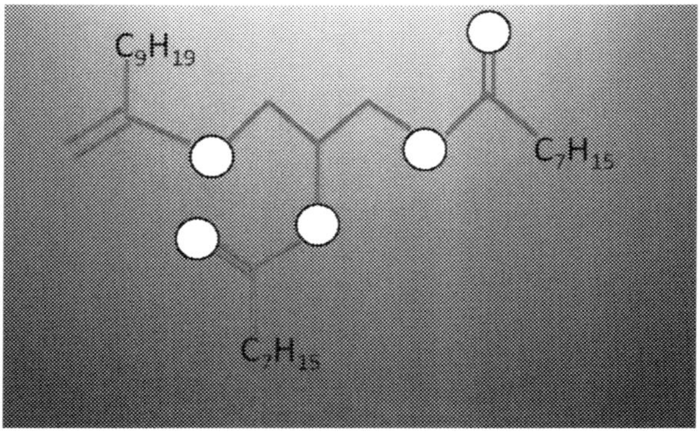

FIGURE 3.1.9.1 Chemical structure of caprylic/capric triglyceride.

coconut or palm oil. In saponification we utilize strong alkalis to react with glycerol and fatty acids to make a soapy compound. Steam hydrolysis with high pressure and heat is also used to split the glycerol from the unsaturated fats. After the separation of capric/caprylic acid we continue for the last procedure called esterification to add the glycerol group back to the unsaturated fats (Stephanie Greenwood, 2015).

Properties of capric triglyceride

Physically, caprylic acid is an oily liquid, minimally soluble in water, and having a slightly unpleasant rancid-like smell and taste. Caprylic acid has excellent shelf life because it has great stability and resistance to oxidation.

The three main forms in which caprylic triglyceride is extensively used are dispersing agent, solvent, and emollient. When applied on the skin, it lubricates the skin by making a light and nongreasy barrier and it is very quickly absorbed. This property of caprylic acid is referred to as its emollient nature. In the form of dispersing agent, it can help evenly disperse many active ingredients like vitamins, pigments, etc. in a solution to make them readily absorbable by the epidermis layer of the skin.

Uses of capric triglyceride

Caprylic acid has widespread uses in the health promoting and disease control sector, and in cosmetics and other industrial products due to some of its unusual but brilliant properties. Caprylic acid is taken as a dietary supplement. There are some studies that show the use of caprylic acid in weight management by burning excess calories in the body. Caprylic acid is also used as part of a ketogenic diet to treat children with intractable epilepsy.

Caprylic acid also works as an antioxidant for skin and also boosts the antioxidants in skin products. It is also used in the form of an antimicrobial pesticide for surface sanitization in the food and dairy industry, and as disinfectant in some healthcare sectors and services. Caprylic acid has an oily texture, hence it is used in many cosmetic products that require slipperiness, easy spreadability, and smoothness after touch. Lack of color and order with high stability makes it a valuable ingredient in many cosmetic products. It has high resistance to oxidation. Some products in which caprylic acid is easily found are facial and eye creams, moisturizers, lipsticks, antiaging creams, sunscreen lotions, face foundations, and lip and eyeliner (Swaminathan and Jicha, 2014). Caprylic acid is also widely used in the commercial production of esters and in perfume and dye manufacturing.

Alzheimer's disease and capric triglyceride-based medical food

The world's population is speedily aging, and the number of individuals with dementia is likely to grow from 35 million to 65 million by 2030 (Korolev, 2014). The German psychiatrist and neuropathologist Dr. Alois Alzheimer was the first to describe a dementing condition later known as AD.

Dementia is a clinical disorder having symptoms related to progressive decline of intellectual function. It is associated with damaged memory, language, thinking, attention, reasoning, orientation, and visuospatial capacity. These cognitive changes cause change in emotional and social behavior of the person suffering with AD, hence they are not able to show improvement in social skills, work, and relationships.

Nowadays, medicinal food has shown a new hope. Food supplements containing caprylic acid are prescribed to patients with AD in addition to mainstream medicines. A normal healthy adult's brain consumes glucose as an energy source. However, in some conditions like starvation, the brain can also metabolize ketone bodies (KBs) as an energy source. KBs are supplements for glucose metabolism in the brain and a competent fuel for cells, so they are present in the circulation. Caprylic triglycerides induce ketosis without any modification to the diet.

Axona is a famous medicinal food supplement that is being used for the treatment of mild-to-moderate AD, which has caprylic acid as its main constituent. Axona is "Generally Recognized as Safe" by the Food and Drug Administration (Henderson, 2009). Accera, Inc. is the company that makes Axona. The company claims that during digestion, caprylic triglycerides are broken down into KBs and offer an alternative source of energy for the brain. In AD, the brain's ability to use a normal energy source such as glucose is decreased, hence ketones help, and this was the company's idea behind the production of Axona.

Mechanism of action and biochemistry

Reduced brain glucose utilization and mitochondrial dysfunction are the hallmarks of AD. In AD, dysfunction of glucose transporter 1 decreases the availability of neuronal glucose levels so the neuron cannot uptake glucose and is not capable of producing enough energy. Caprylic triglyceride or MCFA is metabolized into KBs that can serve as an alternate energy substrate for neuronal metabolism. Monocarboxylic acid transporter 1 transports KBs from blood to neuron. It has also been studied by Bough et al. (2006) that 3-hydroxybutyrate, a KB, induced brain-derived neuropathic factors. These factors are very useful in the treatment of several neurodegenerative disorders. KBs also upregulate mitochondrial biogenesis in the CNS and increase mitochondrial glutathione levels and glutathione peroxidase activity.

The mechanism of ketogenesis and ketolysis is given in Fig. 3.1.9.2.

In a natural condition when body glucose or stored glycogen is exhausted, insulin triggers a signal to the lipolysis of endogenous adipose, which increases the level of circulating free fatty acid and results in ketogenesis. Ketogenesis generally takes place in liver mitochondria where excess amounts of acetyl-CoA are produced. It produces energy via a tricarboxylic acid (TCA) cycle. The remaining amount of acetyl-CoA is converted into HMG-CoA (beta-hydroxyl beta-methyl glutaryl CoA) by hydroxymethyl glutaryl CoA

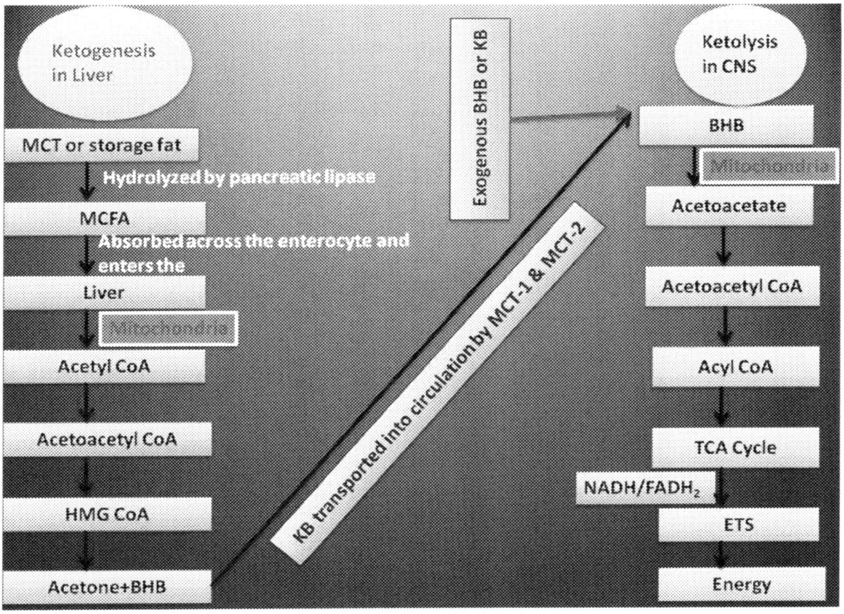

FIGURE 3.1.9.2 Mechanism of ketogenesis and ketolysis.

synthase. The HMG-CoA lyase metabolizes HMG-CoA into acetyl-CoA and acetoacetate. Acetoacetate is converted into BHB (beta-hydroxybutyrate) by phosphatidylcholine-dependent mitochondrial BHB dehydrogenase. BHB enters into the blood circulation and becomes a source of energy for extra hepatic tissues (Kashiwaya et al., 2000).

KBs reconverted into acetyl-CoA produce energy through the TCA cycle. Monocarboxylate transporters (MCT-1 and -2) transport KBs out from mitochondria into the circulation. The exogeneous KBs (BHBs) also induce the same effect. They rapidly metabolize in the CNS mitochondria by a ketolysis process and generate energy. The minimum level of BHBs in the blood is 0.5 mmol/L. As the level of BHBs increases, it produces symptoms like constipation, stomach pain, diarrhea, etc.

Case studies related to caprylic triglyceride-based treatment in Alzheimer's disease

AD is a progressive neurological disorder that mainly affects people over 65 years of age. It is the most frequent type of dementia among older people. Current medicines used for AD include cholinesterase inhibitors (donepezil, tacrine, galantamine, and rivastigmine) and *N*-methyl-D-aspartate receptor antagonists only. Development and clinical studies of new AD therapies are needed. Here are some case studies focusing on the use of caprylic acid in AD.

Croteau et al. (2018) conducted a study to analyze whether brain ketone uptake increases in response to MCTs in patients with AD as it would increase in healthy adults. They gave 30 g/day of two different MCT supplements, having caprylic and capric acids in specific concentrations, to the patients for 1 month. Analysis of brain uptake for ketone and glucose was done using positron emission tomography. Results show that brain uptake and ketone consumption were doubled on both MCT supplements. The brain ketone uptake pattern was also the same in healthy young patients. Results also report that MCT-based supplements increased total brain energy metabolism by increasing ketone supply without any effects on brain glucose utilization.

Newport et al. (2015) demonstrated that exogenous ketone monoester is able to increase circulating KBs. In their investigation they gave a ketone monoester supplement dosage of 21.5 g three times per day to a patient (a 63-year-old Caucasian male patient with sporadic AD) and slowly increased the dosage to 28.7 g three times per day. They found a dose—response fashion in the serum BHB levels and dosages of ketone monoester supplement. The patient was positive for ApoE4 and witnessed a continuous decline in cognition for 12 years.

In a second trial they formulated a new ketogenic formula (165 mL/daily of a 4:3 mixture of MCT and coconut oil) and it was given to the patient for 75 days. After 75 days they found that the patient's Mini-Mental State Examination (MMSE) score was 20 (increments of eight points). They continuously treated the patient for 20 months with the ketogenic formula and subsequently measured the Alzheimer's Disease Assessment Scale—Cognitive Subscale of the patient, which improved by six points. The ability to carry out day-to-day activities was also improved by 14 points.

The results of the research study done by Ota et al. (2019) suggest that dietary changes are necessary for the effectiveness of exogenous KB-inducing agent. To obtain acute cognitive benefits, dietary changes are required otherwise it may only bring about physiological improvement in AD. They conducted their experiment with 20 Japanese patients (11 males and nine females) with mild-to-moderate AD having an average age of 73.4 years. All the patients were treated with 20-g emulsified MCT dosage for 12 week and observations were made after 2 h; however, there was no cognitive improvement but KB level in the plasma was increased. After this, observational research was further continued for the next 12 week on 19 patients from the study. The protocol required that patients consume the same 20-g MCT dosage each day with meals. The result of this study reported progressive improvement in the working memory, processing speed, and short-term memory of 16 out of 19 patients only on chronic consumption of ketogenic formula in mild-to-moderate AD cases.

Ohnuma et al. (2016) analyzed Axona for its tolerance, cognitive function, and adverse effects in mild-to-moderate AD. Axona was given for 3 months (40 g in the form of powder containing 20 g of caprylic triglycerides). Twenty-two patients (12 males and 10 females) with mild-to-moderate sporadic AD were chosen for this clinical study. The initial MMSE score was 10—25 and seven patients out of 22 were ApoE4 positive. MMSE, AD assessment-scale score intolerance, and serum ketone concentrations were observed during Axona administration. No severe gastrointestinal adverse effects were observed, so the tolerance of Axona was good but did not find any improvement in cognitive function in the sample of AD patients.

Maynard and Jeff Gelblum (2015) examined eight 50-year-old patients. They examined the MMSE score of patients treated with caprylic triglyceride for 6 months without pharmacotherapy and patients with pharmacotherapy alone for AD. The observations were based on MMSE scores. The finding of their study showed that the patients treated with caprylic triglyceride alone had slower decrease in MMSE scores than the patients treated with pharmacotherapy alone.

Andrew Farah (2014) did an open label study to analyze the effect of capric triglyceride on a 70-year-old male mild AD patient. The study parameters were MMSE, Montreal Cognitive Assessment (MoCA), and 18-fluorodeoxyglucose (18F) positron emission tomography (FDG PET) scans. Capric triglyceride dosage was given for 109 days and it was found that MoCA scores increased up to four points (24—28), MMSE scores increased up to five points (23—28), but FDG PET scans remained the same. The results obtained from this study suggested that capric triglyceride affected cognitive function in a case of mild AD.

Doody et al. (2012) did a clinical effectiveness study on a new formulation of caprylic triglyceride (AC-1204) to test ketosis with predictable levels of beta-hydroxybutyrate in 480 patients. The clinical trial was assessed in a random, double-blind, placebo-controlled, parallel-group, multicenter trial in mild-to-moderate AD patients. Patients having an MMSE score between 14 and 26 were eligible for the trial; they were treated with either AC-1204 or placebo for 26 weeks. The results of this study showed significant improvement in MMSE score and ketosis as a therapeutic intervention in mild-to-moderate AD.

Precautionary advice

Caprylic triglyceride is also denoted by AC-1204. It is the main ingredient of medicated food used for the treatment of AD. It breaks down in the gut by lipases to MCFAs; in hepatic cells, MCFAs oxidizes into acetyl-CoA and acetoacetyl-CoA. The excess amount of acetyl-CoA and acetoacetyl-CoA produces HMG-CoA, which again forms acetoacetate and beta-hydroxybutyrate by HMG-CoA lyase.

The excess amount of KBs in the blood is termed ketoacidosis. It may cause diarrhea, nausea, abdominal distention, hypertension, dizziness, headache, fatigue, and pain. Patients suffering from diabetes or patients with a history of alcohol abuse are at risk for ketoacidosis so they need to consult medical professionals before they use caprylic acid-based medicated food.

Conclusion

Various research results have proposed that dietary administration of medicinal foods may control the development and increase of various diseases, including AD. Caprylic triglyceride is one of the dietary enhancements that shows positive results in the treatment of AD. Caprylic triglyceride-based nourishments (Axona) have been recommended to patients with mild-to-moderate AD recently, dependent on clinical preliminary outcomes. Caprylic triglycerides prompt ketosis without the need to change the diet; these types of health products have the advantage of being economic, simple to apply, socially acceptable, and above all they have no side effects. But many more studies are needed to prove the effectiveness of these food-based drugs.

References

Andrew Farah, B., 2014. Effects of caprylic triglyceride on cognitive performance and cerebral glucose metabolism in mild Alzheimer's disease: asingle case observation. Front. Aging Neurosci. https://doi.org/10.3389/fnagi.2014.00133.

Anneken, D.J., Both, S., Christoph, R., Fieg, G., Steinberner, U., Westfechtel, A., 2012. Fatty acids. In: Ullmann's Encyclopedia of Industrial Chemistry, 14, pp. 73–116.

Bough, K.J., Wetherington, J., Hassel, B., Pare, J.F., Gawryluk, J.W., Greene, J.G., Shaw, R., Smith, Y., Geiger, J.D., Dingledine, R.J., 2006. Mitochondrial biogenesis in the anticonvulsant mechanism of the ketogenic diet. Ann. Neurol. 60, 223–235.

Croteau, E., et al., 2018. Medium chain triglycerides increase brain energy metabolism in Alzheimer's disease. J. Alzheim. Dis. 64 (2), 551–561.

Greenwood, S., 2015. Capric/Caprylic Triglyceride vs. Fractionated Coconut Oil. http://chemicaloftheday.squarespace.com/qa/2015/2/8/capriccaprylic-triglyceride-vs-fractionated-coconut-oil.html.

Doody, R., Galvin, J., Farlow, M., Shah, R., Doraiswamy, P.M., Ferris, S., 2012. A new 26-week, double-blind, randomized, placebo-controlled, study of AC-1204 (caprylic triglyceride) in mild to moderate Alzheimer's disease: presentation of study design. J. Nutr. Health Aging 16 (9), 868.

Henderson, S.T., Vogel, J.L., Barr, L.J., Garvin, F., Jones, J.J., Costantini, L.C., 2009. Study of the ketogenic agent AC-1202 in mild to moderate Alzheimer's disease: a randomized, double-blind, placebo-controlled, multicenter trial. Nutr. Metabol. 6, 31. https://doi.org/10.1186/1743-7075-6-31.

Kashiwaya, Y., Takeshima, T., Morin, Nakashima, K., Clarke, K., Veech, R.L., 2000. D beta hydroxybutyrate protects neurons in models of Alzheimer's disease. Proc. Natl. Acad. Sci. U.S.A. 97, 5440–5444. https://doi.org/10.1073/pnas.97.10.5440.

Korolev, I.O., 2014. Alzheimer's disease: a clinical and basic science review. MSRJ 4. https://doi.org/10.3402/msrj.v3i0.201333.

Maynard, S.D., Gelblum, J., 2013. Retrospective case study of the efficacy of caprylic triglyceride in patients with mild-to-moderate Alzheimer's disease. Neuropsychiatric Dis. Treat. 9, 1619–1627. https://doi.org/10.2147/NDT.S52331.

Newport, M.T., VanItallie, T.B., Kashiwaya, Y., King, M.T., Veech, R.L., 2015. A new way to produce hyperketonemia: use of ketone ester in a case of Alzheimer's disease. Alzheimer 11, 99–103, 2015.

Ohnuma, T., Toda, A., Kimoto, A., Takebayashi, Y., Higashiyama, R., Tagata, Y., Ito, M., Ota, T., Shibata, N., Arai, H., 2016. Benefits of use, and tolerance of, medium-chain triglyceride medical food in the management of Japanese patients with Alzheimer's disease: a prospective, open-label pilot study. Dovepress 11, 29–36. https://doi.org/10.2147/CIA.S95362, 2016.

Ota, M., Matsuo, J., Ishida, I., Takano, H., Yokoi, Y., Hori, H., Yoshida, S., Ashida, K., Nakamura, K., Takahashi, T., 2019. Effects of a medium-chain triglyceride-based ketogenic formula on cognitive function in patients with mild-to-moderate Alzheimer's disease. Neurosci. Lett. 690, 232–236, 2019.

Swaminathan, A., Jicha, G.A., 2014. Nutrition and prevention of Alzheimer's dementia. Front. Aging Neurosci. 6, 282. https://doi.org/10.3389/fnagi.2014.00282.

Chapter 3.1.10

Berberine

Merve Keskin[1], Gülşen Kaya[2], Fatma Tugce Guragac Dereli[3], Tarun Belwal[4]

[1]*Vocational School of Health Services, Bilecik Seyh Edebali University, Bilecik, Turkey;* [2]*Scientific and Technological Research Center, Inonu University, Malatya, Turkey;* [3]*Department of Pharmacognosy, Faculty of Pharmacy, Suleyman Demirel University, Isparta, Turkey;* [4]*College of Biosystems Engineering and Food Science, Zhejiang University, China*

Introduction

Dementia is a word that describes loss in mental functions. It lowers the quality of life and negatively affects the vital functions of individuals. Alzheimer's disease (AD) is known to be a common cause of this annoying disorder (Mortimer, 1983).

AD is related to the nervous system and known to produce nonreversible degeneration of cognitive functions. The neurodegeneration is related to the deposition of amyloid-β (Aβ) peptides (synthesized from amyloid precursor protein [APP] via β- and γ-secretases) and neurofibrillary components (comprised of hyperphosphorylated tau [τ]-proteins) in the brain (DeTure and Dickson, 2019). The occurrence and progression of this disease have found to be associated with neuronal dysplasia, angiogenic alterations, neuroinflammation, and decreasing cholinergic transmission (Geekiyanage et al., 2012). AD is currently among the predominant cause of death for people 65 and older. Besides, the global burden of the disease increases with each passing year. However, no effective treatment strategy has been found to overcome AD (Shakarami et al., 2020).

The secondary metabolites synthesized by medicinal and aromatic plants have been used as raw materials of diverse drugs. Alkaloids are among the important secondary metabolites used in drug development (Amirkia and Heinrich, 2014). These natural compounds are synthesized mainly from amino acids and carry one or more nitrogen atoms (Ziegler and Facchini, 2008). They are pharmacologically highly active even in minimal doses and most of them are toxic to humans (Debnath et al., 2018).

Naturally Occurring Chemicals against Alzheimer's Disease. https://doi.org/10.1016/B978-0-12-819212-2.00012-8

Berberine and its pharmacological properties

Berberine (BBR), a yellow-colored isoquinoline alkaloid (Fig. 3.1.10.1), is widely found in the rhizomes, stem barks, or roots of several medicinal plants belonging to the Papaveraceae, Ranunculaceae, and Euphorbiaceae families (Grycova et al., 2007). Some of the plants that contain BBR, including *Hydrastis canadensis*, *Berberis vulgaris* L., *Phellodendron amurense*, *Berberis aristata*, *Coptis chinensis*, *Scutellaria baicalensis*, and *Coptis japonica*, have been used as traditional folk remedies to treat infertility and gastrointestinal disorders, especially in Ayurvedic and Chinese medicine (Alzamora et al., 2011; Habtemariam, 2016; Tillhon et al., 2012). Research on BBR to date has revealed that it has several pharmacological activities, e.g., antidiabetic, anti-atherosclerotic, antihypertensive, antihyperlipidemic, antihyperglycemic, antidepressant, antineoplastic, antileishmanial, anti-inflammatory, anti-arrhythmic, anti-oxidant, neuroprotective, anti-Parkinson's, and anti-Alzheimer's (Hou et al., 2011; Hu et al., 2012; Jiang et al., 2011; Lau et al., 2001; Lee et al., 2006, 2007, 2010; Singh et al., 2011; Su et al., 2013; Tang et al., 2006; Tillhon et al., 2012).

Various scientific research studies have shown that BBR may be beneficial in AD by exhibiting neuroprotective activity via anti-oxidant, anti-apoptotic, anti-inflammatory, anticholinesterase, and anti-amyloidogenic effects (Cai et al., 2016). The literature findings are as follows.

Anti-oxidant activity of BBR

Oxidative stress is involved in the pathogenesis of AD by leading to neuron dysfunction and neuronal death (Christen, 2000). In addition, it was determined that oxidative damage and neuroinflammation trigger the formation of neurofibrillary components and Aβ peptides by upregulating β-secretase and γ-secretase (Cai et al., 2016).

Numerous studies have confirmed the antioxidant role of BBR. Kassab and coworkers evaluated the antioxidant activity of BBR by using luminescence and fluorescence techniques. As a result of this study, it was shown that BBR has the ability to inhibit reactive oxygen species (ROS) generation in cell-free assays and human neutrophils (Kassab et al., 2017). BBR not only inhibits ROS-producing oxidase enzymes but also increases the activity of antioxidant enzymes (Thirupurasundari et al., 2009). Bhutada and colleagues demonstrated

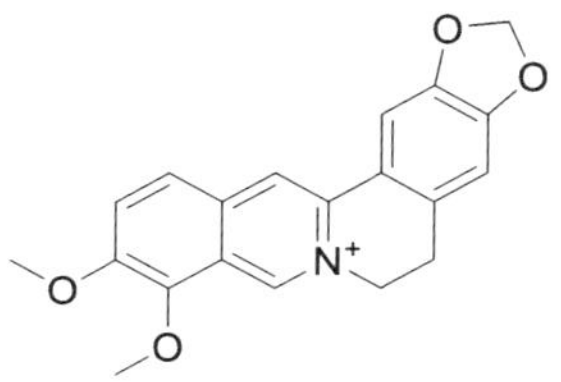

FIGURE 3.1.10.1 The chemical structure of berberine.

that treatment with BBR prevented memory dysfunction in diabetic rats by inhibiting changes in cholinesterase activity and oxidative stress (Bhutada et al., 2011).

Anti-apoptotic activity of BBR

Dysfunction and apoptosis of neuronal cells are among the cardinal characteristics of AD (Zhu et al., 2006). For this reason, the inhibitions of these features may be a promising approach to combat this neurodegenerative disease.

Yu et al. studied the antiapoptotic effect of BBR and found that it showed activity by suppressing the endoplasmic reticulum and mitochondria stress (Yu et al., 2013). Lv et al. determined that BBR exhibited antiapoptotic activity by attenuating mitochondrial impairment (Lv et al., 2012). Wang and colleagues indicated that BBR induced apoptosis in the hepatocellular carcinoma cell lines (Wang et al., 2010). BBR chronic treatment in STZ diabetic rats has been found associated with improving cognitive functions. This beneficial role of BBR is attributed to its anti-apoptotic property (Kalalian-Moghaddam et al., 2013). Findings of a study conducted by Zhou et al. suggested that BBR protected against ischemic brain damage through inhibiting mitochondrial apoptosis and regulating the ROS level (Zhou et al., 2008).

Anti-inflammatory activity of BBR

Alzheimer's patients have chronic brain inflammation, which is manifested by activation of microglial cells, generation of reactive astrocytes, and stimulation of cytokine release. Therefore it is estimated that neuroinflammation causes neuron damage and antiinflammatory drugs could provide prophylaxis against this disease (Rogers et al., 1996).

Current evidence supports the beneficial antiinflammatory activity of BBR. It was found responsible for reduced inflammation through different mechanisms such as modulation of the level of nuclear factor-κB (NF-κB), extracellular signal-regulated kinases, reduction of the expressions of inflammation mediators, and inhibition of the activation of astrocytes (Chen et al., 2017; He et al., 2017; Jia et al., 2012; Lee et al., 2012). In a study performed by Lee and colleagues, it was reported that BBR was useful for improving cognitive functions by reducing inflammatory responses and increasing cholinergic activity (Lee et al., 2012).

Anti-cholinesterase activity of BBR

One of the pathological features of AD is decreased cholinergic transmission (Bowen et al., 1983). Researchers have revealed that BBR may modulate cholinergic stimulation by inhibiting critical enzymes in the development of AD such as monoamine oxidase A, monoamine oxidase B, acetylcholinesterase, and butyrylcholinesterase (Ji and Shen, 2012).

Bhutada et al. demonstrated that treatment with BBR prevents memory disturbance in diabetic rats by improving cholinesterase activity (Bhutada et al., 2011). In another study aimed at investigating the protective activity of BBR on AD, 1-month administration of BBR to rats was found to increase cholinergic activity and prevent memory loss (de Oliveira et al., 2016).

Anti-amyloidogenic activity of BBR

The overexpression and abnormal accumulation of Aβ peptides and also Aβ-dependent τ hyperphosphorylation are among the classical hallmarks of AD. Therefore inhibition of the generation of Aβ peptides represents a significant target for the prevention of AD (Hung et al., 2016).

There are several studies that have examined the anti-amyloidogenic activity of BBR. It has been found to inhibit the formation and aggregation of Aβ and ameliorate cognitive dysfunction in mice (Huang et al., 2017). He and coworkers reported that BBR decreased τ mutations by inactivating NF-κB signal transduction (He et al., 2017). BBR has been demonstrated to reduce Aβ production by altering the trafficking and processing of APP in human neuroglioma cells (Asai et al., 2007). In an experimental AD rabbit model, BBR treatment reduced hippocampal damage and attenuated behavioral symptoms by lowering the activity of β-site APP-cleaving enzyme-1 (Panahi et al., 2013).

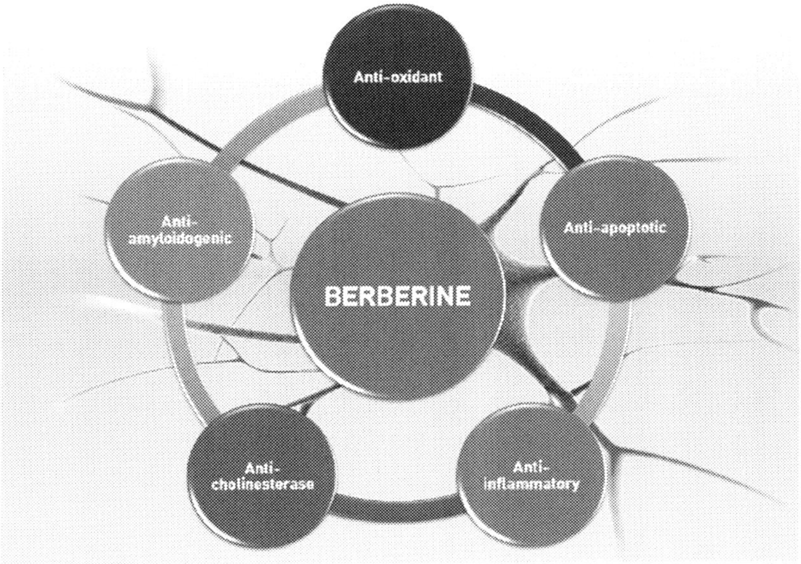

FIGURE 3.1.10.2 Role of berberine in preventing and modulating Alzheimer's disease.

Side effects and toxicological profile of BBR

Studies carried out to investigate the side effects and toxicological profile of BBR are very limited.

The lethal dose 50 of BBR sulfate was found to be 25 mg/kg in mice (Guan et al., 2010). BBR exhibits pharmacological interactions with tolbutamide, warfarin, thiopental, and bilirubin. Therefore BBR-containing plants may be contraindicated in pregnant women and babies with jaundice (Chan, 1993; Tan et al., 2002). The most common side effects of BBR are related to the gastrointestinal system. It can cause gastric lesions in animals (Yin et al., 2008). It also has some toxic effects on dopaminergic neurons and can lead to disturbances in cognitive functions (Shin et al., 2013).

Conclusion

Studies so far have underlined that the administration of BBR may be worthwhile in the prevention and modulation of AD. A summary of the role of BBR is presented in Fig. 3.1.10.2. Nevertheless, future studies should focus deeply on the pharmacokinetic and pharmacodynamic properties of this promising natural compound.

References

Alzamora, R., O'Mahony, F., Ko, W.H., Yip, T.W., Carter, D., Irnaten, M., Harvey, B.J., 2011. Berberine reduces cAMP-induced chloride secretion in T84 human colonic carcinoma cells through inhibition of basolateral KCNQ1 channels. Front. Physiol. 2, 33.

Amirkia, V., Heinrich, M., 2014. Alkaloids as drug leads - a predictive structural and biodiversity-based analysis. Phytochem. Lett. 10, XLVIII–liii.

Asai, M., Iwata, N., Yoshikawa, A., Aizaki, Y., Ishiura, S., Saido, T.C., Maruyama, K., 2007. Berberine alters the processing of Alzheimer's amyloid precursor protein to decrease Abeta secretion. Biochem. Biophys. Res. Commun. 352 (2), 498–502.

Bhutada, P., Mundhada, Y., Bansod, K., Tawari, S., Patil, S., Dixit, P., Umathe, S., Mundhada, D., 2011. Protection of cholinergic and antioxidant system contributes to the effect of berberine ameliorating memory dysfunction in rat model of streptozotocin-induced diabetes. Behav. Brain Res. 220 (1), 30–41.

Bowen, D.M., Allen, S.J., Benton, J.S., Goodhardt, M.J., Haan, E.A., Palmer, A.M., Sims, N.R., Smith, C.C., Spillane, J.A., Esiri, M.M., Neary, D., Snowdon, J.S., Wilcock, G.K., Davison, A.N., 1983. Biochemical assessment of serotonergic and cholinergic dysfunction and cerebral atrophy in Alzheimer's disease. J. Neurochem. 41 (1), 266–272.

Cai, Z.Y., Wang, C.L., Yang, W.M., 2016. Role of berberine in Alzheimer's disease. Neuropsychiatric Dis. Treat. 12, 2509–2520.

Chan, E., 1993. Displacement of bilirubin from albumin by berberine. Biol. Neonate 63 (4), 201–208.

Chen, Q.J., Mo, R., Wu, N.H., Zou, X., Shi, C., Gong, J., Li, J.B., Fang, K., Wang, D.K., Yang, D.S., Wang, K.F., Chen, J., 2017. Berberine ameliorates diabetes-associated cognitive

decline through modulation of aberrant inflammation response and insulin signaling pathway in DM rats. Front. Pharmacol. 8, 334.
Christen, Y., 2000. Oxidative stress and Alzheimer disease. Am. J. Clin. Nutr. 71 (2), 621s–629s.
de Oliveira, J.S., Abdalla, F.H., Dornelles, G.L., Adefegha, S.A., Palma, T.V., Signor, C., da Silva Bernardi, J., Baldissarelli, J., Lenz, L.S., Magni, L.P., Rubin, M.A., Pillat, M.M., de Andrade, C.M., 2016. Berberine protects against memory impairment and anxiogenic-like behavior in rats submitted to sporadic Alzheimer's-like dementia: involvement of acetylcholinesterase and cell death. Neurotoxicology 57, 241–250.
Debnath, B., Singh, W.S., Das, M., Goswami, S., Singh, M.K., Maiti, D., Manna, K., 2018. Role of plant alkaloids on human health: a review of biological activities. Mater Today Chem. 9, 56–72.
DeTure, M.A., Dickson, D.W., 2019. The neuropathological diagnosis of Alzheimer's disease. Mol. Neurodegener. 14 (1), 32.
Geekiyanage, H., Jicha, G.A., Nelson, P.T., Chan, C., 2012. Blood serum miRNA: non-invasive biomarkers for Alzheimer's disease. Exp. Neurol. 235 (2), 491–496.
Grycova, L., Dostal, J., Marek, R., 2007. Quaternary protoberberine alkaloids. Phytochemistry 68 (2), 150–175.
Guan, S., Wang, B., Li, W., Guan, J., Fang, X., 2010. Effects of berberine on expression of LOX-1 and SR-BI in human macrophage-derived foam cells induced by ox-LDL. Am. J. Chin. Med. 38 (6), 1161–1169.
Habtemariam, S., 2016. Berberine and inflammatory bowel disease: a concise review. Pharmacol. Res. 113, 592–599.
He, W., Wang, C., Chen, Y., He, Y., Cai, Z., 2017. Berberine attenuates cognitive impairment and ameliorates tau hyperphosphorylation by limiting the self-perpetuating pathogenic cycle between NF-kB signaling, oxidative stress and neuro-inflammation. Pharmacol. Rep. 69 (6), 1341–1348.
Hou, Q., Tang, X., Liu, H., Tang, J., Yang, Y., Jing, X., Xiao, Q., Wang, W., Gou, X., Wang, Z., 2011. Berberine induces cell death in human hepatoma cells *in vitro* by downregulating CD147. Canc. Sci. 102 (7), 1287–1292.
Hu, J., Chai, Y.S., Wang, Y.G., Kheir, M.M., Li, H.Y., Yuan, Z.Y., Wan, H.J., Xing, D.M., Lei, F., Du, L.J., 2012. PI3K p55γ promoter activity enhancement is involved in the anti-apoptotic effect of berberine against cerebral ischemia-reperfusion. Eur. J. Pharmacol. 674, 132–142.
Huang, M., Jiang, X., Liang, Y., Liu, Q., Chen, S., Guo, Y., 2017. Berberine improves cognitive impairment by promoting autophagic clearance and inhibiting production of β-amyloid in APP/tau/PS1 mouse model of Alzheimer's disease. Exp. Gerontol. 91, 25–33.
Hung, A.S., Liang, Y., Chow, T.C., Tang, H.C., Wu, S.L., Wai, M.S., Yew, D.T., 2016. Mutated tau, amyloid and neuroinflammation in Alzheimer disease-A brief review. Prog. Histochem. Cytochem. 51 (1), 1–8.
Ji, H.F., Shen, L., 2012. Molecular basis of inhibitory activities of berberine against pathogenic enzymes in Alzheimer's disease. Sci. World J. 823201, 2012.
Jia, L., Liu, J., Song, Z., Pan, X., Chen, L., Cui, X., Wang, M., 2012. Berberine suppresses amyloid-beta-induced inflammatory response in microglia by inhibiting nuclear factor-kappaB and mitogen-activated protein kinase signalling pathways. J. Pharm. Pharmacol. 64 (10), 1510–1521.
Jiang, H., Wang, X., Huang, L., Luo, Z., Su, T., Ding, K., Li, X., 2011. Benzenediol-berberine hybrids: multifunctional agents for Alzheimer's disease. Bioorg. Med. Chem. 19 (23), 7228–7235.

Kalalian-Moghaddam, H., Baluchnejadmojarad, T., Roghani, M., Goshadrou, F., Ronaghi, A., 2013. Hippocampal synaptic plasticity restoration and anti-apoptotic effect underlie berberine improvement of learning and memory in streptozotocin-diabetic rats. Eur. J. Pharmacol. 698 (1–3), 259–266.

Kassab, R.B., Vasicek, O., Ciz, M., Lojek, A., Perecko, T., 2017. The effects of berberine on reactive oxygen species production in human neutrophils and in cell-free assays. Interdiscipl. Toxicol. 10 (2), 61–65.

Lau, C.W., Yao, X.Q., Chen, Z.Y., Ko, W.H., Huang, Y., 2001. Cardiovascular actions of berberine. Cardiovasc. Drug Rev. 19 (3), 234–244.

Lee, B., Sur, B., Shim, I., Lee, H., Hahm, D.H., 2012. *Phellodendron amurense* and its major alkaloid compound, berberine ameliorates scopolamine-induced neuronal impairment and memory dysfunction in rats. Korean J. Physiol. Pharmacol. 16 (2), 79–89.

Lee, C.H., Chen, J.C., Hsiang, C.Y., Wu, S.L., Wu, H.C., Ho, T.Y., 2007. Berberine suppresses inflammatory agents-induced interleukin-1beta and tumor necrosis factor-alpha productions via the inhibition of kappaB degradation in human lung cells. Pharmacol. Res. 56 (3), 193–201.

Lee, I.A., Hyun, Y.J., Kim, D.H., 2010. Berberine ameliorates TNBS-induced colitis by inhibiting lipid peroxidation, enterobacterial growth and NF-kB activation. Eur. J. Pharmacol. 648 (1–3), 162–170.

Lee, Y.S., Kim, W.S., Kim, K.H., Yoon, M.J., Cho, H.J., Shen, Y., Ye, J.M., Lee, C.H., Oh, W.K., Kim, C.T., Hohnen-Behrens, C., Gosby, A., Kraegen, E.W., James, D.E., Kim, J.B., 2006. Berberine, a natural plant product, activates AMP-activated protein kinase with beneficial metabolic effects in diabetic and insulin-resistant states. Diabetes 55 (8), 2256–2264.

Lv, X., Yu, X., Wang, Y., Wang, F., Li, H., Wang, Y., Lu, D., Qi, R., Wang, H., 2012. Berberine inhibits doxorubicin-triggered cardiomyocyte apoptosis via attenuating mitochondrial dysfunction and increasing Bcl-2 expression. PloS One 7 (10), e47351.

Mortimer, J.A., 1983. Alzheimer's disease and senile dementia: prevalence and incidence. In: Reisberg, B. (Ed.), Alzheimer's Disease. The Free Press, New York, pp. 141–148.

Panahi, N., Mahmoudian, M., Mortazavi, P., Hashjin, G.S., 2013. Effects of berberine on β-secretase activity in a rabbit model of Alzheimer's disease. Arch. Med. Sci. 9 (1), 146–150.

Rogers, J., Webster, S., Lue, L.F., Brachova, L., Civin, W.H., Emmerling, M., Shivers, B., Walker, D., McGeer, P., 1996. Inflammation and Alzheimer's disease pathogenesis. Neurobiol. Aging 17 (5), 681–686.

Shakarami, A., Tarrah, H., Mahdavi-Hormat, A., 2020. A CAD system for diagnosing Alzheimer's disease using 2D slices and an improved AlexNet-SVM method. Optik 212, 164237.

Shin, K.S., Choi, H.S., Zhao, T.T., Suh, K.H., Kwon, I.H., Choi, S.O., Lee, M.K., 2013. Neurotoxic effects of berberine on long-term L-DOPA administration in 6-hydroxydopamine-lesioned rat model of Parkinson's disease. Arch Pharm. Res. (Seoul) 36 (6), 759–767.

Singh, T., Vaid, M., Katiyar, N., Sharma, S., Katiyar, S.K., 2011. Berberine, an isoquinoline alkaloid, inhibits melanoma cancer cell migration by reducing the expressions of cyclooxygenase-2, prostaglandin E(2) and prostaglandin E(2) receptors. Carcinogenesis 32 (1), 86–92.

Su, T., Xie, S., Wei, H., Yan, J., Huang, L., Li, X., 2013. Synthesis and biological evaluation of berberine-thiophenyl hybrids as multi-functional agents: inhibition of acetylcholinesterase, butyrylcholinesterase, and Aβ aggregation and antioxidant activity. Bioorg. Med. Chem. 21 (18), 5830–5840.

Tan, Y.Z., Wu, A.C., Tan, B.Y., Wu, J.H., Tang, L.F., Li, F., Li, J.H., Li, Y.X., Zhang, D.J., 2002. Study on the interactions of berberine displace other drug from their plasma proteins binding sites. J. Chin. Pharmacol. Bull. 18, 576–578.

Tang, L.Q., Wei, W., Chen, L.M., Liu, S., 2006. Effects of berberine on diabetes induced by alloxan and a high-fat/high-cholesterol diet in rats. J. Ethnopharmacol. 108 (1), 109–115.

Thirupurasundari, C.J., Padmini, R., Devaraj, S.N., 2009. Effect of berberine on the antioxidant status, ultrastructural modifications and protein bound carbohydrates in azoxymethane-induced colon cancer in rats. Chem. Biol. Interact. 177 (3), 190–195.

Tillhon, M., Guaman Ortiz, L.M., Lombardi, P., Scovassi, A.I., 2012. Berberine: new perspectives for old remedies. Biochem. Pharmacol. 84 (10), 1260–1267.

Wang, N., Feng, Y., Zhu, M., Tsang, C.M., Man, K., Tong, Y., Tsao, S.W., 2010. Berberine induces autophagic cell death and mitochondrial apoptosis in liver cancer cells: the cellular mechanism. J. Cell. Biochem. 111 (6), 1426–1436.

Yin, J., Gao, Z., Liu, D., Liu, Z., Ye, J., 2008. Berberine improves glucose metabolism through induction of glycolysis. Am. J. Physiol. Endocrinol. Metab. 294 (1), E148–E156.

Yu, W., Sheng, M., Xu, R., Yu, J., Cui, K., Tong, J., Shi, L., Ren, H., Du, H., 2013. Berberine protects human renal proximal tubular cells from hypoxia/reoxygenation injury via inhibiting endoplasmic reticulum and mitochondrial stress pathways. J. Transl. Med. 11, 24.

Zhou, X.Q., Zeng, X.N., Kong, H., Sun, X.L., 2008. Neuroprotective effects of berberine on stroke models *in vitro* and *in vivo*. Neurosci. Lett. 447 (1), 31–36.

Zhu, X., Raina, A.K., Perry, G., Smith, M.A., 2006. Apoptosis in Alzheimer disease: a mathematical improbability. Curr. Alzheimer Res. 3 (4), 393–396.

Ziegler, J., Facchini, P.J., 2008. Alkaloid biosynthesis: metabolism and trafficking. Annu. Rev. Plant Biol. 59, 735–769.

Chapter 3.1.11

Hypericin and pseudohypericin

Koula Doukani[1], Ammar Sidi Mohammed Selles[2], Hasna Bouhenni[1]
[1]*Faculty of Nature and Life Sciences, University of Ibn Khaldoun, Tiaret, Algeria;* [2]*Institute of Veterinary Sciences, University of Ibn Khaldoun, Tiaret, Algeria*

Introduction

Alzheimer's disease (AD) is a multifactorial disease characterized by an inexorable progressive loss of cognitive function associated with the presence of amyloid plaques in the hippocampal area of the brain. This disease is the most common form of dementing illness among middle-aged and older adults. But some early forms of the disease are linked to a specific genetic defect. Although the etiology is unknown, genetic factors clearly play a role in 10%−15% of cases (Francis et al., 1999; Alzheimer's Association, 2010). Patients with an early form show short-term memory loss, inability to learn new information, mood swings, difficulty finding words, forgetting names, and losing objects. Frustration, hostility, and irritability are common emotional characteristics of AD patients. In severe cases, patients become completely incontinent, memory is completely lost, and the sense of time and place disappears. In addition, they become totally dependent on others and ultimately require comprehensive care (Rao et al., 2012). Until now, the drugs available to treat AD are only symptomatic; they possess limited efficiency and many undesirable effects. However, efforts to find a cure for AD have been disappointing (Bredesen, 2009; Xu et al., 2009).

Several medicinal plants are used as an alternative treatment for AD in different cultures to improve memory, such as *Valeriana officinalis*, *Punica granatum* L., *Salvia officinalis*, *Myristica fragrans*, *Bacopa monnieri* Linn, *Centella asiatica* Linn, *Evolvulus alsinoides* Linn, and *Hypericum perforatum* (Zerrouki et al., 2016; Akram and Nawaz, 2017).

Chen et al. (2011) and Kasper et al. (2010) reported that *H. perforatum* L. extracts were effective and well tolerated in the treatment of mild-to-moderate depressive episodes and in the short-term treatment of the symptoms of mild depressive disorders. In addition, *H. perforatum* extracts can be effective in the treatment of generalized anxiety disorder, somatoform disorder, sleep disorder, obsessive−compulsive disorder, and seasonal

Naturally Occurring Chemicals against Alzheimer's Disease. https://doi.org/10.1016/B978-0-12-819212-2.00013-X

affective disorder (Can et al., 2009). Moreover, Öztürk (1996) noted that *H. perforatum* has been extracted without special side effects. This plant is a rich source of natural products: flavonoids, proanthocyanidins, naphthodianthrones, and acylphloroglucinols (Mennini and Gobbi, 2004). Among naphthodianthrones, hypericin (4,5,7,4′,5′,7-hexa-hydroxy-2,2′-dimethyl-meso-naphthodianthron) (Jacobson et al., 2001) and pseudohypericin (Kitanov, 2001) are characteristic constituents of the genus *Hypericum* (Karioti and Bilia, 2010).

Description of *Hypericum perforatum* L.

The genus *Hypericum* L. (Hypericaceae) contains 484 species that have been classified into 36 taxonomic sections in all the continents except Antarctica (Bejaoui et al., 2017). *H. perforatum* L. is the most studied taxon. It is used for the extraction of several bioactive compounds known for their photodynamic, antidepressive, and antiviral activities (Patocka, 2003; Vinterhalter et al., 2006; Bejaoui et al., 2017).

Hypericum perforatum L. (HP) (commonly called as St. John's Wort (SJW), klamath weed, tipton weed, goat weed, and enola weed) (Fig.3.1.11.1) (Muenscher, 1946) is a five-petalled, yellow-flowered perennial weed, erect, multistemmed and grows to 80 cm in height. It is a native of Europe but is widely distributed in Asia, Australia, North Africa, North and South America, and throughout the temperate areas of the world (Barnes et al., 2001; Vattikuti and Ciddi, 2005; Belwal et al., 2019).

FIGURE 3.1.11.1 *Hypericum perforatum* L. (http://luirig.altervista.org).

Bioactive constituents

St. John's Wort (SJW) is one of the most widely studied medicinal plants because of its chemical composition and pharmacological properties (antidepressant, antimicrobial, anticancer, antiinflammatory, wound healing, etc.) (Mennini and Gobbi, 2004; Oliviera et al., 2016; Jendželovská et al., 2016).

Several groups of medicinally active compounds have been demonstrated in this plant such as naphthodianthrones, phloroglucinols (hyperforin), flavonoids (such as phenylpropanes, flavonol glycosides, and biflavones), tannins, and related compounds, such as phenylpropanoids and other simple phenolic coumpounds, xanthones, and volatile oils (including α-pinene and cineole) (Barnes et al., 2001; Mennini and Gobbi, 2004; Vattikuti and Ciddi, 2005; Saddiqe et al., 2010; Belwal et al., 2019).

Naphthodianthrones are the most common class of compounds isolated from *H. perforatum*. They include hypericin (as major compound), pseudohypericin, isophypericin, protohypericin, and cyclopseudohypericin (Shrivastava and Dwivedi, 2015).

Hypericin and pseudohypericin previously received most of the attention in pharmacological studies because they contribute to the antidepressant actions of *H. perforatum*. However, the latest research has recognized that hyperforin has emerged as the principal antidepressant of this herb (Anonymous, 2004; Sharopov et al., 2015).

Hypericin and pseudohypericin

Hypericin and pseudohypericin are two natural products, structurally belonging to the chemical class of naphthodianthrones (Fig. 3.1.11.2.)

Naphthodianthrone content varies from 0.05% to 0.30% in *H. perforatum*, but this rate can vary according to the cultivar, altitude, lighting conditions, and period of the year. It occurs mainly in the flowering parts of the plant and is especially located in the dark glands. In general, the content of pseudohypericin (0.03%–0.34%) is higher than that of hypericin (0.03%–0.09%) by two to four times, depending on the variety (Saddiqe et al., 2010; Karioti and Bilia, 2010).

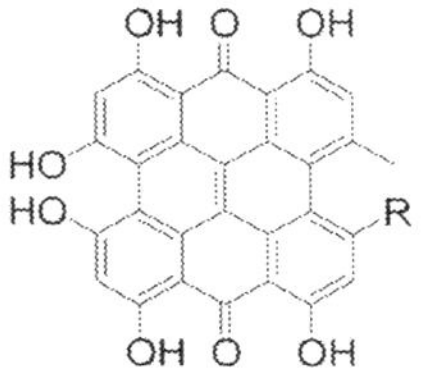

(1) Hypericin: R = CH_3
(2) Pseudohypericin: R = CH_2OH

FIGURE 3.1.11.2 Structures of hypericin and pseudohypericin.

A recent survey on 74 *Hypericum* taxa (approximately 20% of the entire genus) from different continents showed that hypericins were detected only in species of the sections *Hypericum*, *Adenotras*, and *Drosocarpium*. Infrageneric chemotaxonomic surveys have revealed that some *Hypericum* species, such as *Hypericum boissieri*, *Hypericum barbatum*, and *Hypericum rumeliacum*, may contain a two to four fold higher amount of hypericins than *H. perforatum* (Karioti et Bilia, 2010). Besides *Hypericum* sp., hypericin has also been found in some basidiomycetes such as *Dermocybe* spp. (Karioti and Bilia, 2010; Jendželovská et al., 2016) or endophytic fungi, which grow in diverse *Hypericum* sp. or in other plant species (*Thielavia subthermophila*) (Karioti and Bilia, 2010; Jendželovská et al., 2016).

Commonly used is the term total hypericins, which stands for the sum amount of hypericin and pseudohypericin (Kitanov, 2001; Tawaha et al., 2010; Jendželovská et al., 2016). Hypericin; an anthraquinone-derived pigment is responsible for the red color of SJW oils. It is found in the flowers in the form of black dots that are located along the petals. Due to its chemical structure, hypericin is highly photoreactive (Shrivastava and Dwivedi, 2015). It is a bioactive compound that is applicable in several medicinal approaches; its content has been evaluated in in vitro grown *H. perforatum* and in its transgenic clones or in *Hypericum* cultures exposed to various biotechnological applications that are focused on their preservation or stimulation of secondary metabolite production (Jendželovská et al., 2016).

Pharmacological activities of HP

H. perforatum is one of the oldest and most extensively investigated medicinal herbs. It has been used traditionally for a wide range of medicinal applications such as a soothing agent, as an antiphlogistic in the inflammation of bronchi and the urogenital tract, in hemorrhoid treatment, as a healing agent, in the treatment of traumas, burns, scrabs, and ulcers of various kinds, in skin wounds, eczema, tooth extraction, rheumatism, hematomas, lumbago, psoriasis, neuralgia, gastroenteritis, sciatica, snake bite, gout, gallbladder conditions and cystitis, contusions, sprains, diarrhea, hysteria, jaundice, menopausal neurosis, menorrhagia, bedwetting, spinal irritation, shocks, concussions, to strengthen the gums and alleviate halitosis, and for promoting menstruation and periodic fevers. Today, it is largely used in the treatment of depression and AIDS (Barnes et al., 2001; Silva et al., 2005; Vattikuti and Ciddi, 2005; Saddiqe et al., 2010; Sharopov et al., 2015).

It has been known for centuries for its putative medicinal properties, including antidepressant, anxiolytic, diuretic, antibiotic, antiviral, and wound healing effects (Miller, 1998; Khalifa, 2001). Besides, it has been recorded for use traditionally for the treatment of several neurological conditions such as excitability, neuralgia, menopausal neurosis, anxiety, and depression, in addition to its utilization as a nerve tonic (Altun et al., 2013).

H. perforatum possesses antischizophrenic, antibacterial, antifugal, antiretroviral, antimalarial, anticonvulsant, analgesic, spasmolytic, stimulant, hypotensive, apoptotic, sedative, antiinflammatory, antioxidant, antitumor, anticancer, nootropic, antiseptic, and antidiabetic properties (Barnes et al., 2001; Ayan and Çirak, 2008; Shrivastava and Dwivedi, 2015; Oliveira et al., 2016; Ramezani and Zamani, 2017; Belwal et al., 2019). Belwal et al. (2019) reported that *H. perforatum* has also been found helpful in minimizing "the symptoms of attention deficit hyperactivity disorder (ADHD), chronic fatigue syndrome (CFS), irritable bowel syndrome, obsessive–compulsive disorder (OCD), seasonal affective disorder (SAD), somatization disorder, and premenstrual syndrome (PMS). It stops breast tenderness, cramps, and irritability in females with PM."

Hydroalcoholic extracts of *H. perforatum* are the most common and widely used commercial preparations for the treatment of mild-to-moderate depression (Linde et al., 2008). In addition, *H. perforatum* has been reported to be as effective as imipramine, desipramine (Neary and Bu, 1999), maprotiline, amitriptyline, citalopram, fluoxetine, paroxetine, sertraline, and bupropion (Belwal et al., 2019).

The leaves and flowering tops are the two parts of greatest interest for medicinal forms since they contain the main components of medical interest, which are naphthodianthrones (hypericin and pseudohypericin), xanthones, flavonoids (rutin, quercetin, quercitrin), and phloroglucinols, as well as hyperforin and pseudohyperforin (Brockm'oller et al., 1997). SJW extracts or components may have nootropic effects, improving spatial learning and memory, suggesting a potential for the therapy of AD (Widy-Tyszkiewicz et al., 2002; Trofimiuk et al., 2010).

AD causes amyloid plaques in the CNS leading to neurodegeneration, neurotransmitter abnormalities due to cholinergic deficits, as well as other neurotransmitter systems in the CNS, including the glutamatergic and serotonergic systems, inflammation, oxidative damage, apoptosis, and attenuated neuroplasticity (Howes et al., 2017).

The mechanisms of action of the power of hypericin and pseudohypericin are focused on acetylcholinesterase (AChE) enzyme, β-amyloid (Aβ) peptides, and antioxidant and antiinflammatory effects.

Effect on AChE inhibitor

Cholinesterase inhibitors serve as a strategy for the treatment of AD, which is a neurodegenerative disorder. They promote an increase in the level of acetylcholine in the neuronal synaptic area (Mukherjee et al., 2007; Bejaoui et al., 2017).

Several studies have investigated in vitro AChE inhibitors from various *Hypericum* sp.

Božin et al. (2013) investigated the AChE activity of 10 *Hypericum* species extracts (five *H. perforatum*: *Hyperici herba*, *Hypericum maculatum* subsp. *Immaculatum*, *Hypericum olympicum*, *Hypericum richeri* subsp. *Grisebachii*, and *H. barbatum*). All of them have exhibited notable inhibition of AChE activity, with IC_{50} values between 432.74 and over 1500.00 μg of dry extract per mL.

Bejaoui et al. (2017) suggested that the inhibitory activity of *Hypericum humifusum* and *Hypericum perfoliatum* on AChE may be due to the high level of hypericin. These authors noted that methanolic extract has a similar action on AChE with a dose of 4.57 mg GALAEs/g for *H. humifusum* and 3.86 mg GALAEs/g for *H. perfoliatum*, respectively.

Ozkan et al. (2018) studied the AChE inhibitory activities of *Hypericum* extracts from *Hypericum neurocalycinum* and *Hypericum malatyanum*. These authors have recorded that the extracts have a dose-dependent activity for inhibiting AChE. Moreover, *H. neurocalycinum* exerted a stronger AChE inhibitory activity than *H. malatyanum* and the inhibitory activities of the extracts on AChE were lower than that of the standard drug.

Effects on β-amyloid peptides

Several studies report the effect of *H. perforatum* against β-amyloid (Silva et al., 2004; Kraus et al. 2007; Griffith et al., 2010; Hofrichter et al., 2013; Russo et al., 2013) noted that hypericin may interfere with the processes of polymerization of Aβ peptide responsible for the onset of AD.

Kraus et al. (2007) showed that *H. perforatum* extract pretreatment reduced the toxic influence of Aβ (25-35) and (1-40) on microglial cells.

Similarly, the study conducted by the Sgarbossa team found that hypericin at a concentration 10^{-5} M prevented or arrested the aggregation process of Aβ (1-40) peptides. Even so, hypericin was reported to interact also with the early precursors of the β-sheet fibrils and/or protofibrils. More interestingly, hypericin was used to monitor (in vitro) the appearance of early aggregation states of Aβ peptides during the polymerization process (Sgarbossa et al., 2008; Karioti and Bilia, 2010).

However, hypericin at 5 and 15 μM suppresses significantly oligomeric Aβ42 (oAβ42)-induced expression of interleukin-1β (IL)-1β, IL-6, tumor necrosis factor-α, and inducible nitric oxide synthase, and the production of nitric oxide (NO) in microglia without cytotoxicity (Zhang et al., 2016).

These same authors have shown that hypericin improved oAβ42-induced learning and memory impairment in mice in the Morris water maze test. Therefore they concluded that hypericin could be considered as a potential candidate for treating AD (Zhang et al., 2016).

Also the study of Silva et al. (2004) showed the potential neuroprotective action of *H. perforatum* L. in Aβ-induced cell toxicity following the exposure of hippocampal neurons of E18-E19 Wistar rat embryos after treatment to

H. perforatum extracts, and to 25 μM Aβ (25-35) for 48 h. These same authors concluded that *H. perforatum* extracts have neuroprotective activity, which can be of relevance to preventing Aβ peptide neuronal degeneration in AD.

Antiinflammatory

Light-activated pseudohypericin has been reported to inhibit the production of prostaglandin E_2 (Hammer et al., 2007; Karioti and Bilia, 2010), while hypericin has been reported to reduce Croton oil-induced ear edema in mice in a concentration (ID_{50} 0.25 μmol/cm^2) comparable to that of indometacin (ID_{50} 0.26 μmol/cm^2) (Sosa et al., 2007; Karioti and Bilia, 2010).

Moroever, Zhang et al. (2016) found an improvement in the inflammatory response by suppressing MKL1, which is the essential cofactor of p65 during the transcription process, following the administration of hypericin. Likewise, in an Aβ injection AD mouse model, animals orally administrated hypericin (50 mg/kg) for 7 days significantly decreased proinflammatory cytokine expression and NO production in the hippocampus.

Antioxidant activity

Several studies have investigated the in vitro antioxidant activity of various *Hypericum* sp. They have shown that these plants have antioxidant activity due to their richness in flavonoids and polyphenols. All of these studies have shown that these plants have antioxidant activity due to the richness of these plants in flavonoids and polyphenols. However, the effects are variable depending on the plant (Božin et al., 2013; Altun et al., 2013; Bejaoui et al., 2017; Ozkan et al., 2018).

Safety

The most commonly reported adverse reactions to the use of *H. perforatum* extract when doses are higher than those recommended are nausea, rash, fatigue, agitation, and photosensitivity, largely attributed to naphthodianthrones (hypericin and pseudohypericin) (Rodríguez-Landa and Contreras, 2003; Schulz et al., 2006).

Jacobson et al. (2001) showed that subjects treated with hypericin at 0.05 and 0.1 mg/kg had one or more photosensitivity reactions. There were four different types of photosensitivity reactions identified: paresthesias, dermatitis, darkened coloration of exposed skin, and pruritic nodules. All of the photosensitivity reactions were judged to be probably related to hypericin.

Other *H. perforatum* side effects include confusion, restlessness, lethargy, and dryness of the mouth (Klemow et al., 2011), and in some cases psychotic events (Shimizu et al., 2004; Stevinson and Ernst, 2004). However, severe photoallergic reactions (two cases) following ingestion of *H. perforatum* preparations have been reported in the literature (Schulz, 2006).

Nevertheless, Griffith et al. (2010) reported that *H. perforatum* may interfere with the processes of polymerization of Aβ peptide responsible for the onset of AD.

Conclusion

The capacity and potential of *Hypericum* sp. (especially *H. perforatum* L.) for the therapy of AD has been investigated in the literature by clinical and scientific studies of these plants. These plants are characterized by the high content of phenolic, flavonoid, hypericin, pseudohypericin, and hyperforin. The various activities of AChE, Aβ peptides, and antiinflammatory and antioxidant effects improve memory capacity significantly.

However, more advanced studies on the in vivo pharmacological activities of these plants or their extracts as potential substances in the prevention and treatment of AD should be initiated.

References

Akram, M., Nawaz, A., 2017. Effects of Medicinal plants on Alzheimer's disease and memory deficits. Neural Regen. Res. 12 (4), 660–670.

Altun, M.L., Yılmaz, B.S., Ilkay, E.O., Citoglu, G.S., 2013. Assessment of cholinesterase and tyrosinase inhibitory and antioxidant effects of *Hypericum perforatum* L. (St. John's Wort). Ind. Crop. Prod. 43, 87–92.

Alzheimer's Association, 2010. Alzheimer's disease facts and figures. Alzheimers Dement 6, 158–194.

Anonymous, 2004. *Hypericum perforatum*. Alternative Med. Rev. 9 (3), 318–325.

Ayan, A.K., Çirak, C., 2008. Hypericin and pseudohypericin contents in some *Hypericum*. Species growing in Turkey. Pharmaceut. Biol. 46 (4), 288–291.

Barnes, J., Anderson, L.A., Phillipson, J.D., 2001. St John's Wort (*Hypericum perforatum* L.): a review of its chemistry, pharmacology and clinical properties. J. Pharm. Pharmacol. 53, 583–600.

Bejaoui, A., Ben Salem, I., Rokbeni, N., M'rabet, Y., Boussaid, M., Abdennacer Boulila, A., 2017. Bioactive compounds from *Hypericum humifusum* and *Hypericum perfoliatum*: inhibition potential of polyphenols with acetylcholinesterase and key enzymes linked to type-2 diabetes. Pharmaceut. Biol. 55 (1), 906–911.

Belwal, T., Devkota, H.P., Singh, M.K., Sharma, R., Upadhayay, S., Joshi, C., Bisht, K., Gour, J.K., Bhatt, I.D., Rawal, R.S., Pande, V., 2019. 3.40. St. John's Wort (*Hypericum perforatum*). In: Mohammad Nabavi, S., Sanches Silva, A. (Eds.), Nonvitamin and Nonmineral Nutritional Supplements. Academic Press, p. 583.

Božin, B., Kladar, N., Grujic, N., Anackov, G., Isidora, S.I., Neda, G.N., Conic, B.S., 2013. Impact of origin and biological source on chemical composition, anticholinesterase and antioxidant properties of some St. John's Wort Species (*Hypericum* spp., *Hypericaceae*) from the Central Balkans. Molecules 18, 11733–11750.

Bredesen, D.E., 2009. Neurodegeneration in Alzheimer's disease: caspases and synaptic element interdependence. Mol. Neurodegener. 4, 27.

Brockm'oller, J., Reum, T., Bauer, S., Kreb, R., H'ubner, W.D., Roots, I., 1997. Hypericin and pseudohypericin: pharmacokinetics and effects on photosensitivity in humans. Pharmacopsychiatry 30, 94–101.

Can, Ö.D., Öztürk, Y., Demir Özkay, Ü., 2009. A natural antidepressant *Hypericum perforatum* L.: review. Turkiye Klinikleri J. Med. Sci. 29, 708–715.

Chen, X.W., Serag, E.S., Sneed, K.B., Liang, J., Chew, H., Pan, S.Y., Zhou, S.F., 2011. Clinical herbal interactions with conventional drugs: from molecules to maladies. Curr. Mel. Chem. 8, 4836–4850.

Francis, P.T., Palmer, A.M., Snape, M., Wilcock, G.K., 1999. The cholinergic hypothesis pharmacology, biochemistry and behavior of Alzheimer's disease: a review of progress. J. Neurol. Neurosurg. Psychiatry 66, 137–147.

Griffith, T.N., Varela-Nallar, L., Dinamarca, M.C., Inestrosa, N.C., 2010. Neurobiological effects of hyperforin and its potential in Alzheimer's disease therapy. Curr. Med. Chem. 17, 391–406.

Hammer, K.D.P., Hillwig, M.L., Solco, A.K.S., Dixon, P.M., Delate, K., Murphy, P.A., Wurtele, E.S., Birt, D.F., 2007. Inhibition of Prostaglandin E_2 production by anti-inflammatory *Hypericum perforatum* extracts and constituents in RAW264.7 mouse macrophage cells. J. Agric. Food Chem. 55, 7323–7331.

Hofrichter, J., Krohn, M., Schumacher, T., Lange, C., Feistel, B., Walbroel, B., Heinze, H.J., Crockett, S., Sharbel, T.F., Pahnke, J., 2013. Reduced Alzheimer's disease pathology by St. John's Wort treatment is independent of hyperforin and facilitated by ABCC1 and microglia activation in mice. Curr. Alzheimer Res. 10 (10), 1057–1069.

Howes, M.R., Fang, R., Houghton, P.J., 2017. Effect of Chinese herbal medicine on Alzheimer 's disease. Int. Rev. Neurobiol. 135, 29–56.

Jacobson, J.M., Feinman, L., Liebes, L., Ostrow, N., Koslowski, V., Tobia, A., Cabana, B.E., Lee, D.-H., Spritzler, J., Prince, A.M., 2001. Pharmacokinetics, safety, and antiviral effects of hypericin, a derivative of St. John's Wort plant, in patients with chronic hepatitis C virus infection. Antimicrob. Agents Chemother. 45 (2), 517–524.

Jendželovská, Z., Jendželovský, R., Kuchárová, B., Fedoročko, P., 2016. Hypericin in the light and in the dark: two sides of the same coin. Front. Plant Sci. 7, 560.

Karioti, A., Bilia, A.R., 2010. Hypericins as potential leads for new therapeutics. Int. J. Mol. Sci. 11, 562–594.

Kasper, S., Caraci, F., Forti, B., Drago, F., Aguglia, E., 2010. Efficacy and tolerability of *Hypericum* extract for the treatment of mild to moderate depression. Eur. Neuropsycho.Pharmacol. 20, 747–765.

Khalifa, A.E., 2001. *Hypericum perforatum* as nootropic drug: enhancement of retrieval memory of a passive avoidance conditioning paradigm in mice. J. Ethnopharmacol. 76, 49–57.

Kitanov, G.M., 2001. Hypericin and pseudohypericin in some *Hypericum* species. Biochem. Systemat. Ecol. 29, 171–178.

Klemow, M., Bartlow, A., Crawford, J., Kocher, N., Shah, J., Ritsick, M., 2011. Medical attributes of St. John's Wort (*Hypericum perforatum*). In: Benzie, I.F.F., Wachtel Galor, S. (Eds.), Herbal Medicine: Biomolecular and Clinical Aspects, 2nd Ed. CRC Press, Boca Raton, FL. Chapter 11.

Kraus, B., Wolff, H., Heilmann, J., Elstner, E.F., 2007. Influence of *Hypericum perforatum* extract and its single compounds on amyloid-beta mediated toxicity in microglial cells. Life Sci. 81, 884–894.

Linde, K., Berner, M.M., Kriston, L., 2008. St. John's Wort for major depression. Cochrane Database Syst. Rev. 4, Cd000448.

Mennini, T., Gobbi, M., 2004. The antidepressant mechanism of *Hypericum perforatum*. Life Sci. 75, 1021–1027.

Miller, A.L., 1998. St. John's Wort (*Hypericum perforatum*): clinical effects on depression and other conditions. Alternative Medicine review. J. Clin. Therapeutic 3 (1), 18–26.

Muenscher, W.C., 1946. Weeds. The Mac Millan Company, New York.

Mukherjee, P.K., Kumar, V., Mal, M., Houghton, P.J., 2007. Acetylcholinesterase inhibitors from plants. Phytomedicine 14, 289–300.

Neary, J., Bu, Y., 1999. *Hypericum* LI 160 inhibits uptake of serotonin and norepinephrine in astrocytes. Brain Res. 16, 358–363.

Oliveria, A.I., Pinho, C., Sarmento, B., Dias, A.C.P., 2016. Neuroprotective activity of *Hypericum perforatum* and its major components. Front. Plant Sci. 7, 1004.

Ozkan, E.E., Ozden, T.Y., Ozsoy, N., Mat, A., 2018. Evaluation of chemical composition, antioxidant and anti-acetylcholinesterase activities of *Hypericum neurocalycinum* and *Hypericum malatyanum*. South Afr. J. Bot. 114, 104–110.

Öztürk, Y., Aydin, S., Beis, R., Başer, K.H., Berberoǧlu, H., 1996. Effects of *Hypericum perforatum* L. and *Hypericum calycinum* L. extracts on the central nervous system in mice. Phytomedicine 3 (2), 139–146.

Patocka, J., 2003. The chemistry, pharmacology, and toxicology of the biologically active constituents of the herb *Hypericum perforatum* L. J. Appl. Biomed. 1, 61–73.

Ramezani, Z., Zamani, M., 2017. A simple method for extraction and purification of Hypericins from St John's Wort. Jundishapur J. Nat. Pharm. Prod. 12 (1), e13864.

Rao, J.S., Keleshian, V.L., Klein, S., Rapoport, S.I., 2012. Epigenetic modifications in frontal cortex from Alzheimer's disease and bipolar disorder patients. Transl. Psychiatry 2 (7), e132.

Rodriguez-Landa, J.F., Contreras, C.M., 2003. A. review of clinical and experimental observations about antidepressant actions and side effects produced by *Hypericum perforatum* extracts. Phytomedicine 10 (8), 688–699.

Russo, E., Scicchitano, F., Whalley, B.J., Mazzitello, C., Ciriaco, M., Esposito, S., Patane, M., Upton, R., Pugliese, M., Chimirri, S., 2013. *Hypericum perforatum*: pharmacokinetic, mechanism of action, tolerability, and clinical drug-drug interactions. Phytother Res. 28 (5), 643–655.

Saddiqe, Z., Naeem, I., Maimoona, A., 2010. A review of the antibacterial activity of *Hypericum perforatum* L. J. Ethnopharmacol. 131 (3), 511–521.

Schulz, V., 2006. Safety of St. John's Wort extract compared to synthetic antidepressants. Phytomedicine 13, 199–204.

Schulz, H.U., Schürer, M., Bässler, D., Weiser, D., 2006. Investigation of the effect on photosensitivity following multiple oral dosing of two different *Hypericum* extracts in healthy men. Arzneimittelforschung 56, 212–221.

Sgarbossa, A., Buselli, D., Francesco Lenci, F., 2008. In *vitro* perturbation of aggregation processes in b-amyloid peptides: a spectroscopic study. FEBS Lett. 582, 3288–3292.

Sharopov, F.S., Zhang, H., Wink, M., William, N., Setzer, W.N., 2015. Aromatic medicinal plants from Tajikistan (central Asia). Medicines 2, 28–46.

Shimizu, K., Nakamura, M., Isse, K., Nathan, P.J., 2004. First-episode psychosis after taking an extract of *Hypericum perforatum* (St. John's Wort). Hum. Psychopharmacol. 19, 275–276.

Shrivastava, M., Dwivedi, L.K., 2015. Therapeutic potential of *Hypericum perforatum*: A review. IJPSR 6 (12), 4982–4988.

Silva, B.A., Dias, A.C.P., Ferreres, F., Malva, J.O., Oliveira, C.R., 2004. Neuroprotective effect of *H. perforatum* extracts on B-amyloid-induced neurotoxicity. Neurotox. Res. 6, 119–130.

Silva, B.A., Ferreres, F., Malva, J.O., Dias, A.C.P., 2005. Phytochemical and antioxidant characterization of *Hypericum perforatum* alcoholic extracts. Food Chem. 90, 157–167.

Sosa, S., Pace, R., Bornancin, A., Morazzoni, P., Riva, A., Tubaro, A., Della Loggia, R., 2007. Topical anti-inflammatory activity of extracts and compounds from *Hypericum perforatum* L. J. Pharm. Pharmacol. 59, 703–709.

Stevinson, C., Ernst, E., 2004. Can St. John's Wort trigger psychoses? Int. J. Clin. Pharmacol. Ther 42 (9), 473–480.

Tawaha, K., Gharaibeh, M., El-Elimat, T., Alali, F.Q., 2010. Determination of hypericin and hyperforin content in selected Jordanian *Hypericum* species. Ind. Crop. Prod. 32 (3), 241–245.

Trofimiuk, E., Holownia, A., Braszko, J.J., 2010. Activation of CREB by St. John's Wort may diminish deleterious effects of aging on spatial memory. Arch Pharm. Res. (Seoul) 33 (3), 469–477.

Vattikuti, U.M.R., Ciddi, V., 2005. An overview on *Hypericum perforatum* Linn. Nat. Product. Radiance 4 (5), 368–381.

Vinterhalter, B., Ninkovic, S., Cingel, A., Vinterhalter, D., 2006. Shoot and root culture of *Hypericum perforatum* L. transformed with *Agrobacterium rhizogenes* A4M70GUS. Biol. Plantarum 50, 767–770.

Widy-Tyszkiewicz, E., Piechal, A., Joniec, I., Blecharz-Klin, K., 2002. Long term administration of *Hypericum perforatum* improves spatial learning and memory in the water maze. Biol. Pharm. Bull. 25 (10), 1289–1294.

Xu, J., Litterst, C., Georgakopoulos, A., Zaganas, I., Robakis, N.K., 2009. Peptide EphB2/CTF2 generated by the gamma-secretase processing of EphB2 receptor promotes tyrosine phosphorylation and cell surface localization of *N*-methyl-d-aspartate receptors. J. Biol. Chem. 284, 27220–27228.

Zerrouki, K., Djebli, N., Eroglu Ozkan, E., Mat, A., Ozhan, G., 2016. *Hypericum perforatum* improve memory and learning in alzheimer's model: (experimental study in mice). Int. J. Pharm. Pharm. Sci. 8 (8), 49–57.

Zhang, M., Wang, Y., Qian, F., Li, P., Xu, X., 2016. Hypericin inhibits oligomeric amyloid β42-induced inflammation response in microglia and ameliorates cognitive deficits in an amyloid β injection mouse model of Alzheimer's disease by suppressing MKL1. Biochem. Biophys. Res. Commun. 481 (1–2), 71–76.

Chapter 3.1.12

Protopine

Bijo Mathew[1], Della G.T. Parambi[2], Manjinder Singh[3], Omnia M. Hendawy[4], Mohammad M Al-Sanea[2], Rania B. Bakr[2,5]

[1]*Department of Pharmaceutical Chemistry, Amrita School of Pharmacy, Amrita Vishwa Vidyapeetham, Amrita Health Science Campus, Kochi, India;* [2]*College of Pharmacy, Department of Pharmaceutical Chemistry, Jouf University, Sakaka, Saudi Arabia;* [3]*Chitkara College of Pharmacy, Chitkara University, Chandigarh, Punjab, India;* [4]*College of Pharmacy, Department of Pharmacology, Jouf University, Sakaka, Saudi Arabia;* [5]*Department of Pharmaceutical Organic Chemistry, Faculty of Pharmacy, Beni Suef University, Beni Suef, Egypt*

Introduction

Neurodegenerative diseases are incurable and lead to debilitating conditions that become a global burden and are a common cause of dementia. Alzheimer's disease (AD) is a neurodegenerative disease caused by cholinergic dysfunction; it leads to a decrease in cortical cholinergic activity. Among known therapeutic treatments, cholinesterase inhibitors (ChEI) are prime moieties that show promising results by enhancing cholinergic activity and stabilizing cognitive function. However, the positive outcome of ChEI has been constrained by their short half-life and too many adverse events caused by the stimulation of cholinergic systems and hepatotoxicity. A quest for a much better therapy in treating AD with minimal side effects and sustainable action remains in progress (Woodward and Feldman, 2006; Farlow et al., 1993; Knapp, 1994; Rogers et al., 1998).

The search for a better anti-AD drug draws attention to the herbal approach of treatment and shows that natural derivatives can be a better alternative to conventional therapy in treating AD. Various research on phytoconstituents were undertaken and gave promising results in reversing the ill effects of neurodegeneration. Several scientific pieces of literature have focused on herbal plants and their efficacy in improving the symptoms of AD. Phytoconstituents from the medicinal plants help to restore the chemical balance by their inhibitory action of neurotransmitters. Some plants reported to have such ac-

Naturally Occurring Chemicals against Alzheimer's Disease. https://doi.org/10.1016/B978-0-12-819212-2.00014-1

tivity are *Bacopa monnieri* (Chatterji et al., 1965; Uabundit et al., 2010), *Centella asiatica* (Dhanasekaran et al., 2009), *Ginkgo biloba* (Bars, 1997), *Withania somnifera*, (Jayaprakasam et al., 2010) and *Panax ginseng* (Peng et al., 2002).

Protopine is a naturally occurring alkaloid of the isoquinoline class that has a 10-membered nitrogen-containing ring along with a carbonyl group (Wada et al., 2007). Protopine is a compound that is unparalleled compared with other molecules under research as a potent acetylcholinesterase with antioxidant, neuroprotective, and hepatoprotective action. This moiety can be considered as a secondary metabolite that is derived from isoquinoline during a biosynthetic process such as the reticulin pathway; it is included in many phytopreparations (Johns et al., 1969). Protopine is exploited as a pharmacophore and has potential use in AD. Table 3.1.12.1. lists the details for the different plant species from which protopine is isolated. Most are isolated from the Papaveraceae family and a few are from Fumariaceae. This chapter mainly documents the chemistry and anti-AD potential of protopine.

TABLE 3.1.12.1 Plant species from which protopine is isolated.

Plant botanical name	Family	References
Argemone platyceras	Papaveraceae	Siatka et al. (2017a,b)
Argemone mexicana	Papaveraceae	Chen et al. (2012), Capasso et al. (1997)
Corydalis humosa	Papaveraceae	Tao et al. (2005)
Corydalis yanhusuo	Papaveraceae	Lu et al. (2012)
Corydalis yanhusuo	Papaveraceae	Kim et al. (1999)
Fumaria schrammii	Fumariaceae	Vrancheva et al. (2016)
Fumaria indica	Fumariaceae	Tripathi et al. (1988)
Rhizoma corydalis	Papaveraceae	Sun et al. (2014)
Dactylicapnos scandens Hutch	Papaveraceae	Yan et al. (2004)
Eschscholzia californica	Papaveraceae	Vacek et al. (2010)
Corydalis bulbosa	Papaveraceae	Kaneko & Naruto (1969)
Chelidonium majus	Papaveraceae	Slavík et al. (1965)

Synthesis of protopine

The first chemical synthesis of protopine was performed by scientist Haworth along with cryptopine (Shamma, 1972a,b). Margni successfully synthesized 5,6,7,8-tetrahydro-7-methyldibenz[*c*,*g*]azecin-14(13*H*)-one, which is the parent molecular framework of protopine, using Perkin's method (Margni et al., 1970). This process includes tetrahydroprotoberberines as key intermediates. Other studies report the facile synthesis of the alkaloid through a multistep synthetic sequence starting from the ring enlargement of indeno[2,1-*a*]benzazepines (Knapp, 1994) through singlet oxygen oxygenation. The final step synthesis is the formation of a methylene group from the amide carbonyl group of the resultant 10-membered keto-lactam (Shamma, 1972a,b; Orito et al., 1980) (Fig. 3.1.12.1).

Crystal structure of protopine hydrochloride (salt of protopine)

Analysis of the crystal structure of protopine hydrochloride is relevant regarding its pharmacological and biosynthetic aspects. The crystal data of the compound were studied in detail and various details regarding the geometry and lattice parameters were documented. The study revealed that the salt is in a *trans* fused state (Dostál et al., 2001). Also, a detailed x-ray study of the free protopine molecule was reported in the literature and evidenced strong electrostatic interaction between the nitrogen atom in the 10-membered rings; it illustrated the position of the carbonyl carbon as geometrically opposite across the ring. This interaction is significant regarding the stability of 10-membered heterocyclic rings (Mottus et al., 1953). On the other hand, protopine hydrochloride (salt of protopine) exists as *cis* and *trans* isomers in a solution phase, as revealed by nuclear magnetic resonance studies (Takahashi et al., 1985).

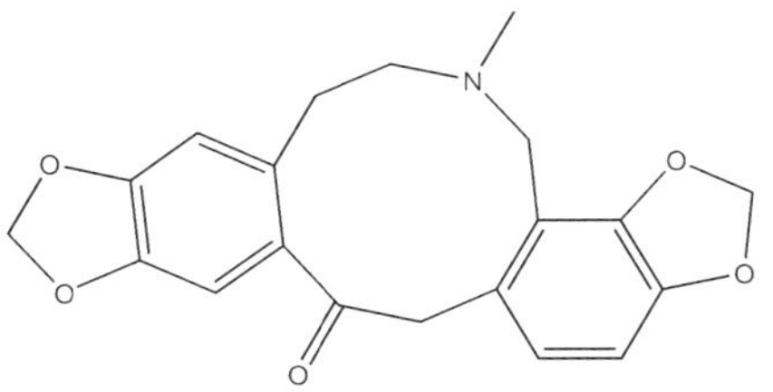

15-methyl-7,9,19,21-tetraoxa-15-azapentacyclo[15.7.0.04,12.06,10.018,22]tetracosa-1(17),4,6(10),11,18(22),23-hexaen-3-one

FIGURE 3.1.12.1 IUPAC name and structure of protopine.

Biological activities of protopine

Herbs containing protopine are used for various purposes in traditional medicine. Most of the pharmacological activities of protopine have been reported, including acetylcholinesterase inhibition (anti-AD) and hepatoprotective and anticancer activity.

Acetylcholinesterase inhibition/anti-Alzheimer's activity

AD has received tremendous attention in research owing to the increased global statistics of the disease over past decades. It is considered to be one of the most chronic classes of neurodegenerative disease. It progresses with time, affecting the quality of life of individuals and causing brain degeneration and death. Most patients are aged greater than 60 years. It is associated with the age-related death of astrocytes (brain cells) and the shrinkage of brain mass as the disease progresses to further stages.

A Chinese medicinal plant, *Dactylicapnos scandens Hutch*, also been ascertained as an inhibitor of serotonin transporter. Protopine isolated from this plant was subjected to 5-hydroxy-DL-tryptophan(5-HTP)-tail suspension and induced a head twitch response during a test. This study opens up further possibilities for the role of protopine as an antidepressant in mild and moderate states of depression (Xu et al., 2006).

Another study reported the promising effect of protopine on focal cerebral ischemic injury in a rat model. This was the first study that investigated the activity of protopine on cerebral ischemia. The authors induced focal cerebral ischemia by the intraluminal filament technique. The results were promising, and pretreatment with the drug decreased serum lactate dehydrogenase activity and the cerebral infarction ratio. The results also implied an improvement in the ischemia-induced neurological deficit score and histological changes of the brain in a dose-dependent manner. Detailed investigations revealed that protopine can also stimulate the activity of the enzyme superoxide dismutase in serum and can decrease total calcium. In rats with middle cerebral artery occlusion with ischemic brain tissue, the moiety can decrease terminal deoxynucleotidyl transferase-mediated dUTP nick end labeling—positive cells. This study suggested the possibility of developing a protopine-based drug for neuroprotective activities. A related study implied the antidementia activity of protopine in lipopolysaccharide-stimulated Raw 264.7 cells; the compound was active in reducing prostaglandin E2, cyclooxygenase-2, and nitric oxide (NO) production with no cytotoxicity. This study provided insight for correlating the antiinflammatory effect of protopine along with a decrease in the NO level for anti-AD activity, because NO and oxidative stress—associated free radicals are suspected to be involved in AD pathophysiology (Xiao et al., 2007).

The effect of protopine on the aggregation of platelet was investigated in detail using in vitro, in vivo, and ex vivo methods. This study analyzed the effect of protopine specifically on the metabolic system of cyclic adenosine monophosphate (AMP) in platelets. The study illustrated that protopine increased cyclic AMP content, activated enzyme adenylate cyclase, and thus affected platelet aggregation (Shiomoto et al., 1990).

Other pharmacological activity

Hepatoprotective activity

The hepatoprotective activity of protopine alkaloid was reported for the plant *Fumaria indica* and its hepatoprotective activity was investigated. The results were compared with the standard drug silymarin. The authors considered various factors such as lipid peroxidation (malondialdehyde content), reduced histological damage, and changes in serum enzymes such as Serum glutamic pyruvic transaminase (SGPT), serum glutamic-oxaloacetic transaminase (SGOT), Alkaline phosphatase (ALP), reduced glutathione, and the metabolite bilirubin. Protopine at doses of 10−20 mg provided significant results for which the isolate was found to be effective as a hepatoprotectant comparable to the action of the standard drug silymarin (Rathi et al., 2008). Another work examined whether protopine affected the expression of CYP1A enzymes in primary cultures of human hepatoma HepG2 cells and human hepatocytes (Vrba et al., 2011). This study also proved that the induction of CYP1A1 expression by protopine was either by mild or negligible activation of the aryl hydrocarbon receptor. Western blotting and ethoxyresorufin-O-deethylase (EROD) assay also confirmed that 2,3,7,8-Tetrachlorodibenzo-p-dioxin (TCDD), CYP1A messenger RNA levels induced by protopine in both HepG2 cells and human hepatocytes, could not elevate the CYP1A protein level. This result provides valuable insight for the safe use of protopine regarding the possible induction of CYP1A enzymes. Janbaz and colleagues reported the similar activity of protopine to prevent hepatotoxicity in rodents induced by paracetamol and CCl_4. The authors suggested an inhibitory effect on microsomal drug metabolizing enzymes indicating an antihepatotoxic action of protopine (Janbaz et al., 2002).

Antioxidant and anticancer activity

It is well-known that phytochemicals have potent antitumor and antioxidant activity (Dev et al., 2016). The role of protopine as a microtubule-stabilizing agent was illustrated by Chen and group. It can cause mitotic arrest and apoptosis in human prostate cancer cell lines. In that study, protopine acted through an antiproliferative manner, resulting in the apoptosis of cancer cells.

The results suggest that the mechanism of apoptosis is through the pathway by toning the CDK-1 activity and BCl-2 family of proteins (Chen et al., 2012). A related study identified protopine as the anticancer component for the herb *Corydalis yanhusuo*, which is used in cancer treatment in Chinese traditional medicine. In that work, the authors suggested that protopine can alter the expression of adhesive factors and thus inhibit human umbilical vein endothelial cells the heterotypic cell adhesion between MDA-MB-231 cells (He and Gao, 2014). The authors investigated five major alkaloids present in the herb for anticancer activity; of those, protopine was the only one to show an anticancer effect.

A study revealed that protopine from *Fumaria schleicheri* showed moderate anticonvulsant activity (Prokopenko et al., 2015). Protopine from corydalis exhibited several cardiovascular effects such as antihypertensive action and antiarrhythmic and negative inotropic effects found to have a promiscuous outcome on cation channel currents (Song et al., 2000).

Concluding remarks

AD is known to be a complex degenerative disease affecting the nervous system because of its multicausal and multifactorial pathogenesis. Thus, it is preferable to have a therapy that focuses on polypharmacological aspects to address the complexity of the disease. ChEI is a favorite choice for the treatment of AD. Interestingly, a natural compound, isolated protopine, was shown to exhibit cholinesterase inhibitory action as well as neuroprotective, antioxidant, and hepatoprotective action. This noncholinergic effect of protopine can be adapted for better management of the neurodegenerative disease by interfering with the major factors causing AD. Preclinical and clinical trial data show that more exciting research needs to be carried out to determine the efficacy and safety profile of protopine.

References

Bars, P.L.L.A., 1997. Placebo-controlled, double-blind, randomized trial of an extract of ginkgo biloba for dementia. J. Am. Med. Assoc. 278 (16), 1327–1332.

Capasso, A., Piacente, S., Pizza, C., Tommasi, N.D., Jativa, C., Sorrentino, L., 1997. Isoquinoline Alkaloids from Argemone Mexicana reduce morphine withdrawal in Guinea pig isolated ileum. Planta Med. 63 (04), 326–328.

Chatterji, N., Rastogi, R.,P., Dhar, M.L., 1965. Chemical examination of *Bacopa monniera* Wettst: parti-isolation of chemical constituents. Indian J. Chem. 3, 24.

Chen, C.H., Liao, C.H., Chang, Y.L., Guh, J.H., Pan, S.L., Teng, C.M., 2012. Protopine, a novel microtubule-stabilizing agent, causes mitotic arrest and apoptotic cell death in human hormone-refractory prostate cancer cell lines. Canc. Lett. 315 (1), 1–11.

Dev, S, Dhaneshwar, R.S., Mathew, B., 2016. Discovery of Camptothecin Based Topoisomerase I Inhibitors: Identification Using an Atom Based 3D-QSAR, Pharmacophore Modeling, Virtual Screening and Molecular Docking Approach. Comb. Chem. High Throughput. Screen 19, 752–763.

Dhanasekaran, M., Holcomb, L.A., Hitt, A.R., Tharakan, B., Porter, J.W., Young, K.A., Manyam, B.V., 2009. Centella asiaticaextract selectively decreases amyloidβlevels in Hippocampus of Alzheimers disease animal model. Phytother Res. 23 (1), 14–19.

Dostál, J., Žák, Z., Nečas, M., Slavík, J., Potáček, M., 2001. Protopine hydrochloride. Acta Crystallogr. Sect. C Cryst. Struct. Commun. 57 (5), 651–652.

Farlow, M., Hershey, L.A., Sadowsky, C.H., Gracon, S.I., Lewis, K.W., Dolan-Ureno, J., 1993. Tacrine in Alzheimers disease-reply. J. Am. Med. Assoc. 269 (22), 2849–2850.

He, K., Gao, J.L., 2014. Protopine inhibits heterotypic celladhesion in MDA-MB-231 cells through down-regulation of multi-adhesive factors. Afr. J. Tradit., Complementary Altern. Med. 11 (2), 415–424.

Janbaz, K.H., Saeed, S.A., Gilani, A.H., 2002. Protective effect of rutin on Paracetamol- and CCl4-induced hepatotoxicity in Rodents. Fitoterapia 73 (7–8), 557–563.

Jayaprakasam, B., Padmanabhan, K., Nair, M.G., 2010. Withanamides InWithania somniferafruit protect PC-12 cells from Î²-amyloid responsible for Alzheimers disease. Phytother Res. 24, 859–863.

Johns, S., Lamberton, J., Tweeddale, H., Willing, R., 1969. Alkaloids of *zanthoxylum conspersipunctatum* (rutaceae): the structure of a new alkaloid isomeric with protopine. Aust. J. Chem. 22 (10), 2233.

Kaneko, H., Naruto, S., 1969. Constituents of corydalis species. VI. Alkaloids from Chinese corydalis and the identity of d-corydalmine with d-corybulbine. J. Org. Chem. 34 (9), 2803–2805.

Kim, S., Hwang, S., Jang, Y., Park, M., Markelonis, G., Oh, T., Kim, Y., 1999. Protopine from corydalis ternata has anticholinesterase and antiamnesic activities. Planta Med. 65 (3), 218–221.

Knapp, M.J., 1994. A 30-week randomized controlled trial of high-dose tacrine in patients with Alzheimers disease. The tacrine study group. J. Am. Med. Assoc. 271 (13), 985–991.

Lu, Z., Sun, W., Duan, X., Yang, Z., Liu, Y., Tu, P., 2012. Chemical constituents from Corydalis yanhusuo. Zhongguo Zhongyao Zazhi 37 (2), 235–237.

Margni, A.L., Giacopello, D., Deulofeu, V., 1970. Synthesis of 5, 6, 7, 8-tetrahydro-7-methyldibenz [c, g] azecin-14 (13 H)-one: the structural skeleton of the protopine alkaloids. J. Chem. Soc. C Org. 18, 2578–2580.

Mottus, E., Schwarz, H., Marion, L., 1953. Interaction of keto- and tertiary amino-function in medium rings. Can. J. Chem. 31 (11), 1144–1151.

Orito, K., Kurokawa, Y., Itoh, M., 1980. Cheminform Abstract: on the synthetic approach to the protopine alkaloids. Chemischer Informationsdienst 11, 24.

Peng, Q., Buz'Zard, A., Lau, B., 2002. Pycnogenol® protects neurons from amyloid-β peptide-induced apoptosis. Mol. Brain Res. 104 (1), 55–65.

Prokopenko, Y., Tsyvunin, V., Shtrygol', S., Georgiyants, V., 2015. In vivo anticonvulsant activity of extracts and protopine from the fumaria schleicheri herb. Sci. Pharm. 84 (3), 547–554.

Rathi, A., Srivastava, A., Shirwaikar, A., Singh Rawat, A., Mehrotra, S., 2008. Hepatoprotective potential of fumaria indica pugsley whole plant extracts, fractions and an isolated alkaloid protopine. Phytomedicine 15 (6–7), 470–477.

Rogers, S.L., Farlow, M.R., Doody, R.S., Mohs, R., Friedhoff, L.T., 1998. A 24-week, double-blind, placebo-controlled trial of donepezil in patients with Alzheimers disease. Neurology 50 (1), 136–145.

Shamma, M., 1972a. Preface. Organic Chemistry the Isoquinoline Alkaloids - Chemistry and Pharmacology.

Shamma, M., 1972b. The protopines. Organic Chemistry The Isoquinoline Alkaloids - Chemistry and Pharmacology, pp. 344–358.

Shiomoto, H., Matsuda, H., Kubo, M., 1990. Effects of protopine on blood platelet aggregation. Ii. Effect on metabolic system of adenosine 3',5'-Cyclic monophosphate in platelets. Chem. Pharmaceut. Bull. 38 (8), 2320–2322.

Siatka, T., Adamcová, M., Opletal, L., Cahlíková, L., Jun, D., Hrabinová, M., Chlebek, J., 2017a. Cholinesterase and prolyl oligopeptidase inhibitory activities of alkaloids from *Argemone platyceras* (Papaveraceae). Molecules 22 (7), 1181.

Siatka, T., Adamcová, M., Opletal, L., Cahlíková, L., Jun, D., Hrabinová, M., Kuneš, J., Chlebek, J., 2017b. Cholinesterase and prolyl oligopeptidase inhibitory activities of alkaloids from *Argemone platyceras* (Papaveraceae). Molecules 22 (7), 1181.

Slavík, J., Slavíková, L., Brabenec, J., 1965. Alkaloide der mohngewächse (papaveraceae) XXX. Über weitere alkaloide aus der wurzel von *Chelidonium Majus* L. Collect. Czech Chem. Commun. 30 (11), 3697–3704.

Song, L., Ren, G., Chen, Z., Chen, Z., Zhou, Z., Cheng, H., 2000. Electrophysiological effects of protopine in cardiac myocytes: inhibition of multiple cation channel currents. Br. J. Pharmacol. 129 (5), 893–900.

Sun, M., Liu, J., Lin, C., Miao, L., Lin, L., 2014. Alkaloid profiling of the traditional Chinese medicine rhizoma corydalis using high performance liquid chromatography-tandem quadrupole time-of-flight mass spectrometry. Acta Pharm. Sin. B 4 (3), 208–216.

Takahashi, H., Iguchi, M., Onda, M., 1985. Utilization of protopine and related alkaloids. XVII. Spectroscopic studies on the ten-membered ring conformations of protopine and .ALPHA.-Allocryptopine. Chem. Pharmaceut. Bull. 33 (11), 4775–4782.

Tao, J., Zhang, X., Ye, W., Zhao, S., 2005. Chemical constituents from Corydalis humosa. Zhong Yao Cai 28 (7), 556–557.

Tripathi, Y., Pandey, V., Pathak, N., Biswas, M., 1988. A seco-phthalideisoquinoline alkaloid from fumaria indica seeds. Phytochemistry 27 (6), 1918–1919.

Uabundit, N., Wattanathorn, J., Mucimapura, S., Ingkaninan, K., 2010. Cognitive enhancement and neuroprotective effects of Bacopa monnieri in alzheimers disease model. J. Ethnopharmacol. 127 (1), 26–31.

Vacek, J., Walterová, D., Vrublová, E., Šimánek, V., 2010. The Chemical and biological properties of protopine and allocryptopine. Heterocycles 81 (8), 1773.

Vrancheva, R.Z., Ivanov, I.G., Aneva, I.Y., Dincheva, I.N., Badjakov, I.K., Pavlov, A.I., 2016. Alkaloid profiles and acetylcholinesterase inhibitory activities of fumaria species from Bulgaria. Z. Naturforsch. C Biosci. 71 (1–2), 9–14.

Vrba, J., Vrublova, E., Modriansky, M., Ulrichova, J., 2011. Protopine and allocryptopine increase MRNA levels of cytochromes P450 1A in human hepatocytes and HepG2 cells independently of AhR. Toxicol. Lett. 203 (2), 135–141.

Wada, Y., Kaga, H., Uchiito, S., Kumazawa, E., Tomiki, M., Onozaki, Y., Orito, K., 2007. On the synthesis of protopine alkaloids. J. Org. Chem. 72 (19), 7301–7306.

Woodward, M., Feldman, H.H., 2006. Drug treatment: cholinesterase inhibitors. Severe Dementia 131–149.

Xiao, X., Liu, J., Hu, J., Li, T., Zhang, Y., 2007. Protective effect of protopine on the focal cerebral ischaemic injury in rats. Basic Clin. Pharmacol. Toxicol. 101 (2), 85–89.

Xu, L., Chu, W., Qing, X., Li, S., Wang, X., Qing, G., Fei, J., Guo, L., 2006. Protopine inhibits serotonin transporter and noradrenaline transporter and has the antidepressant-like effect in mice models. Neuropharmacology 50 (8), 934–940.

Yan, T.Q., Yang, Y.F., Ai, T.M., 2004. Determination of protopine and isocorydine in root of Dactylicapnos scandens by HPLC. Zhongguo Zhongyao Zazhi 29 (10), 961–963.

Chapter 3.1.13

Spinosin

Jessica Pandohee, Mohamad Fawzi Mahomoodally
Department of Health Sciences, Faculty of Medicine and Health Sciences, University of Mauritius, Réduit, Mauritius

Introduction

Alzheimer's disease (AD) is one of the main causes of dementia and affects mainly senior people in their sixties or older. As the name suggests, AD affects the nerve cells, resulting in the degeneration of neurons in the brain. As time goes on, the disease attacks the limited neurons in the brain, slowly leading to decades of incapability and eventually death. A report from the Global Burden of Diseases, Injuries, and Risk Factors published in 2019 (Nichols et al., 2019) stated that there were 43.8 million individuals with dementia in 2016, that more women than men lived with the disease, and that a total of 2.4 million deaths made dementia the fifth leading cause of death around the world. In the United States alone, 5.8 million Americans were reported to have AD, accompanied with a staggering increase in deaths of 145% from 2000 to 2017 compared with a decrease in deaths resulting from stroke, heart disease, and prostate cancer (2019 Alzheimer's disease, 2019).

Although AD is mainly detected in individuals aged over 60 years, the illness is not considered to be part of the normal aging process. Depending on how advance the disease is, AD patients face a dark reality including a loss of physical ability, personality, and state of mental health. The development of AD has been classified into three main steps. In the early stage, the onset of AD appears as forgetfulness, losing track of time, possible difficulty with communication, and difficulty making decisions and personal finances, among others. In the middle stages of AD, the limitations become clearer and more restricting. Patients often have difficulty with speech, comprehension, personal care, preparing food, or being alone safely without considerable support. The last stage is the most deteriorated phase, in which patients are unaware of time and place; unable to recognize relatives and friends; unable to eat without assistance, and have mobility, bowel, and bladder incontinence. Symptoms vary among individuals and can be chronic or progressive, which in the end

Naturally Occurring Chemicals against Alzheimer's Disease. https://doi.org/10.1016/B978-0-12-819212-2.00015-3

leads to death owing to the gradual deterioration and reduction in cognitive functions.

AD and dementia were declared a major health issue globally by the World Health Organization (Global action, 2017). AD is such a health threat because the challenges arising from it are multifaceted. Because there is no cure for AD or treatment to reduce the rate of the disease progress or reverse damage done to the brain, the only approach to enable AD patients to have a life as comfortable as possible is to minimize the effect and consequences of AD's symptoms, often with the constant help of a caregiver or health officer. The public health impact from AD is tremendous and was estimated to have cost US $818 billion from direct medical, social, and informal care expenditures in 2015 (Global action, 2017). AD is a burden to the patient, because the latter spends much time in a state of disability and dependence. Moreover, as the baby boom generation gets older, and with the aging population throughout the world, the health care expenditures will only increase dramatically (Taylor and Sloan, 2000; Lin et al., 2019; Sopina et al., 2019). Indirectly, AD leads to a loss of income as the patient slowly loses the capability to control the body. In low-income countries, where family care predominates, the loss of income by both the patient and relatives negatively affects productivity for economies (Sopina et al., 2019).

Causes and mechanism of Alzheimer's disease

Only recently has there been accumulating evidence that AD begins decades before the symptoms become apparent (Jack et al., 2009; Braak et al., 2011; Gordon et al., 2018). When AD commences in an individual, the latter does not notice the subtle changes happening in the brain. The clinical manifestations appear only when it is too late for the patient and people around to reverse the increasing gravity of the symptoms. Until now, the definite causes of AD have remained unknown. Significant bodies of research have shown the link between the importance of the amyloid precursor protein (APP) locus situated on chromosome 21 on early-onset AD (Rovelet-Lecrux et al., 2006; Tanzi, 2012; Campion et al., 1999) and the peptide amyloid beta (Aβ) (Haass et al., 1995; Bird, 2008). Chromosomal (APP locus) is believed to contribute to less than 1% of AD cases. Roughly 25% is believed to result from familial causes, whereas the origins of 75% of cases are unknown and might involve several factors such as environmental interactions (Bird, 2008).

Postmortem biopsies of AD patients have showed the vast growth of plaques and tangles in the human brain, which have been characterized as Aβ. Aβ is series of 36–43 amino acids abnormally produced by the *trans*-membrane protein APP. At high concentrations, Aβ aggregates to form amyloid fibrils, which deposit outside neurons and in the walls of small blood vessels in the brain. This can interfere with communication at the neuron's synapses. An excess of the tau protein (microtubule-associated protein) has been observed in the brains of AD patients, and this accumulation, which results from misfolding of native proteins, is believed to lead to neurofibrillary degeneration as well as block the transport of nutrients inside neurones.

Diagnosis of Alzheimer's disease

According to the *Diagnostic and Statistical Manual of Mental Disorders*, Fifth Edition, AD is diagnosed by the progression of several defects displayed by memory impairment and cognitive defects affecting social and/or professional performance, which can be observed through a sudden and noteworthy decrease from a previous level of function (Association, 2013). In addition, the cognitive deficits should not have been caused by other nervous issues leading to progressive deficits in memory or cognition, systemic conditions associated with dementia, or health issues resulting from AD-causing specific substances (Association, 2013). Although the definite diagnosis of AD can be made only with a brain biopsy or an autopsy, diagnosis during the lifetime of an individual usually involves months of tests combining brain imaging and clinical assessments.

Physical and neurological examinations include amnestic presentation (impairment in learning and remembering), nonamnestic presentation (for example word-finding), visuospatial presentation (impaired face recognition), and executive dysfunction (impaired reasoning, judgment, and problem solving). The overall aim is to check for signs of memory impairment. A large part of the AD diagnosis process involves excluding other diseases that would have similar symptoms as AD. Medical officers evaluate other possible causes of memory loss, and other diseases such as medication-induced dementia; metabolic, endocrine, or nutritional disorder; brain tumors; hematoma; and depression need to be ruled out (Bature et al., 2017).

Laboratory assessments involves electrophysiologic methods, computed tomography, regional cerebral blood flow, positron emission tomography, magnetic resonance imaging, and an examination of blood fluids and nonneural tissues (McKhann et al., 1984; Richards and Hendrie, 1999). A healthy adult usually has an endogenous potential such as P300 that reflects the speed of cognition. However, the latency of P300 changes with age and 50%−80% of AD patients show an increase in latency of P300 that can be detected as an increase in slow-wave activity. Computed tomography gives a snapshot of the brain and allows measurements of the widths and sizes of the human brain areas. This allows the exclusion of other diseases such as hematoma, brain tumor, and dementia associated with vascular disease, because the volume of the ventricular system and width of the third ventricle in AD are increased whereas the gyri are narrowed and sulci are widened. AD can also be differentiated from cerebrovascular diseases by measuring regional cerebral blood flow, because autoregulation in cerebrovascular patients is decreased but is preserved in AD patients. Examination of the cerebrospinal fluid and blood for

AD biomarkers is also under development. However, definitive assessment of AD through this avenue is still impossible. As a result, AD diagnosis still remains expensive.

Current therapies for Alzheimer's disease

Currently, no cure or treatment for AD is available, which means that patients have no choice but to live with the worsening symptoms, which eventually lead to death. In some cases, short-term improvement has been achieved through pharmacologic therapy, providing 6–18 months of relief. Cholinesterase inhibitors and memantine are the main approved drugs for short-term alleviation of AD symptoms (Anand and Singh, 2013). Cholinesterase inhibitors block the enzyme that breaks down acetylcholine, which is believed to improve brain function. A benefit of cholinesterase inhibitors is improvements in cognition, behavior, and body function, which indicates its usefulness in the early stages of AD (Richards and Hendrie, 1999; Yiannopoulou and Papageorgiou, 2013). Another drug that has been approved is memantine, a glutamatergic agent aimed at preserving glutamatergic receptors *N*-methyl-D-aspartate (Robinson and Keating, 2006; Farrimond et al., 2012). However, there are limited studies that prove the safety and clinical efficacy of glutamatergic agents.

Still under clinical trials, immunotherapy targeting the removal of Aβ from the brain is showing promising results in animal models. Initially active immunotherapy in a phase I human clinical trial showed that a significant percentage of patients developed antibodies to Aβ after multiple doses of Aβ42 in adjuvant (AN1792 and QS-21) (Delrieu et al., 2012). Unfortunately, further evaluations of the vaccine efficacy revealed serious adverse effects. Many patients developed meningoencephalitis (Gilman et al., 2005). Therefore, there is still plenty of research developing active therapy to remove the risk for meningoencephalitis (Pride et al., 2008).

Spinosin

Spinosin is a plant metabolite that is extracted from the seeds of the fruit *Ziziphus jujuba* var. *spinosa*. It has been characterized as 2″-β-O-glucopyranosyl swertisin, a C-glycoside flavonoid with the molecular formula $C_{28}H_{38}O_{15}$. Unlike other well-known flavonoids, spinosin has not received as much academic and Western attention for its health benefits although it is widely used in many countries in Asia including China, Japan, and Korea. The seeds of *Z. jujuba* var. *spinosa* contain various metabolites with biological activity, such as flavonoids, triterpenoid saponins, and alkaloids, and have been used in traditional Chinese medical practice as a tranquilizer and to treat insomnia. The first known report of spinosin was in 1987 by Yuan (Yuan, 1987), who described the sedative and hypnotic properties of the seeds. Since then, there have been several chemical investigations to characterize the seeds and spinosin and other derivatives (Xie et al., 2011; Kim et al., 2014; Wang et al., 2013).

To shine more light on the health benefits of spinosin, various research groups across the Asian continent have been investigating the link between spinosin and diseases such as AD (Fanxing et al., 2019; Jung et al., 2014; Sang Yoon et al., 2015; Lee et al., 2016), sleep initiation, and maintenance disorder (He et al., 2012; Wang et al., 2008), as well as genes such as 5-hydrotryptamine receptor 1A (Jung et al., 2014; Wang et al., 2012), brain-derived neurotrophic factor (Fanxing et al., 2019; Lee et al., 2016), and glutamate ionotropic receptor kainate type subunit 3 (Jiao et al., 2017). There has been increasing evidence supporting spinosin's pharmacological activities. All of these studies suggest that spinosin is a potential chemical against AD.

Potential of spinosin against Alzheimer's disease

The frontal cortex, hippocampus, and septum are rich with 5-hydroxytryptamine 1A (5-HT_{1A}) receptors and are the zones of the brain responsible for learning and memory. Antagonism of the 5-HT_{1A} receptor reverses cognitive impairment supported by cholinergic or glutamatergic blockade. Therefore, compounds that could act as a 5-HT_{1A} receptor antagonist would be a potentially competitive solution to treat cognitive deficit–associated diseases such as AD.

One of the earliest pharmacological investigations of spinosin was carried out by Wang et al. (2008) in 2008 and showed early signs of the mechanism of action of spinosin on the mouse brain. First, the effects of spinosin are dose-dependent. At a dose of 45 mg/kg pentobarbital, spinosin provided an increased sleep time and reduced sleep latency, whereas at the subhypnotic dose of 28 mg/kg pentobarbital, spinosin increased the rate of sleep onset and exhibited a synergistic effect with 5-hydroxytryptophan. They also observed an augmentative effect with 5-hydroxytryptophan on both sleep latency and sleep time in mice treated with hypnotic doses of pentobarbital and inhibited *para*-chlorophenylalanine-induced suppression of pentobarbital-induced hypnosis. This work (Wang et al., 2008) concluded that spinosin potentiated the hypnotic effect of pentobarbital-induced sleep via a serotonergic mechanism.

Moreover, the same group of researchers later (Wang et al., 2010) hypothesized that spinosin could potentiate pentobarbital-induced sleep via serotonin-1A (5-HT_{1A}) receptors. The results demonstrated that spinosin significantly augmented pentobarbital (35 mg/kg, intraperitoneally [i.p.]) induced sleep in rats, reflected by reduced sleep latency and increased total sleep time, nonrapid eye movement (NREM) sleep time, and rapid eye movement (REM) sleep time. With regard to NREM sleep duration, spinosin mainly increased slow-wave sleep (SWS). In addition, spinosin (15 mg/kg, intragastrically) significantly antagonized 5-HT_{1A} agonist 8-OH-DPAT

(0.1 mg/kg, i.p.) induced reductions in total sleep time, NREM sleep, REM sleep, and SWS in pentobarbital-treated rats. Drummond and Brown (2001) and Graves et al. (2003) linked sleep with learning and memory function, suggesting that sleep deprivation may modify the molecular mechanism of memory storage. The significance that spinosin can help with sleep mechanism and molecules such as 5-hydroxytryptophan and 5-HT_{1A} receptors means that further testing is required to understand the role of spinosin in the brain chemistry fully and how it can help in AD.

After the contribution of Wang et al. (2008), Wang et al. (2012) to the role of spinosin and its mechanism of action, Jung et al. (2014) evaluated the memory-ameliorating effect of spinosin on mice and showed that a dose of 10 or 20 mg/kg of spinosin enabled a prolonged latency time in the passive avoidance task, an increased percentage of spontaneous alternation in the Y-maze task, and a lengthened swimming time in the target quadrant in the Morris water maze task. Moreover the memory-ameliorating effects of spinosin were mediated, in part, by 5-HT_{1A} receptor signaling, and a significant increase in the expression levels of phosphorylated extracellular signal-regulated kinases and cyclic adenosine monophosphate response element-binding proteins in the hippocampus was reported (Jung et al., 2014). These findings indicate that spinosin may be a key flavonoid in the treatment of AD.

Another approach that could potentially lead to treatment of AD involves adult neurogenesis. Adult neurogenesis is the formation of functional, mature neurons from neural stem cells; it has an important role in learning, memory, emotion, stress, depression, and response to injury. This process usually occurs in specific regions of the brain including the subventricular zone of the lateral ventricle and the subgranular zone (SGZ) of the hippocampal dentate gyrus (DG). Lee et al. (2016) investigated whether spinosin affected cognitive performance in healthy adult rodents and reported that subchronic application of low-dose spinosin over 2 weeks significantly increased the latency time in the passive avoidance task. Another finding was that spinosin increased the proliferation of neural stem cell (NSC) and the survival of newborn cells in the hippocampal DG region. Bromodeoxyuridine (BrdU) immunostaining showed that subchronic administration of spinosin significantly increased the number of BrdU-incorporated cells in the SGZ compared with the administration vehicle and that smaller doses (2.5 mg/kg) also increased the number of BrdU-positive cells but was not significant.

More interestingly, Ko et al. (2015) tested spinosin as a therapeutic strategic against AD and completed the task by characterizing the ameliorating effects of spinosin on memory impairment or the pathological changes induced through $A\beta_{1\text{-}42}$ oligomer (AβO) in mice. The authors showed that among a cohort of previously AβO-infected (50-μM) mice treated with spinosin at a dose of 5, 10, and 20 mg/kg for 7 days, a subchronic dose of 20 mg/kg of spinosin significantly improved AβO-induced cognitive impairment, and that spinosin treatment did not affect general locomotor activity. Furthermore, spinosin treatment reduced the number of activated microglia and astrocytes observed after AβO injection, and spinosin rescued the AβO-induced decrease in choline acetyltransferase expression levels. These findings may be among the first reports of the removal of AβO in an infected brain, confirming the role of spinosin as a neuroprotective and antiinflammatory agent.

Fanxing et al. (2019) also used a mouse model of AD to investigate the use of spinosin to protect neurons to achieve the recovery of learning and memory and therefore the enhancement of cognitive impairment. Mice with AD induced by AβO intracerebroventricular injection were administered 10 and 100 μg/kg per d spinosin over the course of a week. The authors showed that mice treated with both 10 and 100 μg/kg spinosin demonstrated improved AβO-induced cognitive impairment though long-term memory evaluated by the Morris maze and accumulation of AβO in the hippocampus. Treatment with spinosin also showed a decrease in the level of malondialdehyde in the hippocampus. Their results also showed the upregulation of brain-derived neurotrophic factor and reversal of the level of Bcl-2 in the hippocampus and cerebral cortex of the infected mice at 100 μg/kg per d only, which suggested that spinosin may have antiapoptic activity.

Future directions and conclusion

There has been an increasing amount of evidence supporting the potential of spinosin against AD. It is obvious that spinosin reverses cognitive impairment, improves memory and learning, and reduces the $A\beta_{1\text{-}42}$ oligomer in the brain of AD-induced mice. Although the entire mechanism of action of spinosin on the brain remains unknown, advances in showing that a naturally occurring compound such as spinosin can help fight against AD provide hope for individuals with AD, as well as healthy individuals who might avoid the onset of AD simply by consuming spinosin-containing food.

References

2019 Alzheimer's disease, 2019. Facts and figures. Alzheimer's Dement. 15 (3), 321–387.

Anand, P., Singh, B., 2013. A review on cholinesterase inhibitors for Alzheimer's disease. Arch Pharm. Res. (Seoul) 36 (4), 375–399.

Association, A.P., 2013. Diagnostic and Statistical Manual of Mental Disorders (DSM-5®) (Arlington, Texas.).

Bature, F., et al., 2017. Signs and symptoms preceding the diagnosis of Alzheimer's disease: a systematic scoping review of literature from 1937 to 2016. BMJ Open 7 (8), e015746.

Bird, T.D., 2008. Genetic aspects of Alzheimer disease. Genet. Med. 10, 231.

Braak, H., et al., 2011. Stages of the pathologic process in Alzheimer disease: age categories from 1 to 100 years. J. Neuropathol. Exp. Neurol. 70 (11), 960–969.

Campion, D., et al., 1999. Early-onset autosomal dominant Alzheimer disease: prevalence, genetic heterogeneity, and mutation spectrum. Am. J. Hum. Genet. 65 (3), 664–670.

Delrieu, J., et al., 2012. Clinical trials in Alzheimer's disease. Immunother. Approaches 120 (s1), 186−193.

Drummond, S.P.A., Brown, G.G., 2001. The effects of total sleep deprivation on cerebral responses to cognitive performance. Neuropsychopharmacology 25 (1), S68−S73.

Fanxing, X., et al., 2019. Neuroprotective effects of spinosin on recovery of learning and memory in a mouse model of Alzheimer's disease. Biomol. Ther. 27 (1), 71−77.

Farrimond, L.E., Roberts, E., McShane, R., 2012. Memantine and cholinesterase inhibitor combination therapy for Alzheimer's disease: a systematic review. BMJ Open 2 (3), e000917.

Gilman, S., et al., 2005. Clinical effects of Aβ immunization (AN1792) in patients with AD in an interrupted trial. Neurology 64 (9), 1553.

Global Action Plan on the Public Health Response to Dementia 2017-2025, in, 2017. World Health Organization, Geneva. Licence: CC BY-NC-SA 3.0 IGO.

Gordon, B.A., et al., 2018. Spatial patterns of neuroimaging biomarker change in individuals from families with autosomal dominant Alzheimer's disease: a longitudinal study. Lancet Neurol. 17 (3), 241−250.

Graves, L.A., et al., 2003. Sleep deprivation selectively impairs memory consolidation for contextual fear conditioning. Learn. Mem. 10 (3), 168−176.

Haass, C., et al., 1995. The Swedish mutation causes early-onset Alzheimer's disease by β-secretase cleavage within the secretory pathway. Nat. Med. 1 (12), 1291−1296.

He, B., et al., 2012. A UFLC-MS/MS method for simultaneous quantitation of spinosin, mangiferin and ferulic acid in rat plasma: application to a comparative pharmacokinetic study in normal and insomnic rats. J. Mass Spectrom. 47 (10), 1333−1340.

Jack Jr., C.R., et al., 2009. Serial PIB and MRI in normal, mild cognitive impairment and Alzheimer's disease: implications for sequence of pathological events in Alzheimer's disease. Brain 132 (5), 1355−1365.

Jiao, L., et al., 2017. Degradation kinetics of 6‴-p-coumaroylspinosin and identification of its metabolites by rat intestinal flora. J. Agric. Food Chem. 65 (22), 4449−4455.

Jung, I.H., et al., 2014. Ameliorating effect of spinosin, a C-glycoside flavonoid, on scopolamine-induced memory impairment in mice. Pharmacol. Biochem. Behav. 120, 88−94.

Kim, W.I., et al., 2014. Quantitative and pattern recognition analyses of magnoflorine, spinosin, 6‴-feruloyl spinosin and jujuboside A by HPLC in Zizyphi Semen. Arch. Pharm. Res. 37 (9), 1139−1147.

Ko, S.Y., et al., 2015. Spinosin, a C-glucosylflavone, from *Zizyphus jujuba* var. spinosa ameliorates Aβ1-42 oligomer-induced memory impairment in mice. Biomol. Ther. 23 (2), 156−164.

Lee, Y., et al., 2016. Spinosin, a C-glycoside flavonoid, enhances cognitive performance and adult hippocampal neurogenesis in mice. Pharmacol. Biochem. Behav. 145, 9−16.

Lin, P.-J., et al., 2019. Family and caregiver spillover effects in cost-utility analyses of Alzheimer's disease interventions. Pharmacoeconomics 37 (4), 597−608.

McKhann, G., et al., 1984. Clinical diagnosis of Alzheimer's disease. Neurology 34 (7), 939.

Nichols, E., et al., 2019. Global, regional, and national burden of Alzheimer's disease and other dementias, 1990-2016: a systematic analysis for the Global Burden of Disease Study 2016. Lancet Neurol. 18 (1), 88−106.

Pride, M., et al., 2008. Progress in the active immunotherapeutic approach to Alzheimer's disease: clinical investigations into AN1792-associated meningoencephalitis. Neurodegener. Dis. 5 (3−4), 194−196.

Richards, S.S., Hendrie, H.C., 1999. Diagnosis, management, and treatment of Alzheimer disease: a guide for the internist. JAMA Intern. Med. 159 (8), 789−798.

Robinson, D.M., Keating, G.M.J.D., 2006. Memantine 66 (11), 1515−1534.

Rovelet-Lecrux, A., et al., 2006. APP locus duplication causes autosomal dominant early-onset Alzheimer disease with cerebral amyloid angiopathy. Nat. Genet. 38 (1), 24–26.

Sang Yoon, K., et al., 2015. Spinosin, a C-glucosylflavone, from *Zizyphus jujuba* var. spinosa ameliorates Aβ1-42 oligomer-induced memory impairment in mice. Biomol. Ther. 23 (2), 156–164.

Sopina, E., et al., 2019. Long-term medical costs of Alzheimer's disease: matched cohort analysis. Eur. J. Health Econ. 20 (3), 333–342.

Tanzi, R.E., 2012. The genetics of Alzheimer disease. Acta Neuropathol. 2 (10).

Taylor Jr., D.H., Sloan, F.A., 2000. How much do persons with Alzheimer's disease cost medicare? J. Am. Geriatr. Soc. 48 (6), 639–646.

Wang, L.-E., et al., 2008. Spinosin, a C-glycoside flavonoid from semen *Zizhiphi* Spinozae, potentiated pentobarbital-induced sleep via the serotonergic system. Pharmacol. Biochem. Behav. 90 (3), 399–403.

Wang, L.E., et al., 2010. Potentiating effect of spinosin, a C-glycoside flavonoid of Semen *Ziziphi* spinosae, on pentobarbital-induced sleep may be related to postsynaptic 5-HT1A receptors. Phytomedicine 17 (6), 404–409.

Wang, L.-E., et al., 2012. Augmentative effect of spinosin on pentobarbital-induced loss of righting reflex in mice associated with presynaptic 5-HT1A receptor. J. Pharm. Pharmacol. 64 (2), 277–282.

Wang, B., et al., 2013. New spinosin derivatives from the seeds of *Ziziphus mauritiana*. Nat. Prod. Bioprosect. 3 (3), 93–98.

Xie, Y.-Y., et al., 2011. A novel spinosin derivative from semen *Ziziphi* spinosae. J. Asian Nat. Prod. Res. 13 (12), 1151–1157.

Yiannopoulou, K.G., Papageorgiou, S.G., 2013. Current and future treatments for Alzheimer's disease. Ther. Adv. Neurol. Disord. 6 (1), 19–33.

Yuan, C.L., 1987. [Sedative and hypnotic constituents of flavonoids in the seeds of *Ziziphus spinosae*]. Zhango Yao Tong Bao 12 (9), 34–36.

Chapter 3.1.14

Nobiletin

Hari Prasad Devkota[1], Anjana Adhikari-Devkota[1], Amina Ibrahim Dirar[1], Tarun Belwal[2]

[1]*Graduate School of Pharmaceutical Sciences, Kumamoto University, Kumamoto, Japan;* [2]*College of Biosystems Engineering and Food Science, Zhejiang University, China*

Introduction

Flavonoids, belonging to the group of polyphenols, are one of the most widely studied phytochemicals or plant secondary metabolites. They are found abundantly in fruits and vegetables. Flavonoids possess several beneficial health properties, and consequently many studies have attempted to explore their pharmacological properties including antioxidant, antiinflammatory, and anticancer activities, among others (Kumar and Pandey, 2013; Socci et al., 2017). Many epidemiological studies have also reported the health-beneficial, disease-preventive, and therapeutic effects of the dietary intake of flavonoids (Pan et al., 2010). Similarly, recent studies have correlated the consumption of foods rich in flavonoids with the enhancement of cognitive performance (Bakoyiannis et al., 2019). Elsewhere, other studies have reported that flavonoids delay the progression of Alzheimer's disease (AD), along with their restorative effects on neurodegenerative disorders and AD-like pathological symptoms (Ayaz et al., 2019; Williams and Spencer, 2012).

Nobiletin is one of the most widely studied polymethoxyflavonoids (PMFs), a group of flavonoids commonly found in *Citrus* plants (Tapas et al., 2008) (Fig. 3.1.14.1). Nobiletin has received great interest in recent years due to its diverse pharmacological properties including antioxidant (Wang et al., 2018), anticancer (Chen et al., 2014), antiinflammatory (Zhang et al., 2016) and disease-preventive activities against neurodegenerative diseases (Nakajima et al., 2014). A Scopus database search using the term "Nobiletin" for the past 20 years (from 2000 to 2018) showed a constant growth in the number of publications each year (Fig. 3.1.14.2).

AD, a neurodegenerative disease, is the common form of dementia in elderly people and is characterized by a progressive decline in cognitive function and elevated levels of β-amyloid (Aβ) plaques (Selkoe, 2001).

Naturally Occurring Chemicals against Alzheimer's Disease. https://doi.org/10.1016/B978-0-12-819212-2.00016-5

FIGURE 3.1.14.1 Pictures of fruits of *Citrus* plants and structure of nobiletin.

Although various publications have reported the promising effects of both synthetic and natural compounds in the treatment of AD, only a few have been available in the market as therapeutic agents. Along with other natural compounds, flavonoids have also received great attention in slowing the progression and enhancing the treatment of AD. This chapter discusses the available research results related to the role of nobiletin in the prevention and treatment of AD, with a critical analysis on the mechanism of action.

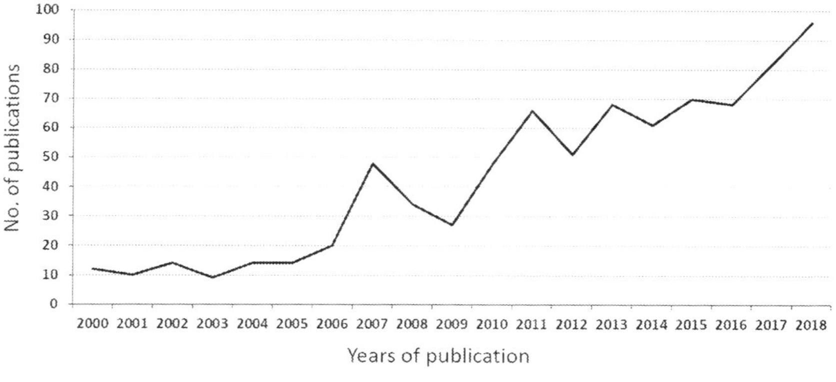

FIGURE 3.1.14.2 Trends in scientific publications regarding nobiletin from 2000 to 2018 (retrieved on August 10, 2019).

Chemistry and sources of nobiletin

Citrus plants belonging to the Rutaceae family are one of the richest sources of flavonoids, e.g., nobiletin, tangeretin, hesperidin, neohesperidin, rutin, narirutin, naringenin, quercetin, etc., which are well known for their beneficial pharmacological activities (Cheigh et al., 2012; Ihara et al., 2012). *Citrus* fruits are widely used all over the world as food ingredients and in traditional medicines. For example, in Japan, the dried peels obtained from ripe fruits of *Citrus reticulata* and *Citrus unshiu* are used as the crude drug "Chinpi" and dried peel from the ripe fruits of *Citrus aurantium* is used as crude drug "Touhi." Similarly, dried immature fruits of *Citrus aurantium* are used as crude drug "Kijitsu" (The Ministry of Health, Labour and Welfare of Japan, 2016). Polymethoxyflavonoids, containing two or more methoxy groups in their basic structure (Li et al., 2009), are abundantly found in the peel of *Citrus* fruits, particularly in sweet orange (*Citrus sinensis*), Satsuma mandarin (*Citrus unshiu*), and mandarin (*Citrus reticulata*) (Nguyen et al., 2017). PMFs are reported for their many biological activities such as antioxidant, antineuroinflammatory, anticancer, neuroprotective, antiviral, and cardioprotective activities among others (Chen et al., 2017; Nguyen et al., 2017). Among these PMFs, nobiletin is one of the most widely studied and has widespread biological activities. Nobiletin has been isolated from the peel of ripened and unripe fruits of various *Citrus* plants (Adhikari-Devkota et al., 2019), *Citrus* flowers (Miyashita et al., 2018), and juice (Wada et al., 2006). It has also been isolated from plants belonging to families other than Rutaceae, such as *Commiphora myrrha* (Burseraceae) (Abdel-Hady et al., 2019), *Calophyllum inophyllum* (Callophyllaceae) (Ponguschariyagul et al., 2018), *Phyllanthus emblica* (Phyllanthaceae) (Nguyen et al., 2018), *Gymnotheca chinensis* (Saururaceae) (Xiao et al., 2016), and *Drynaria bonii* (Polypodaceae) (Trinh et al., 2015), among others.

Metabolism and distribution in the human body

While considering the pharmacological activities of any target molecules in AD, one of the critical factors is their ability to cross the blood–brain barrier (BBB)—often a neglected but crucial area of research. Although reported to be useful in AD, many molecules are not evaluated for their ability to cross the BBB. Asakawa et al. synthesized the ^{11}C-labeled nobiletin and visualized the distribution of nobiletin in the brain using positron emission tomography (PET). A solution of ^{11}-C-nobiletin was administered to 8-week-old male rats intravenously through the tail vain and PET scans of different organs were recorded. The highest accumulation of nobiletin in the brain was observed at 5 min after administration, gradually decreasing for 60 min. However, this study clearly indicated the distribution of nobiletin in the brain. The highest accumulation of nobiletin among different organs was observed in the liver (Asakawa et al., 2011). Similarly, Okuyama et al. studied the permeation of different polymethoxyflavonoids in mouse brain and distribution of nobiletin, and other polymethoxyflavonoids in brain tissues were analyzed by high-performance liquid chromatography (Okuyama et al., 2017).

Pharmacological effects of nobiletin in the prevention and treatment of AD

Antineuroinflammatory and neuroprotective activities

Neuroinflammation is an inflammatory response of the brain and spinal cord which is triggered by various factors such as an initiating stimulus, toxins, pathogens, infections, protein accumulation, aging, genetic conditions, and other disease- and environmental-related factors. When activated by these factors, cells of the central nervous system (CNS) including microglia, endothelial cells, astrocytes, T-cells, and neurons release various inflammatory mediators such as cytokines (e.g., interleukins-IL-1β, IL-6), chemokines, tumor necrosis factor (TNF-α), secondary messengers (nitric oxide, prostaglandins), and reactive oxygen species (DiSabato et al., 2016; Shabab et al., 2017). Although acute neuroinflammation is a protective response, its progression to chronic inflammation is also reported to be associated with the progression of various neurodegenerative diseases (Ransohoff, 2016). These neurodegenerative diseases include AD (Heneka et al., 2015), Parkinson's disease (Collins et al., 2012), amyotrophic lateral sclerosis (ALS), and multiple sclerosis. Various other neurological disorders including dementia, depression, and schizophrenia (Chen et al., 2016; Shabab et al., 2017; Smith et al., 2012) are also reported to be associated with chronic neurodegeneration.

Various studies have reported the potent neuroprotective and antineuroinflammatory activities of polymethoxyflavonoids including nobiletin. For example, Cui et al. reported that nobiletin exerts its antineuroinflammatory activity in lipopolysaccharide (LPS)-induced BV-2 cells through suppression of microglial activation. The molecular mechanism revealed that nobiletin inhibited the release of TNF-α and IL-1β along with phosphorylation of mitogen-activated protein kinases (MAPKs) including extracellular signal-regulated kinase (ERK), c-Jun NH2-terminal kinase (JNK), and p38 (Cui et al., 2010). Similarly, Ho and Kuo reported on the collective effects of nobiletin, hesperidin, and tangeretin for antineuroinflammatory activity of tangerine peel in BV-2 microglial cells (Ho and Kuo, 2014). Adhikari-Devkota et al. evaluated the antineuroinflammatory activity of unripe peel of *Citrus* "Hebesu" peels extract, fractions, and isolated pure compounds including hesperidin, tangeretin, nobiletin, 5-hydroxy-6,7,8,3′,4′-pentamethoxyflavone, and 3,5,6,7,8,3′,4′-heptamethoxyflavone (Adhikari-Devkota et al., 2019) in BV-2 cells. Among them, nobiletin, 5-hydroxy-6,7,8,3′,4′-pentamethoxyflavone, and 3,5,6,7,8,3′,4′-heptamethoxyflavone (HMF) showed potent antineuroinflammatory activities in LPS-induced BV-2 cells, however tangeretin showed weak activity. The presence of two methoxy groups in the B ring of polymethoxyflavonoids was reported as an essential requirement in structure—activity relationship studies. Yasuda et al. have reported the protective role of nobiletin against ischemia-reperfusion (I/R) injury and the improvement of functional outcome in cerebral I/R model rats. This study explored the antiinflammatory and antiapoptotic effects of nobiletin and revealed its neuroprotective mechanism. Nobiletin markedly reduced infarct volume and suppressed brain edema, suppressed neutrophil invasion into the ischemic region, and decreased apoptotic brain cell death in the ischemic hemisphere. Furthermore, the motor functional deficit was improved upon treatment with nobiletin in ischemic rats (Yasuda et al., 2014).

Neuroprotective effect against hydrogen peroxide-induced oxidative stress

Oxidative stress refers to the excess load of reactive oxygen species (ROS) when the body fails to counterbalance the damaging effect with endogenous antioxidants. Oxidative stress has a detrimental effect on human health. ROS are involved in the aging process and the pathogenesis of a wide range of serious diseases including neurodegenerative diseases (Gackowski et al., 2008; Zuo and Motherwell, 2013). Herein, a study reported the mechanism of nobiletin and *Citrus unshiu* immature peel on hydrogen peroxide (H_2O_2)-induced oxidative stress in HT22 murine hippocampal neuronal cells. The study reported that nobiletin as well as *C. unshiu* (water extract) had a protective effect against H_2O_2-induced cell death in HT22 neurons via mitogen-activated protein kinases and apoptotic pathways (Cho et al., 2015).

Cholinesterase inhibitory activities

One of the strategies for treatment of AD is the use of acetylcholinesterase (AChE) inhibitors, which increase the level of acetylcholine in the synapses (Huang and Mucke, 2012; Osborn and Saunders, 2010). Nobiletin and five of its derivatives were evaluated for their AChE-inhibitory activity where nobiletin, 8-demethoxynobiletin, and 6-demethoxynobiletin showed potent activity. Furthermore, the inhibitory activity was correlated structurally to the methoxyl substitution at the 5 and 3′ positions of nobiletin (Kimura et al., 2011). Recently, Lee et al. investigated the neuroprotective effect of nobiletin and *Citrus aurantium* extract against amyloid β_{1-42} ($A\beta_{1-42}$)-induced spatial learning and memory impairment in mice. This study also reported AChE activity in the hippocampus and cortex. Nobiletin and *C. aurantium* extract exerted a profound effect on AChE activity in the cortex and hippocampus and also reduced the spatial learning deficits. The study also reported the neuroprotective effect by regulating antiapoptotic mechanisms to downregulation of Bax and cleavage of caspase-3 protein expression and to the upregulation of Bcl-2 and Bcl-2/Bax expression in the cortex and hippocampus of Aβ-treated mice (Lee et al., 2019).

Effects in cognitive impairment

Nakajima et al. reported the beneficial effect of nobiletin on cognitive impairment using olfactory bulbectomy (OBX) mice as a model. Also, its effect on cholinergic neurodegeneration was estimated by AChE staining in the brain. It was found that nobiletin rescued the OBX-induced decrease in the density of AChE-positive fibers in the hippocampus by 19%–32% (Nakajima et al., 2007). Nagase et al. also reported similar findings where nobiletin exerted a dose-dependent activity on increasing the action on spontaneous alternation behavior in OBX mice (Nagase et al., 2005).

Nagase et al. (2005) also evaluated the neuroprotective role of six *Citrus* flavonoids including nobiletin and 5-demethylnobiletin isolated from the peel of *Citrus depressa*. In several neurological disorders including AD, cAMP response element (CRE) transcription is dysregulated (Rouaux et al., 2004; Vitolo et al., 2002). The study reported that nobiletin potentially triggered activation of the cAMP/CREB signaling pathway coupled with CRE-dependent transcription. In addition, nobiletin rescued impaired memory in olfactory-bulbectomized mice after 11 days of oral administration (Nagase et al., 2005). In a different study, Nagase et al. explored the neurotrophic action mechanism of nobiletin in PC12D cells. PC12 cells, from a rat pheochromocytoma cell line, are widely used as a suitable cellular model for biochemical studies of neuronal differentiation. The study revealed that nobiletin induces neurite outgrowth by activating a cAMP/PKA/MEK/Erk/MAP kinase-dependent but not TrkA-dependent signaling pathway coupling with CRE-mediated gene transcription (Nagase et al., 2005). Takito et al. (2016) examined the effect of nobiletin and nerve growth factor (NGF) on the CRE-dependent transcription in PC12D cells. The combination of nobiletin and NGF has shown noticeable enhancement on CRE-dependent transcription as compared to NGF alone, which suggested cross-talk signaling in activating the CRE-dependent transcription in PC12D cells (Takito et al., 2016).

Matsuzaki et al. investigated the effects of nobiletin on phosphorylation of GluR1 receptor, the subunit of α-amino-3-hydroxy-5-methyl-D-aspartate (AMPA) receptors, and the receptor-mediated synaptic transmission in the hippocampus, a region implicated in memory formation. Nobiletin upregulated synaptic transmission via the postsynaptic AMPA receptors at least partially by stimulation of PKA-mediated phosphorylation of GluR1 receptor in the hippocampus (Matsuzaki et al., 2008).

Effects on amyloid-β protein (Aβ)

One of the essential therapeutic targets for the treatment of AD is amyloid-β protein (Aβ), which is catalyzed by beta-site amyloid precursor protein (APP) cleaving enzyme 1 (BACE1) (Vassar et al., 2009). Some studies have reported that BACE1 inhibitors may produce mechanism-based side effects which depend on the level of BACE1 inhibition (Yan, 2017). Therefore, various studies have been conducted to discover BACE1 inhibitors from natural products. Recently, Youn et al. explored the BACE1-inhibitory activity of some polymethoxyflavones including nobiletin. Moreover, their in silico docking studies revealed that nobiletin interacts by forming hydrogen bonding with important amino acid residues (ALA157, VAL336 and THR232, SER10 and THR232) (Youn et al., 2017). Onozuka et al. reported that nobiletin improved Aβ-mediated memory deficits and reduced the Aβ load in the hippocampus of transgenic animals model of AD (Onozuka et al., 2008). Kimura et al. reported the effects of nobiletin on intracellular and extracellular β -amyloid accumulation in human induced pluripotent stem (iPS) cell-derived AD model neurons. The results indicated that nobiletin significantly regulated the neprilysin mRNA levels and also reduced the intraneural content of amyloid β. Nobiletin was also found to reduce the levels of amyloid β_{1-42} released in a cellular medium. The authors suggested that nobiletin suppressed both intracellular and extracellular amyloid β levels (Kimura et al., 2018). Similarly, Matsuzaki et al. studied the effects of nobiletin in memory deterioration in AD by restoring the amyloid β-impaired CERB phosphorylation (Matsuzaki et al., 2006). Another study reported the neuroprotective effect of "Chotosan," a traditional Kampo prescription, against Aβ-induced neurotoxicity in PC12 cells. The study reported 10 major phenolic acids from "Chotosan" ethyl acetate fraction with pronounced neuroprotective effects on $A\beta_{25-35}$-induced neurotoxicity. Among these compounds, nobiletin (at a dose of 25 μM) displayed moderate inhibitory activities with inhibition ratios ranging from 23.8% to 37.5% on $A\beta_{1-42}$ aggregation (Wei et al., 2016).

A recent study reported the neuroprotective properties of flavonoids against the amyloid β (Aβ) protein associated with Alzheimer's disease. Marsh et al. studied the in vitro neuroprotective effect of nobiletin along with other flavonoids via incubation with human $A\beta_{1-42}$ for 48 h. The study also investigated the effects on neuronal PC12 cell viability. Promising antiamyloid activity was reported for the other flavonoids bearing several hydroxyl substituents in the B ring of the flavone scaffold. However, nobiletin showed less potent antifibrillar and neuroprotective activity, which was suggested to be due to the polymethoxyl groups in the B ring (Marsh et al., 2017). Future studies using various polymethoxylated and polyhydroxylated flavonoids are necessary to confirm these results further.

Studies in humans

Based on previous results of the effectiveness of nobiletin in AD in transgenic mice, Seki et al. performed a pilot clinical study to examine the safety and feasibility of nobiletin-rich *Citrus reticulata* (NChinpi) coadministration with donepezil in donepezil-preadministered patients. A total of six patients with mild to moderate AD were selected for the NChinpi treatment group and five patients were selected for the donepezil treatment group. The treatment was continued for 1 year and after 1 year, the baseline cognitive assessment was assessed with Mini-Mental State Examination and Japanese Version of AD assessment Cognitive Subscale. After 1 year of treatment, there were no significant changes in the NChinpi-treated group as compared to baseline, however the donepezil-administered group showed cognition impairment. NChinpi was also found to be safe, with no adverse effects and digestive symptoms (Seki et al., 2013).

A recent observational study was performed to evaluate the effectiveness of "Yokukansankachimpihange," a Kampo formulation combined with nobiletin-rich *Citrus reticulate*, with donepezil on improving the behavioral and psychological symptoms of dementia (BPSD). A total of 46 patients with dementia were selected for the study and grouped in the donepezil + "Yokukansankachimpihange" group (23 patients) and donepezil group (23 patients). The Frequency-Weighted Behavioral Pathology in Alzheimer's Disease Rating Scale (BEHAVE-AD-FW) was used to evaluate the BPSD, while the Mini-Mental State Examination and the Digit Symbol test of WAIS-R were used to evaluate impairment of global cognitive function and executive function. The study reported a positive clinical effect on enhancing behavioral abnormalities of the combined donepezil + "Yokukansankachimpihange" group and no effect was observed on cognitive functions (Meguro and Yamaguchi, 2018).

Conclusions

Various studies have reported the potential effects of nobiletin in the prevention and treatment of Alzheimer's disease through various mechanisms (Fig. 3.1.14.3). However, there remains a need for detailed clinical studies to bring such results to clinical practice. Although in vitro studies have been reported to be successful in such experiments, detailed factors related to absorption, metabolism, and excretion of nobiletin should be considered when used in human.

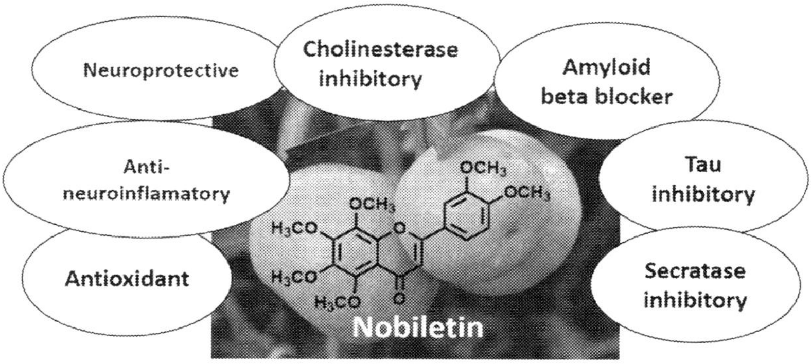

FIGURE 3.1.14.3 Schematic representation of the pharmacological activities of nobiletin in AD.

References

Abdel-Hady, H., El-Wakil, E.A., Morsi, E.A., 2019. Characterization of ethyl acetate and methanol extracts of *Commiphora myrrha* and evaluating in vitro anti-diabetic and anti-obesity activities. J. Appl. Pharm. Sci. 9 (9), 38–44. https://doi.org/10.7324/JAPS.2019.90906.

Adhikari-Devkota, A., Kurauchi, Y., Yamada, T., Katsuki, H., Watanabe, T., Devkota, H.P., 2019. Anti-neuroinflammatory activities of extract and polymethoxyflavonoids from immature fruit peels of Citrus 'Hebesu.'. J. Food Biochem. https://doi.org/10.1111/jfbc.12813.

Asakawa, T., Hiza, A., Nakayama, M., Inai, M., Oyama, D., Koide, H., et al., 2011. PET imaging of nobiletin based on a practical total synthesis. Chem. Commun. 47 (10), 2868–2870. https://doi.org/10.1039/c0cc04936k.

Ayaz, M., Sadiq, A., Junaid, M., Ullah, F., Ovais, M., Ullah, I., et al., 2019. Flavonoids as prospective neuroprotectants and their therapeutic propensity in aging associated neurological disorders. Front. Aging Neurosci. https://doi.org/10.3389/fnagi.2019.00155.

Bakoyiannis, I., Daskalopoulou, A., Pergialiotis, V., Perrea, D., 2019. Phytochemicals and cognitive health: are flavonoids doing the trick? Biomed. Pharmacother. https://doi.org/10.1016/j.biopha.2018.10.086.

Cheigh, C.I., Chung, E.Y., Chung, M.S., 2012. Enhanced extraction of flavanones hesperidin and narirutin from Citrus unshiu peel using subcritical water. J. Food Eng. 110 (3), 472–477. https://doi.org/10.1016/j.jfoodeng.2011.12.019.

Chen, C., Ono, M., Takeshima, M., Nakano, S., 2014. Antiproliferative and apoptosis-inducing activity of nobiletin against three subtypes of human breast cancer cell lines. Anticancer Res. 34 (4), 1785–1792.

Chen, W.W., Zhang, X., Huang, W.J., 2016. Role of neuroinflammation in neurodegenerative diseases (Review). Mol. Med. Rep. 13 (4), 3391–3396. https://doi.org/10.3892/mmr.2016.4948.

Chen, X.M., Tait, A.R., Kitts, D.D., 2017. Flavonoid composition of orange peel and its association with antioxidant and anti-inflammatory activities. Food Chem. 218, 15–21. https://doi.org/10.1016/j.foodchem.2016.09.016.

Cho, H., Jung, S., Lee, G., Cho, J., Choi, I., 2015. Neuroprotective effect of *Citrus unshiu* immature peel and nobiletin inhibiting hydrogen peroxide-induced oxidative stress in HT22 murine hippocampal neuronal cells. Pharmacogn. Mag. 11 (44), S284–S289. https://doi.org/10.4103/0973-1296.166047.

Collins, L.M., Toulouse, A., Connor, T.J., Nolan, Y.M., 2012. Contributions of central and systemic inflammation to the pathophysiology of Parkinson's disease. Neuropharmacology 62 (7), 2154–2168. https://doi.org/10.1016/j.neuropharm.2012.01.028.

Cui, Y., Wu, J., Jung, S.C., Park, D.B., Maeng, Y.H., Hong, J.Y., et al., 2010. Anti-neuroinflammatory activity of nobiletin on suppression of microglial activation. Biol. Pharm. Bull. 33 (11), 1814–1821. https://doi.org/10.1248/bpb.33.1814.

DiSabato, D.J., Quan, N., Godbout, J.P., 2016. Neuroinflammation: the devil is in the details. J. Neurochem. 139, 136–153. https://doi.org/10.1111/jnc.13607.

Gackowski, D., Rozalski, R., Siomek, A., Dziaman, T., Nicpon, K., Klimarczyk, M., et al., 2008. Oxidative stress and oxidative DNA damage is characteristic for mixed Alzheimer disease/vascular dementia. J. Neurol. Sci. 266 (1–2), 57–62. https://doi.org/10.1016/j.jns.2007.08.041.

Heneka, M.T., Carson, M.J., Khoury, J. El, Landreth, G.E., Brosseron, F., Feinstein, D.L., et al., 2015. Neuroinflammation in Alzheimer's disease. Lancet Neurol. 14 (4), 388–405. https://doi.org/10.1016/S1474-4422(15)70016-5.

Ho, S.C., Kuo, C.T., 2014. Hesperidin, nobiletin, and tangeretin are collectively responsible for the anti-neuroinflammatory capacity of tangerine peel (Citri reticulatae pericarpium). Food Chem. Toxicol. 71, 176–182. https://doi.org/10.1016/j.fct.2014.06.014.

Huang, Y., Mucke, L., 2012. Alzheimer mechanisms and therapeutic strategies. Cell. https://doi.org/10.1016/j.cell.2012.02.040.

Ihara, H., Yamamoto, H., Ida, T., Tsutsuki, H., Sakamoto, T., Fujita, T., et al., 2012. Inhibition of nitric oxide production and inducible nitric oxide synthase expression by a polymethoxyflavone from young fruits of *Citrus unshiu* in rat primary astrocytes. Biosci. Biotechnol. Biochem. 76 (10), 1843–1848. https://doi.org/10.1271/bbb.120215.

Kimura, J., Nemoto, K., Onoue, S., Yamakuni, T., Yokosuka, A., Mimaki, Y., et al., 2011. Inhibitory effects of Citrus polymethoxyflavones, nobiletin and its analogues, on acetylcholinesterase activity. Pharmacometrics 81 (1–2), 23–26.

Kimura, J., Shimizu, K., Kajima, K., Yokosuka, A., Mimaki, Y., Oku, N., Ohizumi, Y., 2018. Nobiletin reduces intracellular and extracellular β-amyloid in iPS cell-derived Alzheimer's disease model neurons. Biol. Pharm. Bull. 41 (4), 451–457. https://doi.org/10.1248/bpb.b17-00364.

Kumar, S., Pandey, A.K., 2013. Chemistry and biological activities of flavonoids: an overview. Sci. World J. https://doi.org/10.1155/2013/162750.

Lee, H.J., Lee, S.K., Lee, D.R., Choi, B.K., Le, B., Yang, S.H., 2019. Ameliorating effect of *Citrus aurantium* extracts and nobiletin on β-amyloid (1-42)-induced memory impairment in mice. Mol. Med. Rep. 20 (4), 3448–3455. https://doi.org/10.3892/mmr.2019.10582.

Li, S., Pan, M.H., Lo, C.Y., Tan, D., Wang, Y., Shahidi, F., Ho, C.T., 2009. Chemistry and health effects of polymethoxyflavones and hydroxylated polymethoxyflavones. J. Funct. Foods 1 (1), 2–12. https://doi.org/10.1016/j.jff.2008.09.003.

Marsh, D.T., Das, S., Ridell, J., Smid, S.D., 2017. Structure-activity relationships for flavone interactions with amyloid β reveal a novel anti-aggregatory and neuroprotective effect of 2′,3′,4′-trihydroxyflavone (2-D08). Bioorg. Med. Chem. 25 (14), 3827–3834. https://doi.org/10.1016/j.bmc.2017.05.041.

Matsuzaki, K., Miyazaki, K., Sakai, S., Yawo, H., Nakata, N., Moriguchi, S., et al., 2008. Nobiletin, a citrus flavonoid with neurotrophic action, augments protein kinase A-mediated phosphorylation of the AMPA receptor subunit, GluR1, and the postsynaptic receptor response to glutamate in murine hippocampus. Eur. J. Pharmacol. 578 (2–3), 194–200. https://doi.org/10.1016/j.ejphar.2007.09.028.

Matsuzaki, K., Yamakuni, T., Hashimoto, M., Haque, A.M., Shido, O., Mimaki, Y., et al., 2006. Nobiletin restoring β-amyloid-impaired CREB phosphorylation rescues memory deterioration in Alzheimer's disease model rats. Neurosci. Lett. 400 (3), 230–234. https://doi.org/10.1016/j.neulet.2006.02.077.

Meguro, K., Yamaguchi, S., 2018. Decreased behavioral abnormalities after treatment with combined Donepezil and Yokukansankachimpihange in Alzheimer disease: an observational study. The Osaki-Tajiri project. Neurol. Ther. 7 (2), 333–340. https://doi.org/10.1007/s40120-018-0109-9.

Miyashita, T., Adhikari-Devkota, A., Hori, K., Watanabe, M., Watanabe, T., Devkota, H.P., 2018. Flavonoids from the flowers of citrus "Hebesu.". Nat. Prod. Commun. 13 (7).

Nagase, H., Omae, N., Omori, A., Nakagawasai, O., Tadano, T., Yokosuka, A., et al., 2005a. Nobiletin and its related flavonoids with CRE-dependent transcription-stimulating and neuritegenic activities. Biochem. Biophys. Res. Commun. 337 (4), 1330–1336. https://doi.org/10.1016/j.bbrc.2005.10.001.

Nagase, H., Yamakuni, T., Matsuzaki, K., Maruyama, Y., Kasahara, J., Hinohara, Y., et al., 2005b. Mechanism of neurotrophic action of nobiletin in PC12D cells. Biochemistry 44 (42), 13683–13691. https://doi.org/10.1021/bi050643x.

Nakajima, A., Ohizumi, Y., Yamada, K., 2014. Anti-dementia activity of nobiletin, a citrus flavonoid: a review of animal studies. Clin. Psychopharmacol. Neurosci. https://doi.org/10.9758/cpn.2014.12.2.75.

Nakajima, A., Yamakuni, T., Haraguchi, M., Omae, N., Song, S.Y., Kato, C., et al., 2007. Nobiletin, a citrus flavonoid that improves memory impairment, rescues bulbectomy-induced cholinergic neurodegeneration in mice. J. Pharmacol. Sci. 105 (1), 122–126. https://doi.org/10.1254/jphs.SC0070155.

Nguyen, T.A.T., Duong, T.H., Le Pogam, P., Beniddir, M.A., Nguyen, H.H., Nguyen, T.P., et al., 2018. Two new triterpenoids from the roots of *Phyllanthus emblica*. Fitoterapia 130, 140–144. https://doi.org/10.1016/j.fitote.2018.08.022.

Nguyen, V.S., Li, W., Li, Y., Wang, Q., 2017. Synthesis of citrus polymethoxyflavonoids and their antiproliferative activities on Hela cells. Med. Chem. Res. https://doi.org/10.1007/s00044-017-1871-4.

Okuyama, S., Miyazaki, K., Yamada, R., Amakura, Y., Yoshimura, M., Sawamoto, A., et al., 2017. Permeation of polymethoxyflavones into the mouse brain and their effect on MK-801-induced locomotive hyperactivity. Int. J. Mol. Sci. 18 (3) https://doi.org/10.3390/ijms18030489.

Onozuka, H., Nakajima, A., Matsuzaki, K., Shin, R.W., Ogino, K., Saigusa, D., et al., 2008. Nobiletin, a citrus flavonoid, improves memory impairment and Aβ pathology in a transgenic mouse model of Alzheimer's disease. J. Pharmacol. Exp. Therapeut. 326 (3), 739–744. https://doi.org/10.1124/jpet.108.140293.

Osborn, G.G., Saunders, A.V., 2010. Current treatments for patients with Alzheimer disease. J. Am. Osteopath. Assoc. 110 (9 Suppl. 8).

Pan, M.H., Lai, C.S., Ho, C.T., 2010. Anti-inflammatory activity of natural dietary flavonoids. Food Funct. 1 (1), 15–31. https://doi.org/10.1039/c0fo00103a.

Ponguschariyagul, S., Sichaem, J., Khumkratok, S., Siripong, P., Lugsanangarm, K., Tip-pyang, S., 2018. Caloinophyllin A, a new chromanone derivative from *Calophyllum inophyllum* roots. Nat. Prod. Res. 32 (21), 2535–2541. https://doi.org/10.1080/14786419.2018.1425845.

Ransohoff, R.M., 2016. How neuroinflammation contributes to neurodegeneration. Science. https://doi.org/10.1126/science.aag2590.

Rouaux, C., Loeffler, J.P., Boutillier, A.L., 2004. Targeting CREB-binding protein (CBP) loss of function as a therapeutic strategy in neurological disorders. Biochem. Pharmacol. 68, 1157–1164. https://doi.org/10.1016/j.bcp.2004.05.035.

Seki, T., Kamiya, T., Furukawa, K., Azumi, M., Ishizuka, S., Takayama, S., et al., 2013. Nobiletin-rich *Citrus reticulata* peels, a kampo medicine for Alzheimer's disease: a case series. Geriatr. Gerontol. Int. 13 (1), 236–238. https://doi.org/10.1111/j.1447-0594.2012.00892.x.

Selkoe, D.J., 2001. Alzheimer's disease: genes, proteins, and therapy. Physiol. Rev. https://doi.org/10.1152/physrev.2001.81.2.741.

Shabab, T., Khanabdali, R., Moghadamtousi, S.Z., Kadir, H.A., Mohan, G., 2017. Neuroinflammation pathways: a general review. Int. J. Neurosci. https://doi.org/10.1080/00207454.2016.1212854.

Smith, J.A., Das, A., Ray, S.K., Banik, N.L., 2012. Role of pro-inflammatory cytokines released from microglia in neurodegenerative diseases. Brain Res. Bull. https://doi.org/10.1016/j.brainresbull.2011.10.004.

Socci, V., Tempesta, D., Desideri, G., De Gennaro, L., Ferrara, M., 2017. Enhancing human cognition with cocoa flavonoids. Front. Nutr. 4 (19) https://doi.org/10.3389/fnut.2017.00019.

Takito, J., Kimura, J., Kajima, K., Uozumi, N., Watanabe, M., Yokosuka, A., et al., 2016. Nerve growth factor enhances the CRE-dependent transcriptional activity activated by nobiletin in PC12 cells. Can. J. Physiol. Pharmacol. 94 (7), 728–733. https://doi.org/10.1139/cjpp-2015-0394.

Tapas, A., Sakarkar, D., Kakde, R., 2008. Flavonoids as nutraceuticals: a review. Trop. J. Pharmaceut. Res. 7 (3), 1089–1099. https://doi.org/10.4314/tjpr.v7i3.14693.

Trinh, P.T.N., Tri, M.D., An, N.H., An, P.N., Minh, P.N., Dung, L.T., 2015. Phenolic compounds from the rhizomes of *Drynaria bonii*. Chem. Nat. Compd. 51 (3), 476–479. https://doi.org/10.1007/s10600-015-1318-4.

Vassar, R., Kovacs, D.M., Yan, R., Wong, P.C., 2009. The β-secretase enzyme BACE in health and Alzheimer's disease: regulation, cell biology, function, and therapeutic potential. J. Neurosci. 29, 12787–12794. https://doi.org/10.1523/JNEUROSCI.3657-09.2009.

Vitolo, O.V., Sant'Angelo, A., Costanzo, V., Battaglia, F., Arancio, O., Shelanski, M., 2002. Amyloid β-peptide inhibition of the PKA/CREB pathway and long-term potentiation: reversibility by drugs that enhance cAMP signaling. Proc. Natl. Acad. Sci. U.S.A. 99 (20), 13217–13221. https://doi.org/10.1073/pnas.172504199.

Wada, K., Uehara, M., Takara, K., Tome, Y., Yano, M., Ishii, T., Ohta, H., 2006. Quantitative analysis of nobiletin in shiikuwasa (Citrus depressa Hayata) juice. Food Preserv. Sci. 32 (1), 29–33. https://doi.org/10.5891/jafps.32.29.

Wang, M., Meng, D., Zhang, P., Wang, X., Du, G., Brennan, C., et al., 2018. Antioxidant protection of nobiletin, 5-demethylnobiletin, tangeretin, and 5-Demethyltangeretin from citrus peel in *Saccharomyces cerevisiae*. J. Agric. Food Chem. 66 (12), 3155–3160. https://doi.org/10.1021/acs.jafc.8b00509.

Wei, M., Chen, L., Liu, J., Zhao, J., Liu, W., Feng, F., 2016. Protective effects of a Chotosan fraction and its active components on β-amyloid-induced neurotoxicity. Neurosci. Lett. 617, 143–149. https://doi.org/10.1016/j.neulet.2016.02.019.

Williams, R.J., Spencer, J.P.E., 2012. Flavonoids, cognition, and dementia: actions, mechanisms, and potential therapeutic utility for Alzheimer disease. Free Radic. Biol. Med. https://doi.org/10.1016/j.freeradbiomed.2011.09.010.

Xiao, S.J., Guo, D. Le, Xu, D.L., Zhang, M.S., Chen, F., Ding, L.S., Zhou, Y., 2016. Chemical constituents of Gymnotheca chinensis. Chin. Tradit. Herb. Drugs 47 (10), 1665–1669. https://doi.org/10.7501/j.issn.0253-2670.2016.10.007.

Yan, R., 2017. Physiological functions of the β-site amyloid precursor protein cleaving enzyme 1 and 2. Front. Mol. Neurosci. https://doi.org/10.3389/fnmol.2017.00097.

Yasuda, N., Ishii, T., Oyama, D., Fukuta, T., Agato, Y., Sato, A., et al., 2014. Neuroprotective effect of nobiletin on cerebral ischemia-reperfusion injury in transient middle cerebral artery-occluded rats. Brain Res. 1559, 46–54. https://doi.org/10.1016/j.brainres.2014.02.007.

Youn, K., Yu, Y., Lee, J., Jeong, W.S., Ho, C.T., Jun, M., 2017. Polymethoxyflavones: novel β-secretase (BACE1) inhibitors from citrus peels. Nutrients 9 (9), 1–12. https://doi.org/10.3390/nu9090973.

Zhang, L., Zhang, X., Zhang, C., Bai, X., Zhang, J., Zhao, X., et al., 2016. Nobiletin promotes antioxidant and anti-inflammatory responses and elicits protection against ischemic stroke in vivo. Brain Res. 1636, 130–141. https://doi.org/10.1016/j.brainres.2016.02.013.

Zuo, L., Motherwell, M.S., 2013. The impact of reactive oxygen species and genetic mitochondrial mutations in Parkinson's disease. Gene. https://doi.org/10.1016/j.gene.2013.07.085.

Section 3.2

Plants and their extracts

Chapter 3.2.1

Ginkgo biloba

Ashutosh Paliwal[1], Pooja Pandey[1], Kushagra Pant[2], Manoj Kumar Singh[3], Vipul Chaudhary[4], Jalaj Kumar Gour[5], Ashwini Kumar Nigam[6], Vimlendu Bhushan Sinha[7]

[1]*Department of Biotechnology, Kumaun University Nainital, Bhimtal Campus, Bhimtal, Uttarakhand, India;* [2]*Directorate of Cold Water Fisheries Research, ICAR, Bhimtal, Uttarakhand, India;* [3]*Center for Noncommunicable Diseases (NCD), National Centre for Disease Control, Delhi, India;* [4]*Department of Biotechnology, Deenbandhu Chhotu Ram University of Science and Technology, Sonepat, Haryana, India;* [5]*Department of Biochemistry, Faculty of Science, University of Allahabad, Prayagraj, Uttar Pradesh, India;* [6]*Department of Zoology, Udai Pratap College, Varanasi, Uttar Pradesh, India;* [7]*Department of Biotechnology, School of Engineering and Technology, Sharda University, Greater Noida, Uttar Pradesh, India*

Introduction

This decade can be described as the decade of technology. It is true that technology is an elementary tool in this rapidly changing world. However, as we know, every story has two sides. The good part of this technology is that it reduces work pressure with time saving, but on the other hand technology may induce various neurological disorders. In this world the older population is more affected by neural disorders, such as dementia, which are attributed to progressive neurodegenerative disorders that have now become a social problem worldwide. The global population's health is a social issue (Hugo and Ganguli, 2014; Zhang et al., 2008). Cognitive decline can be both moderate impairment (MCI) and dementia (Howieson, 2016; Jørgensen et al., 2016; Mormino and Papp, 2016). A small but evident diminution in cognitive function is seen during mild/moderate cognitive impairment (MCI), referred to as the evidence stage before dementia (Budson and Solomon, 2012; Fernández-Blázquez et al., 2016; Petersen, 2016). MCI is categorized into: MCI due to Alzheimer's disease (AD) and MCI due to other causes, wherein dementia is described as psychic retardation syndrome (Damiani et al., 2014; Ihl et al., 2015; Wang et al., 2016). The causes include Alzheimer disease, mixed dementia, and vascular dementia (Montine et al., 2014; Altamura et al., 2016). Alzheimer is a type of neurodegenerative disorder which starts with moderate memory problems and within a year develops into multiple cognitive and functional impairment (Brooker et al., 2014; Aygün and Güngör, 2015; Wood,

Naturally Occurring Chemicals against Alzheimer's Disease. https://doi.org/10.1016/B978-0-12-819212-2.00017-7

2016). Ischemic stroke and hemorrhagic stroke are mainly responsible for vascular dementia, which is characterized by vascular disease (cerebral) with low cerebral insertion, often leading to impairment of memory, behavior, and cognition (Tsivgoulis et al., 2014; O'Brien and Thomas, 2015). A patient suffering from both types of dementias, i.e., Alzheimer's disease and vascular dementia, is treated as having mixed dementia (Moore et al., 2014; Bogolepova, 2015; Kim et al., 2016). A study by Prince et al. (2015) revealed that more than 46 million suffer from dementia around the world and this number is forecast to increase to 131.5 million by 2050 and will also have a huge impact on the global economy. Surprisingly, there is still no therapeutic treatment available for dementia (Kennedy and Sud, 2014), with only symptomatic treatment being possible (Peirson et al., 2015). In the series of available treatments for dementia, Chen et al. (2016) reported that the inhibitors of cholinesterase increase the concentration of neurotransmitters which can improve memory. The excitatory amino acid receptor is inhibited by the NMDA receptor which reduces neurotoxicity (Newport et al., 2015), and other commercially available medications/drugs such as dopamine-blocking agents, benzodiazepines, and serotonergic agents cause other symptoms (Tanaka et al., 2015).

Ginkgo biloba

Since ancient time before pharmaceutical industries had evolved, people were totally dependent on plants. In the 18th century when many forests were present and many medicinal plants available with known medicinal values, people were totally dependent on plants for their medical needs. India is a country situated in the Asian subcontinent, with a rich flora and the Indian tradition health system relies on plants, with many communities being totally dependent on plants/herbs. The Indian Himalayan region is blessed with various medicinal plants/herbs which are not only used by local communities but also their subproducts in demand from pharmaceutical industries. There are many well-explored medicinal plants available in the Indian Himalayan region such as *Withania somnifera*, *Taxus baccata*, *Purnella vulgaris*, *Catharanthus roseus*, etc. Drug discovery for the treatment of Alzheimer's disease is going on, but synthetic drugs may exhibit cytotoxic effects apart from suppressing the indications of disease-causing symptoms. Herbal medicine could be the best option to reduce this toxicity, as herbal formulations not only reduce the risk of cytotoxicity but also have the potential to sensitize chemical drugs. Therefore, keeping this in mind to cure severe neurodegenerative disorders, i.e., Alzheimer's disease, a herbal medication is essential, and *Ginkgo biloba* could be a potential candidate to treat AD. *Ginkgo biloba* leaves have been used since ancient times for medication purposes. During the 1970s, Dr. Willmar Schwabe Pharmaceuticals (Karlsruhe, Germany) devised a better method for formulation of *Ginkgo biloba* and obtained stable concentrated extracts of the leaf tissues (Le Bars, 2003). Extract of *Ginkgo biloba* (GBE) yields 24% flavonoids and 6% terpenoids and 5%−10% organic acids (Fig. 3.2.1.1A and B). Herbal decoctions of *G. biloba* are now used in treatment and as well as in prevention of neurodegenerative dementias related to senescence, AD, peripheral vascular diseases, and neurosensory issues (e.g., tinnitus) (Luo, 2001).

Scientific classification of *G. biloba*

Kingdom: Plantae
Division: Gingkgophyta
Class: Ginkgoopsida
Order: Ginkgoales
Family: Ginkgoaceae
Genus: *Ginkgo*
Species: *biloba*

Over the last few years various herbal decoctions have been used for the treatment/inhibition of dementia with the aim that it may cause a delay in its progression. For the last few decades *Ginkgo biloba* extracts have been widely used in the treatment of Alzheimer's and cognitive disorders (Weinmann et al., 2010). It has been also recognized as an antidementia drug. The most probable mechanism behind its action against AD may be its antioxidant and anti-apoptotic properties and its inhibiting potential against activation of caspase-3 and amyloid-β aggregation (Luo et al., 2002). *G. biloba* has gained a great deal of attention in recent years, where a number of articles and studies have shown that *G. biloba* can stop or slow cognitive deterioration and behavioral disturbances in patients affected by Alzheimer's disease. Although *Ginkgo biloba* is a globally accepted and trusted herbal remedy for dementia treatment, its efficacy remains controversial. DeKosky et al. (2008) and Vellas et al. (2012) concluded that there were insignificant outcomes of *G. biloba* in the primary

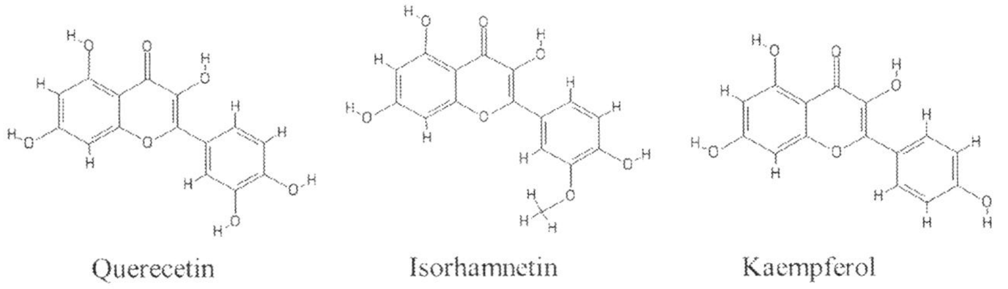

FIGURE 3.2.1.1A Chemical structures of flavonoids present in *Ginkgo biloba*.

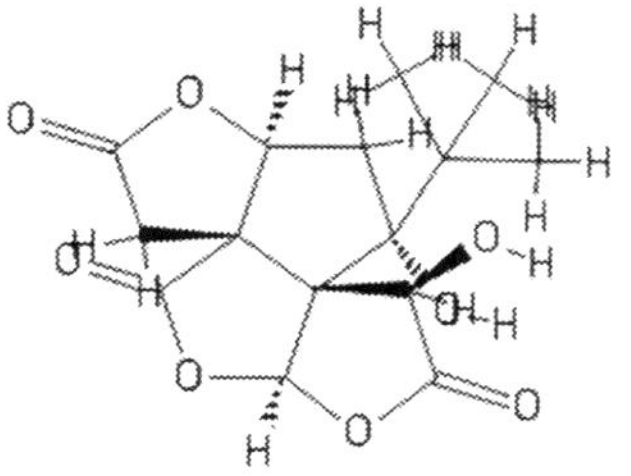

Bilobalide

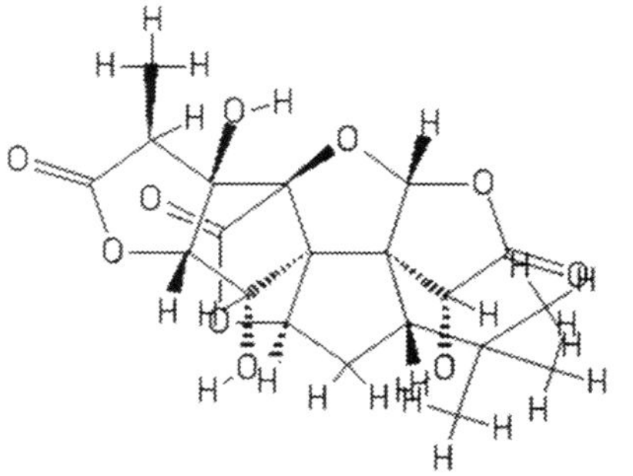

Ginkgolides

FIGURE 3.2.1.1B Chemical structure of terpenoids present in *Ginkgo biloba*.

prevention of dementia and also in Alzheimer's disease. Other workers have confirmed the potent role of *G. biloba* in the prevention/inhibition of dementia progression (Weinmann et al., 2010; Ihl, 2012) and also confirmed the effective role of *Ginkgo biloba* extract (GBE) versus cholinesterase inhibitors (ChEIs) in the prevention of Alzheimer's disease (Wettstein, 2000; Schulz, 2003; Mazza et al., 2006; Yancheva et al., 2009) (Fig. 3.2.1.2).

There are various scientific reports which favor the medicinal potential of *Ginkgo biloba* leaf extracts (Fig. 3.2.1.3). It is a well-known herbal tree in China as it has been used in the Chinese traditional medicinal system for the last 5000 years. Extracts of *G. biloba* are generally used for treating pulmonary diseases. Halpern (1998) reported an interesting fact about *G. biloba* that it was the only plant which survived in Nagasaki and Hiroshima after they were each subject to an atomic bomb during the Second World War. Different research groups from France and Germany have published reports on the efficacy of GBE against cerebral insufficiency or neurasthenia, which is characterized by symptoms including memory impairment, mood depression, confused state, loss of sensation, fatigue, dizziness, and lack of motivation. Oken et al. (1998) reported the antioxidative potential of flavonoids and antiplatelet-activating property of terpenoids found in *Ginkgo biloba* (Shi et al., 2009). It may be possible that both mechanisms could play a pivotal role in the therapeutic treatment of AD. Published literature confirms the importance of free radicals in oxidative and peroxidase neuronal damage but also extends to amyloid deposition in AD, which is attributed as the cause of memory loss and cognitive symptoms. The anti-PAF effect can be used in combating proinflammatory effects and further progression of vascular lesions derived due to activation of platelets.

FIGURE 3.2.1.2 A twig of the *Ginkgo biloba* plant.

Alzheimer's disease

The first thing that comes to mind when talking about AD is dementia. Alzheimer's is a form of dementia where the person affected has a mild to severe impairment of cognitive function, and loss of memory leading to behavioral disturbances, affecting their daily activities. A person suffering from AD will initially face changes in their thinking and reasoning, such as forgetting something they read or confusing the names of new people they meet, etc. It is a progressive disease which worsens with time and in the later stages the patients finds it increasingly difficult to perform day to day activities such as dressing, eating with utensils, and is unable to communicate that they are

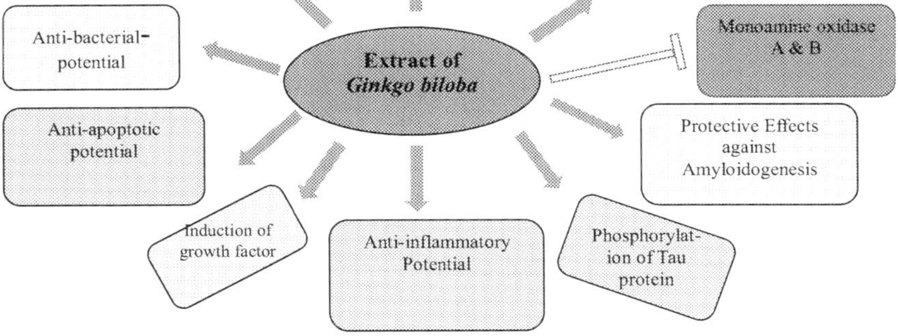

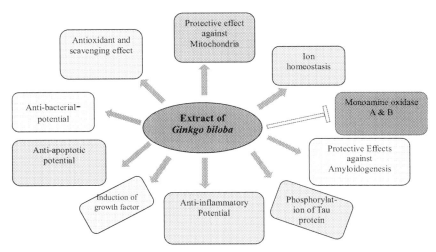

FIGURE 3.2.1.3 Broad-spectrum analysis of the activity of *G. biloba* extract.

thirsty or hungry. The dementia generally worsens with age, with some studies suggesting that people over the age of 65 are most likely to become affected, however, there have been cases in younger people also. The main cause of AD is the death of brain cells over time. According to AD International, 50 million people worldwide had dementia in 2017; in the United States alone there are 5.8 million people living with AD today and, the number could be as high as 14 million. In India, the number is around four million, however this might be an underestimation as mild cases may not be documented for data purposes.

AD is a neurodegenerative disorder followed by cognitive impairments and memory loss. Its main molecular characteristic is the deposition/accumulation of amyloids in senile plaques, cholinergic deficits, intracellular neurofibrillary tangles, and neuron loss in brain regions which are essential for both cognition and memory (Shi et al., 2008). Oxidative stress, inflammation, excitotoxicity, and neuronal apoptosis are some of the pathological indicators which are generally seen in the brains of those suffering with Alzheimer's disease (Magaki et al., 2007; Culpan et al., 2007; Koutsilieri and Riederer, 2007; Shi et al., 2008; Gella and Durany, 2009; Cameron and Landreth, 2009; Bertoni-Freddari et al., 2009). Some agents which have protective potential include antioxidants, antiinflammatory drugs, estrogens, cholinergic agents, and neurotrophic factors that could be used for the treatment/prevention of AD, however there is no scientific report/research group that has confirmed the therapeutic potential of these agents (Zimmermann et al., 2002; Klegeris et al., 2005; Cole and Frautschy, 2007; Shi et al., 2008; Kovács, 2009). *Ginkgo biloba* possesses the potential to combat the progression of AD.

Major constituents of *Ginkgo biloba* and their modes of action

It is known that GBE is effective against cognitive disease and dementia, and there are certain conceivable mechanisms by which *G. biloba* prevents the neurodegenerative effects of AD—or we can say that there are some compounds produced by *G. biloba* in response to AD characteristics. Several studies have confirmed the biological role of *Ginkgo biloba*, these properties are described next.

Antioxidant activity

It has been suggested that Alzheimer's disease is primarily induced by oxidative stress (Gella and Durany, 2009). The free radical scavenging potential of *G. biloba* is responsible for its efficient action against oxidative stress and this fact is supported by various in vivo and in vitro studies (Smith et al., 2004). For example, pretreating cerebellar granule cells with GBE efficaciously reduced the oxidative impairment caused by $H_2O_2/FeSO_4$ (Wei et al, 2000). In another study documented by Wu et al. (2006) and Smith et al.

(2004), two models of AD, i.e., Aβ-expressing neuroblastoma cell line N2a and Aβ-expressing transgenic *Caenorhabditis elegans*, an H_2O_2 reactive oxygen species (ROS) was attenuated by GBE. There are some reports of increasing the level of protein and action of some regulatory enzymes being upregulated, such as superoxide dismutase (SOD), catalase (Colak et al, 1998), glutathion reductase (GSH) (Sasaki et al., 2002), and γ-glutamyl cysteinyl synthetase by *G. biloba* (Bridi et al., 2001; Rimbach et al., 2001; Ahlemeyer and Krieglstein, 2003).

The antioxidative potential of *G. biloba* extract is mainly due to the presence of flavonoids. The proportion of flavonoid present in extract increases the antioxidative effect by different possible pathways such as upregulating antioxidant proteins like GSH and SOD, chelation of metal ions (prooxidant), and direct scavenging of ROS (Smith et al., 2003).

Protective effects on mitochondrial function

Patients suffering from cognitive disorders including Alzheimer's disease generally face problems in proper functioning of mitochondria (Castellani et al., 2002). The protective effect of *G. biloba* on mitochondria indicated that GBE directly affected mitochondria. It may impart its antioxidant effect because the mitochondrial chain is a pronounced source and ROS target. During the research into SH-SY5Y cells by Shi et al. (2009), they reported preventive outcomes of *G. biloba* on dysfunctioning of mitochondria elicited by amyloid β peptide (Aβ) which results in a reduction of ROS generation. Interestingly this preventive impression was only found for ginkgolide B and not quercetin (Shi et al., 2009). GBE protective effects for mitochondria have been illustrated in a handful of in vivo studies. Utilizing two animal models for aging, it has been reported that this herbal treatment is effective for decreasing cytochrome-C oxidase (COX) activity, mitochondrial ATP, and GSH mitochondrial content in the hippocampi of aged SAMP8 mice and ovariectomized rats (Shi et al., 2009). Despite the scientific reports, the basic mechanism behind the protective effect of GBE on mitochondria remains to be clearly elucidated.

Antiapoptotic effect

Apoptosis, programmed cell death, is associated with various neurodegenerative disorders including AD (Yuan and Yankner 2000; Smith et al., 2004; Bertoni-Freddari et al., 2009). GBE contains multifactorial antiapoptotic activity which might act synergistically on signaling pathways for apoptosis (Smith et al., 2004). *G. biloba* has exhibited its role in various mechanisms such as maintaining the state of mitochondrial membrane, precluding cytochrome *c* release from mitochondria, apoptosome organization obstruction, and the apoptotic caspase cascade for intensifying transcription of some antiapoptotic proteins. This helps in the reduction of proapoptotic caspase-12, inactivates apoptotic *c-Jun* N-terminal kinase (JNK), which causes "turning off" of the targets located downstream (*c-Jun*) and helps in inhibiting cleavage of protease caspase-3 and blocks apoptosis execution for the prevention of nuclear DNA fragmentation which is a hallmark of apoptosis (Smith et al., 2004; Shi et al., 2009).

Antiapoptotic property may be partly induced by flavonoids which are having antioxidative potential. Flavonoids prevent cell apoptosis induced by various oxidants (Smith et al., 2003; Gohil et al., 2002), and reports suggests that quercetin, being important in GBE, does not have inhibition potentials for JNK activity apoptosis induced due to hydrogen peroxide and 4-hydroxy-2-nonenal (Spencer et al., 2003). Induction of peroxidation leads to inactivation of the JNK-*c-Jun*/AP-1 pathway which might be responsible for the antiapoptotic effect shown by quercetin (Ishikawa et al., 2000). Kong et al. (2000) studied the outcomes of a low concentration of quercetin, activation promotion of the MAPK pathway which leads to the expression of downstream but essential defensive and survival actions like *c-Fos*, *c-Jun*, phase II detoxifying enzymes, GSH *S*-transferase, and quinone reductase. Although various studies have supported the role of flavonoids in the antiapoptotic potential of *G. biloba* there are some scientific reports available which confirm that terpene present in extracts also contributes to the antiapoptotic potential of GBE, e.g., bilobalide, ginkgolide B, and ginkgolide J. Terpenes from *G. biloba* speed up the rate of cell death in chick embryonic neurons (Ahlemeyer et al., 1999; Defeudis, 2002). Ahlemeyer and Krieglstein (2003) documented the reverse apoptotic role of bilobalide which was induced by staurosporine treatment (12 h) of chick neurons. It was found that in the case of mixed cultures, neurons and astrocytes originating from neonatal rat hippocampus and bilobalide were able to rescue the neurons from being deprived by serum-induced apoptosis (Ahlemeyer and Krieglstein, 2003). However, ginkgolide A was reported as failing in blocking the apoptotic damage in serum-deprived/staurosporine-treated neurons (Ahlemeyer and Krieglstein, 2003). Attenuation in the level of apoptotic markers such as c-myc, p53, and Bax suggested the potential of bilobalide in blocking neuronal apoptosis at an early stage (Defeudis, 2002).

Similarly, blocking of early signaling pathways in apoptosis indicates the antiapoptotic potential of ginkgolides. Chan et al. (2007) reported the inhibition of apoptotic cell (ethanol-induced) death via restraint of JNK activation and caspase-3 by ginkgolide B. This finding was also supported by Shi et al. (2009). Despite these studies there are some contrasting data available which deny the protective role of *G. biloba* on apoptosis, i.e., bilobalide cannot save adult rat hippocampal neurons for apoptosis induced by 2,2′-azobis-2-amidinopropane (Defeudis, 2002). The report of Xin et al. (2000) also denied the protective role of *G. biloba* in apoptosis by showing the results that terpens present in an extract failed to block apoptosis induced by hydroxyl radicals in rat cerebellar granule cells. However, many studies confirm the antiapoptotic potency and presence of various active compounds and apoptotic action in GBE constituents (Shi et al., 2009). Determination of the effective dose for specific/effective action against apoptosis is still under research and not yet confirmed (Shi et al., 2009).

Antiinflammatory effect

Patients affected with AD also suffer with inflammation (De Toledo, 2006). Levels of cytokines, phase reactants (acute), and inflammatory mediators were reported to be high in cases affected with AD (De Toledo, 2006). Braquet et al. (1987) and Chan et al. (2007) reported *G. biloba* as having an antiinflammatory effect and this might be credited to the collective actions of ginkgolides and flavonoids (Chan et al., 2007). Platelet-activating factor (PAF)—antagonist activity is primarily responsible for its antiinflammatory potential. Various studies have confirmed the role of PAF as a cytokine regulator in inflammatory responses (Bonavida et al., 1995; Maclennan et al., 2002). Intracerebroventricular PAF administration in rodents increases proinflammatory mediator leukotriene synthesis, especially leukotriene C4 (Hynes et al., 1991; Maclennan et al., 2002). Synthesis of PAF can be initiated by stimulation of neurotransmitters for, e.g., *N*-methyl-D-aspartic acid (NMDA) and glutamic acid, which further executes excellent role in neuronal functioning related to brain development (Aihara et al., 2000). A higher concentration of PAF in the brain also causes damage giving rise to neurodegenerative disorders like Alzheimer's disease (Farooqui et al., 2006; Bate et al., 2008). Ginkgolides show competitive behavior with PAF (Chan et al., 2007). Intracerebroventricular administration of BN-52,021 (ginkgolide B) in rats importantly weakens PAF-mediated cerebrospinal fluid increase (Hynes et al., 1991; Maclennan et al., 2002). Kunievsky et al. (1992) and Maclennan et al. (2002) reported a reduction in the exhibition of eicosanoid and thromboxane B in rat brain induced by PAF. To support these findings, Shi et al. (2009) documented an inhibiting property of ginkgolide B against a PAF-mediated reduction in the viability of SH-SY5Y cells. Additionally, the collective inhibitory effect of ginkgolides A and B on proinflammatory cytokines, tumor necrosis factor-α, and IL-1 production in LPS-stimulated rat microglial cultures was also recorded (Du et al., 1998; Maclennan et al., 2002). Simultaneously, flavonoid content of *G. biloba* might inhibit lipooxygenase, an important enzyme in leukotriene synthesis (Chan et al., 2007), whereas quercetin might be involved in antagonist activity of PAF (Shi et al., 2009).

Preventive effects on amyloidogenesis and Aβ aggregation

Aβ plaque accumulation has been described as the main characteristic of AD (Gerlai, 2001). In recent years, prevention of Aβ-induced toxicity by *G. biloba* has gained increasing attention. There are various reports which support the role of *G. biloba* in prevention of neurotoxicity, induced by Aβ by different mechanisms such as ROS accumulation, uptake of glucose, dysfunction of mitochondria, AKT activation, JNK, and apoptosis (Bastianetto et al., 2000; Smith et al., 2003; Shi et al., 2009). Additionally, *G. biloba* prevents amyloidogenesis (Ramassamy, 2006). It was reported that hippocampal slices EGb761 (*G. biloba*) can accelerate amyloid precursor protein (APP) secretions, pointing toward the α-secretase pathway, thereby increasing the release of the soluble form of APP (sAPPα) (Colciaghi et al., 2004; Ramassamy, 2006). By using transgenic mice model system capability of GBE against α-secretase pathway was assessed in vivo (Stackman et al., 2003; Ramassamy, 2006). Simons et al. (1998) and Ramassamy (2006) studied the inhibiting potential of *G. biloba* on brain Aβ by lowering the level of circulating free cholesterol, further affecting APP processing and amyloidogenesis. The major compounds of *G. biloba* which are accountable for inhibition of amyloidogenesis have not yet been explored. Luo et al. (2002) and Ramassamy (2006) proposed that GBE had inhibitory potential for Aβ fibril formation, which is considered for neurotoxicity of Aβ and is mainly due to the β-sheet structure of Aβ-fibrils (Pike et al., 1991; Simmons et al., 1994; Soto et al., 1998; Ramassamy, 2006). In this way we can conclude that β-sheet structure inhibition formation of Aβ fibrils can execute and prevent Aβ toxicity. Transition metal ions like iron, zinc, and copper have the potential to influence the aggregation state for Aβ. *G. biloba* is capable of inhibiting Aβ fibril formation by its chelation property (Luo et al., 2002; Ramassamy, 2006). Watanabe et al. (2001) documented the role of *G. biloba* for Aβ fibril formation by transthyretin upregulation (Tsuzuki et al., 2000). Different flavonoid compounds like bilobalide and ginkgolide J from GBE have been reported as having inhibitory potential on Aβ aggregation (Luo et al., 2002; Ramassamy, 2006).

Some other mechanisms

To overcome or facilitate the effect of neurotoxicity in Alzheimer's disease there are some mechanisms reported in GBE including ionic homeostasis, phosphorylation modulation of tau protein, and growth factor induction for synthesis. Ca^{2+} dyshomeostasis is another mechanism which is thought to be important in mediating neurotoxicity in AD (Shirwany et al., 2007; Berrocal et al., 2009). Liu et al. (2002) and Ahlemeyer and Krieglstein (2003) investigated microtubule-associated tau protein being hyperphosphorylated in the brain of an AD survivor and also aggregated as neurofibrillary tangles of paired helical filaments, which were mainly readable for AD pathogenesis. Using mRNA microarrays, Watanabe et al. (2001) elucidated GBE ability to upregulate the gene expression of microtubule-associated tau protein and neural protein phosphatase type 1, a serine/threonine protein phosphatase (Ahlemeyer and Krieglstein, 2003; Watanabe et al., 2001). A subject affected with neurodegenerative disorders were observed with altered nerve growth factor level (NGF) (Pierre et al., 1999; Peng et al., 2009). Pierre et al. (1999) also reported that GBE is capable of enhancing the mRNA expression of NGF in mouse cortex. The studies of Zheng et al. (2000) and Ahlemeyer and Krieglstein (2003) exhibited the effect of Egb761, bilobalide, a terpenoid constituent, to enhance mRNA and protein expression of vascular endothelial growth factor and glial-derived neurotrophic factors.

Possible mechanism of actions

Competitive suppression of platelet-activating factor (PAF) by *Ginkgo biloba* extract is due to its antiinflammatory effect. *Ginkgo* comprises ginkgolide A and B antagonists, which especially inhibit PAF binding with the membrane receptor and may also exert neuroprotective and antithrombotic effects. Various scientific reports have concluded that flavonoid and ginkgolide B collectively exert some activities including inhibition of lipoprotein formation, aggregation of platelets, and adherence of platelets which result in lessening the risk of vascular injury and atherosclerosis. Moreover, PAF antagonism can also help in the prevention of cyclosporin-induced nephrotoxicity, along with decreased activity associated with coronary blood flow, which finally affects myocardial contractility. This process may also help to bring about an understanding of the beneficial effects for bronchospasm, hypersensitivity reaction, and circulatory diseases (Natural Medicines Comprehensive Database, 1999; Micromedex Healthcare Series, 2003). Flavonoid glycosides can also exert some relative antioxidant effects and cause a reduction in endothelial cell injury, which ultimately impairs the development of atherosclerosis. GBE probably can offer some protection to intestinal mucosa against injury, mainly by decreasing neutrophil infiltration, stimulating choline uptake and lipid peroxidation, and preventing muscarinic receptors. The repressive effect of *Ginkgo* on monoamine oxidase activity requires further study for understanding its role.

Conclusion

After successfully reviewing various scientific reports, GBE is thought to be very effective for the treatment of cognitive disorders. AD is a very serious condition, especially as there is no effective medication available for this disease, however there are some conventional medications available to lessen the effects of AD although no cure has been found. In recent years, *G. biloba* has gained a lot of attention due to its broad range of medicinal values as its extract contain various phytocompounds which show various activities such as inhibition of platelet-activating factor (PAF), antioxidant potential, antibacterial potential, and many others. Many research reports for GBE use in cerebrovascular insufficiency, impairment of memory in aged people, AD, dementia, depression of resistant type, and artery and venous insufficiency have been documented. Various scientific literature helps in understanding the antioxidant potential of *G. biloba* and a meta-analysis has unraveled its safety use aspects. Long-term usage of ginkgo has been aimed at preventive intervention and may also be standardized first for studying its effects on preventing cognitive decline or dementia. Another study supported the safe effectiveness of GBE for Alzheimer's dementia and the vascular type of dementia, and also mixed dementias. Various in vivo and in vitro preclinical studies support the fact that *Ginkgo biloba* extract may prove to be efficient in the treatment/prevention of AD and also other age-related diseases. Overall, a better understanding of the mechanisms underlying the neuroprotective effects of *G. biloba* may finally contribute to designing therapeutic strategies for future clinical practices.

References

Ahlemeyer, B., Krieglstein, J., 2003. Neuroprotective effects of *Ginkgo biloba* extract. Cell. Mol. Life Sci. 60 (9), 1779–1792.

Ahlemeyer, B., Möwes, A., Krieglstein, J., 1999. Inhibition of serum deprivation-and staurosporine-induced neuronal apoptosis by *Ginkgo biloba* extract and some of its constituents. Eur. J. Pharmacol. 367 (2–3), 423–430.

Aihara, M., Ishii, S., Kume, K., Shimizu, T., 2000. Interaction between neurone and microglia mediated by platelet-activating factor. Gene Cell. 5 (5), 397–406.

Altamura, C., Scrascia, F., Quattrocchi, C.C., Errante, Y., Gangemi, E., Curcio, G., Rossini, P.M., 2016. Regional MRI diffusion, white-matter hyperintensities, and cognitive function in Alzheimer's disease and vascular dementia. J. Clin. Neurol. 12 (2), 201–208.

Aygün, D., Güngör, İ.L., 2015. Why is Alzheimer's disease confused with other dementias? Turk. J. Med. Sci. 45 (5), 1010–1014.

Bastianetto, S., Ramassamy, C., Doré, S., Christen, Y., Poirier, J., Quirion, R., 2000. The *Ginkgo biloba* extract (EGb 761) protects hippocampal neurons against cell death induced by β-amyloid. Eur. J. Neurosci. 12 (6), 1882–1890.

Bate, C., Tayebi, M., Williams, A., 2008. Ginkgolides protect against amyloid-β 1–42-mediated synapse damage in vitro. Mol. Neurodegener. 3 (1), 1.

Berrocal, M., Marcos, D., Sepúlveda, M.R., Pérez, M., Ávila, J., Mata, A.M., 2009. Altered Ca^{2+} dependence of synaptosomal plasma membrane Ca^{2+}-ATPase in human brain affected by Alzheimer's disease. FASEB J. 23 (6), 1826–1834.

Bertoni-Freddari, C., Fattoretti, P., Casoli, T., Di Stefano, G., Balietti, M., Giorgetti, B., Perretta, G., 2009. Neuronal apoptosis in Alzheimer's disease. Ann. N.Y. Acad. Sci. 1171 (1), 18.

Bogolepova, A.N., 2015. A modern concept of mixed dementia. Zh. Nevrol. Psikhiatr. Im. S S Korsakova 115 (5), 120–126.

Bonavida, B., Mencia-Huerta, J.M., 1995. Platelet-activating factor and the cytokine network in inflammatory processes. Clin. Rev. Allergy Immunol. 12 (4), 381–395.

Braquet, P.I.E.R.R.E., 1987. The ginkgolides: potent platelet-activating factor antagonists isolated from *Ginkgo biloba* L. chemistry, pharmacology and clinical applications. Drugs Future 12, 643–699.

Bridi, R., Crossetti, F.P., Steffen, V.M., Henriques, A.T., 2001. The antioxidant activity of standardized extract of *Ginkgo biloba* (EGb 761) in rats. Phytother. Res. 15 (5), 449–451.
Brooker, D., Fontaine, J.L., Evans, S., Bray, J., Saad, K., 2014. Public health guidance to facilitate timely diagnosis of dementia: Alzheimer's cooperative valuation in Europe recommendations. Int. J. Geriatr. Psychiatr. 29 (7), 682–693.
Budson, A.E., Solomon, P.R., 2012. New criteria for Alzheimer's disease and mild cognitive impairment: implications for the practicing clinician. Neurol. 18 (6), 356.
Cameron, B., Landreth, G.E., 2009. Inflammation, microglia, and Alzheimer's disease. Neurobiol. Dis. 37 (3), 503–509.
Castellani, R., Hirai, K., Aliev, G., Drew, K.L., Nunomura, A., Takeda, A., Smith, M.A., 2002. Role of mitochondrial dysfunction in Alzheimer's disease. J. Neurosci. Res. 70 (3), 357–360.
Chan, P.C., Xia, Q., Fu, P.P., 2007. *Ginkgo biloba* leave extract: biological, medicinal, and toxicological effects. J. Environ. Sci. Health C 25 (3), 211–244.
Chan, W.H., Hsuuw, Y.D., 2007. Dosage effects of ginkgolide B on ethanol-induced cell death in human hepatoma G2 cells. Ann. N. Y. Acad. Sci. 1095 (1), 388–398.
Chen, Y.D., Zhang, J., Wang, Y., Yuan, J.L., Hu, W.L., 2016. Efficacy of cholinesterase inhibitors in vascular dementia: an updated meta-analysis. Eur. Neurol. 75 (3–4), 132–141.
Colak, Ö., Sahin, A., Alataş, Ö., İnai, M., Yaşar, B., Kiper, H., 1998. The effect of *Ginkgo biloba* on the activity of catalase and lipid peroxidation in experimental strangulation ileus. Int. J. Clin. Lab. Res. 28 (1), 69–71.
Colciaghi, F., Borroni, B., Zimmermann, M., Bellone, C., Longhi, A., Padovani, A., Di Luca, M., 2004. Amyloid precursor protein metabolism is regulated toward alpha-secretase pathway by *Ginkgo biloba* extracts. Neurobiol. Dis. 16 (2), 454–460.
Cole, G.M., Frautschy, S.A., 2007. The role of insulin and neurotrophic factor signaling in brain aging and Alzheimer's Disease. Exp. Gerontol. 42 (1–2), 10–21.
Culpan, D., Cornish, A., Love, S., Kehoe, P.G., Wilcock, G.K., 2007. Protein and gene expression of tumour necrosis factor receptors I and II and their promoter gene polymorphisms in Alzheimer's disease. Exp. Gerontol. 42 (6), 538–544.
Damiani, G., Silvestrini, G., Trozzi, L., Maci, D., Iodice, L., Ricciardi, W., 2014. Quality of dementia clinical guidelines and relevance to the care of older people with comorbidity: evidence from the literature. Clin. Interv. Aging 9, 1399.
De Toledo, M., 2006. Inflammation and Alzheimer's disease. Rev. Neurol. 42, 433–438.
Defeudis, F.V., 2002. Bilobalide and neuroprotection. Pharmacol. Res. 46 (6), 565–568.
DeKosky, S.T., Williamson, J.D., Fitzpatrick, A.L., Kronmal, R.A., Ives, D.G., Saxton, J.A., Kuller, L.H., 2008. *Ginkgo biloba* for prevention of dementia: a randomized controlled trial. JAMA 300 (19), 2253–2262.
Du, Z.Y., Li, X.Y., 1998. Effects of ginkgolides on interleukin-1, tumor necrosis factor-alpha and nitric oxide production by rat microglia stimulated with lipopolysaccharides in vitro. Arzneim. Forsch. 48 (12), 1126–1130.
Farooqui, A.A., Horrocks, L.A., 2006. Phospholipase A_2-generated lipid mediators in the brain: the good, the bad, and the ugly. Neuroscientist 12 (3), 245–260.
Fernandez-Blazquez, M.A., Ávila-Villanueva, M., Maestú, F., Medina, M., 2016. Specific features of subjective cognitive decline predict faster conversion to mild cognitive impairment. J. Alzheimers Dis. 52 (1), 271–281.
Gella, A., Durany, N., 2009. Oxidative stress in Alzheimer disease. Cell Adh. Migr. 3 (1), 88–93.
Gerlai, R., 2001. Alzheimer's disease: β-amyloid hypothesis strengthened! Trends Neurosci. 24 (4), 199.

Gohil, K., Packer, L., 2002. Global gene expression analysis identifies cell and tissue specific actions of *Ginkgo biloba* extract, EGb 761. Cell. Mol. Biol. (Noisy-le-Grand, France) 48 (6), 625–631.

Halpern, G., Halpern, G.M., 1998. Ginkgo: A Practical Guide. Avery.

Howieson, D.B., 2016. Cognitive decline in presymptomatic Alzheimer disease. JAMA Neurol. 73 (4), 384–385.

Hugo, J., Ganguli, M., 2014. Dementia and cognitive impairment: epidemiology, diagnosis, and treatment. Clin. Geriatr. Med. 30 (3), 421–442.

Hynes, N., Bishai, I., Lees, J., Coceani, F., 1991. Leukotrienes in brain: natural occurrence and induced changes. Brain Res. 553 (1), 4–13.

Ihl, R., 2012. *Gingko biloba* extract EGb 761®: clinical data in dementia. Int. Psychogeriatr. 24 (S1), S35–S40.

Ihl, R., Bunevicius, R., Frölich, L., Winblad, B., Schneider, L.S., Dubois, B., 2015. WFSBP task force on mental disorders in primary care; WFSBP task force on dementia. World Federation of Societies of Biological Psychiatry guidelines for the pharmacological treatment of dementias in primary care. Int. J. Psychiatry Clin. Pract. 19, 2–7.

Ishikawa, Y., Kitamura, M., 2000. Anti-apoptotic effect of quercetin: intervention in the JNK-and ERK-mediated apoptotic pathways. Kidney Int. 58 (3), 1078–1087.

Jørgensen, K., Hasselbalch, S.G., Waldemar, G., 2016. The risk of dementia and cognitive decline can be reduced. Ugeskr Laeger 178 (7). V11150887-V11150887.

Kennedy, S., Sud, D., 2014. A guide to prescribing anti-dementia medication. Nurs. Times 110, 16–18.

Kim, H.J., Cha, J., Lee, J.M., Shin, J.S., Jung, N.Y., Kim, Y.J., Lee, J.H., 2016. Distinctive resting state network disruptions among Alzheimer's disease, subcortical vascular dementia, and mixed dementia patients. J. Alzheimers Dis. 50 (3), 709–718.

Klegeris, A., McGeer, P.L., 2005. Non-steroidal anti-inflammatory drugs (NSAIDs) and other anti-inflammatory agents in the treatment of neurodegenerative disease. Curr. Alzheimer Res. 2 (3), 355–365.

Kong, A.N.T., Yu, R., Chen, C., Mandlekar, S., Primiano, T., 2000. Signal transduction events elicited by natural products: role of MAPK and caspase pathways in homeostatic response and induction of apoptosis. Arch Pharm. Res. 23 (1), 1–16.

Koutsilieri, E., Riederer, P., 2007. Excitotoxicity and new antiglutamatergic strategies in Parkinson's disease and Alzheimer's disease. Parkinsonism Relat. Disord. 13, S329–S331.

Kovács, T., 2009. Therapy of Alzheimer disease. Neuropsychopharmacol. Hung 11, 27–33.

Kunievsky, B.A.R.U.C.H., Yavin, E.P.H.R.A.I.M., 1992. Platelet-activating factor stimulates arachidonic acid release and enhances thromboxane B2 production in intact fetal rat brain ex vivo. J. Pharmacol. Exp. Therapeut. 263 (2), 562–568.

Le Bars, P.L., 2003. Magnitude of effect and special approach to *Ginkgo biloba* extract EGb 761® in cognitive disorders. Pharmacopsychiatry 36 (S 1), 44–49.

Liu, F., Iqbal, K., Grundke-Iqbal, I., Gong, C.X., 2002. Involvement of aberrant glycosylation in phosphorylation of tau by cdk5 and GSK-3β. FEBS Letters 530 (1–3), 209–214.

Luo, Y., 2001. *Ginkgo biloba* neuroprotection: therapeutic implications in Alzheimer's disease. J. Alzheimers Dis. 3 (4), 401–407.

Luo, Y., Smith, J.V., Paramasivam, V., Burdick, A., Curry, K.J., Buford, J.P., Butko, P., 2002. Inhibition of amyloid-β aggregation and caspase-3 activation by the *Ginkgo biloba* extract EGb761. Proc. Natl. Acad. Sci. U.S.A. 99 (19), 12197–12202.

Maclennan, K.M., Darlington, C.L., Smith, P.F., 2002. The CNS effects of Ginkgo biloba extracts and ginkgolide B. Prog. Neurobiol. 67 (3), 235–257.

Magaki, S., Mueller, C., Dickson, C., Kirsch, W., 2007. Increased production of inflammatory cytokines in mild cognitive impairment. Exp. Gerontol. 42 (3), 233–240.

Mazza, M., Capuano, A., Bria, P., Mazza, S., 2006. Ginkgo biloba and donepezil: a comparison in the treatment of Alzheimer's dementia in a randomized placebo-controlled double-blind study. Eur. J. Neurol. 13 (9), 981–985.

Micromedex Healthcare Series, 2003. MICROMEDEX, Inc., Englewood, Colorado (Edition Expires [3/2003]).

Montine, T.J., Koroshetz, W.J., Babcock, D., Dickson, D.W., Galpern, W.R., Glymour, M.M., Manly, J.J., 2014. Recommendations of the Alzheimer's disease-related dementias conference. Neurology 83 (9), 851–860.

Moore, A., Patterson, C., Lee, L., Vedel, I., Bergman, H., 2014. Fourth Canadian consensus Conference on the Diagnosis and Treatment of Dementia: recommendations for family physicians. Can. Fam. Phys. 60 (5), 433–438.

Mormino, E.C., Papp, K.V., 2016. Cognitive decline in preclinical stage 2 Alzheimer disease and implications for prevention trials. JAMA Neurol. 73 (6), 640–642.

Natural Medicines Comprehensive Database, 1999. Therapeutic Research Faculty, pp. 377–380.

Newport, D.J., Carpenter, L.L., McDonald, W.M., Potash, J.B., Tohen, M., Nemeroff, C.B., APA Council of Research Task Force on Novel Biomarkers and Treatments, 2015. Ketamine and other NMDA antagonists: early clinical trials and possible mechanisms in depression. Am. J. Psychiatry 172 (10), 950–966.

O'Brien, J.T., Thomas, A., 2015. Vascular dementia. Lancet 386, 1698–1706.

Oken, B.S., Storzbach, D.M., Kaye, J.A., 1998. The efficacy of *Ginkgo biloba* on cognitive function in Alzheimer disease. Arch. Neurol. 55 (11), 1409–1415.

Peirson, L., Fitzpatrick-Lewis, D., Morrison, K., Ciliska, D., Kenny, M., Ali, M.U., Raina, P., 2015. Prevention of overweight and obesity in children and youth: a systematic review and meta-analysis. CMAJ Open 3 (1), E23.

Peng, S., Garzon, D.J., Marchese, M., Klein, W., Ginsberg, S.D., Francis, B.M., Fahnestock, M., 2009. Decreased brain-derived neurotrophic factor depends on amyloid aggregation state in transgenic mouse models of Alzheimer's disease. J. Neurosci. 29 (29), 9321–9329.

Petersen, R.C., 2016. Mild cognitive impairment. Continuum lifelong learning. Neurology 22 (2), 404–418.

Pierre, S., Jamme, I., Droy-Lefaix, M.T., Nouvelot, A., Maixent, J.M., 1999. *Ginkgo biloba* extract (EGb 761) protects Na, K-ATPase activity during cerebral ischemia in mice. Neuroreport 10 (1), 47–51.

Pike, C.J., Walencewicz, A.J., Glabe, C.G., Cotman, C.W., 1991. In vitro aging of ß-amyloid protein causes peptide aggregation and neurotoxicity. Brain Res. 563 (1–2), 311–314.

Prince, M., Wimo, A., Guerchet, M., Gemma-Claire, A., Wu, Y.T., Prina, M., 2015. World Alzheimer Report 2015: The Global Impact of Dementia. Alzheimer's Disease International (ADI), London.

Ramassamy, C., 2006. Emerging role of polyphenolic compounds in the treatment of neurodegenerative diseases: a review of their intracellular targets. Eur. J. Pharmacol. 545 (1), 51–64.

Rimbach, G., Gohil, K., Matsugo, S., Moini, H., Saliou, C., Virgili, F., Packer, L., 2001. Induction of glutathione synthesis in human keratinocytes by *Ginkgo biloba* extract (EGb761). Biofactors 15 (1), 39–52.

Sasaki, K., Hatta, S., Wada, K., Ueda, N., Yoshimura, T., Endo, T., Haga, M., 2002. Effects of extract of *Ginkgo biloba* leaves and its constituents on carcinogen-metabolizing enzyme activities and glutathione levels in mouse liver. Life Sci. 70 (14), 1657–1667.

Schulz, V., 2003. Ginkgo extract or cholinesterase inhibitors in patients with dementia: what clinical trials and guidelines fail to consider. Phytomedicine 10, 74–79.

Shi, C., Xu, X.W., Forster, E.L., Tang, L.F., Ge, Z., Yew, D.T., Xu, J., 2008. Possible role of mitochondrial dysfunction in central neurodegeneration of ovariectomized rats. Cell Biochem. Funct. 26 (2), 172–178.

Shi, C., Zhao, L., Zhu, B., Li, Q., Yew, D.T., Yao, Z., Xu, J., 2009. Protective effects of *Ginkgo biloba* extract (EGb761) and its constituents quercetin and ginkgolide B against β-amyloid peptide-induced toxicity in SH-SY5Y cells. Chem. Biol. Interact. 181 (1), 115–123.

Shirwany, N.A., Payette, D., Xie, J., Guo, Q., 2007. The amyloid beta ion channel hypothesis of Alzheimer's disease. Neuropsychiatr. Dis. Treat. 3 (5), 597.

Simmons, L.K., May, P.C., Tomaselli, K.J., Rydel, R.E., Fuson, K.S., Brigham, E.F., Brems, D.N., 1994. Secondary structure of amyloid beta peptide correlates with neurotoxic activity in vitro. Mol. Pharmacol. 45 (3), 373–379.

Simons, M., Keller, P., De Strooper, B., Beyreuther, K., Dotti, C.G., Simons, K., 1998. Cholesterol depletion inhibits the generation of β-amyloid in hippocampal neurons. Proc. Natl. Acad. Sci. U.S.A. 95 (11), 6460–6464.

Smith, J.V., Luo, Y., 2003. Elevation of oxidative free radicals in Alzheimer's disease models can be attenuated by *Ginkgo biloba* extract EGb 761. J. Alzheimers Dis. 5 (4), 287–300.

Smith, J.V., Luo, Y., 2004. Studies on molecular mechanisms of *Ginkgo biloba* extract. Appl. Microbiol. Biotechnol. 64 (4), 465–472.

Soto, C., Sigurdsson, E.M., Morelli, L., Kumar, R.A., Castaño, E.M., Frangione, B., 1998. β-sheet breaker peptides inhibit fibrillogenesis in a rat brain model of amyloidosis: implications for Alzheimer's therapy. Nat. Med. 4 (7), 822.

Spencer, J.P., Rice-Evans, C., Williams, R.J., 2003. Modulation of pro-survival Akt/protein kinase B and ERK1/2 signaling cascades by quercetin and its in vivo metabolites underlie their action on neuronal viability. J. Biol. Chem. 278 (37), 34783–34793.

Stackman, R.W., Eckenstein, F., Frei, B., Kulhanek, D., Nowlin, J., Quinn, J.F., 2003. Prevention of age-related spatial memory deficits in a transgenic mouse model of Alzheimer's disease by chronic *Ginkgo biloba* treatment. Exp. Neurol. 184 (1), 510–520.

Tanaka, A.J., Cho, M.T., Millan, F., Juusola, J., Retterer, K., Joshi, C., Wilkins, A., 2015. Mutations in SPATA5 are associated with microcephaly, intellectual disability, seizures, and hearing loss. Am. J. Hum. Genet. 97 (3), 457–464.

Tsivgoulis, G., Katsanos, A.H., Papageorgiou, S.G., Dardiotis, E., Voumvourakis, K., Giannopoulos, S., 2014. The role of neurosonology in the diagnosis of vascular dementia. J. Alzheimers Dis. 42 (s3), S251–S257.

Tsuzuki, K., Fukatsu, R., Yamaguchi, H., Tateno, M., Imai, K., Fujii, N., Yamauchi, T., 2000. Transthyretin binds amyloid β peptides, Aβ1–42 and Aβ1–40 to form complex in the autopsied human kidney—possible role of transthyretin for Aβ sequestration. Neurosci. Lett. 281 (2–3), 171–174.

Vellas, B., Coley, N., Ousset, P.J., Berrut, G., Dartigues, J.F., Dubois, B., Touchon, J., 2012. Long-term use of standardised *Ginkgo biloba* extract for the prevention of Alzheimer's disease (GuidAge): a randomised placebo-controlled trial. Lancet Neurol. 11 (10), 851–859.

Wang, D.C., Black, S.E., Zukotynski, K.A., Zukotynski, K.A., 2016. Diagnosing dementia. CMAJ (Can. Med. Assoc. J.) 188, 603.

Watanabe, C.M., Wolffram, S., Ader, P., Rimbach, G., Packer, L., Maguire, J.J., Gohil, K., 2001. The in vivo neuromodulatory effects of the herbal medicine *Ginkgo biloba*. Proc. Natl. Acad. Sci. U.S.A. 98 (12), 6577–6580.

Wei, T., Ni, Y., Hou, J., Chen, C., Zhao, B., Xin, W., 2000. Hydrogen peroxide-induced oxidative damage and apoptosis in cerebellar granule cells: protection by *Ginkgo biloba* extract. Pharmacol. Res. 41 (4), 427–433.

Weinmann, S., Roll, S., Schwarzbach, C., Vauth, C., Willich, S.N., 2010. Effects of *Ginkgo biloba* in dementia: systematic review and meta-analysis. BMC Geriatr. 10 (1), 14.

Wettstein, A., 2000. Cholinesterase inhibitors and Ginkgo extracts—are they comparable in the treatment of dementia?: comparison of published placebo-controlled efficacy studies of at least six months' duration. Phytomedicine 6 (6), 393–401.

Wood, H., 2016. Meta-analysis finds high reversion rate from MCI to normal cognition. Nat. Rev. Neurol. 12 (4), 189–190.

Wu, Y., Wu, Z., Butko, P., Christen, Y., Lambert, M.P., Klein, W.L., Luo, Y., 2006. Amyloid-β-induced pathological behaviors are suppressed by *Ginkgo biloba* extract EGb 761 and ginkgolides in transgenic *Caenorhabditis elegans*. J. Neurosci. 26 (50), 13102–13113.

Xin, W., Wei, T., Chen, C., Ni, Y., Zhao, B., Hou, J., 2000. Mechanisms of apoptosis in rat cerebellar granule cells induced by hydroxyl radicals and the effects of EGb761 and its constituents. Toxicology 148 (2–3), 103–110.

Yancheva, S., Ihl, R., Nikolova, G., Panayotov, P., Schlaefke, S., Hoerr, R., GINDON Study Group., 2009. *Ginkgo biloba* extract EGb 761®, donepezil or both combined in the treatment of Alzheimer's disease with neuropsychiatric features: a randomised, double-blind, exploratory trial. Aging Ment. Health 13 (2), 183–190.

Yuan, J., Yankner, B.A., 2000. Apoptosis in the nervous system. Nature 407 (6805), 802.

Zhang, H.Y., Zheng, C.Y., Yan, H., Wang, Z.F., Tang, L.L., Gao, X., Tang, X.C., 2008. Potential therapeutic targets of huperzine A for Alzheimer's disease and vascular dementia. Chem. Biol. Interact. 175 (1–3), 396–402.

Zheng, S.X., Zhou, L.J., Chen, Z.L., Yin, M.L., Zhu, X.Z., 2000. Bilobalide promotes expression of glial cell line-derived neurotrophic factor and vascular endothelial growth factor in rat astrocytes. Acta Pharmacol. Sinica 21 (2), 151–155.

Zimmermann, M., Colciaghi, F., Cattabeni, F., Di, M.L., 2002. *Ginkgo biloba* extract: from molecular mechanisms to the treatment of Alzheimer's disease. Cell. Mol. Biol. 48 (6), 613–623.

Chapter 3.2.2

Panax ginseng c.a. Meyer

Amit Bahukhandi, Shashi Upadhyay, Kapil Bisht
G.B. Pant National Institute of Himalayan Environment, Almora, Uttarakhand, India

Introduction

Ginseng (meaning shaped like a man) is a common and respected traditional herbal plant in Korea, northeastern China, and far eastern Siberia, and belongs to the family Araliaceae (Yun, 2001). Many medicinally important plant species are known as ginseng, but, *Panax ginseng* (PG) is one of the most important herbal healing plants which helps to maintain physical activity. The genus name *Panax* is derived from the Greek words, "pan" meaning all and "axos" means medicine, and so *Panax* means "all healing." PG is commonly known as "Korean ginseng" and is also known as the lord or king of herbs. It is taken by inhabitants orally to enhance their thinking, concentration, memory, work efficiency, and physical stamina, to prevent muscle damage from exercise, and to increase athletic endurance, etc. It is also used in other diseases such as anxiety, general fatigue, and Alzheimer's, etc.

Taxonomy and distribution

PG is a deciduous perennial plant belonging to the family Araliaceae (Table 3.2.2.1) and is distributed in 35 countries, mainly in Asia, particularly South Korea and China (Bag and So, 2013). Genus *Panax* having a total of 13 species (Yun, 2001). Among the ginseng species, *P. ginseng* (Korean ginseng), *P. notoginseng* (Chinese ginseng), *P. japonicum* (Japan ginseng), and *P. quinquefolius* (American ginseng) are the most common.

Bioactive constituents

PG contains many bioactive compounds including amino acids, alkaloids, phenols, proteins, polypeptides, vitamins (B1 and B2) and ginsenoides are major active compound (Qi et al., 2011). More than 100 ginsenosides were isolated from different portions, i.e., rhizome, root, stem, leaves, fruits, and flowers of the species (Shin et al., 2015). According to the chemical structure of sapogenins, ginsenosides can be divided into three categories, the

Naturally Occurring Chemicals against Alzheimer's Disease. https://doi.org/10.1016/B978-0-12-819212-2.00046-3

TABLE 3.2.2.1 Systematic classification.

Kingdom:	Plantae
(Unranked):	Angiosperms
(Unranked):	Eudicots
(Unranked):	Asterids
Order:	Apiales
Family:	Araliaceae
Genus:	*Panax*
Species:	*Panax ginseng*

protopanaxadiol (PPD), protopanaxatriol (PPT), and oleanolic acid types (Ma et al., 2017; Qiu et al., 2017; Yu et al., 2017). The root portion of the species was used as medicine more than 2000 years ago. It shows various properties such as anticancer, antidiabetic, antiaging, and antiinflammatory (Qiu et al., 2017; Yu et al., 2017; Ogawa and Kawasaki, 2018), etc. Due to their potential uses and biochemical properties they were recommended as an important food resource by the Chinese government in 2012, and can be used by numerous food industries (Wang et al., 2016). Nowadays, the demand for the species is constantly increasing due to its medicinal properties, with several countries around the world, i.e., China, Korea, Japan, etc. having started intensive cultivation of the species (Baeg and So, 2013; Liu et al., 2017). In addition, a few studies have indicated that ginsenosides can be influenced by several environmental factors such as light, temperature, humidity, etc. and also by the plant type, harvesting methods, storage condition, etc. (Shi et al., 2007). There are several products, such as tonics, cosmetics, healthcare products, herbal teas, candy, honey, etc., formulated from PG (Shin et al., 2015).

Mechanism of action

PG is considered as an adaptogen, which suggests it has varied actions and effects on the body that support nonspecific resistance to biochemical and physical stressors, improve vitality and longevity, and enhance mental capacity (Blumenthal, 2003; Rochester, 2003; Kiefer and Pantuso, 2003). Reviews suggest that PG has immunomodulating activity by affecting the hypothalamic–pituitary–adrenal (HPA) axis (Fleming, 1998). Exposure of ginsenoside under in vitro conditions showed enhancement in natural killer (NK) cell activity and increased immune cell phagocytosis (Blumenthal, 2003). According to a WHO review report, in male impotence the ginseng saponins are thought to decrease serum prolactin, thereby increasing libido (WHO, 1999).

Pharmacological studies

Alzheimer's disease (AD) is a chronic neurodegenerative disease and usually occurs after 65 years of age (Kim et al., 2018). This is a very expansive disease that is present all over the world and there are very limited therapeutic drugs available. It is clinically characterized by learning and memory impairments, as well as deterioration of other cognitive and noncognitive mental functions (Hardy and Allsop, 1991). A study reported that administration of ginseng did not show sufficient monotherapeutic effects on Alzheimer's disease progression, although it can ameliorate cognition deficits in patients when combined with conventional Alzheimer's disease drugs. Two major components, ginsenosides and gintonin, may be responsible for the ginseng extract-mediated Alzheimer's disease improvements through diverse molecular mechanisms in mouse models of Alzheimer's disease (Kim et al., 2018). Various pharmacological studies on PG have been performed by researchers. The essential oil of the species inhibited acetylcholinesterase (AChE), butyrylcholinesterase (BChE), and β-secretase, which are enzymes related to the treatment and prevention of Alzheimer's disease (Kawamoto et al., 2019). Likewise, Kawamoto et al. (2019) reported the most potent activity with 51.3% inhibition at 500 μg/mL against β-secretase, however, 70.4% acetylcholinesterase and 84.4% butyrylcholinesterase inhibition occurred at 50–500 μg/mL, respectively. This species is traditionally used as a vital energy-reinforcing agent, with a high safety profile and few adverse reactions. Since ancient times the root of PG has been a popular and widely used traditional herbal medicine in Korea, China, and Japan. It has now become popular as a functional health food and is used globally as a natural medicine. Evidence is accumulating in the literature of the physiological and pharmacological effects of PG on neurodegenerative diseases. Possible ginseng- or ginsenosides-mediated neuroprotective mechanisms mainly involve maintaining homeostasis, antiinflammatory, antioxidant, antiapoptotic, immune-stimulatory activities, and reduced neurotherapeutic efficacies in neurodegenerative diseases and neurological disorders such as Parkinson's disease, Alzheimer's disease, Huntington's disease, amyotrophic lateral sclerosis, and multiple sclerosis (Cho, 2012). Hu et al. (2011) reported that PG extracts can obviate cell death, curb the overproduction of reactive oxygen species (ROS), elevate the Bax/Bcl-2 ratio, stimulate the release of cytochrome C, and activate caspase-3 expression in 1-methyl-4-phenylpyridinium (MPP) (+)-treated SH-SY5Y human neuroblastoma cells. Van Kampen et al. (2003) reported that the oral administration of PG extract G115 significantly and dramatically blocked tyrosine hydroxylase (TH) (+) cell loss in the SN and reduced the appearance of locomotor dysfunction in 1-methyl-4-phenyl-1,2,3,6- tetrahydropyridine

(MPTP)/MPP(+)-induced C57BL/6 mice and Sprague–Dawley (SD) rats. Thus, PG extracts appear to provide protective effects against neurotoxicity in in vitro and in vivo models of Parkinson's disease. However, some research has indicated that the species acts as an antidepressant (Jin et al., 2019). It is also used as an invigorant to combat memory lapses/loss by improving blood and oxygen flow to the brain and is considered to stimulate mental activity. The influence of feeding single doses of the ginsenosides Re, Rg1, and Rg3 at 50 μM in conditioned medium of CHO 2B7 cells resulted in reductions of 32.2%, 19.4%, and 69.3%, respectively, of Aβ42 after 3 h of treatment and for Rg3 the apparent IC50 was found to be < 25 μM. This was supported by further evidence from the administration of 25 mg/kg of ginsenosides which resulted in a 20%–30% reduction in Aβ42 in vivo studies in a Tg2576 mouse model after 18 h (Abdel-Salam, 2019).

In addition to all these studies, previous attempts have indicated that PG has been widely used to treat diabetes (Cho et al., 2006). The results of clinical studies demonstrated that ginseng could improve the immune response in diabetic patients (Kiefer and Pantuso, 2003). Xie et al. (2004) concluded that this antidiabetic effect of ginseng is related to ginsenosides in diabetic rats and suggested that such a type of action is due to the presence of phenolic, flavonoids, triterpenoid saponins, ginsenoides, and polysaccharides, etc., and different mechanisms are involved in suppressing blood glucose levels (El-Khayat et al., 2011). Likewise, Kim (2012) reported that ginsenosides can inhibit ROS production, stimulate NO production, increase blood circulation, ameliorate vasomotor tone, and adjust lipid profile. Additionally, several studies indicate that ginsenosides have a multitude of activities in both physiological and/or pathologic conditions concerned with cardiovascular diseases.

Dosage

PG roots can be taken orally, either chewed, or taken as a powder, liquid extract, decoction, or infusion. The level of ginsenosides can vary depending on the steeping time and type of preparation. The ginsenoside concentration can vary from approximately 64% to 77%. Crude preparations of 1–2 g dried root powder can be taken daily for up to 3 months, according to recommendations by the German Commission E (WHO, 1999). A decoction can be prepared by simmering 3–9 g dried root in 720–960 mL (24–32 oz) water for 45 min. A fluid extract (1:2 concentration) prepared from crude root can be dosed at 1–6 mL daily (Scaglione et al., 1996). An infusion can be made by pouring 150–250 mL (5–8 oz) of boiling water over 1–2 g root, steeping for 10 min while covered, and then straining before drinking. A dosage of *Panax ginseng* extract standardized to 4% ginsenosides is 200 mg per day, in divided doses, yielding 8 mg ginsenosides daily. Other reports suggest significantly higher doses of 80–240 mg ginsenosides daily might be warranted in some cases (Blumenthal, 2003).

Side effects

PG is safe when applied to the skin as part of a multi-ingredient product in the short term. However, it is unsafe when taken directly or orally mainly during pregnancy. The most common side effect is insomnia. Some reports have shown that users may experience menstrual problems, breast pain, increased heart rate, high or low blood pressure, headache, loss of appetite, diarrhea, itching, rash, dizziness, mood changes, vaginal bleeding, and other side effects (WHO, 1999; Blumenthal, 2003).

Conclusion and recommendations

The main bioactive compounds of PG are ginsenosides (ginseng saponins). Many studies have shown the pharmacological and physiological importance of PG. Traditionally it has been used for physical vitality. Current in vivo and in vitro studies have shown its beneficial effects in a wide range of pathological conditions such as Alzheimer's, cardiovascular diseases, cancer, immune deficiency, and hepatotoxicity. In general, antioxidant, antiinflammatory, antiapoptotic, and immunostimulant activities are mostly behind the possible ginseng-mediated protective mechanisms. Also, research shows that it has many side effects, hence it should be used or taken under medical observation and consultation as it may be unsafe for some users.

References

Abdel-Salam, O.M., 2019. Use of herbal products/alternative medicines in neurodegenerative diseases (Alzheimer's disease and Parkinson's disease). In: Pathology, Prevention and Therapeutics of Neurodegenerative Disease. Springer, Singapore, pp. 279–301.

Baeg, I.H., So, S.H., 2013. The world ginseng market and the ginseng (Korea). J. Ginseng Res. 37 (1), 1–7.

Blumenthal, M., 2003. The ABC Clinical Guide to Herbs. Theime, New York, NY, pp. 211–225.

Cho, I.H., 2012. Effects of *Panax ginseng* in neurodegenerative diseases. J. Ginseng Res. 36 (4), 342.

Cho, W.C., Chung, W.S., Lee, S.K., Leung, A.W., Cheng, C.H., Yue, K.K., 2006. Ginsenoside Re of *Panax ginseng* possesses significant antioxidant and antihyperlipidemic efficacies in streptozotocin-induced diabetic rats. Eur. J. Pharmacol. 550 (1–3), 173–179.

El-Khayat, Z., Hussein, J., Ramzy, T., Ashour, M., 2011. Antidiabetic antioxidant effect of *Panax ginseng*. J. Med. Plants Res. 5 (18), 4616–4620.

Fleming, T., 1998. Physician Desk References for Herbal Medicine, First ed. Medical Economics Company, Montvale, NJ.

Hardy, J., Allsop, D., 1991. Amyloid deposition as the central event in the aetiology of Alzheimer's disease. Trends Pharmacol. Sci. 12, 383–388.

Hu, S., Han, R., Mak, S., Han, Y., 2011. Protection against 1-methyl-4-phenylpyridinium ion (MPP+)-induced apoptosis by water extract of ginseng (*Panax ginseng* CA Meyer) in SH-SY5Y cells. J. Ethnopharmacol. 135 (1), 34–42.

Jin, Y., Cui, R., Zhao, L., Fan, J., Li, B., 2019. Mechanisms of Panax ginseng action as an antidepressant. Cell Prolif. 52 (6), e12696.

Kawamoto, H., Takeshita, F., Murata, K., 2019. Inhibitory effects of essential oil extracts from Panax plants against β-secretase, cholinesterase, and amyloid aggregation. Nat. Prod. Commun. 14 (10), 1934578X19881549.

Kiefer, D., Pantuso, T., 2003. Panax ginseng. Am. Fam. Physician 68, 1539–1542.

Kim, J.H., 2018. Pharmacological and medical applications of *Panax ginseng* and ginsenosides: a review for use in cardiovascular diseases. J. Ginseng Res. 42 (3), 264–269. https://doi.org/10.1016/j.jgr.2017.10.004.

Kim, H.J., Jung, S.W., Kim, S.Y., Cho, I.H., Kim, H.C., Rhim, H., Kim, M., Nah, S.Y., 2018. *Panax ginseng* as an adjuvant treatment for Alzheimer's disease. J. Ginseng Res. 42 (4), 401–411.

Kim, J.H., 2012. Cardiovascular diseases and *Panax ginseng*: a review on molecular mechanisms and medical applications. J. Ginseng Res. 36 (1), 16.

Liu, Z., Wang, C.Z., Zhu, X.Y., Wan, J.Y., Zhang, J., Li, W., Raun, C.C., Yuan, C.S., 2017. Dynamic changes in neutral and acidic ginsenosides with different cultivation ages and harvest seasons: identification of chemical characteristics for *Panax ginseng* quality control. Molecules 22 (5), 734.

Ma, G.D., Chiu, C.H., Hsu, Y.J., Hou, C.W., Chen, Y.M., Huang, C.C., 2017. Changbai Mountain ginseng (*Panax ginseng* CA Mey) extract supplementation improves exercise performance and energy utilization and decreases fatigue-associated parameters in mice. Molecules 22 (2), 237.

Ogawa, K.O., Kawasaki, K., 2018. Panax ginseng for frailty-related disorders: a review. Front. Nutr. 5, 140.

Qi, L.W., Wang, C.Z., Yuan, C.S., 2011. Isolation and analysis of ginseng: advances and challenges. Nat. Prod. Rep. 28, 467–495.

Qiu, S., Yang, W.Z., Yao, C.L., Shi, X.J., Li, J.Y., Lou, Y., Duan, Y.N., Wu, W.Y., Guo, D.A., 2017. Malonylginsenosides with potential antidiabetic activities from the flower buds of *Panax ginseng*. J. Nat. Prod. 80 (4), 899–908.

Rochester, V.T., 2003. Medical Herbalism: The Science and Practice of Herbal Medicine. Healing Arts Press, p. 570.

Scaglione, F., Cattaneo, G., Alessandria, M., Cogo, R., 1996. Efficacy and safety of the standardized ginseng extract G115 for potentiating vaccination against the influenza syndrome and protection against the common cold. Drugs Exp. Clin. Res. 22, 65–72.

Shi, W., Wang, Y., Li, J., Zhang, H., Ding, L., 2007. Investigation of ginsenosides in different parts and ages of *Panax ginseng*. Food Chem. 102 (3), 664–668.

Shin, B.K., Kwon, S.W., Park, J.H., 2015. Chemical diversity of ginseng saponins from *Panax ginseng*. J. Ginseng Res. 39 (4), 287–298.

Van Kampen, J., Robertson, H., Hagg, T., Drobitch, R., 2003. Neuroprotective actions of the ginseng extract G115 in two rodent models of Parkinson's disease. Exp. Neurol. 184 (1), 521–529.

Wang, H.P., Zhang, Y.B., Yang, X.W., Yang, X.B., Xu, W., Xu, F., Cai, S.Q., Wang, Y.P., Xu, Y.H., Zhang, L.X., 2016. High-performance liquid chromatography with diode array detector and electrospray ionization ion trap time-of-flight tandem mass spectrometry to evaluate ginseng roots and rhizomes from different regions. Molecules 21 (5), 603.

World Health Organization, 1999. Radix Ginseng. WHO Monographs on Selected Medicinal Plants, 1. World Health Organization, Geneva, Switzerland, pp. 168–182.

Xie, J.T., Mehendale, S.R., Wang, A., Han, A.H., Wu, J.A., Osinski, J., Yuan, C.S., 2004. American ginseng leaf: ginsenoside analysis and hypoglycemic activity. Pharmacol. Res. 49 (2), 113–117.

Yu, S., Zhou, X., Li, F., Xu, C., Zheng, F., Li, J., Zhao, H., Dai, Y., Liu, S., Feng, Y., 2017. Microbial transformation of ginsenoside Rb1, Re and Rg1 and its contribution to the improved anti-inflammatory activity of ginseng. Sci. Rep. 7 (1), 138.

Yun, T.-K., 2001. Brief introduction of *Panax ginseng* C.A. Meyer. J. Korean Med. Sci. 16, S3–S5.

Chapter 3.2.3

Melissa officinalis (lemon balm)

Koula Doukani[1], Ammar Sidi Mohammed Selles[2], Hasna Bouhenni[1]
[1]*Faculty of Nature and Life Sciences, University of Ibn Khaldoun, Tiaret, Algeria;* [2]*Institute of Veterinary Sciences, University of Ibn Khaldoun, Tiaret, Algeria*

Introduction

Alzheimer's disease (AD) is considered as a major neurodegenerative disorder and the main cause of dementia in the elderly (Andrade et al., 2019).

Various allopathic medicines are prescribed in the treatment of AD but they exert side effects. Therefore, herbal medicines could be a good source of drugs for treatment of AD and memory deficit with fewer or no side effects (Akram and Nawaz, 2017).

Medicinal plants have played an excellent role in the treatment and prevention of disease in human (Zarei et al., 2015). According to the World Health Organization (WHO), the majority of the world's population uses traditional medicines in primary health care (Taiwo, 2007). This use is mainly related to the unwanted side effects of synthetic drugs, although herbal products are safe and effective (Kumar et al., 2012).

The discovery of new herbal medicines with neurobiological activity is sought. However, the explanations of their mechanisms of action are among the most intensive areas of scientific development. In addition, plant-based substances as a component of a healthy diet are an attractive alternative in the prevention and treatment of dementia (Ozarowski et al., 2016).

In traditional medicine, several thousand plant species have been used worldwide and constitute a potential reservoirs for the discovery of new drugs (Limen-Ben Amor et al., 2009). However, more than 150 plants are used to improve learning and memory, supporting the direct or indirect traditional beliefs in various preparations (Adams et al., 2007).

Fermino et al. (2015) report according to Carolus Linnaeus in 1753 in the French Pharmacopoeia that among these plant species lemon balm L. (Lemon balm, bee balm, honey balm) is one of the main species of medicinal plants cultivated mainly in natural flora especially in the Mediterranean region, and

Naturally Occurring Chemicals against Alzheimer's Disease. https://doi.org/10.1016/B978-0-12-819212-2.00047-5

native to southern Europe and North Africa, and east to the Caucasus and in the north of Iran. Its wild types are in all Mediterranean countries and the southern part of the Alps (Bagdat and Cosge, 2006).

Traditional herbal medicine suggests the use of *Melissa officinalis* (MO) for all complaints proceeding from a nervous system disorder (Scholey and Stough, 2011) and to improve memory and concentration (Dehbani et al., 2019).

Taxonomy

Belonging to the Lamiaceae family, *M. officinalis* (lemon balm) is a perennial herb with a lemon scent (Fig. 3.2.3.1) (Jalal et al., 2015). Known by nine synonyms, *Melissa officinalis* L. also has an infraspecific taxon of species, *Melissa officinalis* subsp. Inodora Bornm (The Plant List, 2013).

Synonyms: "*Melissa officinalis* subsp. altissima (Sm.) Ar- cang., *Melissa officinalis* var. altissima (Sm.) K.Koch, *Melissa officinalis* var. cordifolia (Pers.) K.Koch, *Melissa officinalis* var. foliosa Briq., *Melissa officinalis* var. graveolens (Host) Nyman, *Melissa officinalis* var. hirsuta K. Koch, *Melissa officinalis* subsp. officinalis, *Melissa officinalis* var. romana (Mill.) Woodv. and *Melissa officinalis* var. villosa Benth" (The Plant List, 2013).

The taxonomical classification of this plant is as follows:

Kingdom: Plantae
Subkingdom: Tracheobionta
Division: Magnoliophyta
Subdivision: Spermatophyta
Class: Magnoliopsida
Subclass: Asteridae
Order: Lamiales
Family: Lamiaceae.
Genus: Melissa
Species: *Melissa officinalis* L.

FIGURE 3.2.3.1 *Melissa officinalis* (www.hear.org).

Cultivation

Lemon balm is considered by some gardeners as a weed, this is a consequence of its ease of growing and spreading, hence it is easier to cultivate by beginners (Miraj et al.,2017). For the best performance of the cultivation of this plant, fertile sandy and loamy soils, well drained and a pH varying from 5 to 7 are necessary. However, for better growth and development it requires sunny days. Nevertheless, it also grows in cloudy climates. When the plants grow in semi-shade, they make larger leaves than those grow up toward sunny conditions. A moderate temperature of 15−35°C and rainfall between 500 and 600 mm well distributed over the season are required for good growth. It is particularly sensitive to drought (Moradkhani et al., 2010; Verma et al., 2015).

Phytochemical profile

The leaf of *Melissa officinalis* is a rich source of compounds including flavonoids (quercitrin, rhamnocitrin, luteolin), polyphenolic compounds (hydroxycinnamic acid derivatives, especially caffeic acid, rosmarinic acid, and protocatechuic acid), monoterpene glycosides, monoterpenoid aldehydes, triterpenes (oleanolic acids and ursolic), tannins, and sesquiterpenes (Miraj et al., 2016, 2017; Sofowora et al., 2013; Gurčík et al., 2005). The yield of *Melissa* essential oil varies from 0.02% to 0.37% of the dry weight. Citral, neral, citronellal, and geraniol are the major constituents of this essential oil (Fermino et al., 2015; Moradkhani et al., 2010; Carnat et al., 1998).

Traditional uses and pharmacology

The medicinal properties of *M. officinalis* (leaf extract or essential oil) give it a great role in traditional medicine, aromatherapy, and the food industry (Moaca et al., 2018).

Used formerly in traditional medicine, lemon balm was incorporated into the preparation of tonic solutions, called "elixir of life," by French monks and nuns, and Paracelsus, a famous doctor and chemist in the period from 1493 to 1541. John Evelyn (1620−1706) described this plant as "the rule of the brain, strengthening the mind and withdrawing from melancholy." The Hebrew names for the essential oil is "bal-smin" or "leader of oils". Meanwhile Avicenna suggests that lemon balm strengthens the heart. Today, lemon balm is used in various industries (medicine, cosmetics, food, etc.) in many countries (Bagdat and Cosge, 2006).

Several studies have reported folk medicinal use of MO for memory enhancement, menstrual-inducing, cardiotonic, anxiolytic agent, nervousness and stress-induced headaches, antigas, fever-reducing, antiseptic, antimicrobial, antifungal, antispasmodic, antiparasitic, carminative, hypotensive, diaphoretic, sedative-hypnotic, hepatoprotective, effects and as surgical dressing for wounds (Moradkhani et al., 2010; Verma et al., 2015; Jalal et al., 2015; Shakeri et al., 2016; Moradpour et al., 2017). Besides, it is effective in the treatment of indigestion, colic, rheumatism, vertigo, malaise, insomnia, nausea, epilepsy, anemia, migraine, syncope, depression, hysteria, and Alzheimer's psychosis (Jalal et al., 2015; Miraj et al., 2016, 2017). In addition, there are studies indicating that extract of lemon balm presents a cytotoxic effect on several cancers, including breast and colon cancers (Zarei et al., 2015; Miraj et al., 2016). It has been reported that it contains natural products that can inhibit protein biosynthesis in cancer cells (Aldal'in, 2018).

Furthermore, it can be used for its antitumor, antiherpes, anti-HIV, antiviral, and antidiabetic activities (Basar and Zaman, 2013; Jalal et al., 2015; Miraj et al., 2016, 2017).

Additionally, extract of lemon balm possesses an antioxidant effect (Zarei et al., 2015), and antinociceptive and antiinflammatory activities by binding to muscarinic and nicotinic receptors (Moaca et al., 2018). Recently, it has been shown that it is potentially useful in treating hyperthyroidism (Basar and Zaman, 2013); this results from its effect on the function of the pituitary gland by improving the hormonal levels of TSH (thyroid-stimulating hormone), T3 (triiodothyronine), and T4 (thyroxine) (Zarei et al., 2015). Likewise, lemon balm causes a decrease in total lipids by improving the amount of high-density lipoproteins and decreasing hepatic cholesterol synthesis from where it can be used in the treatment of cardiovascular diseases. It can also be used in the treatment of respiratory, mental, and central nervous system diseases (Moaca et al., 2018).

Moreover, it was also found to treat externally diseases such as gout, herpes, and sores, and to act as an insect repellent (Aldal'in, 2018).

Currently, in modern botanical medicine, lemon balm is being researched for its effects as a mood- and concentration-enhancing plant, and for its calming, sleep, and cognition-improving effects (Bazzari and Bazzari, 2018).

Treatment

Neurodegenerative disorders including Alzheimer's disease are a greater public health problem around the world (Cole and Vassar, 2007; Craig et al., 2011). Several drugs have been employed in the treatment of Alzheimer's disease (AD), including two different modes of action that have been observed for these drugs (Winslow et al., 2011):

Acetylcholinesterase inhibitors (AChEI)

Acetylcholinesterase inhibitors (AChEI); including one of the first generation (tacrine) and three of the second generation (donepezil, rivastigmine, and galantamine) have been adopted for the treatment of mild to moderate AD (Weinstock, 1999; Farlow et al., 2008; Bishara et al., 2015).

These agents vary in their pharmacological activities (Bishara et al., 2015). Pharmacologically, tacrine has several mechanisms of action, among which are the increase in cerebral blood flow and the inhibition of the production of amyloid-β (Summers, 2000). However, donepezil has a relative selectivity for acetylcholinesterase (AchE) compared to butyryl cholinesterase (BuChE) allowing it to have a mixed competitive and noncompetitive AchEI effect (Tsuno, 2009). Meanwhile, rivastigmine (rivastigmine tartrate) is a selective inhibitor of AChE and BuChE of the brain compared to peripheral tissues by inhibition of pseudo-irreversible carbamate (Onor et al., 2007) and galantamine selectively inhibits AChE as well as having an effect on the allosteric modulator of nicotinic acetylcholine receptors (nAChRs) (Weinstock, 1999; Bishara et al., 2015).

N-methyl-D-aspartate (NMDA) receptor antagonists

N-methyl-D-aspartate (NMDA) receptor antagonists like memantine have been indicated in moderately severe to severe disease (Bishara et al., 2015). This is the only glutamate receptor ligand that is approved for treatment of AD (Johnson et al., 2015).

Unfortunately, several adverse effects can be observed following the use of these drugs. Table 3.2.3.1 summarizes these adverse effects.

The limited chemical drugs along with their reported adverse effects (diarrhea, vomiting, fatigue, nausea, weight loss, loss of appetite) are the main reasons for seeking natural effective treatments by scientists (Mahboubi, 2019). Medicinal plants play an important role in the treatment and prevention of diseases in human health (Changizi-Ashtiyani et al., 2013). The discovery of new herbal medicines with neurobiological activity associated with the determination of their mechanisms of action is currently a very interesting scientific field (Ozarowski et al., 2016).

On the other hand, different herbal prescriptions are used in the prevention and treatment of AD in traditional systems (Mahboubi, 2019). This long-standing use has been described in many traditional medicines up to the present time (traditional Chinese medicine, Iranian traditional medicine, and European herbal medicine) (Perry and Howes, 2011; Shi et al., 2017; Mahboubi, 2019).

Recently, many new compounds have been isolated from plants to improve the treatment of dementia with fewer side effects than conventional drugs and they have been developed by scientists. These substances are considered to be potential anti-AD drugs (Sun et al., 2013).

The richness of medicinal plants in several compounds have resulted in various pharmacological activities and wide use in the treatment of many diseases such as AD (Anekonda and Reddy, 2005).

TABLE 3.2.3.1 Adverse effects of conventional drugs in the treatment of AD.

	Adverse effects	References
Tacrine	Cholinergic "side effects": nausea, myalgia, diarrhea, tremor, dyspepsia, emesis, excessive urination, rhinitis	Davis and Powchik (1995); Summers (2000)
Donepezil	Diarrhea, headache, common cold, nausea, anorexia, hallucinations, aggressive agitation, behavior, abnormal dreams and nightmares, dizziness, syncope, abdominal disturbance, insomnia, vomiting, pruritus, rash, muscle cramps, fatigue, urinary incontinence, pain	Schneider (2013); Bishara et al. (2015)
Rivastigmine	Anorexia, dizziness, nausea, vomiting, diarrhea, agitation, confusion, malaise, weight loss, headache, somnolence, tremor, abdominal pain and dyspepsia, sweating fatigue, anxiety, asthenia	Schneider (2013); Bishara et al. (2015)
Galantamine	Nausea, vomiting, fatigue, decreased appetite, anorexia, hallucination, somnolence, depression, lethargy, syncope, dizziness; tremor, headache, bradycardia, hypertension, abdominal pain and discomfort, diarrhea, dyspepsia, sweating, muscle spasms, asthenia, malaise, weight loss, falls	Schneider (2013); Bishara et al. (2015)
Memantine	Drug hypersensitivity, headache, somnolence, dizziness, hypertension, dyspnea, balance disorders, constipation, elevated liver function tests	Bishara et al. (2015)

Several plants have been utilized in traditional medicine for the treatment of dementia and Alzheimer's disease, among which mention may be made of *Ginkgo biloba* L., *Panax ginseng* CA Meyer, genus *Turmeric*, genus *Glycyrrhiza*, *Camellia sinensis* Kuntze (Tewari et al., 2018), *Melissa officinalis*, and *Salvia officinalis* (Akhondzadeh and Abbasi, 2006).

For a long time, *Melissa officinalis* (lemon balm) has been used in the treatment of many diseases such as headaches, neurological diseases, gastrointestinal diseases, and rheumatoid problems (Jun et al., 2012; Wichtl, 2004).

Alternative treatment by *Melissa officinalis*

Popular and formerly known as a medicinal plant, *Melissa officinalis* has been used for the prevention and treatment of nervous disturbances in phytotherapy (Ozarowski et al., 2016; WHO, 2004). In popular medicine, *M. officinalis* has been used for the treatment of dementia and amnesia which are strongly associated with Alzheimer's disease (Ellis, 2005; Cummings, 2000).

The daily oral dose of dried herb for adults is 1.5−4.5 g, as an infusion in 150 mL water, 2−4 mL of 45% ethanol extract (1:1), three times a day, and 2−6 mL tincture (1:5 in 45% of ethanol), three times a day (Mahboubi, 2019; Edwards et al., 2015).

According to Iranian traditional medicine, the typical dose for *M. officinalis* was 40 g of dry leaves, 80 g of fresh leaves. and 9 g of dry seeds (Mahboubi, 2019). Doses of its extract at 600−1600 mg are used in clinical studies. Topical agents containing 1% *M. officinalis* extract are used for treatment of herpes virus lesions (Mahboubi, 2019; Koytchev et al., 1999).

The efficacy and safety of *Melissa officinalis* extract to moderate Alzheimer's disease patients were examined using a fixed dose (60 drops/day). After 4 months, *Melissa officinalis* extract presented a significantly better effect on cognitive function than placebo (Akhondzadeh et al., 2003).

Mechanism of action

The pathology of Alzheimer's disease involves β-amyloidosis that causes amyloid plaques in the central nervous system (CNS), forming neurofibrillary tangles, and leading to neurodegeneration. Neurotransmitter abnormalities are caused by cholinergic deficits, as well as other neurotransmitter systems of the central nervous system, including glutamatergic and serotonergic systems. Inflammatory mechanisms, apoptosis, and attenuated neuroplasticity oxidative damage have also been implied in the pathology of AD (Howes et al., 2017).

Mechanisms of action of *M. officinalis* potency in memory functions and Alzheimer disease are summarized in Fig. 3.2.3.2.

Effect of *M. officinalis* on AchE

One pathological factor for AD is deficiency in acetylcholine neurotransmitters and reduced acetylcholine in the cholinergic system with a critical role of AChE (Mahboubi, 2019). The compounds of *M. officinalis* improve cognitive functions such as memory by binding to acetylcholine and by inhibition of acetylcholinesterase (AChE) (Rostami et al., 2010). It has been reported that *Melissa officinalis* (lemon balm) increases cognitive function and decreases

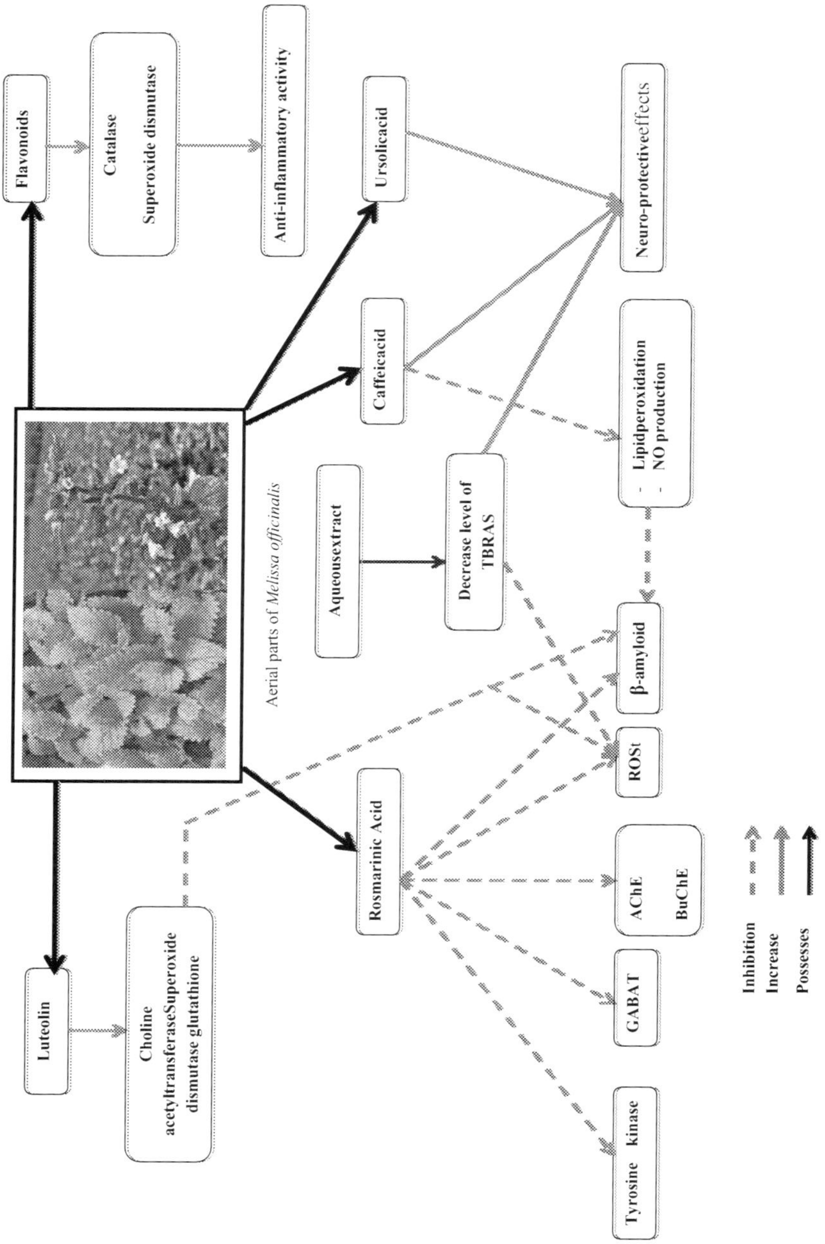

FIGURE 3.2.3.2 Proposition of the various mechanisms of action of *M. officinalis* in Alzheimer's disease.

agitation in patients with mild to moderate AD (Akhondzadeh and Abbasi, 2006). This effect is in relation to acetylcholinesterase inhibitors (AChE), which raise the level and time of action of acetylcholine in brain synapses (Ellis, 2005; Cummings, 2000). This activity is due to the link in the activity of ACh receptor in the central nervous system with both muscarinic and nicotinic binding properties (Akhondzadeh and Abbasi, 2006). The effect of cholinergic can reduce the effects of AD and decrease its cognitive symptoms (on memory and learning deficits), and also cognitive impairments in patients with schizophrenia (Ellis, 2005; Cummings, 2000). Soodi et al. (2014) revealed that intraperitoneal injection of extract of *M. officinalis* decreased AChE enzyme in the hippocampus of rats.

In addition, lemon balm consistently contains high levels of rosmarinic acid. This compound is an inhibitor of AChE and BuChE (Kim et al., 2015a) and is known for its antioxidative and anti-inflammatory properties (Vladimir-Kneževic et al., 2014; Orhan et al., 2008). Rosmarinic acid, by improving hippocampal functions and inhibiting γ-aminobutyric acid (GABA) transaminase, could prevent the progression of dementia (Awad et al., 2009). Increasing levels of GABA may suggest additional therapeutic benefits to accompany symptoms such as anxiety, insomnia, and aggressive behavior (Jembrek and Vlainic, 2015).

Additionally, rosmarinic acid can also inhibit the activity of tyrosine kinase (Baluchnejadmojarad et al., 2013; Sebestík et al., 2011), which is expressed extensively in the hippocampus and is involved in the induction of long-term potentiation (LTP) which is the basis of learning and memory (Baluchnejadmojarad et al., 2013; Jelic et al., 2007).

Effects of *M. officinalis* on β-amyloid (Aβ)

Alzheimer's disease is a weakening neurodegenerative disorder typified by increased β-amyloid (Aβ) deposition and neuronal dysfunction causing impaired learning and memory function (Baluchnejadmojarad et al., 2013).The neurodegenerative process in AD is attributed to oxidative stress. β-Amyloid significantly increases the free radicals and their products and enhances the production of peroxynitrite ($ONOO^-$) and peroxidation of membrane lipid, which are associated with degeneration of neurons and cell death in AD (Mahboubi, 2019). Baluchnejadmojarad et al. (2013) showed that rosmarinic acid reversed the Aβ-induced increase in malondialdehyde and nitrite production but could not improve superoxide dismutase activity, suggesting that the beneficial effect of rosmarinic acid does not occur primarily via its antioxidant capacity. Likewise, Lopez et al. (2009) reported that *M. officinalis* has a direct free radical scavenging activity.

Rosmarinic acid acts by repressing the formation of fibrils from Aβ and disrupting preformed Aβ fibrils in vitro (Ono et al., 2004). Moreover, this compound inhibits the production of reactive oxygen species directly or indirectly by preventing Aβ fibril formation in PC12 cells (Iuvone et al., 2006). The total extract of *M. officinalis* can protect PC12 cells against cell death induced by hydrogen peroxide and oxidative stress (Lopez et al., 2009). Aliev et al. (2008) reported that antioxidants can ameliorate disease progression. These same authors attributed the antioxidant activity of *M.officinalis* to the neuroprotective effect of *M. officinalis*, which is opposed to Aβ toxicity.

These same authors noted that rosmarinic acid protects the cells against β-amyloid induced toxicity, inhibiting the phosphorylated p38 mitogen excited protein kinase and reducing the membrane lipid peroxidation in a dose-dependent manner. Rosmarinic acid inhibits the p38 MAP kinase pathway, caspase-3 activity, and DNA fragmentation, which results in inhibition of tau protein hyperphosphorylation and protection against β-amyloid induced apoptosis (Mahboubi, 2019; Iuvone et al., 2006).

Additionally, caffeic acid, by inhibiting lipid peroxidation and production of NO, improves the cognitive damage producing Aβ in the brain (Kim et al., 2015b).

Among the active components of *M. officinalis*, luteolin works by increasing various enzymes in the hippocampal tissue (choline acetyl transferase, superoxide dismutase, and glutathione peroxidase) that improves the memory deficit induced by Aβ. Also, it reverses the increased activity of acetylcholine esterase by increasing the amount of acetylcholine and reducing malondialdehyde in the hippocampal homogenate. Moreover, in an experimental study on AD rats, luteolin could protect against memory deficits induced by Aβ by regulating the cholinergic system and inhibiting oxidative damages (Beheshti and Shahmoradi, 2018; Yu et al., 2015).

Antioxidant activity of *M. officinalis*

M. officinalis has powerful antioxidant effects (Zarei et al., 2015). These effects have been related to the high amounts of potent antioxidant and phenolic compounds, hydroxycinnamic derivatives, rosmarinic acid, benzodioxole, chlorogenic acid, caffeic acid, linoleic acid, urosolic acid, and carnosic acid that are present in the extract (Mahboubi, 2019; Zarei et al., 2015). According to the literature, the most important component of lemon balm responsible for its antioxidant activity is rosmarinic acid (Barros et al. 2013).

Lemon balm extract could inhibit the onset of lipid peroxidation by inhibiting the generation of early reactive chemical species. This reaction can explain the blocking of the common final route in the process of peroxidation of polyunsaturated fatty acids. Lemon balm infusion increases plasma levels of catalase, superoxide dismutase, and glutathione peroxidase and a marked decrease in plasma DNA damage, myeloperoxidase, and lipid peroxidation. Likewise, the increase in the antioxidant potential of the extract is due to its chelating activity of iron (II) (Miraj et al., 2017; Dastmalchi et al., 2008).

Martins et al. (2012) reported that manganese increases the level of thiobarbituric acid reactive substances (TBARS) as a marker of oxidative stress in the hippocampus and striatum. These authors also reported that mice treated with the aqueous extract of *Melissa officinalis* had decreased levels of cerebral manganese and concluded that the aqueous extract of *M. officinalis* has powerful antioxidant and neuroprotective characteristics, validating its effectiveness in attenuation of oxidative stress caused by Mn in the brain of mice.

Ursolic acid is one of the components of lemon balm, and plays a role in antioxidant activity. This triterpenoid eliminates hydroxyl radicals and increases the activities of antioxidant enzymes (superoxide dismutase, catalase, and glutathione peroxidase) (Liu, 1995).

Anti-inflammatory activity

M. officinalis is a rich source of monoterpene aldehydes linked to geraniol and nerol (geranial and neral). These compounds have anti-inflammatory activity (Gbenou et al., 2013; Shen et al., 2015). Likewise, *Melissa officinalis* has the capacity to stop the activity of monooxygenases, lipooxygenases, and cyclooxygenases which is due to its phenolic compounds such as flavonoids (Fermino et al., 2015; Svobodová et al., 2003). Nevertheless, competition in some cases and allosteric inhibitions in others have been observed following oxidoreductases, and hydrolases such as the hyaluronate lyase that catalyzes the degradation of hyaluronic acid (Fermino et al., 2015; Havsteen, 2002). Other authors report that the anti-inflammatory activity is probably due to flavonoids which have an effective role in facilitating the synthesis of prostaglandins (Miladi-Gorgi et al., 2005; Anjaneyulu and Chopra, 2003).

In addition, different authors have cited rosmarinic acid as one of the most common compounds. This molecule has an anti-inflammatory activity, and an inhibiting activity of 5-lipoxygense, 3R-hydroxysteroid dehydrogenase, and lipid peroxidation (Nakazawa and Ohsawa, 1998).

Safety

No side effects have so far been reported for this herb when used topically or orally at recommended doses for up to 30 days. In otherwise healthy adults and when consumed in amounts found in foods, lemon balm has been assigned generally regarded as safe (GRAS) status in the United States with a maximum level of 0.5% in baked goods (Miraj et al., 2017).

Several studies have reported that the administration of *M. officinalis* extracts in young healthy persons has no effects or symptoms of toxicity. Meanwhile, it improves and actively amends mood and cognitive ability (Obulesu and Rao, 2011; Akhondzadeh et al., 2003; Kennedy et al., 2002; Wake et al., 2000; Schulz et al., 1998).

Akhondzadeh et al. (2003), indicated that the effects associated with *Melissa* were usually those supposed from cholinergic stimulation, and are identical to those reported with cholinesterase inhibitors. Although we did not take account agitation to be a side effect, the frequency of agitation was higher in the placebo group, and this showed an additional advantage of *Melissa officinalis* in the management of agitated patients with Alzheimer's disease. This is in harmony with the results of a double-blind, placebo-controlled study that tested the effects of *Melissa officinalis* essential oil aromatherapy on ratings of agitation and quality of life in 71 patients suffering from severe dementia (Akhondzadeh et al., 2003).

Conclusion

M. officinalis is one of the most ancient and best known herbaceous aromatic plants. Several activities are characteristic of this plant or its extracts. Antioxidant, sedative, neuroprotective, antianxiety, and hypnotic properties are included in these activities. Expectations for the potential of *M. officinalis* for diseases like Alzheimer's therapy have been met by an impressive literature on the clinical and scientific studies of this plant. The plant is recognized as safe and side effects are very rare and generally mild when they do occur, which allows for its use in many treatments.

A comparison between *M. officinalis* treatment and pharmacological treatments should be performed. Such studies should include identification of the active principal in order to improve the validation of the clinical trial. Further large-scale studies are necessary to determine the effectiveness of this plant in the treatment of AD.

References

Adams, M., Gmunder, F., Hamburger, M., 2007. Plants traditionally used in age-related brain disorders- a survey of ethnobotanical literature. J. Ethnopharmacol. 113 (3), 363–381.

Akhondzadeh, S., Abbasi, S.H., 2006. Herbal medicine in the treatment of Alzheimer's disease. Am. J. Alzheimers Dis. Other Demen. 21 (2), 113–118.

Akhondzadeh, S., Noroozian, M., Mohammadi, M., Ohadinia, S., Jamshidi, A.H., Khani, M., 2003. *Melissa officinalis* extract in the treatment of patients with mild to moderate Alzheimer's disease: a double blind, randomised, placebo controlled trial. J. Neurol. Neurosurg. Psychiatry 74 (7), 863–866.

Akram, M., Nawaz, A., 2017. Effects of medicinal plants on Alzheimer's disease and memory deficits. Neural. Regen. Res. 12 (4), 660–670.

Aldal'in, H.K., 2018. Phytochemical analysis of methanolic extract of Jordanian *Melissa officinalis* L. Int. J. Biol. 10 (4), 58–62.

Aliev, G., Obrenovich, M.E., Reddy, V.P., Shenk, J.C., Moreira, P.I., Nunomura, A., Zhu, X., Smith, M.A., Perry, G., 2008. Antioxidant therapy in Alzheimer's disease: theory and practice. Mini Rev. Med. Chem. 8 (13), 1395–1406.

Andrade, S., Ramalho, M.J., Loureiro, J.A., Pereira, M.D.C., 2019. Natural compounds for Alzheimer's disease therapy: a systematic review of preclinical and clinical studies. Review. Int. J. Mol. Sci. 20 (9), 2313.

Anekonda, T.S., Reddy, P.H., 2005. Can herbs provide a new generation of drugs for treating Alzheimer's disease? Brain Res. Rev. 50 (2), 361–376.

Anjaneyulu, M., Chopra, K., 2003. Quercetin, a bioflavonoid, attenuates thermal hyperalgesia in a mouse model of diabetic neuropathic pain. Prog. Neuro-Psychopharmacol. Biol. Psychiatry 27 (6), 1001–1005.

Awad, R., Muhammad, A., Durst, T., Trudeau, V.L., Arnason, J.T., 2009. Bioassay-guided fractionation of lemon balm (*Melissa officinalis* L.) using an in vitro measure of GABA transaminase activity. Phytother. Res. 23 (8), 1075–1081.

Bagdat, R.B., Cosge, B., 2006. The essential oil of lemon balm (*Melissa officinalis* L.), its components and using fields. J. Fac. Agric. OMU 21 (1), 116–121.

Baluchnejadmojarad, T., Roghani, M., Kazemloo, P., 2013. Rosmarinic acid mitigates learning and memory disturbances in amyloid β(25–35)-induced model of Alzheimer's disease in rat. J. Basic Clin. Pathophysiol. 2 (1), 7–14.

Barros, L., Duenãs, M., Dias, M.I., Sousa, M.J., Santos-Buelga, C., Ferreira, I.C., 2013. Phenolic profile of cultivated, in vitro cultured and commercial samples of *Melissa officinalis L.* infusions. Food Chem. 136 (1), 1–8.

Basar, S.N., Zaman, R., 2013. An overview of Badranjboya (*Melissa officinalis*). Int. Res. J. Biol. Sci. 2 (12), 107–109.

Bazzari, A.H., Bazzari, F.H., 2018. Medicinal plants for Alzheimer's disease: an updated review. J. Med. Plants Stud. 6 (2), 81–85.

Beheshti, S., Shahmoradi, B., 2018. Therapeutic effect of *Melissa officinalis* in an amyloid-β rat model of Alzheimer's disease. J. Herb. Med. Pharmacol. 7 (3), 193–199.

Bishara, D., Sauer, J., Taylor, D., 2015. The pharmacological management of Alzheimer's disease. Prog. Neurol. Psychiatr. 19 (4), 9–16.

Carnat, A.P., Carnat, A., Fraisse, D., Lamaison, J.L., 1998. The aromatic and polyphenolic composition of lemon balm (*Melissa officinalis* L. subsp. *officinalis*) tea. Pharm. Acta Helv. 72 (5), 301–305.

Changizi-Ashtiyani, S., Najafi, H., Jalalvandi, S., Hosseinei, F., 2013. Protective effects of *Rosa canina* L. fruit extracts on renal disturbances induced by reperfusion injury in rats. Iran J. Kidney Dis. 7 (4), 290–298.

Cole, S.L., Vassar, R., 2007. The Alzheimer's disease secretase enzyme, BACE1. Mol. Neurodegener. 2, 22.

Craig, L.A., Hong, N.S., McDonald, R.J., 2011. Revisiting the cholinergic hypothesis in the development of Alzheimer's disease. Neurosci. Biobehav. Rev. 35 (6), 1397–1409.

Cummings, J.L., 2000. Cholinesterase inhibitors: a new class of psychotropic compounds. Am. J. Psychiatry 157 (1), 4–15.

Dastmalchi, K., Dorman, H.D., Oinonen, P.P., Darwis, Y., Laakso, I., Hitunen, R., 2008. Chemical composition and in vitro antioxidative activity of a lemon balm (*Melissa officinalis L.*) extract. LWT Food Sci. Technol. 41 (3), 391–400.

Davis, K.L., Powchik, P., 1995. Tacrine. Lancet 345 (8950), 625–630.

Dehbani, Z., Komaki, A., Etaee, F., Shahidi, S., Taheri, M., Komaki, S., Faraji, N., 2019. Effect of a hydro-alcoholic extract of *Melissa officinalis* on passive avoidance learning and memory. J. Herbmed. Pharmacol. 8 (2), 120–125.

Edwards, S.E., da Costa Rocha, I., Williamson, E.M., Heinrich, M., 2015. Phytopharmacy: An evidence-based guide to herbal medicinal products. John Wiley & Sons, USA, p. 414.

Ellis, J.M., 2005. Cholinesterase inhibitors in the treatment of dementia. J. Am. Osteopath. Assoc. 105 (3), 145–158.

Farlow, M.R., Miller, M.L., Pejovic, V., 2008. Treatment options in Alzheimer's disease: maximizing benefit, managing expectations. Dement. Geriatr. Cogn. Disord. 25 (5), 408–422.

Fermino, B.L., Khalil, N.M., Bonini, J.S., Pereira, R.P., da Rocha, J.B.T., da Silva, W.C.F.N., 2015. Anxiolytic properties of *Melissa officinalis* and associated mechanisms of action: a review. Afr. J. Pharm. Pharmacol. 9 (3), 53–59.

Gbenou, J.D., Ahounou, J.F., Akakpo, H.B., Laleye, A., Yayi, E., Gbaguidi, F., Baba- Moussa, L., Darboux, R., Dansou, P., Moudachirou, M., Kotchoni, S.O., 2013. Phytochemical composition of *Cymbopogon citratus* and *Eucalyptus citriodora* essential oils and their anti-inflammatory and analgesic properties on Wistar rats. Mol. Biol. Rep. 40 (2), 1127–1134.

Gurčík, Ľ., Dúbravská, R., Miklovičová, J., 2005. Economics of the cultivation of *Salvia officinalis* and *Melissa officinalis*. Agric. Econ. Czech 51, 348–356.

Havsteen, B.H., 2002. The biochemistry and medical significance of the flavonoids. Pharmacol. Ther. 96 (2–3), 67–202.

Howes, M.-J.R, Fang, R., Houghton, P.J., 2017. Effect of Chinese herbal medicine on Alzheimer's disease. Int. Rev. Neurobiol. 135, 29–56.

Iuvone, T., De Filippis, D., Esposito, G., D'Amico, A., Izzo, A.A., 2006. The spice sage and its active ingredient rosmarinic acid protect PC12 cells from amyloid-beta peptide-induced neurotoxicity. J. Pharmacol. Exp. Ther. 317 (3), 1143–1149.

Jalal, Z., El Atki, Y., Lyoussi, B., Abdellaoui, A., 2015. Phytochemistry of the essential oil of *Melissa officinalis* L. growing wild in Morocco: preventive approach against nosocomial infections. Asian Pac. J. Trop. Biomed. 5 (6), 458–461.

Jelic, D., Mildner, B., Kostrun, S., Nujic, K., Verbanac, D., Culic, O., Antolovic, R., Brandt, W., 2007. Homology modeling of human Fyn kinase structure: discovery of rosmarinic acid as a new Fyn kinase inhibitor and in silico study of its possible binding modes. J. Med. Chem. 50 (6), 1090–1100.

Jembrek, M.J., Vlainic, J., 2015. GABA Receptors: pharmacological potential and pitfalls. Curr. Pharm. Des. 21 (34), 4943–4959.

Johnson, J.W., Glasgow, N.G., Povysheva, N.V., 2015. Recent insights into the mode of action of memantine and ketamine. Curr. Opin. Pharmacol. 20, 54–63.

Jun, H.J., Lee, J.H., Jia, Y., Hoang, M.H., Byun, H., Kim, K.H., Lee, S.J., 2012. *Melissa officin*alis essential oil reduces plasma triglycerides in human apolipoprotein E2 transgenic mice by inhibiting sterol regulatory element-binding protein-1c-dependent fatty acid synthesis. J. Nutr. 142 (3), 432–440.

Kennedy, D.O., Scholey, A.B., Tildesley, N.T.J, Perry, E.K., Wesnes, K.A., 2002. Modulation of mood and cognitive performance following acute administration of *Melissa officinalis* (lemon balm). Pharmacol. Biochem. Behav. 72 (4), 953–964.

Kim, G.-D., Park, Y.S., Jin, Y.-H., Park, C.-S., 2015a. Production and applications of rosmarinic acid and structurally related compounds. Appl. Microbiol. Biotechnol. 99 (5), 2083–2092.

Kim, J.H., Wang, Q., Choi, J.M., Lee, S., Cho, E.J., 2015b. Protective role of caffeic acid in an $A\beta_{25-35}$-induced Alzheimer's disease model. Nutr. Res. Pract. 9 (5), 480–488.

Koytchev, R., Alken, R.G., Dundarov, S., 1999. Balm mint extract (Lo-701) for topical treatment of recurring herpes labialis. Phytomedicine 6 (4), 225–230.

Kumar, H., More, S.V., Han, S.-D., Choi, J.-Y., Choi, D.-K., 2012. Promising therapeutics with natural bioactive compounds for improving learning and memory — a review of randomized trials. Molecules 17 (9), 10503–10539.

Limen-Ben Amor, I., Boubaker, J., Sgaier, M.B., Skandrani, I., Bhouri, W., Neffati, A., Kilani, S., Bouhlel, I., Ghedira, K., Chekir-Ghedira, L., 2009. Phytochemistry and biological activities of *Phlomis* species. J. Ethnopharmacol. 125 (2), 183–202.

Liu, J., 1995. Pharmacology of oleanolic acid and ursolic acid. J. Ethnopharmacol. 49 (2), 57–68.

Lopez, V., Martin, S., Gomez-Serranillos, M.P., Carretero, M.E., Jager, A.K., Calvo, M.I., 2009. Neuroprotective and neurological properties of *Melissa officinalis*. Neurochem. Res. 34 (11), 1955–1961.

Mahboubi, M., 2019. *Melissa officinalis* and rosmarinic acid in management of memory functions and Alzheimer disease. Asian Pac. J. Trop. Biomed. 9 (2), 47–52.

Martins, E.N., Pessano, N.T., Leal, L., Roos, D.H., Folmer, V., Puntel, G.O., Rocha, J.B.T, Aschner, M., Ávila, D.S., Puntel, R.L., 2012. Protective effect of *Melissa officinalis* aqueous extract against Mn-induced oxidative stress in chronically exposed mice. Brain Res. Bull. 87, 74–79.

Miladi-Gorgi, H., Vafaee, A., Rashidipoor, A., Taherian, A.A., Jarrahi, M., Emami Abarghoee, M., Sadeghi, H., 2005. The role of opioid receptors on anxiolytic effects of the aqueous extract of *Melissa officinalis* in mice. Persian. Razi J. Med. Sci. 12 (47), 145–153.

Miraj, S., Azizi, N., Kiani, S., 2016. A review of chemical components and pharmacological effects of *Melissa officinalis* L. Der Pharm. Lett. 8 (6), 229–237.

Miraj, S., Kopaei, R., Kiani, S., 2017. *Melissa officinalis L*: a review study with an antioxidant prospective. J. Evid. Based Complement. Alternat. Med. 22 (3), 385–394.

Moaca, E.A., Farca, C., Ghitu, A., Coricovac, D., Popovici, R., Caraba-Meita, N.L., Ardelean, F., Antal, D.S., Dehelean, C., Avram, S., 2018. A comparative study of *Melissa officinalis* leaves and stems ethanolic extracts in terms of antioxidant, cytotoxic, and antiproliferative potential. Evid. Based Complement. Alternat. Med. 1–12.

Moradkhani, H., Sargsyan, E., Bibak, H., Naseri, B., Sadat-Hosseini, M., Fayazi-Barjin, A., Meftahizade, H., 2010. *Melissa officinalis* L., a valuable medicine plant: a review. J. Med. Plants Res. 4, 2753–2759.

Moradpour, M., Ghoran, S.H., Asghari, J., 2017. Phytochemical investigation of *Melissa officinalis* L. flowers from Northern part of Iran (Kelardasht). J. Med. Plants Stud. 5 (3), 176–181.

Nakazawa, T., Ohsawa, K., 1998. Metabolism of rosmarinic acid in rats. J. Nat. Prod. 61 (8), 993–996.

Obulesu, M., Rao, D.M., 2011. Effect of plant extracts on Alzheimer's disease: An insight into therapeutic avenues. J. Neurosci. Rural Pract. 2 (1), 56–61.

Ono, K., Hasegawa, K., Naiki, H., Yamada, M., 2004. Curcumin has potent anti-amyloidogenic effects for Alzheimer's beta-amyloid fibrils in vitro. J. Neurosci. Res. 75 (6), 742–750.

Onor, M.L., Trevisiol, M., Aguglia, E., 2007. Rivastigmine in the treatment of Alzheimer's disease: an update. Clin. Interv. Aging 2 (1), 17–32.

Orhan, I., Kartal, M., Kan, Y., Sener, B., 2008. Activity of essential oils and individual components against acetyl- and butyrylcholinesterase. Z. Naturforsch. C Biosci. 63 (7–8), 547–553.

Ozarowski, M., Mikolajczak, P.L., Piasecka, A., Kachlicki, P., Kujawski, R., Bogacz, A., Bartkowiak-Wieczorek, J., Szulc, M., Kaminska, E., Kujawska, M., Jodynis-Liebert, J., Gryszczynska, A., Opala, B., Lowicki, Z., Seremak-Mrozikiewicz, A., Czerny, B., 2016. Influence of the *Melissa officinalis* leaf extract on long-term memory in scopolamine animal model with assessment of mechanism of action. Evid. Based Complement. Alternat. Med. 1–17.

Perry, E., Howes, M.-J.R., 2011. Medicinal plants and dementia therapy: herbal hopes for brain aging? CNS Neurosci. Ther. 17 (6), 683–698.

Rostami, S., Momeni, Z., Behnam-Rassouli, M., Ghayour, N., 2010. Comparison of antioxidant effect of *Melissa officinalis* leaf and vitamin C in lead acetate induced learning deficits in rat. Daneshvar Med. 17 (86), 47–54.

Schneider, L.S., 2013. Alzheimer disease pharmacologic treatment and treatment research. Continuum (Minneapolis, Minn.) 19 (2 Dementia), 339–357.

Scholey, A., Stough, C., 2011. Neurocognitive effects of herbal extracts. In: Benton, D. (Ed.), Lifetime Nutritional Influences on Cognition, Behaviour and Psychiatric Illness. Woodhead Publishing, pp. 272–297.

Schulz, V., Hansel, R., Tyler, V., 1998. Rational Phytotherapy: A physician's guide to herbal medicine. Springer-Verlag, New York, p. 306.

Sebestík, J., Marques, S.M., Falé, P.L., Santos, S., Arduíno, D.M., Cardoso, S.M., Oliveira, C.R., Serralheiro, M.L., Santos, M.A., 2011. Bifunctional phenolic-choline conjugates as antioxidants and acetylcholinesterase inhibitors. J. Enzym. Inhib. Med. Chem. 26 (4), 485–497.

Shakeri, A., Sahebkar, A., Javadi, B., 2016. *Melissa officinalis* L. - A review of its traditional uses, phytochemistry and pharmacology. J. Ethnopharmacol. 188, 204–228.

Shen, Y., Sun, Z., Guo, X., 2015. Citral inhibits lipopolysaccharide-induced acute lung injury by activating PPAR-γ. Eur. J. Pharmacol. 747, 45–51.

Shi, J., Ni, J., Lu, T., Zhang, X., Wei, M., Li, T., Liu, W., Wang, Y., Shi, Y., Tian, J., 2017. Adding Chinese herbal medicine to conventional therapy brings cognitive benefits to patients with Alzheimer's disease: a retrospective analysis. Complement. Alternat. Med. 17 (533), 1–7.

Sofowora, A., Ogunbodede, E., Onayade, A., 2013. The role and place of medicinal plants in the strategies for disease prevention. Afr. J. Tradit. Complement. Alternat. Med. 10 (5), 210–229.

Soodi, M., Naghdi, N., Hajimehdipoor, H., Choopani, S., Sahraei, E., 2014. Memory-improving activity of *Melissa officinalis* extract in naive and scopolamine-treated rats. Res. Pharm. Sci. 9 (2), 107–114.

Summers, W.K., 2000. Tacrine (THA, Cognex®). J. Alzheim. Dis. 2 (2), 85–93.

Sun, Z.-K., Yang, H.-Q., Chen, S.-D., 2013. Traditional Chinese medicine: a promising candidate for the treatment of Alzheimer's disease. Transl. Neurodegener. 2 (6), 1–7.

Svobodová, A., Psotová, J., Walterová, D., 2003. Natural phenolics in the prevention of UV-induced skin damage. A review. Biomed. Pap. Med. Fac. Univ. Palacky Olomouc Czech Repub. 147 (2), 137–145.

Taiwo, A.E, 2007. Alterações Comportamentais decorrentes da administração de *Mellissa officinalis* emratos. Faculdade de Ciência da Saúde. Universidade de Brasília.

Tewari, D., Stankiewicz, A.M., Mocan, A., Sah, A.N., Tzvetkov, N.T., Huminiecki, L., Horbañczuk, J.O., Atanasov, A.G., 2018. Ethnopharmacological approaches for dementia therapy and significance of natural products and herbal drugs. Front. Aging Neurosci. 10 (3), 1–24.

The Plant list, 2013. http://www.theplantlist.org. Version 1.1.

Tsuno, N., 2009. Donepezil in the treatment of patients with Alzheimer's disease. Expert Rev. Neurother. 9 (5), 591–598.

Verma, P.P.S., Singh, A., Rahaman, L.U., Bahl, J.R., 2015. Lemon balm (*Melissa officinalis* L.) an herbal medicinal plant with broad therapeutic uses and cultivation practices: a review. Int. J. Recent Adv. Multidisc. Res. 2 (1), 928–933.

Vladimir-Kneževic, S., Blažekovic, B., Kindl, M., Vladic, J., Lower-Nedza, A.D., Brantner, A.H., 2014. Acetylcholinesterase inhibitory, antioxidant and phytochemical properties of selected medicinal plants of the *Lamiaceae* family. Molecules 19 (1), 767–782.

Wake, G., Court, J., Pikering, A., Lewis, R., Wilkins, R., Perry, E., 2000. CNS acetylcholine receptor activity in European medicinal plants traditionally used to improve failing memory. J. Ethnopharmacol. 69 (2), 105–114.

Weinstock, M., 1999. Selectivity of cholinesterase inhibition: clinical implications for the treatment of Alzheimer's disease. CNS Drugs 12, 307–323.

WHO, WHO Monographs on Selected Medicinal Plants - Volume 2, WHO, Geneva, Switzerland

Wichtl, M., 2004. Herbal Drugs and Phytopharmaceuticals: A Handbook for Practice on a Scientific Basis. Medpharm Press GmbH Scientific Publishers, Germany, p. 704.

Winslow, B.T., Onysko, M.K., Stob, C.M., Hazlewood, K.A., 2011. Treatment of Alzheimer disease. Am. Fam. Physician 83 (12), 1403–1412.

Yu, T.X., Zhang, P., Guan, Y., Wang, M., Zhen, M.Q., 2015. Protective effects of luteolin against cognitive impairment induced by infusion of Aß peptide in rats. Int. J. Clin. Exp. Pathol. 8 (6), 6740–6747.

Zarei, A., Changizi-Ashtiyani, S., Taheri, S., Hosseini, N., 2015. A brief overview of the effects of *Melissa officinalis L.* Extract on the function of various body organs. Zahedan J. Res. Med. Sci. 17 (7), 1–6.

Chapter 3.2.4

Bacopa monnieri (Brahmi)

Tanuj Joshi[1], Abhishek Gupta[2], Prashant Kumar[3], Anita Singh[1], Aadesh Kumar[1]

[1]*Department of Pharmaceutical Sciences, Faculty of Technology, Kumaun University, Nainital, Uttarakhand, India;* [2]*Hygia Institute of Pharmaceutical Education and Research, Lucknow, Uttar Pradesh, India;* [3]*School of Pharmacy, Bharat Institute of Technology, Meerut, Uttar Pradesh, India*

Introduction

Neurodegenerative diseases like Alzheimer's disease (AD), Parkinson's disease, and other associated disorders are a combination of disorders of the central nervous system. They are characterized by progressive loss of neurons present in the brain and spinal cord, influencing motor abilities (Ramanan and Saykin, 2013; Brettschneider et al., 2015; Medina and Evans, 2015).

As per the reports, it is estimated that until 2020, neurodegenerative diseases will be the eighth leading cause of death in developed countries. And until the mid-century, neurodegenerative diseases will be the second leading cause of death in the world, overtaking cancer (Menken et al., 2000).

Currently, available drug treatments have proven to slow down the progression of the disease and provide symptomatic relief, but they fail to achieve a remarkable cure (Connolly and Lang, 2014; Kumar and Ekavali, 2015). The cause and pathogenesis are totally uncertain in the case of AD, a well-known neurodegenerative disorder in the brain, which is most common in elderly patients. Bacopa is widely used in the cases of forgetfulness and long-term memory loss especially in elderly patients. AD is the main reason behind dementia in 60%–80% of cases in elderly patients (Hebert et al., 2003).

Initially, AD is portrayed by the impairment of recent memory, followed by impairment of cognitive abilities, vocabulary, and perception due to the engagement of hippocampus and median temporal lobe (Markowitsch and Staniloiu, 2012).

Subsequently, due to engagement of other parts of the central nervous system, there may be evident disturbances in sleep, psychological changes, and impaired judgment, as well as pyramidal and extrapyramidal motor signs (Alzheimer's Association, 2010).

Naturally Occurring Chemicals against Alzheimer's Disease. https://doi.org/10.1016/B978-0-12-819212-2.00020-7

The findings of the World Alzheimer's Report (2015) state that the worldwide incidences of dementia had been risen from 30 million (2010) to 46.8 million. The frequency of dementia in India was 33.6 out of 1000 cases, including 54% of AD. These statistical data suggest the significant rise in the overall worldwide burden of AD, while the treatment is limited (World Alzheimer Report, 2015).

The pathogenesis of AD is uncertain. AD represents an epidemic that without an efficient therapy can create tremendous economical burden. Past studies of AD neuropathology gave valuable but limited insights. Recent research has provided genetic, molecular, and cellular data that are creating opportunities to both interpret and handle AD. The most significant research includes the presence of toxic oligomeric species of Aβ and tau in the brain and how oligomers of Aβ and tau initiate and drive pathogenesis. There is much still to learn about pathogenesis and the discovery of disease-modifying treatments for fighting AD (Chen and Mobley, 2019).

In general, aging is the progressive slow death of the organs, tissues, and cells in the body. And the brain is the most susceptible organ to aging, due to lack of regeneration of the neurons. According to Ayurveda, medha governs the core cognitive functioning of the brain, which starts decreasing in later stages of life, and as a result, decision-making capacity becomes imperative, leading to profound dementia. Hence, for prolonged life and healthy functioning of brain, rejuvenating therapies were introduced in Ayurveda. Further, aging is linked with degenerative changes like functioning of hormones, damage to proteins, lipids, and DNA, resulting in abnormal functioning (Dahanukar and Thatte, 1989).

Over the decades, it has been seen that there is high interest in cognitive enhancement, especially focused on the cognition enhancement in the younger population involving cognitively intact individuals in the studies (Sahakian and Morein-Zamir, 2007). In the context of pharmaceutical cognition enhancers, methylphenidate and modafinil (Provigil) have taken center stage (Repantis et al., 2010). Herbal extracts are supposed to have various active components influencing numerous neuronal, metabolic, and hormonal activities in the body, responsible for various behavioral processes (Scholey et al., 2005).

Taxonomical classification of *B. monnieri*

The taxonomy of *B. monnieri* is in the kingdom Plantae; subkingdom Viridiplantae; in Kingdom Streptophyta; superdivision Embryophyta; division Tracheophyta; subdivision Spermatophytina; class Magnoliopsida; subclass Asteridae; superorder Asteranae; order Lamiales; family Plantaginaceae; genus Bacopa; species *monnieri*. The image of *B. monnieri* could be identified with Fig. 3.2.4.1 (ITIS, 2015).

Nomenclature: *B. monnieri* is native in India, Bangladesh, and Southern Asia, and it also grows in Australia, Europe, and Africa. The vernacular names of *B. monnieri* are Indian pennywort and water hyssop (English) (PMP, 2014).

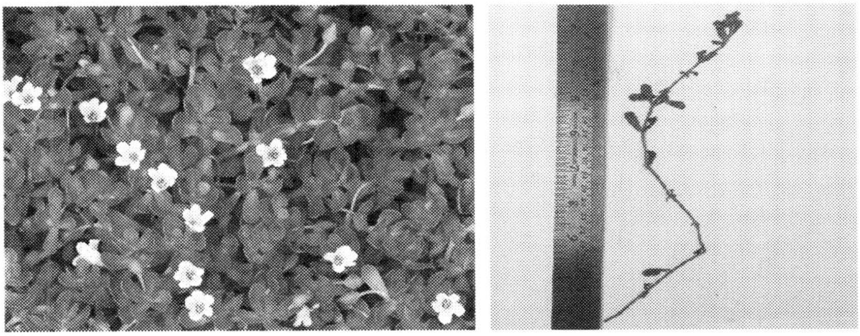

FIGURE 3.2.4.1 *B. monnieri*, with manual measurement of the single twig.

B. monnieri is classified as a memory enhancer and rejuvenator, and also for increasing longevity. Due to its significant use in age-related memory and cognitive impairment, enhancing life-span, and providing nutrition, it is recommended in various psychiatric disorders, like hallucination, schizophrenia, and obsessive compulsive disorders (Diggavi, 1998).

Bacopa contains bacoside A (64.28%), bacoside B (27.11%), betullic acid, D-mannitol, β-sitosterol, and stigmastanol. Bacoside A is levo-rotatory and bacoside B is dextro-rotatory. Bacoside A is a major component in this plant, including two sets of saponins. One set was derived from pseudojujubogenin, which upon acid hydrolysis furnishes four triterpenoid transformation products, viz., Bacogenins A1, A2, A3, and A4 (Chatterjee et al., 1963; Basu et al., 1967).

Saponins upon acid hydrolysis yield two triterpenoids with triene side chains as the major transformation products. These triterpenoids were designated as Bacogenins A4 (trans) and (cis) and are associated with memory impairment activities, like antianxiety (Singh et al., 1975). Bacopa has also been used to treat numerous inflammatory conditions such as asthma, bronchitis, dropsy, and rheumatism. List of all major chemical constituents is summarised in Table 3.2.4.1 (Channa et al., 2006).

Bacopa is used in Ayurveda for memory enhancement, to increase cognitive ability, and for lack of concentration and anxiety (Aguiar and Borowski, 2013).

Bacopa inhibits the release of IL-6 and TNF-α from LPS (lipopolysaccharide)-activated microglia and also inhibits the enzyme activity of MMP-3 and caspase 1 and 3. *Bacopa* also has the potential to treat neurodegenerative diseases and disorders like depression, anxiety, and schizophrenia (Michelle et al., 2017).

TABLE 3.2.4.1 List of major chemical constituents in plant *B. monnieri*.

S.No	Chemical category/ metabolites	Chemical constituents
1.	Primary metabolites	Carbohydrates, saponins
2.	Secondary metabolites	Bacoside A, bacoside B, betullic acid, D-mannitol stigmasterol, β-sitosterol, stigmastanol, and triterpenoid saponins

Brahmi Rasayan, an Ayurvedic preparation, which has *Bacopa* as the major component, showed antiinflammatory actions in large oral doses in animal models of inflammation (Jain et al., 1994).

Brahmi ghrita, a popular Ayurvedic formulation, had proven to significantly improved latency during preclinical study in rats (Achilya et al., 2004). Recently, using neuroimaging, scientists have investigated the correlation between brain neurometabolite levels, comprising molecular or cellular changes that may be related to aging (Chavhan, 2013).

Magnetic resonance spectroscopy (MRS) has been used extensively to study cognitive impairment, AD (Mandal et al., 2012), and depression (Auer et al., 2000), as well as schizophrenia (Bertolino et al., 1998). Further MRS can be employed for early diagnosis of disease (Mountford et al., 2010).

Moreover, oxidative stress has a role in aging and neurodegenerative disorders. Based on *in vitro* and *in vivo* studies, *Bacopa* can be utilized as a therapeutic strategy against oxidative damage and cognitive decline in the elderly (Tamara et al., 2015).

Mechanism of action of *B. monnieri* in Alzheimer's disease (AD)

B. monnieri plays an important role in neuroprotection and in improving cognition. Different mechanisms of action have been proposed for *B. monnieri* by which it may exert beneficial effect in AD. The phytoconstituents that are mainly held responsible for the activity of *B. monnieri* are bacoside A, bacoside B, and bacopaside IX (Howes and Houghton, 2012). Although, in an *in silico* study using computer software, it was found that the parent triterpinoid and saponins found in *Bacopa* get converted to other active metabolites that may be ultimately responsible for the action. These metabolites may show a better pharmacokinetic profile and bioactivity than the parent compound (Ramasamy et al., 2015). The action of *Bacopa* mainly revolves around its ability to enhance cognition and memory. Also, it possesses antistress and antianxiety activity (Singh, 2015).

Bacopa produces its action through several mechanisms that might provide a beneficial effect in AD. Some of these mechanisms are as follows: an antioxidant action, by modulation of the cholinergic function, by prevention of β-amyloid accumulation (Aβ), and by an effect on the nitric oxide pathway (Howes and Houghton, 2012). Bacosides found in *B. monnieri* may be responsible for the ability of *B. monnieri* to increase nerve impulse transmission. Restoration of activity of synapse, increase in kinase activity, and neuronal synthesis leads to repair of damaged neurons by bacosides. Antioxidant activity shown by bacosides is also held responsible for therapeutic activity of *B. monnieri* in AD. Bacosides have demonstrated antioxidant activity in the striatum, frontal cortex, and the hippocampus in experimental studies. Studies on animals have shown that *Bacopa* extract modulates expression of enzymes that help in scavenging and generation of reactive oxygen species in the brain (Devishree et al., 2017). Some researchers have found that *Bacopa* provides protection against oxidative damage by maintenance of ionic balance of membrane and mitochondrial functional integrity. Also, a decrease in lipoxygenase activity, decrease in lipid peroxides, and reduction in concentration of divalent metal ions have been found (Dhawan, 2014). Due to destruction of the cholinergic neurons the levels of acetylcholine are reduced, and this reduction in levels of acetylcholine leads to a decrease in cholinergic transmission. The preceding theory has been used by researchers to explain the generation of AD. Thus, drugs that can elevate the levels of acetylcholine can help in the treatment of AD. Acetylcholinesterase (AChE) is an enzyme that leads to degradation of acetylcholine. Thus, by inhibition of this enzyme, levels of acetylcholine can be elevated, which may result in therapeutic benefits in AD. Various experiments have demonstrated that *Bacopa* might produce useful effect in AD by inhibiting the enzyme AChE (Ahirwar and Tembhre, 2016). Researchers have found that *B. monnieri* activates the enzyme choline acetyltransferase and enhances blood flow in brain (Singh, 2015). Also *Bacopa* stabilizes the functional and structural integrity of membranes, and this may lead to its neuroprotective effect in AD. Regulation of translation of mRNA and surface expression of receptors like gamma amino butyric acid receptor (GABAR), α-amino-3-hydroxy-5-methyl-4-isoxazolepropionic acid receptor (AMPAR), and N-methy-D-aspartate receptor (NMDAR) in various regions of the brain are formed by bacosides present in *B. monnieri* (Lal and Baraik, 2019).

Experiments on *B. monnieri* related to Alzheimer's disease (AD) using *in vitro* and *in vivo* methods

B. monnieri has shown significant effect on AD and other diseases that lead to neurodegeneration. Beneficial effects are produced by it through improvement in memory, intelligence, cognition, and neuroprotection (Malve, 2017). Many actions of *Bacopa* have been studied that may support its role as a neuroprotector in the aging brain. It may facilitate neurotransmission in the brain and help in the repair of neurons that are damaged (Ray and Ray, 2015). In AD, *B. monnieri* has shown remarkable effect that is showcased by various studies. *B. monnieri* leads to maintenance of ionic equilibrium, restoration of selenium and zinc, and amelioration of adenosine triphosphatases in the brain (Roy and Awasthi, 2017). Studies have shown that *B. monnieri* enhances acquisition, retention, and motor learning and retards extinction of behavior that has been newly adopted. Anterograde memory is enhanced by *B. monnieri*. It reduced anterograde amnesia induced by sodium nitrite and scopolamine by improvements in hypoxic conditions and acetylcholine respectively in experimental studies. Retrograde amnesia produced by BN52021 (an antagonist of receptors for platelet-activating factor) was also reversed by bacosides probably due to enhancement in synthesis of platelet-activating factor. Diazepam-induced amnesia is also significantly reversed by treatment with *B. monnieri*. Retention of memory and improvement of acquisition was produced by treatment of *B. monnieri* in a study in which cognitive impairment was produced by phenytoin. This effect was produced due to elevation in cerebral levels of glutamate. A preparation of *Bacopa* is also available commercially, and it is called Keen Mind. Keen Mind is a preparation that can be useful for maintaining cognitive and brain health in the elderly (Al-Snafi, 2013). In a study, neuroprotective effect of *B. monnieri* was observed by combining extracts of rosemary and supercritical carbon dioxide extracts of *B. monnieri*. In the study, *B. monnieri* and rosemary extracts were evaluated for antioxidant and neuroprotective effects, individually as well as in combination with each another. Embryonic mouse hypothalamus cell lines and human glial cells were used for the evaluation of *B. monnieri* and rosemary extract. The combined extracts showed better antioxidant activity than either extract used individually. Also, with regard to neuroprotective effect and inhibitory effects on expression of phobophobia, it was found to be similar when the effects of the individual drugs were compared with the combination. However, with respect to inhibition of amyloid protein production and brain-derived neurotrophic factor's increased production in hypothalamic cells, the combination showed a better effect than the individual drugs. The study proved that combination of *Bacopa* and rosemary proves to be better in comparison to individual extracts in producing neuroprotective effect (Ramachandran et al., 2014). In a study, effect of *B. monnieri* (ethanoilc extract) was evaluated for anticholinesterase activity in various brain regions of rats. Anticholinesterase activity of different extent was shown by the extract (100 mg/kg) in various brain areas. The Km (Michaelis Menten constant) values for acetylcholinesterase were increased by the extract in enzyme kinetic studies that were performed *in vivo*. The Km values were found to increase in various brain areas. The hippocampal area showed lower enhancements in Km values compared to the cerebral cortex by the extract. However, there was no change in the maximum velocity (Vmax) when comparison was done between the vehicle-treated group and the group that was administered the extract. Thus, competitive inhibition of acetylcholinesterase (AChE) was shown by the extract in different regions of the rat brain. This study reflects that due to its anticholinesterase potential, *B. monnieri* may prove beneficial for AD treatment (Tembhre et al., 2015). An *in vitro* and *in silicio* study was performed by researchers to study the effect of bacosides (bacoside A and bacopaside X), aglycones (pseudojujubogenin, jujubogenin), and their derivatives like bacogenin A1, ebelin lactone. AChE inhibitory assays and radioligand receptor binding assays were performed for determining *in vitro* potential of the aforementioned compounds of *B. monnieri*. *In silico* studies were performed using ADMET descriptors and discovery studio molecular properties for evaluating activities of the aforementioned compounds. Auto dock software was used for performing docking of the compounds into AChE and M_1, D_1, D_2, $5HT_{2A}$, $5HT_{1A}$ receptors. It was found in the experiment that bacopaside X and bacoside A were affinitive toward D_1 receptors, whereas ebelin lactone showed affinity toward M_1 and $5HT_{2A}$ receptors. No compound showed inhibitory activity against AChE. $5HT_{2A}$ and M_1 receptor stimulation are involved with cognition and memory. In *in silico* studies, parent bacosides did not show docking into any of the receptors or AChE. Their molecular properties for qualifying as drugs acting on the central nervous system were also not found to be good. Aglycones and derivatives of aglycones on the other side demonstrated good activities on the central nervous system. Better binding affinity was shown by these compounds. Intestinal absorption of these compounds as well as their entry into the brain was also found to be good. Due to the stimulation of aforementioned receptors ($5HT_{2A}$ and M_1), strongest binding energy, and highest penetration into the blood—brain barrier, ebelin showed better effect than its parent molecules. The study thus suggests that in comparison to the parent molecules like bacosides, derivatives of aglycone like ebelin might show pharmacologic activity. This means the phytoconstituents of *B. monnieri* may get converted to active forms within the body to show their pharmacologic actions (Ramasamy et al., 2015). *In vivo* studies carried out on murine brain had shown that *B. monnieri* enhanced activities of endogenous antioxidant enzymes on one side and reduced damage due to oxidants in striatum, prefrontal cortex, and hippocampus on the other side. *B. monnieri* had shown improvements in olfactory bulbectomized mice suffering from dysfunction in cognition by preserving effect on cholinergic neuron and increasing signaling related to synaptic plasticity. Also, improvements were observed in spatial memory and long-term memory defect induced by fear and recognition of objects. In a study on C57/B16 mice, a decrease in amyloid fibrils formation, decrease in reactive oxygen species, decrease in two types of amyloidogenic proteins, and inhibitory effects on the potentiation of AD by β amyloid protein were observed by the extract of *B. monnieri*. In brains of mice, damage due to oxidative species was induced with 3-nitropropionic acid (3-NPA). This oxidative damage induced by 3-NPA was reduced by extract of

B. monnieri as demonstrated by results of various antioxidant assays (Chakraborty and Chakraborty, 2018; Chaudhari et al., 2017; Dhanasekaran et al., 2007). *B. monnieri*'s rhizomes and aerial parts were subjected to ethanolic extraction, and the extract was found to possess nontropic activity in experimental studies. This is due to membrane dephosphorylation induced by bacosides, and it subsequently leads to elevation in ribonucleic acid (RNA) and protein levels in a specific area of the brain. Alternatively, it has been suggested that *B. monnieri* increases cognition by modulating the cholinergic system and increasing protein kinase activity in the hippocampus. In another study, it was shown that decrease in cognition produced by colchicines and ibotenic acid was reversed by treatment with stem and leaf extract of *B. monnieri* that was rich in bacoside. Also in the same study, a decrease in the levels of acetylcholine, decrease in binding with muscarinic receptors in frontal cortex and hippocampus, and decrease in choline acetyltransferase activity were reversed. *Bacopa* extract in hippocampus, striatum, and frontal cortex of rat showed antioxidant activity. In a different study, toxicity produced by nitric oxide was inhibited by methanolic extract of *B. monnieri*. Also, deoxyribonucleic acid (DNA) cleavage produced by hydrogen peroxide was inhibited by the extract of *B. monnieri in vitro* (Dastmalchi et al., 2007). Researchers found that bacopaside I in oral dose of 3, 10, 30 mg/kg for 6 days exhibited antioxidant, cerebral ATP-enhancing action and neuroprotective action after induction of cerebral ischemia in rats. Infarct volume and neurologic deficits were significantly decreased, and Ca^{2+}, Mg^{2+}, ATPase activity, $Na^{+}K^{+}$ ATPase, nitric oxide, energy charge, total adenine nucleotides, and ATP content of brain were increased by bacopaside. The antioxidant activities of endogenous enzymes were improved, whereas the elevation in malondialdehyde was inhibited by bacopaside I in brain. In a mice model of AD, pretreatment with *B. monnieri* extract in 20, 40, and 80 mg/kg doses was given for 2 weeks by oral route. Also, treatment was given for 1 week after inducing AD. Reduction in latency time was observed in water maze test due to treatment with Bacopa. The reduction in cholinergic neurons was also mitigated by *B. monnieri* extract (Aguiar and Borowski, 2013). In another study, *B. monnieri* extract showed a neuroprotective in rat hippocampus by combating oxidative destruction induced by aluminum (Ansari et al., 2007).

Clinical studies on *B. monnieri* with respect to Alzheimer's disease (AD)

Different experiments performed by researchers have established importance of *Bacopa* treatment in AD. A formulation (KeenMind-CDRI 08) containing *Bacopa* extract was evaluated for its effects in a study that was cross over, double blind, and placebo controlled. The percentage of bacoside was kept no less than 55% in KeenMind-CDRI 08. A multitasking framework was used for assessing the cognitive effects of KeenMind-CDRI 08 in 17 healthy participants. The multitasking framework comprised of tasks like stroop, mental arithmetic test, letter search, and visual tracking. The four tasks were presented to the participants simultaneously, each of them in one of the quadrants of the computer screen in a particular order. The accuracy and speed of solving these tasks by the participants was counted. The level of the tasks was set at medium and 20 min was given for the test. The mood of the participants was evaluated by the help of State Trait Anxiety Inventory (STAI) and the Bond-Lader VAS. Also, salivary samples were collected from the participants for measuring the levels of cortisol in them through luminescence immunoassay. The results of the study showed that *B. monnieri* produced adaptogenic and nootropic actions in the participants to a certain extent. Studies on a larger population are required to assess the actual magnitude of these actions of *Bacopa* (Benson et al., 2014). In another study, effect of *B. monnieri* was studied in patients suffering from AD. The effect of *Bacopa* on these patients regarding cognitive functions was noted. The participants taken in the study were newly designated AD patients in the age group of 60−65 years. The study was a nonrandomized, uncontrolled, prospective, and open label clinical trial. The baseline scores of all patients were recorded. After recording the baseline scores with respect to cognition the patients were administered Bacognize, which is a 300 mg standardized extract of *B. monnieri*. It was administered twice daily for a duration of 6 months. The scores were again recorded for all the patients at the end of the study. The cognitive functions of AD patients used showed improvements after receiving treatment with Bacognize for 6 months (Goswami et al., 2011). A clinical experiment was conducted on *B. monnieri* extract that was standardized for its effect on depression, anxiety, and cognition in the elderly. Placebo control, randomization, and double binding were the different criteria of the study. Two groups were used in the study: one was given placebo and the other treatment with extract of *B. monnieri*. It was administered in tablet form (300 mg/kg), and placebo was also given in a similar way. The subjects belonged to the age group of 65 years or more with no signs of clinical dementia. Various tests were used to assess cognition. *B. monnieri* was found useful in improving cognition in elderly (Calabrese et al., 2008). A study was conducted by researchers in Australia to evaluate the improvement in memory produced by *B. monnieri* in elderly, healthy humans. The study was placebo controlled, randomized, and double blind. Participants selected in the study were healthy and above the age of 55. The participants were divided into groups. One of the groups received Bacomind (300 mg/kg) and the other group received placebo in a similar manner. At baseline and after 12 weeks of treatment, different tests were done to assess memory. Results showed that *B. monnieri* significantly enhanced retention and acquisitions in the elderly people. This study gave support to the traditional uses of *Bacopa* as a memory enhancer (Morgan and Stevens, 2010). A clinical trial was performed in India that highlighted *B. monnieri's* role regarding impairment of memory due to aging. It was found that the extract of *B. monnieri* significantly produced beneficial effects on learning and memory during the 12 weeks of treatment in the study. Thus the study showed that the extract of *B. monnieri* provided a beneficial effect against memory impairment due to aging (Raghav et al., 2006). Seventy-six adults in the age group of 40–65 years of age were used in a study to analyze the chronic effects of *Bacopa* with respect to memory. Various functions of the memory were tested, and anxiety levels were measured in the participants. In the test concerning retention of new information, *Bacopa* showed a significant effect. In the follow-up tests, it was seen that there was no effect on the learning rate. The results showed that *Bacopa* decreased information forgetting complication in the participants (Roodenrys et al., 2002). A clinical trial was conducted by researchers in elderly participants who were healthy for evaluation of *Bacopa* with respect to cholinergic functions, monoaminergic functions, attention, working memory, and cognitive processing. Treatments were given with *Bacopa* and placebo to both male and female participants according to their respective groups for duration of 12 weeks. The results of the study demonstrated that *Bacopa* showed beneficial effects on cognition and memory due to its inhibitory action on AChE (Peth-Nui et al., 2012). A study was carried out in healthy adults belonging to India in the age group of 35–60 years. The participants were divided into *Bacopa*-treated group and placebo group. At the end of study, it was found that the group treated with *Bacopa* demonstrated a lower state of anxiety in comparison to placebo group. However, there was no significant effect on cognitive parameters by *Bacopa* in comparison to the placebo. Thus, this study proved to be different from other clinical trials on *Bacopa* with respect to the fact that in many clinical trials cognition was actually improved. Many studies involving humans have demonstrated benefits of *Bacopa* on cognitive functions and improvements in symptoms of AD. However conclusive clinical trials carried out for longer duration of time are required (Sathyanarayanan et al., 2013).

Adverse effects associated with *B. monnieri*

Bacopa is safe in therapeutic doses. Its safety profile is backed by years of use in Ayurveda. Also the safety profile of bacosides has been confirmed by different studies. LD_{50} (lethal dose) of the extract of *Bacopa* administered to rats was found to be 17 g/kg for alcoholic extract and 5 g/kg for aqueous extract (Al-snafi, 2013). However, *Bacopa* can produce mild gastrointestinal adverse effects like enhanced frequency of stools, nausea, and abdominal cramps (Howes and Houghton, 2012; Morgan and Stevens, 2010). Thus the adverse effect profile of Bacopa confirms that it is a very safe drug for human use.

Conclusion

The plant *B. monnieri* has prominent potential in the amelioration of cognitive disorders and produces a prophylactic reduction of oxidative damage, NT (Neurotensin) modulation, and cognitive enhancement as desired in AD. Biomedical research on *B. monnieri* is still in its inception, but studies carried out on *B. monnieri* so far have begun to open new areas of research on it. The critical long-term studies conducted on *B. monnieri* in combination with other substances, as per the Ayurvedic system, have resulted in synergistic effects and can be further explored by the clinical investigations. Moreover the social implications of cognition-enhancing drugs are highly promising, but these should be properly tempered with the desired ethical considerations for the new era of neural well being.

References

Achilya, C., Barabde, U., Wadodkar, S., Dorle, A., 2004. Effect of Brahmi Ghrita, a polyherbal formulation on learning and memory paradigms in experimental animals. Indian J. Pharmacol. 36 (3), 159–162.

Aguiar, S., Borowski, T., 2013. Neuropharmacological review of the nootropic herb *B. monnieri*. Rejuvenation Res. 16, 313–326.

Ahirwar, S., Tembhre, M., 2016. Assessment of acetylcholinestrase inhibiton by *Bacopa monneiri* and acephate in hippocampus of chick brain for impediment of Alzheimer's disease. Pharm. & Pharmacol. Int. J. 4 (5), 413–417.

Al-snafi, A.E., 2013. The pharmacology of Bacopa monniera: a review. Int. J. Pharma Sci. Res. 4 (12), 154–159.

Alzheimer's Association, 2010. Alzheimer's disease facts and figures. Alzheimers Dement 6, 158–194.

Ansari, O.A., Tripathi, J.S., Ansari, S., 2007. Evidence based anti-dementing activity of Saraswata ghrita. A nootropic compound from Ayurveda. Int. J. Pharmaceut. Sci. Res. C 4 (11), 4194–4202.

Auer, D.P., Utz, B.P., Kraft, E., Lipinski, B., Schill, J., Holsboer, F., 2000. Reduced glutamate in the anterior cingulate cortex in depression: an in-vivo protonmagnetic resonance spectroscopy study. Biol. Psychiatr. 47 (4), 305–313.

Basu, N., Rastogi, R.P., Dhar, M.L., 1967. Chemical examination of *Bacopa monniera* Wettst: Part III-bacoside B. Indian J. Chem. 5, 84–86.

Benson, S., Downey, L.A., Stough, C., et al., 2014. An acute, double-blind, placebo-controlled cross-over study of 320 mg and 640 mg doses of *B. monnieri* (CDRI08) on multitasking stress reactivity and mood. Phytother. Res. 28 (4), 551–559.

Bertolino, A., Callicott, J., Elman, H.I., 1998. Regionally specific neuronal pathology in untreated patients with schizophrenia: a proton magnetic resonance spectroscopic imaging study. Biol. Psychiatr. 43 (9), 641–648.

Brettschneider, J., Tredici, K.D., Lee, V.M.Y., Trojanowski, J.Q., 2015. Spreading of pathology in neurodegenerative diseases: a focus on human studies. Nat. Rev. Neurosci. 16, 109–120.

Calabrese, C., Gregory, W.L., Leo, M., Kraemer, D., Bone, K., Oken, B., 2008. Effects of astandardized *B. monnieri* extract on cognitive perfomance, anxiey, and depression in the elderly: a randomized double-blind, placebo-controlled trial. J. Alternative Compl. Med. 14, 707–713.

Chakraborty, S., Chakraborty, R., 2018. Prospect of Ayurveda in neuromedicine. Acta Sci. Med. Sci. 2 (8), 63–69.

Channa, S., Dar, A., Anjum, S., Yaqoob, M., Atta, U.R., 2006. Anti-inflammatory activity of *Bacopa monniera* in rodents. J. Ethnopharmacol. 104, 286–289.

Chatterjee, N., Rastogi, R.P., Dhar, M.L., 1963. Chemical examination of *Bacopa monniera* Wettst: Part I—isolation of chemical constituents. Indian J. Chem. 1, 212–215.

Chaudhari, K.S., Tiwari, N.R., Tiwari, R.R., et al., 2017. Neurocognitive effect of nootropic drug brahmi (*B. Monnieri*) in Alzheimer's disease. Ann. Neurosci. 24, 111–122.

Chavhan, G.B., 2013. MRI Made Easy (for Beginners), second ed. Japee Brothers Medical Publishers, New Delhi, India.

Chen, X.-Q., Mobley, W.C., 2019. Alzheimer disease pathogenesis: insights from molecular and cellular biology studies of oligomeric Aβ and tau species. Front. Neurosci. 13, 659.

Connolly, B.S., Lang, A.E., 2014. Pharmacological treatment of Parkinson disease. A review. J. Am. Med. Assoc. 311, 1670–1683.

Dahanukar, S., Thatte, U., 1989. Ayurveda Revisited. Popular Prakashan, Bombay, India, p. 138.

Dastmalchi, K., Damien, D., Vuorela, H., Hiltunen, R., 2007. Plant as potential sources for drug development against Alzheimer's disease. Int. J. Biomed. Pharmaceut. Sci. 1 (2), 83–104.

Devishree, R.A., Kumar, S., Jain, A.R., 2017. Short-term effect of *B. monnieri* on memory; a brief review. J. Pharm. Res. 11, 1447–1450.

Dhanasekaran, M., Tharakan, B., Holcomb, L.A., Hitt, A.R., Young, K.A., Manyam, B.V., 2007. Neuroprotective mechanisms of ayurvedic antidementia botanical. *Bacopa monniera*. Phytother Res. 21, 965–969.

Dhawan, B.N., 2014. Experimental and clinical evaluation of nootropic activity of *Bacopa monniera* Linn. (brahmi). Ann. Natl. Acad. Med. Sci. (India) 50 (1&2), 20–33.

Diggavi, M.V., 1998. Role of Manas in Klaibya (Male Sexual Dysfunctions) and its Management with Medhya Rasāyana and Urşya Basti. M.D. Thesis. Department of Kāyaçikitsā, Institute of Post Graduate Teaching and Research in Āyurveda, Gujarat Āyurveda University, Jamnagar, India.

Goswami, S., Saoji, A., Kumar, N., Thawani, V., Tiwari, M., Thawani, M., 2011. Effect of *B. monnieri* on cognitive functions in Alzheimer's disease patients. Int. J. Collab. Res. Intern. Med. Public Health 3 (4), 285–293.

Hebert, L.E., Scherr, P.A., Bienias, J.L., Bennett, D.A., Evans, D.A., 2003. Alzheimer disease in the US population: prevalence estimates using the 2000 census. Arch. Neurol. 60 (8), 1119–1122.

Howes, M.J.R., Houghton, P.J., 2012. Ethnobotanical treatment strategies against Alzheimer's disease. Curr. Alzheimer Res. 9 (1), 67–85.

Integrated Taxonomic Information System (ITIS), 2015. *B. Monnieri*. Taxonomic Serial No: 33038. Geological Survey, VA, USA.

Jain, P., Khanna, N.K., Trehan, N., Pendse, V.K., Godhwan, i J.L., 1994. Anti-inflammatory effects of ayurvedic preparation brahmi rasayan in rodents. Ind. J. Exp. Bio. 32, 633–636.

Kumar, A., Ekavali, A.S., 2015. A review on Alzheimer's disease pathophysiology and its management: an update. Pharmacol. Rep. 67, 195–203.

Lal, S., Baraik, B., 2019. Phytochemical and pharmacological profile of *B. monnieri*; an ethnomedicinal plant. Int. J. Pharmaceut. Sci. Res. 10 (3), 1001–1013.

Malve, H.O., 2017. Management of Alzheimer's disease: role of existing therapies, traditional medicines and new treatment targets. Indian J. Pharmaceut. Sci. 79 (1), 2–15.

Mandal, P.K., Tripathi, M., Sugunan, S., 2012. Brain oxidative stress: detection and mapping of anti-oxidant marker 'Glutathione' in different brain regions of healthy male/female, MCI and Alzheimer patients using non-invasive magnetic resonance spectroscopy. Biochem. Biophys. Res. Commun. 417 (1), 43–48.

Markowitsch, H.J., Staniloiu, A., 2012. Amnesic disorders. Lancet 380, 1429–1440.
Medina, Y.I., Evans, A.C., 2015. On the central role of brain connectivity in neurodegenerative disease progression. Front. Aging Neurosci. 7, 90–99.
Menken, M., Munsat, T.L., Toole, J.F., 2000. The global burden of disease study: implications for neurology. Arch. Neurol. 57, 418–420.
Michelle, D., Nemetchek, A.A., Stierle, D.B., Stierle, D.I.L., 2017. The Ayurvedic plant *B. monnieri* inhibits inflammatory pathways in the brain. J. Ethnopharmacol. 197, 92–100.
Morgan, A., Stevens, J., 2010. Does *B. monnieri* improve memory performance in older persons. Results of a randomized, placebo-controlled, double-blind trial. J. Alternative Compl. Med. 16, 735–739.
Mountford, C.E., Stanwell, P., Lin, A., Ramadan, S., Ross, B., 2010. Neurospectroscopy: the past, present and future. Chem. Rev. 110 (5), 3060–3086.
Peth-Nui, T., Wattanathorn, J., Muchimapura, S., Tong-Un, T., Piyavhatkul, N., Rangseekajee, P., Ingkaninan, K., Vittaya-areeku, S., 2012. Effects of 12-week *B. monnieri* consumption on attention, cognitive processing, working memory, and functions of both cholinergic and monoaminergic systems in healthy elderly volunteers. Evid. Base Compl. Alternative Med. 1–10.
Philippine Medicinal Plants (PMP), 2014. *B. Monnieri* (Linn.) Wettst.
Raghav, S., Singh, H., Dalai, P.K., Srivastava, J.S., Asthana, O.P., 2006. Randomized controlled trial of standardized *Bacopa monniera* extract in age-associated memory impairment. Indian J. Psychiatr. 48, 238–242.
Ramachandran, C., Quirin, K.W., Escalon, E., Melnick, S.J., 2014. Improved neuroprotective effects by combining *B. Monnieri* and Rosmarinus officinalis supercritical CO_2 extracts. J. Evid. Based Complement. & Altern. Med. 19 (2), 119–127.
Ramanan, V.K., Saykin, A.J., 2013. Pathways to neurodegeneration: mechanistic insights from GWAS in Alzheimer's disease, Parkinson's disease, and related disorders. Am. J. Neuro. Dis. 2, 145–175.
Ramasamy, S., Chin, S.P., Sukumaran, S.D., Buckle, M.J.C., Kiew, L.V., Chung, L.Y., 2015. In-silico and in-vitro analysis of bacoside A, aglycones and its derivatives as the constituents responsible for the cognitive effects of *B. Monnieri*. PLoS One 10 (5), 1–19.
Ray, S., Ray, A., 2015. Medhya Rasayanas in brain function and disease. Med. Chem. 5, 505–511.
Repantis, D., Schlattmann, P., Laisney, O., Heuser, I., 2010. Modafinil and methylphenidate for neuroenhancement in healthy individuals: a systematic review. Pharmacol. Res. 62, 187–206.
Roodenrys, S., Booth, D., Bulzomi, S., Phipps, A., Micallef, C., Smoker, J., 2002. Chronic effects of brahmi (*B. monnieri*) on human memory. Neuropsychopharmacology 27, 279–281.
Roy, S., Awasthi, H., 2017. Herbal medicines as neuroprotective agent: a mechanistic approach. Int. J. Pharm. Pharmaceut. Sci. 9, 1–7.
Sahakian, B., Morein-Zamir, S., 2007. Professor's little helper. Nature 450, 1157–1159.
Sathyanarayanan, V., Thomas, T., Einöther, S.J.L., Dobriyal, R., Joshi, M.K., Krishna, M.S., 2013. Brahmi for the better? New findings challenging cognition and anti-anxiety effects of Brahmi (*Bacopa monniera*) in healthy adults. Psychopharmacology 227, 299–306.
Scholey, A., Kennedy, D., Wesnes, K., 2005. The psychopharmacology of herbal extracts: issues and challenges. Psychopharmacology 179, 705–707.
Singh, S.P., 2015. Brahmi (*B.monnieri*): research on herbal medicines and the UN protocol on biodeversity for profit sharing. J. Geriatr. Care Res. 2 (2), 17–18.
Singh, H.K., Ott, T., Matthies, H., 1975. Effect of intrahippocampal injection of atropine on different phases of a learning experiment. Psychopharmacology (Berl.) 38, 247–258.

Tamara, S., Matthew, P., Con, S., 2015. *B. Monnieri* as an antioxidant therapy to reduce oxidative stress in the aging brain. Evid. Base Compl. Alternative Med. 9. Article ID 615384.

Tembhre, M., Ahirwar, S., Gour, S., Namdeo, A., 2015. Inhibitory potential of acephate and ethanol extract of *B. Monnieri* on AChE in rat cortex and Hippocampus. Int. J. Biosci. Biochem. & Bioinf. 5 (1), 45–53.

World Alzheimer Report, 2015. The Global Impact of Dementia. http://www.alz.co.uk/research/worldreport 2015.

Chapter 3.2.5

Centella asiatica

Arvind Jantwal[1], Sumit Durgapal[1], Jyoti Upadhyay[2], Mahendra Rana[3], Mohd Tariq[4,5], Aadesh Dhariwal[1], Tanuj Joshi[1]

[1]*Department of Pharmaceutical Sciences, Kumaun University, Nainital, Uttarakhand, India;* [2]*School of Health Sciences, University of Petroleum and Energy Studies, Dehradun, Uttarakhand, India;* [3]*Department of Pharmaceutical Sciences, Center for Excellence in Medicinal Plants and Nanotechnology, Kumaun University, Nainital, Uttarakhand, India;* [4]*G.B. Pant National Institute of Himalayan Environment & Sustainable Development, Kosi-Katarmal, Almora, Uttarakhand, India;* [5]*Department of Biotechnology and Microbiology, Meerut Institute of Engineering and Technology, Meerut, Uttar Pradesh, India*

Introduction

The use of plants globally as a source of various new drugs has been utilized by mankind since time immemorial. Humans have used these plants as sources of food, medicine, and fodder for animals. Plants and humans are entangled more than we realize (Vasala and Peter, 2004). Plants have been utilized as a source of supplements for the treatments of various diseases and ailments for thousands of years, based on experience and folk remedies, and continue to draw wide attention for their role in the treatment of mild and chronic diseases. In more recent times, the focus on plant research has increased all over the world and a large body of evidence has been accumulated to highlight the immense potential of medicinal plants used in various traditional systems of medicine and is proving to be integral to human well-being (Gohil, 2009; Moses and Goossens, 2017).

Centella asiatica, also known as Mandukaparni [consisting of dried whole plant of *Centella asiatica* (Linn.) Urban. Syn. *Hydrocotyle asiatica* Linn. (Fam. Apiceae, Umbellifere)], is a prostrate, faintly aromatic, stoloniferous perennial herb, commonly found as a weed in crop fields and other waste land throughout India up to an altitude of 600 m above sea level. It is a greenish-yellow colored herb with a slightly bitter taste found throughout India and is used as a brain tonic, anabolic, alterative, and anxiolytic (Wealth of India, 1992; IP, 2014). *Centella asiatica* has gained interest as a potential plant with promising pharmacological properties. It is used in the treatment of a number of ailments. Its pharmacological activities include wound healing (Shukla, 1999), anticonvulsant, antiulcer, antileprotic/antitubercular, immunomodulatory, diuretic, etc.

Naturally Occurring Chemicals against Alzheimer's Disease. https://doi.org/10.1016/B978-0-12-819212-2.00021-9

(Arora et al., 2002). *C. asiatica* has been extensively utilized in the Ayurvedic system of medicine to treat a variety of ailments (Babu et al., 1995). This plant holds a reputed position in the indigenous system of medicine and is often misinterpreted as *Bacopa monnieri* Brahmi (Wealth of India, 1992) (Fig. 3.2.5.1).

Nomenclature	Botanical Name
Centella asiatica	Syn. *Hydrocotyle asiatica* Linn.
Sanskrit:	Manduki, Darduracchada
English:	Indian pennywort
Hindi:	Brahma Manduki, Brahmi
Urdu/Punjabi:	Brahmi
Bengali:	Jholkhuri, Thalkuri, Thankuni (API Part 1 Vol. IV I[st] Edition)

FIGURE 3.2.5.1 *Centella asiatica* plant.

Distribution

Centella asiatica is distributed throughout tropical and subtropical countries from an elevation of 200–2100 m above sea level, and is native to Asia, Africa, and America (Belwal et al., 2019). In India, it is generally found on wasteland and crop fields as a weed. It is commonly found in moist clayey soils, in large clumps, and hence is often used as a cover crop for plantations to prevent soil erosion (surface runoff) (Wealth of India, 1992).

Chemical constituents

Extensive investigations have been carried out on *C. asiatica* regarding its chemical constituents and composition. The leaves are rich in carotenoids and vitamins B and C. It is also a rich source of amino acids, carbohydrates, phenols, and terpenoids, and for this reason the plant is often used as a vegetable and has found its way into food supplements (Wealth of India, 1992; Belwal et al., 2019). Other constituents of *C. asiatica* include an alkaloid hydrocotylin. Glycosides like asiaticoside A, asiaticoside B, brahmoside, brahminoside, centelloside, madecassoside, thankuni side, glycoside D, and glycoside E, and phenolic compounds like quercitrin, kaempferol, luteolin, and chlorogenic acid and triterpene acid are also present, as are asiatic acid, asiaticoside, madecassic acid, madecassoside, terminolic, centic, centellic, centoic acid, indocentoic acid, isobrahmic, brahmic, betulic, and madasiatic acid. Other phytochemicals include glycerides of oleic, palmitic, stearic, lignoceric, linoleic, and linolenic acids, oligosaccharide centellose, stigmasterol, sitosterol, campsterol, polyacetylenes, carotenoids, vitamins B and C, vellarine, pectic acid, tannins, sugars, inorganic acid, and resins (Belwal et al., 2019; Bhandari et al., 2007; Chopra et al., 1956; Chopra et al., 1992; Datta et al., 1962; Kapoor, 2005; Schaneberg et al., 2003; Singh and Rastogi, 1969) (Fig. 3.2.5.2) (Table 3.2.5.1).

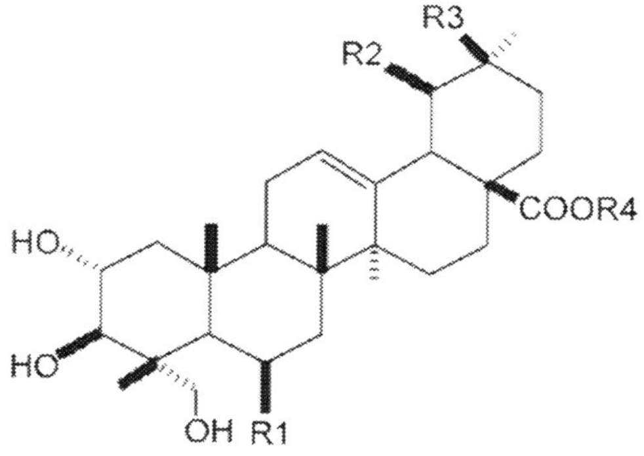

FIGURE 3.2.5.2 Structure of asiatic acid, asiaticoside, asiaticoside B, madecassic acid, madecassoside, and terminolic acid.

TABLE 3.2.5.1

Compound	R1	R2	R3	R4
Asiatic acid	H	CH_3	H	H
Asiaticoside	H	CH_3	H	Glu-Glu-Rha
Asiaticoside B	OH	H	CH_3	Glu-Glu-Rha
Madecassic acid	OH	CH_3	H	H
Madecassoside	OH	CH_3	H	Glu-Glu-Rha
Terminolic acid	OH	H	CH_3	H

Pharmacological activity

Centella asiatica is an effective neuroprotective and has been used as an antidiabetic, antibacterial, cardioprotective, and antiviral drug. Other pharmacological activities are described in Table 3.2.5.2.

Alzheimer's disease

Alzheimer's disease (AD) is a progressive, degenerative neurological disorder characterized by cognitive loss of reasoning and memory. The pathogenesis of AD is still not clear, however it is suggested that it is due to a combination of genetic, environmental/lifestyle factors and the aging process.

Patients initially face difficulty in remembering recent events, followed by forgetfulness and confusion. As the disease progresses the ability to read, write, talk, and eat is lost.

The brain of a diseased AD patient shows three structural abnormalities:

1. Loss of neurons that release acetylcholine → destruction of nucleus basalis (center for Ach release);
2. Beta-amyloid plaques → deposition of protein outside neurons;
3. Neurofibrillary tangles → bundles of filaments inside neurons (consisting of hyperphosphorylated protein "tau").

Acetylcholinesterase inhibitors (Tacrine and Donepezil) have been shown to improve alertness and behavior in around 5% of AD patients some evidence has suggested vitamin E (antioxidant) may have a beneficial effect on AD current research in drug discovery for treatment of AD involves various targets, and is only symptomatic, with the main therapeutic strategies based on the "cholinergic hypothesis" and "amyloid cascade hypothesis" (Orhan et al., 2006; Tortora and Derrickson, 2008).

TABLE 3.2.5.2 Biological activities of *Centella asiatica.*

S. no.	Pharmacological activity	References
1	Antibacterial	Oyedeji and Afolayan, (2005)
2	Anticancer	Verma and Gurmaita, (2019), Hamid et al., (2016), Gohil et al., (2010), Cho et al., (2006), Xu et al., (2012), Babu et al., (1995), Babu and Paddikkala, (1993)
3	Anticonvulsant	Visweswari et al., (2010)
4	Antidiabetic	Mutayabarwa et al., (2003)
5	Antifungal	Dash et al., (2011)
6	Antiinflammatory	Somchit et al., (2004), Yun et al., (2008), Liu et al., (2008)
7	Antioxidant	Hamid et al., (2002), Dewi and Maryani, (2015), Jayashree et al., (2003), Gohil et al., (2010), Zhimin et al., (2017)
8	Antiproliferative	Aizad et al., (2015)
9	Antipsoriatic	Sampson et al., (2001)
10	Antiulcer	Cheng et al., (2004), Cheng and Koo, (2000)
11	Antiviral	Yoosook et al., (2000)
12	Cardioprotective	Gnanapragasam et al., (2004)
13	Cytotoxic	Bunpo et al., (2004), Xu et al., (2012)
14	Hepatoprotective	Antony et al., (2006)
15	Immunostimulant	Wang et al., (2003)
16	Insecticidal	Rajkumar and Jebanesan, (2005)
17	Memory enhancer	Nalini et al., (1992); Veerendra and Gupta, (2002), Veerendra and Gupta, (2003)
18	Sedative	Wijeweera et al., (2006)
19	Wound healing	Bonte et al., (1994), Shukla et al., (1999), Maquart et al., (1999), Gohil et al., (2010), Kumar et al., (1998), Suguna et al., (1996)

In vitro and in vivo studies

Centella asiatica (CA) has been a well-known drug in Asian traditional medicine for over 2000 years and has been used in the Chinese and Ayurvedic medicine systems (Chen et al., 2016; Dey et al., 2017). Asiaticoside is one of the most important phytochemicals present in CA, it protects in vitro ischemia hypoxia (Sun et al., 2015). Studies have shown activity of asiaticoside-6, a chemical constituent of CA, as an anti-Alzheimer's drug protecting against amyloid-β toxicity (Mook-Jung et al., 1999). Juice of CA, liquorice powder, juice of Guduchi, and Sankhpuspi paste when taken with milk, prevented dementia and improved memory (Manyam, 1999). Witter tested CA along with seven other plants for inhibition of amyloid-β1−40 (Aβ40) and methionine amyloid-β1−40 (MAβ40) fibrillation, which causes plaque formation resulting in dementia and AD. It was found that CA extract acted as the strongest inhibiter and interacted with MAβ40, due to asiaticosides (Witter, 2018). In one study, CA extract against streptozotocin-induced cognitive impairment and oxidative stress in Wistar rats was tested, where it was observed that the aqueous extract of CA effectively prevented cognitive defects and oxidative stress caused by intracerebroventricular streptozotocin (Kumar and Gupta, 2002). Aqueous extract of CA was tested on Tg2576 mouse (murine model for AD) with high β-amyloid burden. The result showed improvement of behavioral abnormalities in treated mice. In another study, it was observed that aqueous extract of CA had a protective effect on SH-SY5Y cells from toxicity (caused by H_2O_2) (Somyanath, 2012). Another investigation into the neuroprotective effect of CA against aluminum hexahydrate ($AlCl_3.6H_2O$) was carried out on Wistar rats and the results indicated that CA treatment reduced the alterations caused by $AlCl_3$. This study suggested that the neuroprotective effect was due to CA's antioxidant potential (Amjad and Umesalma, 2015). Matthews in his study observed that CA extract increased cytochrome B, NADH dehydrogenase 1, cytochrome C oxidase 1, and ATP synthase 6 in the hippocampal neurons which resulted in its cognitive effect (Matthews, 2017). Gray demonstrated that aqueous extract of CA (CAW) attenuated amyloid-(Aβ)-induced cognitive deficits in vivo, and prevented Aβ-induced cytotoxicity using an in vitro model. The suggested mechanism of CA aqueous extract is that it affects mitochondrial biogenesis along with activation of the antioxidant response (Gray et al., 2014, 2015).

Dhanasekaran concluded that CA extract decreased the amyloid-β1−40, 1−42 and also reduced Congo red-stained fibrillar amyloid plaques significantly, and the extract acted as an antioxidant and prevented DNA damage (Dhanasekaran et al., 2009). Xu illustrated that the potential mechanism for CA extract in neuroblastoma cells expressing Aβ peptide was implicated in CREB phosphorylation and modulation of ERK/40S ribosomal protein S6 kinase (RSK) signaling pathway (Xu et al., 2008). Chen and coworkers demonstrated the anti-Alzheimer potential of CA extract by the ability of CA extract to improve Aβ-mediated reactive oxygen species production in neural cells (Chen et al., 2016).

Clinical studies

There are a number of in vitro/in vivo studies on neuroprotective activity of CA but only limited clinical studies on CA have been carried out. In a double-blind study on mentally deficient children, it was observed that administration of CA revealed significant improvement in children after 6 months (Rao, 1977).

A randomized double-blind placebo-controlled clinical study was carried out on CA extract containing tannic acid, asiaticoside, and asiatic acid (29.9, 1.09, 48.89 mg/g, respectively). The study was carried out on 28 volunteers with an average age 65.05 ± 3.5. The subjects were given 250, 500, and 750 mg/day for 2 months. The results revealed that working memory, self-rated mood, i.e., overall cognitive functions, had improvement after a high dose of CA (Wattanathorn et al., 2008). In another study on 41 subjects, the effect of CA extract on cognitive functions was tested. The extract was administered in the form of a capsule (one/day) for 2 months. The cognitive activity was recorded using Woodcock-Johnson Cognitive Abilities Test III and it was found that the extract had a positive effect on all test subjects measured at different times (Dev et al., 2009). Tiwari and team studied the effect of CA extract on 60 elderly patients (mean age 65) with mild cognitive deficiency for 6 months. The results indicated that CA extract, when administered twice daily ($500 \text{ mg} \times 2$) resulted in significant cognitive improvement. The results were obtained using Mini Mental State Examination scoring (Tiwari et al., 2008).

Toxicity and interactions

Centella asiatica is a plant with promising pharmacological properties. A literature survey of the toxicity and adverse effects of *Centella asiatica* displayed a scarcity of data. In 2010, Oruganti carried out a study to assess the safety levels for oral doses of *Centella asiatica* on albino rats. After an extensive study for 30 days, it was reported that there was a dose-dependent increase in serum biomarkers. A dose of 1000 mg/kg increased the weight of the spleen and caused a high level of apoptosis in hepatic and renal tissues (Oruganti, 2010).

In a study conducted in 2004, it was established that the oral median lethal dose of dried powder of *Centella asiatica* for mice was over 8 g/kg. In another chronic toxicity study on Wistar rats (male and female) receiving 20, 200, 600, and 1200 mg/kg/day of *C. asiatica* for 6 months, displayed no sign of significant alteration in body weight, blood chemistry, clinical chemistry, or histopathology when compared to the control group (Chivapat, 2004). Chivapat, in 2011, carried out toxicity studies on ECa 233, a standardized extract of

Centella asiatica. The study was carried out on male and female Wistar rats. After a study period of 14 days, no lethality was observed at a dose of 10 g/kg along with absence of any form of pharmacological damage to any organ. Subchronic toxicity of the standardized extract on a group of 24 rats (Wistar rats) showed no difference in average weight, nor did the study animals show any signs of abnormal behavior. On autopsy of the study animals, all the ECa 233-treated animals displayed no damage to organs, there was no difference in organ weight except for the male rat group that was given ECa 233 (10 mg/kg/day) showed a lower relative right adrenal weight when compared to the control group (Chivapat, 2011).

Dandekar and his team carried out studies on the interaction between phenytoin and an Ayurvedic formulation of *Convolvulus pluricaulis* (sankhapushpi) in which *Centella asiatica* is the second most abundant component of the drug. The study was carried out on random-bred, Sprague–Dawley rats of either sex with weight ranging between 100 and 150 g. The study showed that plasma phenytoin levels were lowered significantly, when sankhapushpi was coadministered with phenytoin orally twice daily for 5 days, thereby suggesting the clinical combination of sankhapushpi with phenytoin is not advised (Dandekara et al., 1992)

Laerum and Iversen in 1972 reported asiaticoside (a glycoside terpene from the plant *Centella asiatica*) to be a weak tumor promoter in hairless mouse epidermis. It was also reported to be weakly carcinogenic to the dermis on surface application (Laerum and Iversen, 1972) In a case study carried out by Izu and team they reported the occurrence of allergic contact dermatitis with topical application of creams containing *Centella asiatica* extracts (Izu et al., 1992).

Conclusion

Centella asiatica is a well-known and reputed herbal medicinal plant, and is a plant with promising pharmacological potential for the treatment of a number of ailments. Its pharmacological activities include wound healing, anticonvulsant, antiulcer, antileprotic/antitubercular, immunomodulatory, diuretic, etc. It has a potent protective effect on the central nervous system, which is generally associated with its bioactive compounds including asiatic acid, madecassic acid, asiaticoside, madecassoside, and brahmic acid. The neuroprotective effect of CA has been reported by various mechanisms, the majority of them being an indication of reduction of oxidative stress. Further studies on clinical activities need to be carried out on CA in order to explore its wider possibilities in treating Alzheimer's disease conditions.

Due to the fact that *Centella asiatica* has diversity in its pharmacological activity, the overexploitation of this miraculous plant has become a serious issue and as a result the plant has been listed as a highly threatened plant by the IUCN.

References

Aizad, S., Yahaya, B.H., Zubairi, S.I., 2015. Carboxy-methyl-cellulose (CMC) hydrogel-filled 3-D scaffold: preliminary study through a 3-D antiproliferative activity of *Centella asiatica* extract. In: Ahmad, A., Karim, N.H.A., Hassan, N.I., Zubairi, S.I., Bakar, S.A., Ibarahim, Z., Kartini, H.K., Agustar, H.B., Samsudin (Eds.), AIP Conference Proceedings, vol. 1678. AIP Publishing., p. 050005. No. 1.

Amjad, S., Umesalma, S., 2015. Protective effect of *Centella asiatica* against aluminium-induced neurotoxicity in cerebral cortex, striatum, hypothalamus and hippocampus of rat brain-histopathological, and biochemical approach. J. Mol. Biomarkers Diagn. 6 (1), 1.

Antony, B., Santhakumari, G., Merina, B., Sheeba, V., Mukkadan, J., 2006. Hepatoprotective effect of *Centella asiatica* (L) in carbon tetrachloride-induced liver injury in rats. Indian J. Pharm. Sci. 68 (6), 772.

Arora, D., Kumar, M., Dubey, S.D., 2002. *Centella asiatica* — a review of its medicinal uses and pharmacological effects. J. Nat. Remedies 2/2, 143—149.

Babu, T.D., Paddikkala, J., 1993. Anticancer activity of an active principle from *Centella asiatica*. Amala Res. Bull. 13, 46—49.

Babu, T.D., Kuttan, G., Padikkala, J., 1995. Cytotoxic and anti-tumour properties of certain taxa of Umbelliferae with special reference to *Centella asiatica* (L.) Urban. J. Ethnopharmacol. 48 (1), 53—57.

Belwal, T., Andola, H.C., Atanassova, M.S., Joshi, B., Suyal, R., Thakur, S., Bisht, A., Jantwal, A., Bhatt, I.D., Rawal, R.S., 2019. Gotu kola (*Centella asiatica*). In: Nonvitamin and Nonmineral Nutritional Supplements. Academic Press, pp. 265—275.

Bhandari, P., Kumar, N., Gupta, A.P., Singh, B., Kaul, V.K., 2007. A rapid RP-HPTLC densitometry method for simultaneous determination of major flavonoids in important medicinal plants. J. Sep. Sci. 30 (13), 2092—2096.

Bonte, F., Dumas, M., Chaudagne, C., Meybeck, A., 1994. Influence of asiatic acid madecassic acid, and asiaticoside on human collagen I synthesis. Planta Med. 60 (2), 133—135.

Bunpo, P., Kataoka, K., Arimochi, H., Nakayama, H., Kuwahara, T., Bando, Y., Izumi, K., Vinitketkumnuen, U., Ohnishi, Y., 2004. Inhibitory effects of *Centella asiatica* on azoxymethane-induced aberrant crypt focus formation and carcinogenesis in the intestines of F344 rats. Food Chem. Toxicol. 42 (12), 1987—1997.

Chen, C.L., Tsai, W.H., Chen, C.J., Pan, T.M., 2016. *Centella asiatica* extract protects against amyloid β1—40-induced neurotoxicity in neuronal cells by activating the antioxidative defence system. J. Tradit. Complement. Med. 6 (4), 362—369.

Cheng, C.L., Koo, M.W., 2000. Effects of *Centella asiatica* on ethanol induced gastric mucosal lesions in rats. Life Sci. 67 (21), 2647—2653.

Cheng, C.L., Guo, J.S., Luk, J., Koo, M.W., 2004. The healing effects of Centella extract and asiaticoside on acetic acid induced gastric ulcers in rats. Life Sci. 74 (18), 2237—2249.

Chivapat, S., Chavalittumrong, P., Attawish, A., Boonruad, T., Bansiddhi, J., Phadungpat, S., Punyamong, S., Mingmuang, J., 2004. Toxicity study of *Centella asiatica* (L) urban. J. Thai Tradit. Altern. Med. 2, 3—17.

Chivapat, S., Chavalittumrong, P., Tantisira, M.H., 2011. Acute and sub-chronic toxicity studies of a standardized extract of *Centella asiatica* ECa 233. Thai J. Pharm. Sci. 35, 55—64.

Cho, C.W., Choi, D.S., Cardone, M.H., Kim, C.W., Sinskey, A.J., Rha, C., 2006. Glioblastoma cell death induced by asiatic acid. Cell Biol. Toxicol. 22 (6), 393—408.

Chopra, R.N., Nayar, S.L., Chopra, I.C., 1956. Glossary of Indian Medicinal Plants. Council for Scientific and Industrial Research, New Delhi, p. 58.

Chopra, R.N., Chopra, I.C., Varma, B.S., 1992. Supplement to Glossary of Indian Medicinal Plants. CSIR, New Delhi, India, p. 14.

Dandekara, U.P., Chandraa, R.S., Dalvi, S.S., Joshi, M.V., Gokhalea, P.C., Sharmaa, A.V., Shahb, P.U., Kshirsagar, N.A., 1992. Analysis of a clinically important interaction between phenytoin and Shankhapushpi, an Ayurvedic preparation. J. Ethnopharmacol. 35, 285–288.

Dash, B.K., Faruquee, H.M., Biswas, S.K., Alam, M.K., Sisir, S.M., Prodhan, U.K., 2011. Antibacterial and antifungal activities of several extracts of *Centella asiatica* L. against some human pathogenic microbes. Life Sci. Med. Res. 2011, 1–5.

Datta, T., Basu, U.P., Triterpenoids, F.I., 1962. Thankuniside and thankunic acid: a new triterpene glycoside and acid from *Centella asiatica* Linn Urb. Indian J. Chem. Sec. B 21, 239.

Dev, R.D.O., Mohamed, S., Hambali, Z., Samah, B.A., 2009. Comparison on cognitive effects of *Centella asiatica* in healthy middle age female and male volunteers. Eur. J. Sci. Res. 31 (4), 553–565.

Dewi, R.T., Maryani, F., 2015. Antioxidant and α-glucosidase inhibitory compounds of *Centella asiatica*. Procedia Chem 17, 147–152.

Dey, A., Bhattacharya, R., Mukherjee, A., Pandey, D.K., 2017. Natural products against Alzheimer's disease: pharmaco-therapeutics and biotechnological interventions. Biotechnol. Adv. 35 (2), 178–216.

Dhanasekaran, M., Holcomb, L.A., Hitt, A.R., Tharakan, B., Porter, J.W., Young, K.A., Manyam, B.V., 2009. *Centella asiatica* extract selectively decreases amyloid β levels in hippocampus of Alzheimer's disease animal model. Phytother Res. 23 (1), 14–19.

Gnanapragasam, A., Ebenezar, K.K., Sathish, V., Govindaraju, P., Devaki, T., 2004. Protective effect of *Centella asiatica* on antioxidant tissue defense system against adriamycin induced cardiomyopathy in rats. Life Sci. 76 (5), 585–597.

Gohil, K.J., 2009. Pharmacological Investigation of Clinically Important and Potential Herbal Drug Interactions.

Gohil, K.J., Patel, J.A., Gajjar, A.K., 2010. Pharmacological review on *Centella asiatica*: a potential herbal cure-all. Indian J. Pharm. Sci. 72 (5), 546–556.

Gray, N.E., Morré, J., Kelley, J., Maier, C.S., Stevens, J.F., Quinn, J.F., Soumyanath, A., 2014. Caffeoylquinic acids in *Centella asiatica* protect against amyloid-β toxicity. J. Alzheimers Dis. 40 (2), 359–373.

Gray, N.E., Sampath, H., Zweig, J.A., Quinn, J.F., Soumyanath, A., 2015. *Centella asiatica* attenuates amyloid-β-induced oxidative stress and mitochondrial dysfunction. J. Alzheimers Dis. 45 (3), 933–946.

Hamid, A.A., Shah, Z.M., Muse, R., Mohamed, S., 2002. Characterization of antioxidative activities of various extracts of *Centella asiatica* (L) Urban. Food Chem. 77 (4), 465–469.

Hamid, I.S., Widjaja, N.M.R., Damayanti, R., 2016. Anticancer activity of *Centella asiatica* leaves extract in Benzo (a) pyrene-Induced Mice. Int. J. Pharm. Phytochem. Res. 8 (1), 80–84.

Indian Pharmacopoeia, 2014. Gaziabad. Indian Pharmacopoeia Commission, Uttar Pradesh.

Izu, R., Agurre, A., Gil, N., Diaz-Perez, J.L., 1992. Allergic Contact Dermatitis from a cream containing *Centella asiatica* extract. Contact Dermatitis 26, 192–213.

Jayashree, G., Kurup, M.G., Sudarslal, S., Jacob, V.B., 2003. Anti-oxidant activity of *Centella asiatica* on lymphoma-bearing mice. Fitoterapia 74 (5), 431–434.

Kapoor, L.D., 2005. CRC Handbook of Ayurvedic Medicinal Plants. CRC Press LLC, Florida, pp. 208–209.

Kumar, M.V., Gupta, Y.K., 2002. Effect of different extracts of *Centella asiatica* on cognition and markers of oxidative stress in rats. J. Ethnopharmacol. 79 (2), 253–260.

Kumar, S., Parameshwaraiah, S., Shivakumar, H.G., 1998. Evaluation of topical formulations of aqueous extract of *Centella asiatica* on open wounds in rats. Indian J. Exp. Biol. 36 (6), 569–572.

Laerum, O.D., Iversen, O.H., 1972. Reticuloses and epidermal tumors in hairless mice after topical skin applications of cantharidin and asiaticoside. Canc. Res. 32, 1463–1469.

Liu, M., Dai, Y., Yao, X., Li, Y., Luo, Y., Xia, Y., Gong, Z., 2008. Anti-rheumatoid arthritic effect of madecassoside on type II collagen-induced arthritis in mice. Int. Immunopharmacol. 8 (11), 1561–1566.

Manyam, B.V., 1999. Dementia in Ayurveda. J. Alternative Compl. Med. 5 (1), 81–88.

Maquart, F.X., Chastang, F., Simeon, A., Birembaut, P., Gillery, P., Wegrowski, Y., 1999. Triterpenes from *Centella asiatica* stimulate extracellular matrix accumulation in rat experimental wounds. Eur. J. Dermatol. 9 (4), 289–296.

Matthews, D.G., Gray, N.E., Meshul, C., Caruso, M., Moore, C., Murchison, C.F., Harris, C., Quinn, J.F., Soumyanath, A., 2017. Exploring the effect of *Centella asiatica* on mitochondrial biogenesis in the mouse brain. Alzheimers Dement. 13 (7), P663.

Mook-Jung, I., Shin, J.E., Yun, S.H., Huh, K., Koh, J.Y., Park, H.K., Jew, S.S., Jung, M.W., 1999. Protective effects of asiaticoside derivatives against beta-amyloid neurotoxicity. J. Neurosci. Res. 58 (3), 417–425.

Moses, T., Goossens, A., 2017. Plants for human health: greening biotechnology and synthetic biology. J. Exp. Bot. 68 (15), 4009.

Mutayabarwa, C.K., Sayi, J.G.M., Dande, M., 2003. Hypoglycaemic activity of *Centella asiatica* (L) Urb. East Cent. Afr. J. Pharm. Sci. 6 (2), 30–35.

Nalini, K., Aroor, A.R., Rao, A., Karanth, K.S., 1992. Effect of *Centella asiatica* fresh leaf aqueous extract on learning and memory and biogenic amine turnover in albino rats. Fitoterapia 63 (3), 231–238.

Orhan, G., Orhan, I., Sener, B., 2006. Recent developments in natural and synthetic drug research for Alzheimer's disease. Lett. Drug Des. Discov. 3 (4), 268–274.

Oruganti, M., Roy, B.K., Singh, K.K., Prasad, R., Kumar, S., 2010. Safety assessment of *Centella asiatica* in albino rats. Pharmacogn. J. 2 (16), 5–13.

Oyedeji, O.A., Afolayan, A.J., 2005. Chemical composition and antibacterial activity of the essential oil of *Centella asiatica*. Growing South Africa. Pharm. Biol. 43 (3), 249–252.

Rajkumar, S., Jebanesan, A., 2005. Larvicidal and adult emergence inhibition effect of *Centella asiatica* Brahmi (Umbelliferae) against mosquito *Culex quinquefasciatus* Say (Diptera: Culicidae). Afr. J. Biomed. Res. 8 (1), 31–33.

Rao, V.A., Srinivasan, K., Rao, T.K., 1977. The effect of *Centella asiatica* on the general mental ability of mentally retarded children. Indian J. Psychiatr. 19 (4), 54.

Sampson, J.H., Raman, A., Karlsen, G., Navsaria, H., Leigh, I.M., 2001. In vitro keratinocyte antiproliferant effect of *Centella asiatica* extract and triterpenoid saponins. Phytomedicine 8 (3), 230–235.

Schaneberg, B.T., Mikell, J.R., Bedir, E., Khan, I.A., 2003. An improved HPLC method for quantitative determination of six triterpenes in *Centella asiatica* extracts and commercial products. Pharmazie 58 (6), 381–384.

Shukla, A., Rasik, A.M., Jain, G.K., Shankar, R., Kulshrestha, D.K., Dhawan, B.N., 1999b. In vitro and in vivo wound healing activity of asiaticoside isolated from *Centella asiatica*. J. Ethnopharmacol. 65 (1), 1–11.

Singh, B., Rastogi, R.P., 1969. A reinvestigation of the triterpenes of *Centella asiatica*. Phytochemistry 8, 917.

Somchit, M.N., Sulaiman, M.R., Zuraini, A., Samsuddin, L., Somchit, N., Israf, D.A., Moin, S., 2004. Antinociceptive and antiinflammatory effects of *Centella asiatica*. Indian J. Pharmacol. 36 (6), 377.

Soumyanath, A., Zhong, Y.P., Henson, E., Wadsworth, T., Bishop, J., Gold, B.G., Quinn, J.F., 2012. *Centella asiatica* extract improves behavioral deficits in a mouse model of Alzheimer's disease: investigation of a possible mechanism of action. Int. J. Alzheimers Dis. 2012.

Suguna, L., Sivakumar, P., Chandrakasan, G., 1996. Effect of *Centella asiatica* extract on dermal wound healing in rats. Indian J. Exp. Biol. 34, 1208–1211.

Sun, T., Liu, B., Li, P., 2015. Nerve protective effect of asiaticoside against ischemia-hypoxia in cultured rat cortex neurons. Med. Sci. Mon. 21, 3036.

Tiwari, S., Singh, S., Patwardhan, K., Gehlot, S., Gambhir, I.S., 2008. Effect of *Centella asiatica* on mild cognitive impairment (MCI) and other common age-related clinical problems. Digest J. Nanomater. Biostruct. 3 (4), 215–220.

Tortora, G.J., Derrickson, B.H., 2008. Principles of Anatomy and Physiology. John Wiley & Sons, pp. 539–540.

Vasala, P.A., Ginger Peter, K.V., 2004. Handbook of Herbs and Spices, pp. 1–8.

Veerendra, K.M.H., Gupta, Y.K., 2002. Effect of different extracts of Centella asiatica on cognition and markers of oxidative stress in rats. J. Ethnopharmacol. 79 (2), 253–260.

Veerendra, K.M.H., Gupta, Y.K., 2003. Effect of Centella asiatica on cognition andoxidative stress in an intracerebroventricular streptozotocin model of Alzheimer's disease in rats. Clin. Exp. Pharmacol. Physiol. 30 (5–6), 336–342.

Verma, R., Gurmaita, A., 2019. A review on anticarcinogenic activity of "Centella asiatica". Endangered Species 6, 7.

Visweswari, G., Prasad, K.S., Chetan, P.S., Lokanatha, V., Rajendra, W., 2010. Evaluation of the anticonvulsant effect of *Centella asiatica* (gotu kola) in pentylenetetrazol-induced seizures with respect to cholinergic neurotransmission. Epilepsy Behav. 17 (3), 332–335.

Wang, X.S., Dong, Q., Zuo, J.P., Fang, J.N., 2003. Structure and potential immunological activity of a pectin from *Centella asiatica* (L.) Urban. Carbohydr. Res. 338 (22), 2393–2402.

Wattanathorn, J., Mator, L., Muchimapura, S., Tongun, T., Pasuriwong, O., Piyawatkul, N., Yimtae, K., Sripanidkulchai, B., Singkhoraard, J., 2008. Positive modulation of cognition and mood in the healthy elderly volunteer following the administration of *Centella asiatica*. J. Ethnopharmacol. 116 (2), 325–332.

Wealth of India Vol 3rd, 1992. New Delhi National Institute of Science Communication and Information Resource, CSIR.

Wijeweera, P., Arnason, J.T., Koszycki, D., Merali, Z., 2006. Evaluation of anxiolytic properties of Gotukola – (*Centella asiatica*) extracts and asiaticoside in rat behavioral models. Phytomedicine 13, 668–676.

Witter, S., Witter, R., Vilu, R., Samoson, A., 2018. Medical plants and nutraceuticals for amyloid-β fibrillation inhibition. J. Alzheimers Dis. Rep. (Preprint) 1–14.

Xu, Y., Cao, Z., Khan, I., Luo, Y., 2008. Gotu kola (*Centella asiatica*) extract enhances phosphorylation of cyclic AMP response element binding protein in neuroblastoma cells expressing amyloid beta peptide. J. Alzheim. Dis. 13 (3), 341–349.

Xu, C.L., Wang, Q.Z., Sun, L.M., Li, X.M., Deng, J.M., Li, L.F., Zhang, J., Xu, R., Ma, S.P., 2012. Asiaticoside: attenuation of neurotoxicity induced by MPTP in a rat model of Parkinsonism via maintaining redox balance and up-regulating the ratio of Bcl-2/Bax. Pharmacol. Biochem. Behav. 100 (3), 413–418.

Yoosook, C., Bunyapraphatsara, N., Boonyakiat, Y., Kantasuk, C., 2000. Anti-herpes simplex virus activities of crude water extracts of Thai medicinal plants. Phytomedicine 6 (6), 411–419.

Yun, K.J., Kim, J.Y., Kim, J.B., Lee, K.W., Jeong, S.Y., Park, H.J., Jung, H.J., Cho, Y.W., Yun, K., Lee, K.T., 2008. Inhibition of LPS-induced NO and PGE 2 production by asiatic acid via NF-κB inactivation in RAW 264.7 macrophages: possible involvement of the IKK and MAPK pathways. Int. Immunopharmacol. 8 (3), 431–441.

Zhimin, Q., Xinxin, C., Jingbo, H., Qinmei, L., Qinlei, Y., Junfeng, Z., Xuming, D., 2017. Asiatic acid enhances Nrf2 signaling to protect HepG2cells from oxidative damage through Akt and ERK activation. Biomed. Pharmacother. 88, 252–259.

Chapter 3.2.6

Rosmarinus officinalis L.

Shashi Upadhyay, Kapil Bisht, Amit Bahukhandi, Monika Bisht, Poonam Mehta, Arti Bisht
G.B. Pant National Institute of Himalayan Environment, Almora, Uttarakhand, India

Introduction, botanical description, and distribution

Rosmarinus officinalis L. is a medicinally important plant that belongs to family Lamiaceae and is commonly known as rosemary (Table 3.2.6.1; Andrade et al., 2018). Rosemary is an aromatic, dense bush, branched, evergreen, having whitish-blue flowers, reaching up to 1 m in height with upright stems (Hassani et al., 2016). Leaves are 1−4 cm long and 2−4 mm wide, sessile, leathery, linear to linear-lanceolate, with curved edges, dark green upper side and granulosa and page bottom tomentous, with prominent midrib, and very characteristic smell (Begum et al., 2013; Andrade et al., 2018, Fig. 3.2.6.1).

Rosemary has been widely used not only in cooking, especially to modify and enhance flavors, improving the shelf life of perishable foods and cosmetics, but also in traditional medicine, being a highly appreciated medicinal plant to prevent and cure colds, rheumatism, and pain of muscles and joints (Zhang et al., 2014; Alipour and Saharkhiz, 2016; Andrade et al., 2018). Around the world, rosemary is frequently used as a medicinal plant in the traditional and modern medicines due to presence of several secondary metabolites. In folk medicine, it is widely used as a remedy for curing pain, stimulating hair growth, diuretic, etc. (Al-Sereiti et al., 1999). In addition, species showed antibacterial, antimutagenic properties, antiinflammatory, hypoglycemic, hypolipidemic, hypotensive, antiatherosclerotic, antithrombotic, hepatoprotective, and hypocholesterolemic effects (Tai et al., 2012; Fernandes et al., 2014; Hassani et al., 2016). It has been also reported that the essential oil of rosemary has antibacterial and cytotoxic activities due to presence of three major components, namely 1,8-cineole, α-pinene, and β-pinene (Wang et al., 2012).

Rosemary is a sun-loving herb, native to the south of France and other Mediterranean regions, often used by North African populations for the treatment of several inflammatory and infectious diseases (Chobba et al.,

Naturally Occurring Chemicals against Alzheimer's Disease. https://doi.org/10.1016/B978-0-12-819212-2.00022-0

TABLE 3.2.6.1 Systematic classification.

Kingdom	**:**	**Plantae**
Subkingdom	:	Tracheobionta
Superdivision	:	Spermatophyta
Division	:	Magnoliophyta
Class	:	Magnoliopsida
Subclass	:	Asteridae
Order	:	Lamiales
Family	:	Lamiaceae
Genus	:	*Rosmarinus* L.
Species	:	*Officinalis*

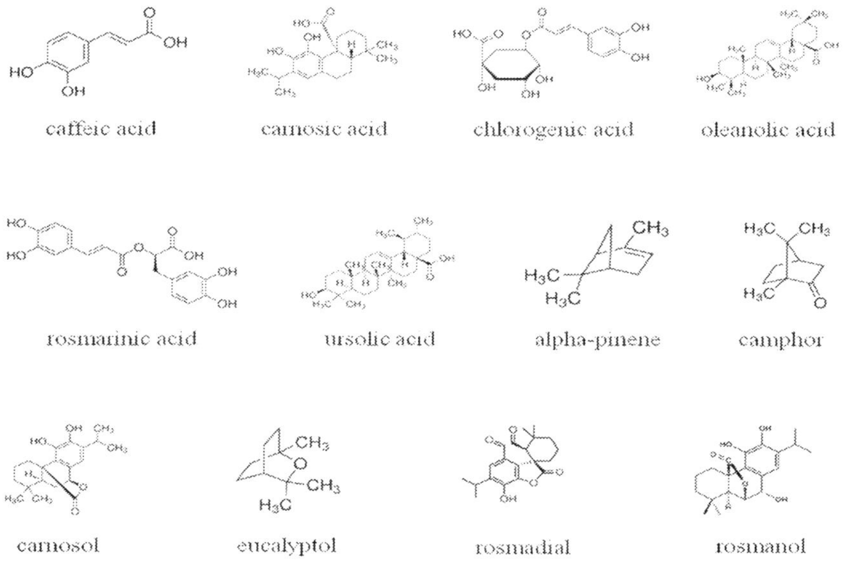

FIGURE 3.2.6.1 Bioactive compounds present in rosemary. *Source: de Oliveira, J.R., Camargo, S.E.A., de Oliveira, L.D., 2019. Rosmarinus officinalis L. (rosemary) as therapeutic and prophylactic agent. J. Biomed. Sci. 26 (5). https://doi.org/10.1186/s12929_019_0499_8.*

2012). It was cultivated in Mediterranean first, then transplanted to China during the Jin Dynasty, and it is cultivated in the entire world now. It grows widely along the north and south coasts of the Mediterranean seas, and also in the sub-Himalayan areas (Tyler et al., 1976; Korb, 1985; Chopra, 1958). It is usually grown for the production of essential oil in Columbia. The province of Murcia (Southeast Spain) is one of the major processors and importers of rosemary. In the United States and Europe, rosemary is a unique species commercially available for use as an antioxidant (Cuvelier et al., 1996).

Bioactive compounds

In the essential oil of rosemary, several biologically active molecules have been reported, mainly monoterpenes, like 1,8-cineole, borneol, pinene, limonene, camphene, 3-octanone, sabinene, myrcene, o-cymene, 1,8-cineole, linalool, myrcenol, camphor, borneol, terpinen-4-ol, α-terpineol, verbinone, piperitone, bornyl acetate, β-caryophyllene cis-b-farnesene, germacrene D, α-bisabolol (Juhás et al., 2009; Machado et al., 2013; Akbari et al., 2015; Borges et al., 2017; Vilela et al., 2016). The oil yield of dried plant (volume/dry weight) obtained by hydrodistillation was 1.9%. Twenty compounds representing 99.93% of the oils were identified. The main constituents of the oils were p-cymene (44.02%), linalool (20.5%), gamma-terpinene (16.62%), thymol (1.81%), beta-pinene (3.61%), alpha-pinene (2.83%), and eucalyptol (2.64%). The oil consisted of monoterpenic hydrocarbons, oxygenated monoterpenes, and sesquiterpene hydrocarbons. Also, the inhibition effect of rosemary oil was investigated against *Alternaria alternata*, *Botrytis cinerea*, and *Fusarium oxysporum*. The experiment was carried out in vitro using disc diffusion to investigate the antifungal action of the oil. Oil tested on potato dextrose agar plates exhibited an inhibitory effect. The extent of inhibition of fungal growth varied depending on the levels of essential oil used in experiment.

Rosemary–drug interactions

The chemical composition of rosemary affects its biologic activities. Among the main components of rosemary that were attributed its pharmacologic activities are 1,8-cineole, camphor, and α-pinene. Biologic activities attributed to 1,8-cineole include antidepressive, antimicrobial, antioxidant, antiallergic, smooth muscle relaxant effect, and antiinflammatory activity; to α-pinene was attributed antioxidant, antifungal, antibacterial, and antiinflammatory activities.

- **Aminophylline:** In vitro evidences revealed that rosemary may increase skin permeability and percutaneous absorption of aminophylline in human skin (Wang et al., 2007).
- **Analgesics:** Based on human evidence, inhalation of the essential oil of rosemary may affect subjective perception of pain, although without reducing pain sensitivity (Gedney et al., 2004).

- **Antianxiety drugs:** In clinical study, inhalation of rosemary essential oil reduced anxiety (Diego et al., 1998; Moss et al., 2003; Burnett et al., 2004; Park and Lee, 2004; McCaffrey et al., 2009).
- **Antibiotics:** On the basis of laboratory study, rosemary essential oil may act antagonistically with ciprofloxacin (van Vuuren et al., 2009). Incorporation of 10 mcg/mL carnosic acid and carnosol into the growth medium caused a 32- and 16-fold potentiation of activity of erythromycin against an erythromycin-effluxing strain, respectively (Oluwatuyi et al., 2004). Rosemary and several of its constituents, including carnosic acid and carnosol, have exhibited antibacterial effects against various Gram-positive and Gram-negative bacteria in vitro, including oral planktonic bacteria (Silva et al., 2008), *Bacillus subtilis*, *Escherichia coli*, *Enterococcus faecalis*, *Pseudomonas aeruginosa*, MRSA, H[2]O[2]-producing lactobacilli, *Bacillus brevis* FMC3, *Bacillus megaterium* DSM32, *Micrococcus luteus* LA 2971, *Mycobacterium smegmatis* RUT, *Listeria monocytogenes* SCOTT A, *Streptococcus thermophilus*, *Pseudomonas fluorescens*, *Yersinia enterocolitica* O:3 P 41797, *Propionibacterium acnes* (ATCC 6919), *Staphylococcus epidermidis*, *Propionibacterium acnes*, and *Staphylococcus aureus* ME/GM/TC resistant (ATCC 33592) (Panizzi et al., 1993; Elgayyar et al., 2001; Abdel-Fatah et al., 2002; Verluyten et al., 2004; Weckesser et al., 2007; Scollard et al., 2009).
- **Anticoagulants/antiplatelet drugs:** Rosemary has shown significant in vitro and in vivo antithrombotic activity in mice (Yamamoto et al., 2005; Naemura et al., 2009). The antithrombotic mechanism may involve a direct inhibitory effect on platelets. In a rat study, oral rosmarinic acid decreased fibronectin and fibrin in the glomerulus (Makino et al., 2002). Theoretically, concurrent use may increase the risk of bleeding.
- **Antidiabetic agents:** Based on animal study, rosemary extract may increase blood sugar levels in both diabetics and nondiabetics (al Hader et al., 1994). However, laboratory studies have indicated that rosemary extracts may theoretically lower glucose levels (Kwon et al., 2006; Rau et al., 2006), a hypothesis substantiated in animal study (Erenmemisoglu et al., 1997; Bakirel et al., 2008).
- **Antihypertensive drugs:** On the basis of in vitro study, water extracts of rosemary may inhibit ACE (Kwon et al., 2006).
- **Antiinflammatory drugs:** On the basis of in vitro study, rosemary may have antiinflammatory activity (Chan et al., 1995). However, in a rat study, injection of 1,8-cineole, a rosemary constituent, produced inflammatory edema in the hind paw (Santos and Rao, 1997).
- **Antineoplastic agents:** On the basis of in vitro study, rosemary may increase the intracellular accumulation of commonly used chemotherapeutic agents, including doxorubicin and vinblastine, in cancer cells that express P-glycoprotein (Plouzek et al., 1999). However, rosemary extract probably does not affect accumulation or efflux of doxorubicin in cells that lack

P-glycoprotein. An increase in the activation of caspase-3 in high-risk pre-B acute lymphoblastic leukemia cells has been observed in vitro during coadministration of carnosol and chemotherapeutic agents high-risk pre-B acute lymphoblastic leukemia has been observed in vitro (Zunino and Storms, 2009). Furthermore, a lower percentage of caspase-3 positive cells progressed to an apoptotic phenotype during coadministration compared to treatment with chemotherapeutics alone.

- **Antiobesity agents:** In mice fed with a high-fat diet, rosemary leaf extract induced a significant reduction of weight and fat mass gain, an effect that may be related to the inhibition of pancreatic lipase activity, as determined in vitro (Harach et al., 2010). Carnosic acid and carnosol from rosemary inhibited the in vitro differentiation of mouse preadipocytes, 3T3-L1 cells, into adipocytes, possibly mediated by the activation of the antioxidant-response element and induction of phase-2 enzymes (Takahashi et al., 2006).
- **Antispasmodic agents:** Rosemary oil produced spasmolytic effects in circular smooth muscle strips of the guinea pig stomach accompanied by agonistic effects on alpha (1) and alpha (2) adrenergic receptors in vitro (Sagorchev et al., 2010). The antispasmodic effects of alcoholic extracts of *R. officinalis* have also been evaluated in isolated guinea pig ileum using acetylcholine and histamine as spasmogens (Forster et al., 1980).
- **Cyclosporine:** Rosemary may potentially interact with cyclosporine (Beaulieu, 2001).
- **Cytochrome P450-metabolized agents:** Results from in vitro and rat study suggest that rosemary may selectively induce P450 enzymes in the liver, particularly CYP 2B, CYP 1A1, CYP 2B1/2, and CYP 2E1 (Offord et al., 1997; Debersac et al., 2001a, b).
- **Diuretics:** In an animal study, rosemary demonstrated diuretic effects, decreasing electrolytes (Haloui et al., 2000). Rosemary has been shown to increase the permeability of furosemide in vitro (Laitinen et al., 2004).
- **Hormonal agents:** On the basis of human evidence, a combination of botanical supplements (i.e., *Curcuma longa*, *Cynara scolymus*, *R. officinalis*, *Schisandra chinensis*, *Silybum marinum*, and *Taraxacum officinalis*) decreased dehydroepiandrosterone, dehydroepiandrosterone sulfate, androstenedione, and estrone sulfate levels in women (Greenlee et al., 2007). On the basis of evidence from mouse study, rosemary may enhance the liver's rate of deactivating estrogen in the body (Zhu et al., 1998).
- **Iron salts:** Rosemary has been shown to decrease iron absorption (Samman et al., 2001).
- **Lithium:** According to case reports, rosemary may precipitate lithium toxicity due to its diuretic properties (Pyevich and Bogenschutz, 2001).
- **Salicylates:** On the basis of in vitro evidence, rosemary may contain high levels of salicylates (Swain et al., 1985).

Impact of rosemary on diseases of central nervous system

Central nervous system (CNS)-related diseases such as depression, Parkinson, Alzheimer, etc., are incurable chronic conditions, and presumably that is why there has been an increasing number of studies in rosemary during recent years, in an attempt to find the curing capacity. Regarding depression, there are several studies reporting a decreasing immobility time and regulation of several neurotransmitters (dopamine, norepinephrine, serotonin, and acetylcholine) and gene expression in mice brain like TH, PC, and MAPK phosphatase (MKP-1) (Sasaki et al., 2013; Machado et al., 2013). These studies contributed to the understanding of the molecular mechanism behind the antidepressant effect of rosemary and its major active compounds. Rosmarinic acid, however, seems to have potential against neurodegenerative diseases. It was found that this compound had cholinergic and neuroprotective effects and inhibited acetylcholinesterase (Ozarowaski et al., 2013). Pengelly et al. (2012) performed a randomized, placebo-controlled, double-blinded, repeated-measures crossover study to investigate possible acute effects of dried rosemary leaf powder on cognitive performance. This work reported significant speed of memory, a potentially useful predictor of cognitive function during aging, using rosemary powder at the dose normally used at culinary consumption. This work expands the value of future studies on effects of low doses of rosemary on memory and cognition (Pengelly et al., 2012).

Side effects

During pregnancy or breastfeeding, rosemary should be avoided in medicinal preparations, although it is safe to use it in cooking in small quantities to seasonal foods. This is not recommended for people suffering with high blood pressure, epilepsy, diverticulitis, chronic ulcers, or colitis. It acts as an emmenagogue, stimulating the flow of menstrual blood. The essential oil of rosemary was once used in folk practice in attempts to induce abortion. As with all essential oils, only small amounts of it should be used, either topically or internally. According to the PDR for herbal medicines the overdose of essential oil of rosemary may cause deep coma, vomiting, spasms, uterine bleeding, gastroenteritis, kidney irritation, and even death (Fig. 3.2.6.2).

Conclusion

Rosemary is an important medicinal plant that has been used to curing many disorders or diseases mainly related with the CNS. Research evidences suggest that it can be used to treat depression, Parkinson disease, Alzheimer disease, etc. Besides this, further studies and clinical trials are needed for knowing the impact of rosemary on CNS-related diseases or disorders.

FIGURE 3.2.6.2 *Rosmarinus officinalis* in flowering stage.

References

Abdel-Fatah, M.K., El-Hawa, M.A., Samia, E.M., Rabie, G., Amer, A.M., 2002. Antimicrobial activities of some local medicinal plants. J. Drug Res. 24, 179–186.

Akbari, J., Saeedi, M., Farzin, D., Morteza-Semnani, K., Esmaili, Z., 2015. Transdermal absorption enhancing effect of the essential oil of *Rosmarinus officinalis* on percutaneous absorption of Na diclofenac from topical gel. Pharm. Biol. 53, 1442–1447. https://doi.org/10.3109/13880209.2014.984855.

al Hader, A.A., Hasan, Z.A., Aqel, M.B., 1994. Hyperglycemic and insulin release inhibitory effects of *Rosmarinus officinalis*. J. Ethnopharmacol. 43, 217–221.

Alipour, M., Saharkhiz, M.J., 2016. Phytotoxic activity and variation in essential oil content and composition of Rosemary (*Rosmarinus officinalis* L.) during different phenological growth stages. Biocatal. Agric. Biotechnol. 7, 271–278.

Al-Sereiti, M.R., Abu-Amer, K.M., Sena, P., 1999. Pharmacology of rosemary (*Rosmarinus officinalis* Linn.) and its therapeutic potentials. Indian J. Exp. Biol. 37, 124–130.

Andrade, J.M., Faustino, C., Garcia, C., Ladeiras, D., Reis, C.P., Rijo, P., 2018. *Rosmarinus officinalis* L.: an update review of its phytochemistry and biological activity. Future Sci. OA 4 (4), FSO283.

Bakirel, T., Bakirel, U., Keles, O.U., Ulgen, S.G., Yardibi, H., 2008. In vivo assessment of antidiabetic and antioxidant activities of rosemary (*Rosmarinus officinalis*) in alloxan-diabetic rabbits. J. Ethnopharmacol. 116, 64−73.

Beaulieu, J.E., 2001. Herbal therapy interactions with immunosuppressive agents. U.S. Pharm. 26, 13−22.

Begum, A., Sandhya, S., Vinod, K.R., Reddy, S., Banji, D., 2013. An in-depth review on the medicinal flora *Rosmarinus officinalis* (Lamiaceae). Acta Sci. Pol. Technol. Aliment. 12 (1), 61−74.

Borges, R.S., Lima, E.S., Keita, H., Ferreira, I.M., Fernandes, C.P., Cruz, R.A.S., Duarte, J.L., Velázquez-Moyado, J., Ortiz, B.L.S., Castro, A.N., Ferreira, J.V., da Silva Hage-Melim, L.I., Carvalho, J.C.T., 2017. Anti-inflammatory and antialgic actions of a nanoemulsion of *Rosmarinus officinalis* L. essential oil and a molecular docking study of its major chemical constituents. Inflammopharmacology. https://doi.org/10.1007/s10787-017-0374-8.

Burnett, K.M., Solterbeck, L.A., Strapp, C.M., 2004. Scent and mood state following an anxiety-provoking task. Psychol. Rep. 95, 707−722.

Chan, M.M., Ho, C.T., Huang, H.I., 1995. Effects of three dietary phytochemicals from tea, rosemary, and turmeric on inflammation-induced nitrite production. Cancer Lett. 96, 23−29.

Chobba, I.B., Bekir, A., Mansour, R.B., Driral, N., Gharsallah, N., Kadri, A., 2012. *In vitro* evaluation of antimicrobial and cytotoxic activities of *Rosarinus officinalis* L. (Lamiaceae) essential oil cultivated from South-West Tunisia. J. Appl. Pharm. Sci. 2 (11), 034−039. https://doi.org/10.7324/JAPS.2012.21107.

Chopra, R.N., 1958. Chopra's Indigenous Drugs of India. UN Dhar and Sons, Calcutta.

Cuvelier, M.E., Richard, H., Berset, C., 1996. Antioxidative activity and phenolic composition of pilot-plant and commercial extracts of sage and roseary. J. Am. Oil Chem. Soc. 73, 645−652.

Debersac, P., Heydel, J.M., Amiot, M.J., Goudonnet, H., Artur, Y., Suschetet, M., Siess, M.H., 2001a. Induction of cytochrome P450 and/or detoxication enzymes by various extracts of rosemary: description of specific patterns. Food Chem. Toxicol. 39 (9), 907−918.

Debersac, P., Vernevaut, M.F., Amiot, M.J., Suschetet, M., Siess, M.H., 2001b. Effects of a water-soluble extract of rosemary and its purified component rosmarinic acid on xenobiotic-metabolizing enzymes in rat liver. Food Chem. Toxicol. 39, 109−117.

Diego, M.A., Jones, N.A., Field, T., Hernandez-Reif, M., Schanberg, S., Kuhn, C., McAdam, V., Galamaga, R., Galamaga, M., 1998. Aromatherapy positively affects mood, EEG patterns of alertness and math computations. Int. J. Neurosci. 96, 217−224.

Elgayyar, M., Draughon, F.A., Golden, D.A., Mount, J.R., 2001. Antimicrobial activity of essential oils from plants against selected pathogenic and saprophytic microorganisms. J. Food Protect. 64, 1019−1024.

Erenmemisoglu, A., Saraymen, R., Ustun, S., 1997. Effect of a *Rosmarinus officinalis* leave extract on plasma glucose levels in normoglycaemic and diabetic mice. Pharmazie 52, 645−646.

Fernandes, R.V.D.B., Borges, S.V., Botrel, D.A., Oliveira, C.R.D., 2014. Physical and chemical properties of encapsulated rosemary essential oil by spray drying using whey protein−inulin blends as carriers. Int. J. Food Sci. Technol. 49 (6), 1522−1529.

Forster, H.B., Niklas, H., Lutz, S., 1980. Antispasmodic effects of some medicinal plants. Planta Med. 40, 309−319.

Gedney, J.J., Glover, T.L., Fillingim, R.B., 2004. Sensory and affective pain discrimination after inhalation of essential oils. Psychosom. Med. 66, 599−606.

Greenlee, H., Atkinson, C., Stanczyk, F.Z., Lampe, J.W., 2007. A pilot and feasibility study on the effects of naturopathic botanical and dietary interventions on sex steroid hormone metabolism in premenopausal women. Cancer Epidemiol. Biomarkers Prev. 16, 1601−1609.

Haloui, M., Louedec, L., Michel, J.B., Lyoussi, B., 2000. Experimental diuretic effects of *Rosmarinus officinalis* and *Centaurium erythraea*. J. Ethnopharmacol. 71, 465–472.
Harach, T., Aprikian, O., Monnard, I., Moulin, J., Membrez, M., Beolor, J.C., Raab, T., Mace, K., Darimont, C., 2010. Rosemary (*Rosmarinus officinalis* L.) leaf extract limits weight gain and liver steatosis in mice fed a high-fat diet. Planta Med. 76, 566–571.
Hassani, F.V., Shirani, K., Hosseinzadeh, H., 2016. Rosemary (*Rosmarinus officinalis*) as a potential therapeutic plant in metabolic syndrome: a review. Naunyn-Schmiedeberg's Arch. Pharmacol. 389 (9), 931–949.
Juhás, Š., Bukovská, A., Čikoš, Š., Czikková, S., Fabian, D., Koppel, J., 2009. Anti-inflammatory effects of *Rosmarinus officinalis* essential oil in mice. Acta Vet. Brno. 78, 121–127. https://doi.org/10.2754/avb200978010121.
Korb, D.F., 1985. Medicinal Plants in Libya. Arab Encyclopedia House, Tripoli, p. 720.
Kwon, Y.I., Vattem, D.A., Shetty, K., 2006. Evaluation of clonal herbs of Lamiaceae species for management of diabetes and hypertension. Asia Pac. J. Clin. Nutr. 15, 107–118.
Laitinen, L.A., Tammela, P.S., Galkin, A., Vuorela, H.J., Marvola, M.L., Vuorela, P.M., 2004. Effects of extracts of commonly consumed food supplements and food fractions on the permeability of drugs across Caco-2 cell monolayers. Pharm. Res. 21, 1904–1916.
Machado, D.G., Cunha, M.P., Neis, V.B., Balen, G.O., Colla, A., Bettio, L.E.B., Oliveira, Á., Pazini, F.L., Dalmarco, J.B., Simionatto, E.L., Pizzolatti, M.G., Rodrigues, A.L.S., 2013. Antidepressant-like effects of fractions, essential oil, carnosol and betulinic acid isolated from *Rosmarinus officinalis* L. Food Chem. 136, 999–1005. https://doi.org/10.1016/j.foodchem.2012.09.028.
Makino, T., Ono, T., Liu, N., Nakamura, T., Muso, E., Honda, G., 2002. Suppressive effects of rosmarinic acid on mesangioproliferative glomerulonephritis in rats. Nephron 92, 898–904.
McCaffrey, R., Thomas, D.J., Kinzelman, A.O., 2009. The effects of lavender and rosemary essential oils on test-taking anxiety among graduate nursing students. Holist. Nurs. Pract. 23, 88–93.
Moss, M., Cook, J., Wesnes, K., Duckett, P., 2003. Aromas of rosemary and lavender essential oils differentially affect cognition and mood in healthy adults. Int. J. Neurosci. 113, 15–38.
Naemura, A., Ura, M., Yamashita, T., Arai, R., Yamamoto, J., 2009. Long-term intake of rosemary and common thyme herbs inhibits experimental thrombosis without prolongation of bleeding time. Thromb. Res. 122, 517–522.
Offord, E.A., Mace, K., Avanti, O., Pfeifer, A.M., 1997. Mechanisms involved in the chemoprotective effects of rosemary extract studied in human liver and bronchial cells. Cancer Lett. 114, 275–281.
de Oliveira, J.R., Camargo, S.E.A., de Oliveira, L.D., 2019. *Rosmarinus officinalis* L. (rosemary) as therapeutic and prophylactic agent. J. Biomed. Sci. 26 (5) https://doi.org/10.1186/s12929_019_0499_8.
Oluwatuyi, M., Kaatz, G.W., Gibbons, S., 2004. Antibacterial and resistance modifying activity of *Rosmarinus officinalis*. Phytochemistry 65, 3249–3254.
Ozarowski, M., Mikolajczak, P.L., Bogacz, A., Gryszczynska, A., Kujawska, M., Jodynis-Liebert, J., Piasecka, A., Napieczynska, H., Szulc, M., Kujawski, R., Bartkowiak-Wieczorek, J., Cichocka, J., Bobkiewicz-Kozlowska, T., Czerny, B., Mrozikiewicz, P.M., 2013. *Rosmarinus officinalis* L. leaf extract improves memory impairment and affects acetylcholinesterase and butyrylcholinesterase activities in rat brain. Fitoterapia 91, 261–271.
Panizzi, L., Flamini, G., Cioni, P.L., Morelli, I., 1993. Composition and antimicrobial properties of essential oils of four Mediterranean Lamiaceae. J. Ethnopharmacol. 39, 167–170.

Park, M.K., Lee, E.S., 2004. The effect of aroma inhalation method on stress responses of nursing students. Taehan Kanho Hakhoe Chi 34, 344–351.

Pengelly, A., Snow, J., Mills, S.Y., Scholey, A., Wesnes, K., Butler, L.R., 2012. Short-term study on the effects of rosemary on cognitive function in an elderly population. J. Med. Food 15 (1), 10–17.

Plouzek, C.A., Ciolino, H.P., Clarke, R., Yeh, G.C., 1999. Inhibition of P-glycoprotein activity and reversal of multidrug resistance *in vitro* by rosemary extract. Eur. J. Cancer 35, 1541–1545.

Pyevich, D., Bogenschutz, M.P., 2001. Herbal diuretics and lithium toxicity. Am. J. Psychiatry. 158, 1329.

Rau, O., Wurglics, M., Paulke, A., Zitzkowski, J., Meindl, N., Bock, A., Dingermann, T., Abdel-Tawab, M., Schubert-Zsilavecz, M., 2006. Carnosic acid and carnosol, phenolic diterpene compounds of the labiate herbs rosemary and sage, are activators of the human peroxisome proliferator-activated receptor gamma. Planta Med. 72, 881–887.

Sagorchev, P., Lukanov, J., Beer, A.M., 2010. Investigations into the specific effects of rosemary oil at the receptor level. Phytomedicine 17, 693–697.

Samman, S., Sandstrom, B., Toft, M.B., Bukhave, K., Jensen, M., Sorensen, S.S., Hansen, M., 2001. Green tea or rosemary extract added to foods reduces nonheme-iron absorption. Am. J. Clin. Nutr. 73, 607–612.

Santos, F.A., Rao, V.S., 1997. Mast cell involvement in the rat paw oedema response to 1, 8-cineole, the main constituent of eucalyptus and rosemary oils. Eur. J. Pharmacol. 331, 253–258.

Sasaki, K., El Omri, A., Kondo, S., Han, J., Isoda, H., 2013. *Rosmarinus officinalis* polyphenols produce anti-depressant like effect through monoaminergic and cholinergic functions modulation. Behav. Brain Res. 238, 86–94.

Scollard, J., Francis, G.A., O'Beirne, D., 2009. Effects of essential oil treatment, gas atmosphere, and storage temperature on Listeria monocytogenes in a model vegetable system. J. Food Protect. 72, 1209–1215.

Silva, M.D., Silva, M.A., Higino, J.S., Pereira, M.S., Carvalho, A.D., 2008. *In vitro* antimicrobial activity and antiadherence of *Rosmarinus officinalis* Linn. against oral planktonic bacteria. Rev. Bras Farmacogn. 18, 236–240.

Swain, A.R., Dutton, S.P., Truswell, A.S., 1985. Salicylates in foods. J. Am. Diet Assoc. 85, 950–960.

Tai, J., Cheung, S., Wu, M., Hasman, D., 2012. Antiproliferation effect of Rosemary (*Rosmarinus officinalis*) on human ovarian cancer cells *in vitro*. Phytomedicine 19 (5), 436–443.

Takahashi, T., Tabuchi, T., Tamaki, Y., Kosaka, K., Takikawa, Y., Satoh, T., 2006. Carnosic acid and carnosol inhibit adipocyte differentiation in mouse 3T3-L1 cells through induction of phase 2 enzymes and activation of glutathione metabolism. Biochem. Biophys. Res. Commun. 382, 549–554.

Tyler, V.E., Brady, L.R., Robbers, J.E., 1976. Pharmacognosy. Lea and Febiger, Philadelphia), p. 171.

van Vuuren, S.F., Suliman, S., Viljoen, A.M., 2009. The antimicrobial activity of four commercial essential oils in combination with conventional antimicrobials. Lett. Appl. Microbiol. 48, 440–446.

Verluyten, J., Leroy, F., De Vuyst, L., 2004. Effects of different spices used in production of fermented sausages on growth of and curvacin A production by *Lactobacillus curvatus* LTH 1174. Appl. Environ. Microbiol. 70, 4807–4813.

Vilela, J., Martins, D., Monteiro-Silva, F., González-Aguilar, G., de Almeida, J.M.M.M., Saraiva, C., 2016. Antimicrobial effect of essential oils of *Laurus nobilis* L. and *Rosmarinus officinallis* L. on shelf-life of minced "Maronesa" beef stored under different packaging conditions. Food Packag. Shelf Life 8, 71–80. https://doi.org/10.1016/j.fpsl.2016.04.002.

Wang, L.H., Wang, C.C., Kuo, S.C., 2007. Vehicle and enhancer effects on human skin penetration of aminophylline from cream formulations: evaluation in vivo. J. Cosmet. Sci. 58, 245–254.

Wang, W., Li, N., Luo, M., Zu, Y., Efferth, T., 2012. Antibacterial activity and anticancer activity of *Rosmarinus officinalis* L. essential oil compared to that of its main components. Molecules 17 (3), 2704–2713.

Weckesser, S., Engel, K., Simon-Haarhaus, B., Wittmer, A., Pelz, K., Schempp, C.M., 2007. Screening of plant extracts for antimicrobial activity against bacteria and yeasts with dermatological relevance. Phytomedicine 14, 508–516.

Yamamoto, J., Yamada, K., Naemura, A., Yamashita, T., Arai, R., 2005. Testing various herbs for antithrombotic effect. Nutrition 21, 580–587.

Zhang, Y., Adelakun, T.A., Qu, L., Li, X., Li, J., Han, L., Wang, T., 2014. New terpenoid glycosides obtained from *Rosmarinus officinalis* L. aerial parts. Fitoterapia 99, 78–85.

Zhu, B.T., Loder, D.P., Cai, M.X., Ho, C.T., Huang, M.T., Conney, A.H., 1998. Dietary administration of an extract from rosemary leaves enhances the liver microsomal metabolism of endogenous estrogens and decreases their uterotropic action in CD-1 mice. Carcinogenesis 19, 1821–1827.

Zunino, S.J., Storms, D.H., 2009. Carnosol delays chemotherapy-induced DNA fragmentation and morphological changes associated with apoptosis in leukemic cells. Nutr. Cancer. 61, 94–102.

Chapter 3.2.7

Valeriana officinalis (valerian)

Manjul Mungali[1], Gauri[2], Alok Tripathi[1], Surabhi Singhal[1]
[1]*Department of Biotechnology, School of Life Sciences & Technology, IIMT University, Meerut, Uttar Pradesh, India;* [2]*S.M.P. Government Girls PG College, Meerut, Uttar Pradesh, India*

Valeriana officinalis

Valerian (*Valeriana officinalis*) is a perennial flowering plant belonging to the Valerianaceae family, commonly found in Europe and Asia. This is generally known as valerian. Valerianaceae family has roughly 250 species, and *V. officinalis* is one of the most commonly used medicinal species. Some other famous subspecies of Valerianaceae family are *Alternifolis, Angustifolia, Latifolia, Collina,* and *Sambucifolia*. In the summer, mature plants bear scented pink and white small flowers that bloom from June to September. From ancient time, this plant has been used as a medicinal herb. The name valerian is derived from the Latin word "valere," which means health or well-being. The most common names for this plant in English are valerian, garden heliotrope, or allheal, and in Sanskrit, it is called balahrivera, in urdu balchar or balchhar, and in tamil, jatamansi (Fig. 3.2.7.1).

Natural chemicals in *V. officinalis*

The extract of *V. officinalis* is from the rhizomes and roots for tinctures, fundamental oils, terpenes, terpene free fractions, and residues from the plant. The chief constituents of valerian plant include flavonoids, sesquiterpenes, caffeic acid derivatives, monoterpenes, lignans, alkaloids, amino acids, and valepotriates (Fernández et al., 2004).

Given subsequently are some primary characteristic chemicals found in this plant as reported and published in a chemical information review document by the National Toxicology Program, NIH, United States, in 2009. According to this database, sesquiterpenes, valerenic acid, and its hydroxy and acetoxy derivatives are the primary characteristic markers of *V. officinalis*. Table 3.2.7.1 shows the chemicals with their molecular formula and weight. Fig. 3.2.7.2 show the chemical structure of each.

Various constituents extracted from *Valeriana* plant can be analyzed by different scientific techniques such as gas chromatography/mass spectrometry

Naturally Occurring Chemicals against Alzheimer's Disease. https://doi.org/10.1016/B978-0-12-819212-2.00023-2

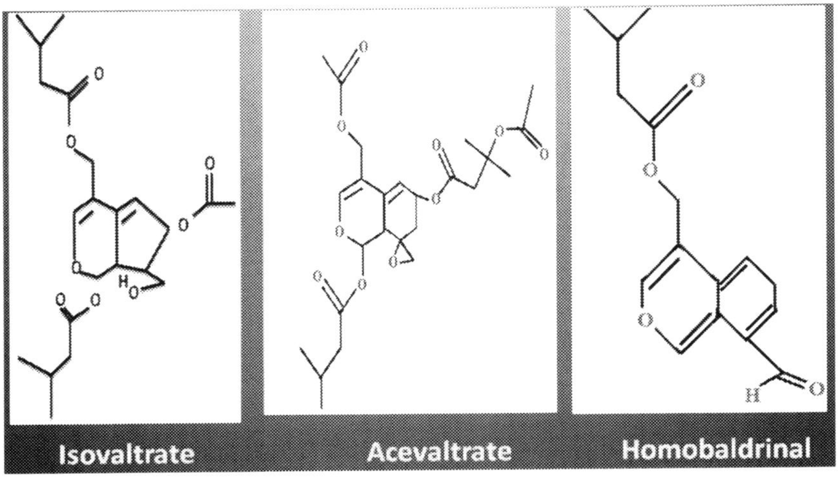

FIGURE 3.2.7.1 Natural chemicals of V. officinalis with molecular formula.

TABLE 3.2.7.1 Natural chemicals found in *V. officinalis* with their molecular formula and weight.

Sr. No	Name of chemical	Molecular formula	Molecular weight
1.	Valerenic acid	$C_{15}H_{22}O_2$	234.33
2.	Valtrate	$C_{22}H_{30}O_8$	422.47
3.	Didrovaltrate	$C_{22}H_{32}O_8$	424.48
4.	Isovaltrate	$C_{22}H_{30}O_8$	422.47
5.	Acevaltrate	$C_{24}H_{32}O_{10}$	480.05
6.	Baldrinal	$C_{12}H_{10}O_4$	218.21
7.	Homobaldrinal	$C_{15}H_{16}O_4$	260.29

(GC/MS), pressure chromatography (HPLC), retention spectrometry, and thin layer chromatography.

There are some discrepancies related to the valerian extraction method and its clinical trials. Many traditional methods utilize water as extraction medium; the most common example is valerian tea. But aqueous extracts of valerian have different modes of action with a different concentration of constituent each time. Ethanol is also used for extraction, and it is considered a precise mechanism for pharmacologic uses.

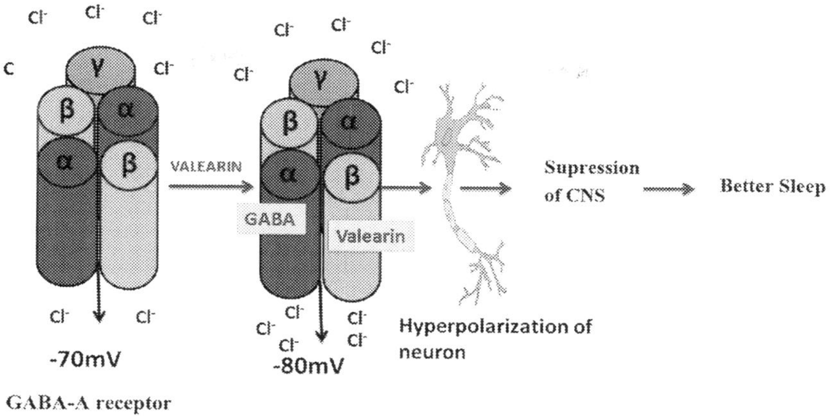

FIGURE 3.2.7.2 Natural chemicals of V. officinalis with molecular formula.

Medicinal uses of *V. officinalis*

Valeriana roots are traditionally used for the management and treatment of nervous anxiety, hysteria, epilepsy, strengthening eyesight, stress, high blood pressure, irritability, convulsions in children, tension headaches, muscle cramps, menopausal restlessness, asthma, colic and irritable bowel syndrome, and restless legs.

Various administration methods of this herb include internally as pressed juice from the fresh plant, tincture, extracts, and other galenical preparations. This herb can be used externally as a bath additive to help relieve nervousness and to induce sleep. Valerian can be boiled and the stem can be inhaled to produce the same effect. The herb has antispasmodic, carminative, diuretic, expectorant, blood pressure lowering, sedative, and sleep-promoting actions.

Valerian roots are specially used for making herbal tea or taken orally for relaxation and sedation purposes. For people having trouble in sleeping, drinking one cup of Valeriana root tea about 40 min before bedtime will help with getting sound sleep. The extract of valerian roots is used for sedative and anxiolytic effects (Del Valle Mojica et al., 2011). Valerian plant was reported to relieve menstrual pain due to its spasmolytic properties. It is also used in the form of dietary supplements to reduce insomnia. It is also reported that valerian enhances the signaling of gamma-aminobutyric acid (GABA), the principle narcotic synapse (Kamal, 2013). The plant also possesses some antimicrobial antioxidant properties.

Although it is a traditionally used herbal drug to treat tension and sleep disorder, its mechanism of action is not fully known. Due to its mood modulation effects, it is believed to be connected with the GABA-ergic system (Scaglione and Zangara, 2004). Clinical experiments have shown that valerian root extracts stimulate GABA transmission; a few studies have also examined the effect of *V. officinalis* in excitatory glutamate-interceded neurotransmission (Del Valle Mojica et al., 2011).

Apart from the sedative, anxiolytic, and spasmolytic properties, Jihua Wang et al., 2010 examined antimicrobial and antioxidant activity of oils from two valerianaceous species, *Nardostachys chinensis* and *V. officinalis*. They used hydro-distillation technique for extraction of essential oils from the roots and rhizomes of these two plants. The chemical composition was analyzed by gas chromatography (GC) and gas chromatography mass spectrometry (GC-MS). Among 20 identified constituents, the three major components were patchoulol (16.75%), α-pinene (14.81%), and β-humulene (8.19%), constituting 88.11% of total extracted oil of *V. officinalis*. They also found that these oils have a wide range of antibacterial activity with MIC values that ranged from 62.5 to 400 μg/mL, and IC_{50} values from 36.93 to 374.72 μg/mL. The oils were also reported to have some antifungal property against *Candida albicans* growth and inhibition of spore germination of *Magnaporthe oryzae*.

According to the chemical information review document published by National Toxicology Program, NIH, United States, in 2009 (CIRD, 2009), *V. officinalis* has no carcinogenicity, cogenotoxicity, nor immunotoxicity.

Mechanism of action of key constituent of *V. officinalis* plant

The traditional medical uses of valerian plant include migraine treatment, pain reliever, anticonvulsant, and sedative. Apart from these, the key effect is the restorative effect on the central nervous system (CNS). The most common use of *V. officinalis* is the management of insomnia and anxiety.

Valeric acid, volatiles, and valepotriates are the key constituents of *Valeriana* rhizome. These compounds are reported for interfering with the brain receptor for the neurotransmitter GABA. The detailed mechanism of action as a mild sedative is unknown. It was observed by many researchers that valerian decreased the removal of GABA, which results in a longer stay of GABA at the brain receptor. The mechanism of action of valerian in the treatment of insomnia is depicted in Figs. 3.2.7.3 and 3.2.7.4.

GABA is the chief inhibitory neurotransmitter of GABA-A receptor present in the CNS. Valerian enhances the binding of GABA to its receptor. It was reported that GABA binds to the alpha (α) subunit of the GABA-A receptor. In the absence of valerian, it is weakly bound to the receptor, which results in diffusion of chloride ions (Cl^-) inside the neurons (−70 mV). In the presence of valerian the GABA firmly binds to the GABA-A receptor and causes more chloride ions (Cl^-) to diffuse inside the neuron, and this results in hyperpolarization of the neuron (−80 mV). In hyperpolarization states, neurons are less active for excitatory postsynaptic potentials, thus suppressing the function of the CNS.

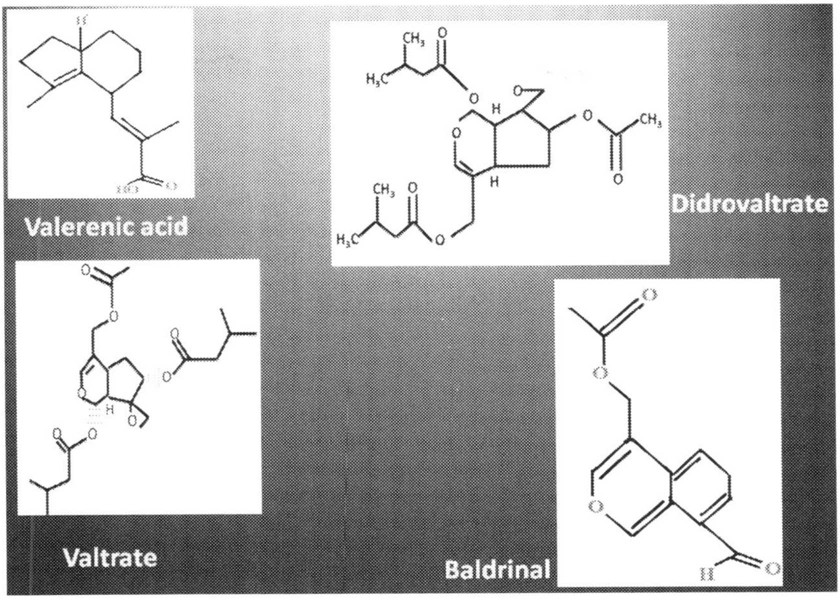

FIGURE 3.2.7.3 Valeriana Plant.

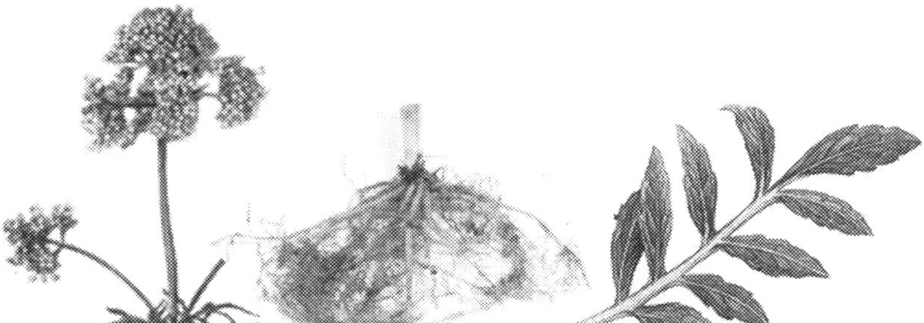

FIGURE 3.2.7.4 Mechanism of action.

V. officinalis and Alzheimer disease

Medicinal plants play a significant role in Alzheimer disease (AD) and memory deficit. These traditional medicines are preventative, protective, nutritive, and curative in nature, so these natural medicines considered safe and harmless for patients. In AD the use of *V. officinalis* is mainly as a neuroprotective and sedative. The roots and rhizomes of *V. officinalis* help in enhancing memory even in elderly people, as well as helping to cure AD-related mental disorders. Extract of *V. officinalis* is sold under various common and trade names including Valerian oil, Valerian rhizome and root oil, Baldrian oil, Baldrianwurzel, Valerianae radix, Baldrion, Alderbrackenwurzel, Laege-baldrian, Nervex, Neurol, Orasedon, Racine de valeriane, Radix valerian, Sedonium, Ticalma, Valeriana radiz, Valerianaheel, Valmane, and ZE 911 (CIRD, 2009).

Clinical studies

Leathwood et al. (1982) studied the effect of valerian root aqueous extract on 128 people. Nine samples were given to each person, three containing placebo, three containing 400 mg valerian extract, and three containing a proprietary over-the-counter valerian preparation. Results showed that valerian produced a significant decrease in subjectively evaluated sleep latency scores. A significant improvement in sleep quality was noticed among people who considered themselves poor or irregular sleepers and smokers.

V. officinalis is a traditionally popular herbal supplement used to treat insomnia and anxiety. Until recently, its mechanism of action has remained unknown. Neurobiologic research has begun to show that the herb, with its active valerenic acid, interacts with the GABA-ergic system. A series of experiments were carried out by Murphy et al. (2009) to confirm this finding. Rats were administered either ethanol (1 mL/kg), diazepam (1 mg/kg), valerian root extract (3 mL/kg), valerenic acid (3 mg/kg), or a solution of valerenic acid and exogenous GABA (75 and 3.6 μg/kg, respectively). Rats were then assessed for the number of entries and time spent on the open arms of an elevated plus maze. Results showed that there was a significant reduction in anxious behavior when subjects exposed to valerian extract or valerenic acid were compared to the ethanol control group. Thus, according to findings, *V. officinalis* is a potential alternative to traditional anxiolytics as measured by the elevated plus maze.

Sudati et al. (2009) examined the protective effect of *V. officinalis* on lipid peroxidation (LPO) induced by different prooxidant agents with neuropathologic importance. Ethanolic extract of valerian (0−60 μg/mL) was tested against quinolinic acid (QA), 3-nitropropionic acid, sodium nitroprusside, iron sulfate ($FeSO_4$), and Fe^{2+}/EDTA induced LPO in rat brain homogenates. In brain homogenates, *V. officinalis* inhibited thiobarbituric acid reactive substances induced by all prooxidants tested in a concentration-dependent manner. Similarly, *V. officinalis* caused a significant decrease on the LPO in cerebral cortex and in deoxyribose degradation. QA-induced ROS production in cortical slices was also significantly reduced by *V. officinalis*. They suggested that *V. officinalis* extract was effective in modulating LPO induced by different prooxidant agents. *V. officinalis* extract, functioning as an antioxidant agent, can be beneficial for reducing insomnia complications linked to oxidative stress.

Sung et al. (2013) investigated the effects of valerian root extract's major component, valerenic acid, on memory function, cell proliferation, neuroblast differentiation, serum corticosterone, and lipid peroxidation in adult and aged mice. For the aging model, D-galactose (100 mg/kg) was administered subcutaneously to 6-week-old male mice for 10 weeks. At 13 weeks of age, valerian root extracts (100 mg/kg) or valerenic acid (340 μg/kg) were administered orally to control and D-galactose-treated mice for 3 weeks. The administration of valerian root extract and valerenic acid significantly improved the preferential exploration of new objects in novel object recognition test and the escape latency, swimming speeds, platform crossings, and spatial preference for the target quadrant in Morris water maze test compared to the D-galactose-treated mice. Cell proliferation and neuroblast differentiation were significantly decreased, while serum corticosterone level and lipid peroxidation in the hippocampus were significantly increased in the D-galactose-treated group compared to that in the control group. The results showed that valerian root extract and valerenic acid enhance cognitive function, promote cell proliferation and neuroblast differentiation, and reduce serum corticosterone and lipid peroxidation in aged mice.

Morin et al. (2005) examined 184 adults (110 women, 74 men, age of 44.3 years) having mild insomnia. Patients were given valerian extract tablets by prescription for 28 days. They concluded a modest hypnotic effect was produced by valerian extract tablets, and it can be used for treatment of mild insomnia, without any side effects.

Safety issues

The medicinal products of *V. officinalis* should be avoided during pregnancy and lactation. Some side effects that have been reported are headaches, excitability, restlessness, dilated pupils, and irregular heartbeat. Because this medicine makes you drowsy, it should not be taking while operating machinery, driving, or performing any activity that requires full alertness. This herb may worsen high blood pressure.

Patients with liver disease are advised not to take valerian. In some individuals, some side effects like headaches and night terrors were reported. The reason may be due to the fact that some people lack a digestive conversion property required to effectively break down valerian. In these individuals, valerian can cause agitation.

Conclusion

Pharmacologic studies indicate that valerian plants possess various biologic activities, like antioxidant, antiinflammatory, anticancer, anticonvulsive, anti-Parkinson, and anti-Alzheimer disease. Valepotriates are major components for the treatment of epilepsy, depressant, Parkinson disease, and AD. Due to the vast medicinal properties, valerian plants are today utilized in the form of many health supplements and medicines either given alone or in combination with other drugs to produced desired effects. Valerian-based treatment produces neuroprotective and sleep-inducing effects in AD patients. There were no side effects found with *V. officinalis* like causing cancer or promoting immunotoxicity in patients. Valerian is not addictive, resulting in no withdrawal symptoms on its discontinuation. There is a strong need for future researches to enlighten the effects of *V. officinalis* in the management of AD.

References

CIRD (Chemical Information Review Document) for Valerian (*Valeriana officinalis* L.), 2009. National Toxicology Program. National Institute of Environmental Health Sciences, National Institutes of Health, U.S Department of Health and Human Services, Research Triangle Park, NC complete web address. http://ntp.niehs.nih.gov/.

Del Valle Mojica, L.M., José, M., Hernández, C., Medina, G.G., Vélez, I.R., Cartagena, N.B., Bianca, A., Torres, H., José, G., Ortíz, 2011. Aqueous and ethanolic *Valeriana officinalis* extracts change the binding of ligands to glutamate receptors. Evid.-Based Complementary Altern. Med. 7. https://doi.org/10.1155/2011/891819. Volume, Article ID 891819.

Fernández, S., Wasowski, C., Paladini Alejandro, C., Marder, M., 2004. Sedative and sleep-enhancing properties of linarin, a flavonoid-isolated from *Valeriana officinalis*. Pharmacol. Biochem. Behav. 77 (2), 399–404. Abstract from PubMed 14751470.

Kamal, patel, 2013. Valeriana Officinalis. https://examine.com/supplements/valeriana-officinalis/.

Leathwood, P.D., et al., 1982. Aqueous extract of valerian root (*Valeriana offcinalis* L.) improves sleep quality in man. Pharmacol. Biochem. Behav. 17 (1), 65–71.

Morin, C.M., Koetter, U., Bastien, C., Catesby Ware, J., Wooten, V., 2005. Valerian-hops combination and diphenhydramine for treating insomnia: a randomized placebo-controlled clinical trial. Sleep 28 (11), 1465–1471. https://doi.org/10.1093/sleep/28.11.1465.

Murphy, K., et al., 2009. *Valeriana officinalis* root extracts have potent anxiolytic effects in laboratory rats. Phytomedicine 17 (Issues 8–9), 674–678. July 2010.

Scaglione, F., Zangara, A., 2004. *Valeriana officinalis* and *Melissa officinalis* extracts normalize brain levels of GABA and glutamate altered by chronic stress sleep. Disord. Manag. 3 (issue 1) https://doi.org/10.23937/2572-4053.1510016. ISSN: 2572-4053.

Sudati, J.H., et al., 2009. In vitro antioxidant activity of *Valeriana officinalis* against different neurotoxic agents. Neurochem. Res. 34, 1372.

Sung, M.N., et al., 2013. *Valeriana officinalis* extract and its main component, valerenic acid, ameliorate d-galactose-induced reductions in memory, cell proliferation, and neuroblast differentiation by reducing corticosterone levels and lipid peroxidation. Exp. Gerontol. 48 (11), 1369–1377. November 2013.

Wang, J., Zhao, J., Liu, H., Zhou, L., Liu, Z., Wang, J., Han, J., Zhu, Y., Yang, F., 2010. Chemical analysis and biological activity of the essential oils of two valerianaceous species from China: Nardostachys chinensis and Valeriana officinalis. Molecules 15 (9), 6411–6422. https://doi.org/10.3390/molecules15096411, 2010.

Further reading

Bos, R., Hendriks, H., Scheffer, J.J.C., Woerdenbag, H.J., 1998. Cytotoxic potential of valerian constituents and valerian tinctures. Phytomedicine 5 (3), 219–225. Abstract from CABA 1998:114166.

Ghafari, S., Esmaeili, S., Aref, H., Naghibi, F., Mosaddegh, M., 2009. Qualitative and quantitative analysis of some brands of valerian pharmaceutical products. Stud. Ethno-Med. 3 (1), 61–64. Abstract from CABA 2009:142048.

Giraldez, 2009. Cytoprotective effect of Valeriana officinalis extract on an in vitro experimental model of Parkinson disease. Neurochem. Res. 34 (2), 215–220. February, Volume.

Huang, B., Qin, L., Chu, Q., Zhang, Q., Gao, L., Zheng, H., 2009. Comparison of headspace SPME with hydrodistillation and SFE for analysis of the volatile components of the roots of *Valeriana officinalis* var. *latifolia*. Chromatographia 69 (5–6), 489–496. Abstract from AGRICOLA 2009:38003.

Malva, J.O., Santos, Sandra, Macedo, Tice, 2004. Neuroprotective properties of *Valeriana officinalis* extracts. Neurotox. Res. 6 (2), 131–140.

Mikell, J.R., Ganzera, M., Khan, I.A., 2001. Analysis of sesquiterpenes in *Valeriana officinalis* by capillary electrophoresis. Pharmazie 56 (12), 946–948 (Abstract from PubMed).

Pilerood, S.A., Prakash, J., 2013. Nutritional and medicinal properties of valerian (*Valeriana officinalis*) herb: a review. International Journal of Food, Nutrition and Dietetics 1. Number 1, Janaury - June 2013.

Pizzorno, J.E., Michael, T., Murray Textbook of Natural Medicine, fourth ed. Elsevier, ISBN 978-1-4377-2333-5.

Santos, M.S., Ferreira, F., Cunha, A.P., Carvalho, A.P., Ribeiro, C.F., Macedo, T., 1994. Synaptosomal GABA release as influenced by valerian root extract–involvement of the GABA carrier. Arch. Int. Pharmacodyn. Ther. 327 (2), 220–231.

Shohet, D., Wills, R.B., Stuart, D.L., 2001. Valepotriates and valerenic acids in commercial preparations of valerian available in Australia. Pharmazie 56 (11), 860–863 (Abstract from PubMed).

Singh, N., Gupta, A.P., Singh, B., Kaul, V.K., 2006. Quantification of valerenic acid in *Valeriana jatamansi* and *Valeriana officinalis* by HPTLC. Chromatographia 63 (3–4), 209–213.

Tortarolo, M., Braun, R., Hubner, G.E., Maurer, H.R., 1982. *In vitro* effects of epoxide-bearing valepotriates on mouse early hematopoietic progenitor cells and human t-lymphocytes. Arch. Toxicol. 51 (1), 37–42.

Ye, J.M., Hu, P.J., Yi, C.Q., Hu, C.Y., Chen, F.M., Qian, W., Xue, C.K., 2007. Valepotriate-induced apoptosis of gastric cancer cell line MKN-45 (Chinese). World Chin. J. Dig. 15 (1), 22–28. Abstract from EMBASE 2007048506.

Chapter 3.2.8

Matricaria recutita

Fatma Tugce Guragac Dereli[1], Tarun Belwal[2]
[1]*Department of Pharmacognosy, Faculty of Pharmacy, Suleyman Demirel University, Isparta, Turkey;* [2]*College of Biosystems Engineering and Food Science, Zhejiang University, China*

Introduction

Alzheimer disease (AD), an untreatable degenerative neurological disease, mainly affects elderly people and results in impaired cognitive abilities in learning, thinking, and memory in the aged population. As the most prevalent cause of dementia, AD sharply decreases the quality of the patient's life and hinders their daily functioning (Singh, 2020).

The pathologic markers used to confirm this disease include the extraneuronal deposition of amyloid-β (Aβ) protein, the intracellular creation of hyperphosphorylated tubulin-binding (tau) protein, decreased cholinergic neurotransmission, and distinct extensive neuroinflammation (Eikelenboom et al., 1989; Sanabria-Castro et al., 2017).

In 2019, an estimated 50 million people were known to live with AD, and this number is expected to triple by 2050 (International, 2019). In addition, it is currently ranked sixth among the top 10 causes of death (Kozlov et al., 2017; Picanco et al., 2018; Singh, 2020). Despite its constantly growing global burden, there is no effective therapeutic strategy available to slow or treat this progressive disease. Current FDA-approved treatment options are palliative and provide only temporary relief of symptoms by changing the level of neurotransmitters in the brain. Therefore, curative treatment is urgently needed to overcome it (Benek et al., 2020; Ramesh et al., 2020; Singh, 2020).

The search for treatment with medicinal plants is as old as human history (Petrovska, 2012). What makes plants so important are the secondary metabolites (SMs) they contain. The classification of these metabolites is based on their chemical structures or synthesis pathways and consists of phenolics, terpenoids, and alkaloids. SMs are involved in the adaptation of plants to their habitats and provide defence against pathogens, herbivores, competing plants,

Naturally Occurring Chemicals against Alzheimer's Disease. https://doi.org/10.1016/B978-0-12-819212-2.00024-4

and pests. These organic compounds are also responsible for a wide variety of medicinal activities, and they represent unique sources for active pharmaceuticals. Several SMs synthesized by plants have been used as raw materials of diverse drugs (Harborne, 2000; Li et al., 2020; Satish et al., 2020).

In several studies, *Matricaria recutita* L. has been found promising in the treatment of pathophysiologic processes of AD, and this anti-AD effect has been attributed to some SMs present in the plant (Alibabaei et al., 2014; Ionita et al., 2018).

M. recutita L

M. recutita L. (synonym is *Matricaria chamomilla* L.) is an annual herbaceous plant that belongs to Asteraceae family (Singh et al., 2011). This medicinal daisy plant, which is also known as chamomile or German chamomile, grows natively in Asia and Europe. It is also cultivated in several other parts of the world because of its cosmetic and therapeutic values (Ortiz et al., 2016).

Chamomile has a very old history as a traditional herbal medicine. Since ancient times, it has been used for the treatment of pain, inflammation, wounds, burns, skin diseases, gout, rheumatism, insomnia, flatulence, colic, fever, hemorrhoids, mastitis, hysteria, neuralgia, and sciatica (Chauhan and Jaya, 2017; Das et al., 1998). It is also used as a memory enhancer in different countries (Adams et al., 2007; Orhan and Aslan, 2009).

Today, chamomile is still among the most used medicinal plants worldwide due to its versatile pharmacologic applications. So far, much preclinical and clinical data support its antimicrobial, antidiarrheal, analgesic, hypocholesterolemic, hypoglycemic, anticancer, antimutagenic, antidiabetic, antidepressant, anxiolytic, wound healing, antiallergic, immunomodulatory, antiinflammatory, antiamnesic, neuroprotective, and antioxidant properties (Al-Dabbagh et al., 2019; Amsterdam et al., 2012; Ionita et al., 2018; Jarrahi et al., 2010; Lee et al., 2010; Mao et al., 2016; Marino et al., 2001; Mehmood et al., 2015; Miguel, 2010; Miraj and Alesaeidi, 2016; Roby et al., 2013; Sebai et al., 2014; Srivastava et al., 2010; Zargaran et al., 2014). The phytochemical analysis on chamomile revealed that flowers include more than 120 SMs, which can be grouped as flavonoids, phenylpropanoids, terpenoids, polyacetylenes, coumarins, spiroethers, tannins, and polysaccharides (Murti et al., 2012).

Chamomile has been shown to be a potential candidate for the treatment of AD by modulating cholinergic neurotransmission; decreasing the expression of interleukin-1β (IL-1β); increasing the level of brain-derived neurotrophic factor (BDNF); inactivating the toxic Aβ proteins; and exhibiting antiinflammatory, antiamnesic, antioxidant, and neuroprotective potential (Alibabaei et al., 2014; Ionita et al., 2018). The literature findings are presented in detail subsequently.

Preclinical studies

To date, many preclinical trials have revealed the anti-AD potential of chamomile.

Ionita and colleagues demonstrated that the hydroalcoholic extract of chamomile has antiamnesic activity via enhancing free radical scavenging activity, decreasing neuroinflammation, and modulating cholinergic activity in scopolamine-induced memory deficit in rodents. The extract has been shown to increase the level of BDNF expression and decrease IL-1β expression (Ionita et al., 2018). It has been proven in previous studies that BDNF dysfunction and excessive IL-1β expression might contribute to AD pathology (Qin et al., 2017; Xian et al., 2015).

Alibabaei et al. investigated the activity of chamomile on learning and memory performance in a scopolamine-induced amnesia rat model. They have established that ethanolic extract prepared from chamomile has repairing activity on cognitive impairment and could be beneficial in AD patients. It is known that reactive oxygen species are related to brain damage and lead to AD. The memory-promoting effect has been attributed to the antioxidant activity of the extract (Alibabaei et al., 2014; Rahnama et al., 2015).

A study conducted by Asgharzade et al. aimed to search the therapeutic efficiency of alcoholic extract of chamomile on the dysfunction of motor coordination induced by scopolamine. The obtained results indicated that the extract protects the rat brain against oxidative stress caused by scopolamine application through its antioxidant feature (Asgharzade et al., 2015).

Similarly, in another study, the extract of chamomile prepared with ethanol has been shown to enhance the motor balance in scopolamine-receiving rats (Rabiei et al., 2015).

Ranpariya and colleagues have found that the methanolic extract of chamomile has a neuroprotective effect in rats against oxidative brain damage induced by AlF_4^-. They estimate that this activity might be due to the ability of chamomile to reduce the glutamate-induced increase in intracellular Ca^{2+} levels (Ranpariya et al., 2011).

The ligulate flowers of chamomile are rich in apigenin (4′, 5, 7-trihydroxyflavone) (Fig. 3.2.8.1), and several reports are available on the

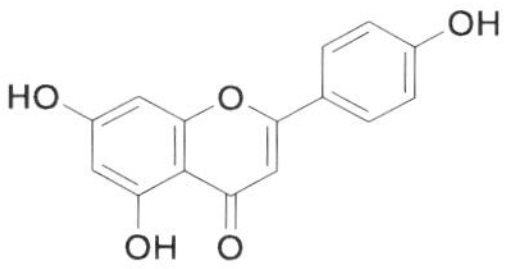

FIGURE 3.2.8.1 The chemical structure of apigenin.

anti-AD activity of this flavone (McKay and Blumberg, 2006). The neurovascular protective, neurotrophic, antiinflammatory, antiapoptotic, anti-amyloidogenic, and antioxidant activities of this polyphenolic compound have been proven (Balez et al., 2016; Liang et al., 1999; Nicholas et al., 2007; Smolinski; Pestka, 2003). The compound was found to be responsible for decreasing the production of adhesion molecules, reducing the deposition of fibrillar amyloid, and enhancement of BDNF signalling (Panes et al., 1996; Zhao et al., 2013). It was also determined that apigenin improved the memory and learning deficits of mice (Liu et al., 2011).

Clinical studies

According to the literature, there is only one clinical study in which apigenin was evaluated for inflammation or cognitive performance. In this experiment, clinical stabilization was noted in all of the patients with AD treated with the formulation containing apigenin, and most of the patients obtained a positive result in the mini-mental test. Given its role against oxidative stress and neuroinflammation, apigenin looks promising for the treatment of AD (Rojas and Dorazco-Barragan, 2010).

Side effects of *M. recutita* L.

Chamomile consumption is generally considered safe (Hausen et al., 1984). Due to sesquiterpene lactones in its composition, there is a risk of developing hypersensitivity reactions to chamomile, but it is low. However, it should be used with caution by individuals allergic to other plants of the Asteraceae family (Mitchell and Dupuis, 1971). The presence of any toxic compound in chamomile has not been reported, but contamination or mixing with different types of daisy may create toxicity. There is no report regarding its safety in pregnant/lactating women or children nor drug–herb interaction (Jo et al., 2016).

Conclusion

Previous preclinical and clinical studies have underlined that chamomile may be worthwhile in the treatment of AD. A summary of the role of chamomile in the treatment of AD is presented in Fig. 3.2.8.2. Therefore, future studies should deeply focus on the pharmacokinetic and pharmacodynamic profile of this promising medicinal plant.

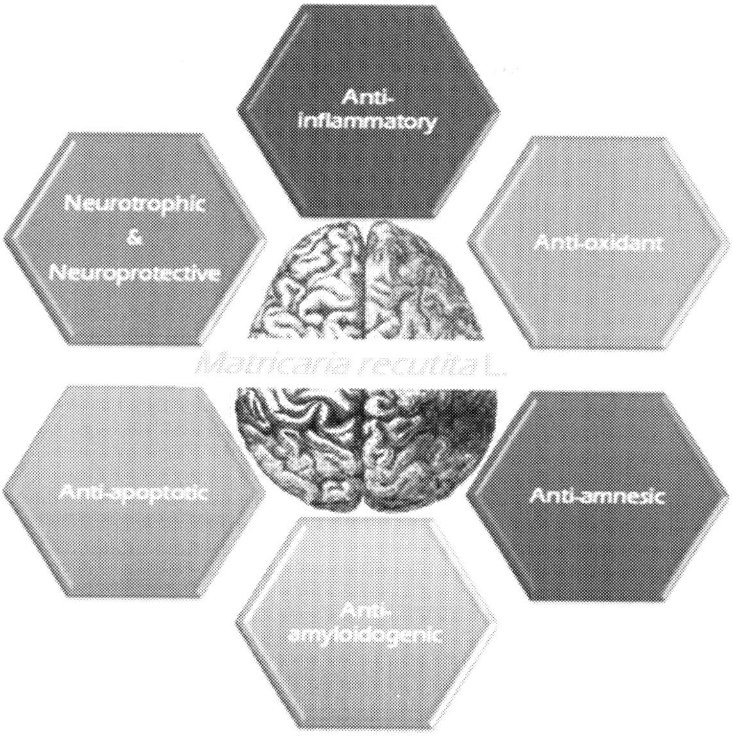

FIGURE 3.2.8.2 Role of *M. recutita* L. in the treatment of AD.

References

Adams, M., Gmunder, F., Hamburger, M., 2007. Plants traditionally used in age related brain disorders- A survey of ethnobotanical literature. J. Ethnopharmacol. 113 (3), 363–381.

Al-Dabbagh, B., Elhaty, I.A., Elhaw, M., Murali, C., Al Mansoori, A., Awad, B., Amin, A., 2019. Antioxidant and anticancer activities of chamomile (*Matricaria recutita* L.). BMC Res. Notes 12 (1), 3.

Alibabaei, Z., Rabiei, Z., Rahnama, S., Mokhtari, S., Rafieian-kopaei, M., 2014. *Matricaria chamomilla* extract demonstrates antioxidant properties against elevated rat brain oxidative status induced by amnestic dose of scopolamine. Biomed Aging Pathol. 4 (4), 355–360.

Amsterdam, J.D., Shults, J., Soeller, I., Mao, J.J., Rockwell, K., Newberg, A.B., 2012. Chamomile (*Matricaria recutita*) may provide antidepressant activity in anxious, depressed humans: an exploratory study. Alternative Ther. Health Med. 18 (5), 44–49.

Asgharzade, S., Rabiei, Z., Rafieian-Kopaei, M., 2015. Effects of *Matricaria chamomilla* extract on motor coordination impairment induced by scopolamine in rats. Asian Pac. J. Trop. Biomed. 5 (10), 829–833.

Balez, R., Steiner, N., Engel, M., Munoz, S.S., Lum, J.S., Wu, Y.Z., Wang, D.D., Vallotton, P., Sachdev, P., O'Connor, M., Sidhu, K., Munch, G., Ooi, L., 2016. Neuroprotective effects of apigenin against inflammation, neuronal excitability and apoptosis in an induced pluripotent stem cell model of Alzheimer's disease. Sci. Rep-Uk 6.

Benek, O., Korabecny, J., Soukup, O., 2020. A Perspective on multi-target drugs for Alzheimer's disease. Trends Pharmacol. Sci. (in press).

Chauhan, E.S., Jaya, A., 2017. Chamomile an ancient aromatic plant-A review. J. Ayu. Med. Sci. 2 (4), 251−255.

Das, M., Mallavarapu, G.R., Kumar, S., 1998. Chamomile (*Chamomilla recutita*): economic botany, biology, chemistry, domestication and cultivation. J. Med. Aromat. Plant Sci. 20, 1074−1109.

Eikelenboom, P., Hack, C.E., Rozemuller, J.M., Stam, F.C., 1989. Complement activation in amyloid plaques in Alzheimer's dementia. Virchows Arch. B Cell Pathol. Incl. Mol. Pathol. 56 (4), 259−262.

Harborne, J.B., 2000. Arsenal for survival: secondary plant products. Taxon 49 (3), 435−449.

Hausen, B.M., Busker, E., Carle, R., 1984. [The sensitizing capacity of composite plants. VII. Experimental studies with extracts and compounds of *Chamomilla recutita* (L.) Rauschert and *Anthemis cotula* L]. Planta Med. 50 (3), 229−234.

International, A.s.D., 2019. World Alzheimer's Report 2019: Attitudes to Dementia. https://www.alz.co.uk/research/WorldAlzheimerReport2019.pdf.

Ionita, R., Postu, P.A., Mihasan, M., Gorgan, D.L., Hancianu, M., Cioanca, O., Hritcu, L., 2018. Ameliorative effects of *Matricaria chamomilla* L. hydroalcoholic extract on scopolamine-induced memory impairment in rats: a behavioral and molecular study. Phytomedicine 47, 113−120.

Jarrahi, M., Vafaei, A.A., Taherian, A.A., Miladi, H., Pour, A.R., 2010. Evaluation of topical *Matricaria chamomilla* extract activity on linear incisional wound healing in albino rats. Nat. Prod. Res. 24 (8), 697−702.

Jo, F., Lp, D., Ca, E., Ec, C., Pharmacy, F., 2016. *Matricaria recutita* and its isolate apigenin: economic value, ethnopharmacology and chemico-biological profiles in retrospect. J. Pharmacogn. Phytochem. 4 (4), 17−31.

Kozlov, S., Afonin, A., Evsyukov, I., Bondarenko, A., 2017. Alzheimer's disease: as it was in the beginning. Rev. Neurosci. 28 (8), 825−843.

Lee, S.H., Heo, Y., Kim, Y.C., 2010. Effect of German chamomile oil application on alleviating atopic dermatitis-like immune alterations in mice. J. Vet. Sci. 11 (1), 35−41.

Li, Y.Q., Kong, D.X., Fu, Y., Sussman, M.R., Wu, H., 2020. The effect of developmental and environmental factors on secondary metabolites in medicinal plants. Plant Physiol. Biochem. (Paris) 148, 80−89.

Liang, Y.C., Huang, Y.T., Tsai, S.H., Lin-Shiau, S.Y., Chen, C.F., Lin, J.K., 1999. Suppression of inducible cyclooxygenase and inducible nitric oxide synthase by apigenin and related flavonoids in mouse macrophages. Carcinogenesis 20 (10), 1945−1952.

Liu, R., Zhang, T.T., Yang, H.G., Lan, X., Ying, J.A., Du, G.H., 2011. The flavonoid apigenin protects brain neurovascular coupling against amyloid-beta(25-35)-induced toxicity in mice. J. Alzheimers Dis. 24 (1), 85−100.

Mao, J.J., Xie, S.X., Keefe, J.R., Soeller, I., Li, Q.S., Amsterdam, J.D., 2016. Long-term chamomile (*Matricaria chamomilla* L.) treatment for generalized anxiety disorder: a randomized clinical trial. Phytomedicine 23 (14), 1735−1742.

Marino, M., Bersani, C., Comi, G., 2001. Impedance measurements to study the antimicrobial activity of essential oils from Lamiaceae and Compositae. Int. J. Food Microbiol. 67 (3), 187–195.

McKay, D.L., Blumberg, J.B., 2006. A review of the bioactivity and potential health benefits of chamomile tea (*Matricaria recutita* L.). Phytother Res. 20 (7), 519–530.

Mehmood, M.H., Munir, S., Khalid, U.A., Asrar, M., Gilani, A.H., 2015. BMC complementary and alternative medicine: antidiarrhoeal, antisecretory and antispasmodic activities of *Matricaria chamomilla* are mediated predominantly through K(+)-channels activation. J. Aust. Tradit. Med. Soc. 21 (2), 126–127.

Miguel, M.G., 2010. Antioxidant and anti-inflammatory activities of essential oils: a short review. Molecules 15 (12), 9252–9287.

Miraj, S., Alesaeidi, S., 2016. A systematic review study of therapeutic effects of *Matricaria recuitta chamomile* (chamomile). Electron. Physician 8 (9), 3024–3031.

Mitchell, J.C., Dupuis, G., 1971. Allergic contact dermatitis from sesquiterpenoids of the Compositae family of plants. Br. J. Dermatol. 84 (2), 139–150.

Murti, K., Panchal, M.A., Gajera, V., Solanki, J., 2012. Pharmacological properties of *Matricaria recutita*: a review. Pharmacologia 3 (8), 348–351.

Nicholas, C., Batra, S., Vargo, M.A., Voss, O.H., Gavrilin, M.A., Wewers, M.D., Guttridge, D.C., Grotewold, E., Doseff, A.I., 2007. Apigenin blocks lipopolysaccharide-induced lethality *in vivo* and proinflammatory cytokines expression by inactivating NF-kappaB through the suppression of p65 phosphorylation. J. Immunol. 179 (10), 7121–7127.

Orhan, I., Aslan, M., 2009. Appraisal of scopolamine-induced antiamnesic effect in mice and *in vitro* antiacetylcholinesterase and antioxidant activities of some traditionally used Lamiaceae plants. J. Ethnopharmacol. 122 (2), 327–332.

Ortiz, M.I., Fernandez-Martinez, E., Soria-Jasso, L.E., Lucas-Gomez, I., Villagomez-Ibarra, R., Gonzalez-Garcia, M.P., Castaneda-Hernandez, G., Salinas-Caballero, M., 2016. Isolation, identification and molecular docking as cyclooxygenase (COX) inhibitors of the main constituents of *Matricaria chamomilla* L. extract and its synergistic interaction with diclofenac on nociception and gastric damage in rats. Biomed. Pharmacother. 78, 248–256.

Panes, J., Gerritsen, M.E., Anderson, D.C., Miyasaka, M., Granger, D.N., 1996. Apigenin inhibits tumor necrosis factor-induced intercellular adhesion molecule-1 upregulation *in vivo*. Microcirculation 3 (3), 279–286.

Petrovska, B.B., 2012. Historical review of medicinal plants' usage. Phcog. Rev. 6 (11), 1–5.

Picanco, L.C.d.S., Ozela, P.F., Brito, M.d.F.d.B., Pinheiro, A.A., Padilha, E.C., Braga, F.S., Silva, C.H.T.d.P.d., Santos, C.B.R.d., Rosad, J.M.C., Hage-Melim, L.I.d.S., 2018. Alzheimer's disease: a review from the pathophysiology to diagnosis, new perspectives for pharmacological treatment. Curr. Med. Chem. 25, 3141–3159.

Qin, X.Y., Cao, C., Cawley, N.X., Liu, T.T., Yuan, J., Loh, Y.P., Cheng, Y., 2017. Decreased peripheral brain-derived neurotrophic factor levels in Alzheimer's disease: a meta-analysis study (N=7277). Mol. Psychiatr. 22 (2), 312–320.

Rabiei, Z., Alibabaei, Z., Rafieian-Kopaei, M., 2015. Determining the antioxidant properties of chamomile and investigating the effects of chamomile ethanol extract on motor coordination disorders in rats. J. Babol. Univ. Medical. Sci. 17 (4), 44–50.

Rahnama, S., Rabiei, Z., Alibabaei, Z., Mokhtari, S., Rafieian-kopaei, M., Deris, F., 2015. Antiamnesic activity of *Citrus aurantium* flowers extract against scopolamine-induced memory impairments in rats. Neurol. Sci. 36 (4), 553–560.

Ramesh, M., Gopinath, P., Govindaraju, T., 2020. Role of Post-translational modifications in Alzheimer's disease. Chembiochem 21 (8), 1052–1079.

Ranpariya, V.L., Parmar, S.K., Sheth, N.R., Chandrashekhar, V.M., 2011. Neuroprotective activity of *Matricaria recutita* against fluoride-induced stress in rats. Pharm. Biol. 49 (7), 696–701.

Roby, M.H.H., Sarhan, M.A., Selim, K.A.H., Khalel, K.I., 2013. Antioxidant and antimicrobial activities of essential oil and extracts of fennel (*Foeniculum vulgare* L.) and chamomile (*Matricaria chamomilla* L.). Ind. Crop. Prod. 44, 437–445.

Rojas, E.D., Dorazco-Barragan, G., 2010. Clinical stabilisation in neurodegenerative diseases: clinical study in phase II. Rev. Neurol. 50 (9), 520–528.

Sanabria-Castro, A., Alvarado-Echeverria, I., Monge-Bonilla, C., 2017. Molecular pathogenesis of Alzheimer's disease: an update. Ann. Neurosci. 24 (1), 46–54.

Satish, L., Shamili, S., Yolcu, S., Lavanya, G., Alavilli, H., Swamy, M.K., 2020. Biosynthesis of secondary metabolites in plants as influenced by different factors. In: Swamy, M. (Ed.), Plant-derived Bioactives. Springer, Singapore, pp. 61–100.

Sebai, H., Jabri, M.A., Souli, A., Rtibi, K., Selmi, S., Tebourbi, O., El-Benna, J., Sakly, M., 2014. Antidiarrheal and antioxidant activities of chamomile (*Matricaria recutita* L.) decoction extract in rats. J. Ethnopharmacol. 152 (2), 327–332.

Singh, O., Khanam, Z., Misra, N., Srivastava, M.K., 2011. Chamomile (*Matricaria chamomilla* L.): an overview. Phcog. Rev. 5 (9), 82–95.

Singh, R.K., 2020. Antagonism of cysteinyl leukotrienes and their receptors as a neuro-inflammatory target in Alzheimer's disease. Neurol Sci ([published online ahead of print]).

Smolinski, A.T., Pestka, J.J., 2003. Modulation of lipopolysaccharide-induced proinflammatory cytokine production *in vitro* and *in vivo* by the herbal constituents apigenin (chamomile), ginsenoside Rb(1) (ginseng) and parthenolide (feverfew). Food Chem. Toxicol. 41 (10), 1381–1390.

Srivastava, J.K., Shankar, E., Gupta, S., 2010. Chamomile: a herbal medicine of the past with a bright future (Review). Mol. Med. Rep. 3 (6), 895–901.

Xian, Y.F., Ip, S.P., Mao, Q.Q., Su, Z.R., Chen, J.N., Lai, X.P., Lin, Z.X., 2015. Honokiol improves learning and memory impairments induced by scopolamine in mice. Eur. J. Pharmacol. 760, 88–95.

Zargaran, A., Borhani-Haghighi, A., Faridi, P., Daneshamouz, S., Kordafshari, G., Mohagheghzadeh, A., 2014. Potential effect and mechanism of action of topical chamomile (*Matricaria chammomila* L.) oil on migraine headache: a medical hypothesis. Med. Hypotheses 83 (5), 566–569.

Zhao, L., Wang, J.L., Liu, R., Li, X.X., Li, J.F., Zhang, L., 2013. Neuroprotective, anti-amyloidogenic and neurotrophic effects of apigenin in an Alzheimer's disease mouse model. Molecules 18 (8), 9949–9965.

Chapter 3.2.9

Galanthus nivalis L. (snowdrop)

Devesh Tewari[1], Tanuj Joshi[2], Archana N. Sah[2]
[1]*Department of Pharmacognosy, School of Pharmaceutical Sciences, Lovely Professional University, Phagwara, Punjab, India;* [2]*Department of Pharmaceutical Sciences, Faculty of Technology, Kumaun University, Nainital, Uttarakhand, India*

Background

Genus *Galanthus* of the family Amaryllidaceae is an important genus in terms of its therapeutic potential. Different species of this genus are native to various countries around the globe. It is more commonly found in Europe including Turkey, Bulgaria, Iran, and the Caucasus mountains (Wendelbo, 1971; Plaitakis and Duvoisin, 1983; Salehi Sourmaghi et al., 2010). *Galanthus* derives its name from the Greek words "gala" and "anthos," meaning milk and flower, respectively. The word *Galanthus* thus means milky white flowers (Davis, 2000; Salehi Sourmaghi et al., 2010). The plant is bulbous with grassy leaves which are narrow and it bears white flowers with erect flowering stalks (Salehi Sourmaghi et al., 2010).

Many pharmacological activities have been reported for various phytoconstituents of different species of this genus. Most of the alkaloids, specifically isoquinoline alkaloids including but not limited to galanthamine, caranine, lycorine, tazettine, narciclasine, montanine, and narwedine, possess a range of therapeutic activities by virtue of their acetylcholinesterase inhibitory, antiviral, antitumor, antimalarial, and immunostimulatory activities (Salehi Sourmaghi et al., 2010).

Distribution and habitat associated with *Galanthus nivalis*

The genus is found in Europe, with the primary center of occurrence being South Central and South Eastern Europe. It is also found in western regions of France in scarce amounts. It is generally found in dispersed areas, mainly in meadows, close to larger river, and also at high altitudes (https://botany.cz/cs/galanthus-nivalis/ accessed on July 7, 2019). The genus is also distributed in the Caucasus and Asia Minor regions (Berkov, 2012). Of the many species, only a few are prevalent, however there are many others that are limited to smaller regions. For instance, *G. nivalis* is innate to the broader area of

Naturally Occurring Chemicals against Alzheimer's Disease. https://doi.org/10.1016/B978-0-12-819212-2.00025-6

Europe, extending from Italy to the Pyrenees, Turkey, the northern part of Greece, and Ukraine. *Galanthus trojanus* is a rare species found in western parts of Turkey (Davis and Özhatay, 2001; Unver 2007).

The plant species was widespread in the Eastern Carpathians long ago, however there was a substantial reduction in its occurrence later due to habitat destruction and the practice of digging out the bulbs directly. Due to the threat of its complete extinction, it is included in the lists of fast-disappearing plants (Budnikov and Kricsfalusy, 1994).

The plant grows on humus soils, mostly in deciduous wet woodlands, flood plains, and is also found in parks. The plant blooms from February until April. It is a perennial herb with an underground bulb, has an approximately 30 cm long straight flower stalk, and always bears a flower that grows from a tremor-green bristle. The flowers of the plant are made from white, elongated outer petals with short inner white petals at the edge with a green border. However, the species shows much variability. The snowdrop plant is categorized as an endangered species and is also listed in the International Endangered Species Convention (CITES) (https://botany.cz/cs/galanthus-nivalis/ accessed on July 7, 2019).

Botany of *Galanthus nivalis*

The family Amaryllidaceae is among 20 families that contain some important alkaloids (Jin, 2013). This family consists of around 1100 species that are bulbous and perennial. The species are further divided into 85 genera that are found in warm, temperate, and tropical regions (Berkov, 2012).

Galanthus nivalis L. (snowdrop) is a bulbous ephemeroid type geophyte of the *Amaryllidaceae* family. It is a spring species and is highly decorative in appearance. Of the many alkaloids this plant produces, the most important are nivalidin and galantamine (Budnikov and Kricsfalusy, 1994).

The entire *Galanthus* species is perennial, herbaceous, and grows from bulbs. They contain an erect, leafless scape with two or three linear leaves. A couple of spathe valves are found above the scape, from which a bell-shaped, solitary, white flower emerges. This flower contains six tepals and among these the three on the outside are greater in size and relatively higher in convexity in comparison to their counterparts lying inside. The flower contains a three-celled ovary which ripens to form a capsule with three cells. Individual seeds, which are white in color, contain an elaiosome which contains substances that are attractive to ants. Moreover, the genus *Galanthus* is closely related to the *Leucojum* L. genus. However, the flowers of *Leucojum* have six equal tepals and several leaves enabling these genera to be distinguished from one another (Berkov, 2012).

Anatomical studies have shown that *G. nivalis* has elongated and somewhat longitudinal cells containing an upper and lower epidermis. Beneath these there are some starch grains containing parenchymal isodiametric cells. The epidermis of the leaf of *G. nivalis* contains elongated and slightly narrowed cells. A regular distribution of the stomata is found throughout the surface of the blade. Parenchyma cells are assimilated beneath and on the upper side of the lower epidermis in the young leaf (Budnikov and Kricsfalusy, 1994).

Phytochemistry and structure–activity relationship

Amaryllidaceous plants have attracted considerable interest due to the presence of diverse alkaloids in this species which possess assorted pharmacological properties. During the early 1950s, phytochemical studies were conducted on the genus *Galanthus*. Of the many alkaloids, two initially identified alkaloids were galanthine (Proskurina and Ordzhonikidze, 1953) and galantamine (Berkov, 2012), which were isolated from the plant *G. voronowii*. Many other phytoconstituents have been reported from this plant, the chemical structures of some of which are presented in Fig. 3.2.9.1.

Nivalin and Razadyne (formerly Reminyl) are hydrobromide salt formulations of galanthamine for the treatment of AD. These formulations are also effective in neurological diseases and poliomyelitis (Berkov, 2012; Heinrich and Teoh, 2004). AChE is accountable for acetylcholine degradation in myoneural junctions and in peripheral/central synapses that utilize acetylcholine as their neurotransmitter. Galanthamine can act on the central nervous system by crossing the blood–brain barrier (Bastida et al., 2006; Heinrich and Teoh, 2004).

Other AChE inhibitors from this plant include epigalanthamine and narwedine, however these are over 130 times less potent than galanthamine (Thomsen et al., 1990). A 10-fold reduction in activity can occur due to the

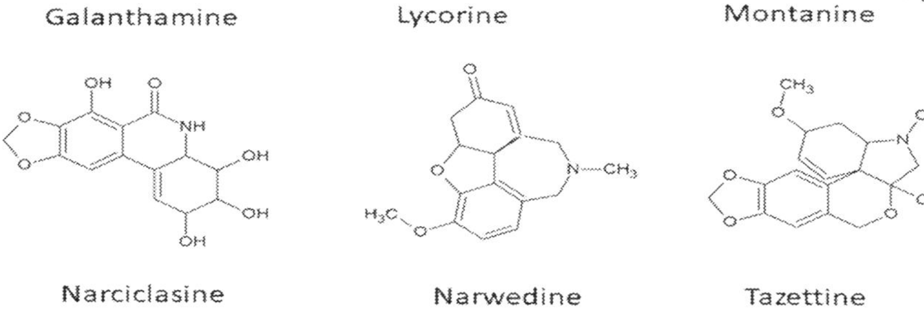

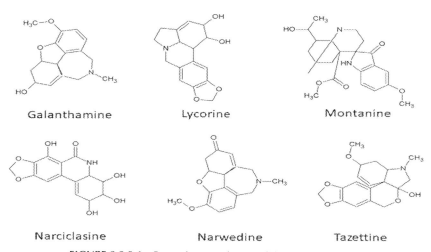

FIGURE 3.2.9.1 Some phytoconstituents of the genus *Galanthus*.

loss of the methyl group located at the N atom. An example of this loss is N-demethylgalanthamine. However, if the hydroxyl group is present at C9 rather than the methoxyl group, as is the case in sanguinine, then the activity increases 10-fold compared with galanthamine. It is also interesting to note that if hydrogenation takes place at C4—C4a, as occurs in lycoramine, it will entirely destroy the inhibitory action of AChE (López et al., 2002).

Hypothalamic—pituitary—adrenal axis activity is modulated by another type of phytochemical constituent present in *Galanthus nivalis*, known as tyramine-type compounds that contain phenolic amine groups (Vera-Avila et al., 1996). The above-mentioned activity of the compounds is probably due to the structural similarity of these compounds with noradrenaline. Trisphaeridine is another constituent of this plant which is effective against retrovirus, however this compound has a low therapeutic index. Some other phytoconstituents of the plant include ismine, arolycoricidine, narciprimine, and arolycoricidine (Berkov, 2012; Kaya et al., 2011).

Apoptosis is started in tumor cells by low concentrations of haemanthamine, which is a phytochemical isolated from *Galanthus nivalis* (McNulty et al., 2007). *Plasmodium falciparum* which is sensitive to chloroquine also responds to this compound. An antihypertensive effect and an effect against retrovirus are other actions attributed to this compound (Bastida et al., 2006; Kaya et al., 2011). Other compounds that possess cytotoxic activity are vittatine and maritidine, which have shown cytotoxic activity against lung carcinoma, colon carcinoma, adenocarcinoma, and carcinoma of renal cells (Bastida et al., 2006; Silva et al., 2008). Various Gram-negative and Gram-positive bacteria are also sensitive toward the action of vittatine (Kornienko and Evidente, 2008).

Importance of snowdrops in neurodegenerative disorders

Galanthus nivalis against AD

The term Alzheimer's disease (AD) was first used by Dr. Alois Alzheimer, a German psychiatrist. He described symptoms including delusions, progressive language and memory impairment, disorientation, and psychosocial impairment, etc. for the first time in a 51-year-old woman. Many symptoms and our understanding regarding AD are due to the work of Alois Alzheimer (Korolev, 2014). AD is the most common cause of dementia, with worsening of intellectual functions being the hallmark of AD. Cognitive functions like language, memory, reasoning, decision-making, attention, and orientation are impaired due to dementia. Alterations in emotional regulation, and changes in personality and social behaviors are also observed with dementia, which impacts the patient's working capacity and social and personal life. Dementia can be reversible or irreversible. Reversible dementias can be cured, and usually occur due to a secondary condition, such as malnutrition, medication (sedative-hypnotics, analgesics, psychotropics, anticholinergics, etc.), depression, endocrine disorders, metabolic disorders, brain tumors, substance abuse, etc. (Gilman, 2010; Korolev, 2014; McKhann et al., 2011; Plassman et al., 2007; Shadlen and Larson, 2010).

Neurodegeneration and/or impairment of vascular processes result in irreversible dementia. Other types of dementias include Lewy body-associated dementia, Parkinson's disease-associated dementia, and dementia of the frontotemporal region of the brain. A comprehensive review of dementia has been recently published by Tewari and coworkers (Tewari et al., 2018). The development of AD involves the presence of various risk factors, including age. After 65 years of age, the risk of developing AD is greatly increased. Onset of AD earlier than 65 years involves a rare genetic mutation in its development, with Down's syndrome being an important factor in AD development (Korolev, 2014).

According to the neuropathology of AD, there is a loss of neurons in the cortex, especially in pyramidal cells. The early stages of AD are characterized by synaptic dysfunction, as a result of which communication within neural circuits is disrupted. These neural circuits are associated with memory and cognition. The medial temporal lobe is the site where degeneration commences, especially in areas like the hippocampus and entorhinal cortex. Impairments in learning and memory (classical symptoms of early AD) result due to the degenerative changes in the above-mentioned brain structures. The spread of AD then occurs to parietal areas and the temporal association cortex. With the progression of disease damage is also seen in areas such as the neocortex and frontal cortex (Korolev, 2014).

Neurodegeneration in limbic and neocortical areas is associated with cognitive dysfunction and behavioral alterations. Not only do AD patients suffer from language, memory, and reasoning problems, they also experience emotional, psychiatric, and personality disturbances (Bozoki et al., 2012; Holtzman et al., 2011; Korolev, 2014). Neuronal disposition of unusual proteins is associated with destruction of neurons in AD. Plaques and tangles are the terms used to describe the pathophysiological lesions found in AD. The above-mentioned abnormal proteins accumulate in the cerebral cortex and along the neural pathway which is responsible for mediating cognitive and memory functions. When the amyloid proteins are deposited extracellularly, they are referred to as senile plaques. During the lifetime of normal brain cells, APP is cleaved to soluble Aβ but, in the case of AD, APP degradation is abnormal and the resulting Aβ precipitates in dense beta sheets to form senile plaques. The inflammatory response that is mounted by microglia- and astrocyte- mediated removal of aggregation of abnormal amyloids is responsible for destroying proximal neurons and neurites (Braak et al., 2011; Holtzman et al., 2011; Korolev, 2014; Norfray; Provenzale, 2004; Querfurth; LaFerla, 2010).

Exhaustive research is required to gain a greater insight into the neuropathology of AD, since clear-cut mechanisms leading to AD are still unknown. AD and its stages can be identified in dead patients by the distribution and presence of amyloid plaques and NFT after performing a postmortem study. In routine clinical examination it is diagnosed by various neurological, neuropsychological, physical, and medical history-related examinations. AD can be diagnosed by keeping in mind the following points: dementia, history of progressive cognitive decline, onset of symptoms from months to years, etc. Still very careful examination is necessary to distinguish AD from other neurodegenerative diseases (Korolev, 2014).

Current treatment therapy for AD provides relief from symptoms of AD and improves the quality of life of patients suffering from AD to a degree, however currently there are no drugs available that can modify AD progression and provide complete relief. The current allopathic drugs that can provide relief in AD are Rivastigmine, tacrine (Cognex), Alantamine (Reminyl) and Donepezil (Aricept). All these drugs are cholinesterase (an enzyme that degrades acetylcholine) inhibitors and lead to improvement in cholinergic transmission. Memantine is another drug which acts as an *N*-methyl-D-aspartate (NMDA) receptor antagonist. It is employed in therapeutics either alone or with other drugs that are cholinesterase inhibitors. Moreover, supportive health care needs to be provided to AD patients by health workers as the complications of AD, including dementia, worsen with time (Bhushan et al., 2018).

Traditionally, many medicinal plants have found uses in the treatment of neurodegenerative diseases. Ayurvedic, Japanese, Chinese, European, and other systems of medicine have their own sets of plants to treat AD. Different plants and their phytochemicals provide relief in cognitive disorders like AD due to their anticholinesterase, antioxidant, antiinflammatory, and other properties. Many such plants including *Gingko biloba*, *Withania somnifera*, *Bacopa monniera*, *Artemisia absinthium*, *Centella asiatica*, etc. have been used traditionally for the treatment of cognitive decline in diseases like AD.

Physostigmine isolated from *Physostigmine venenosum* is a standard anticholinesterase inhibitor that has been used in modern medicine for a very long time. Physostigmine has shown improvement in cognitive functions in several studies. In a model of oxygen deficiency it has shown improvements in cognitive function in rats. Also, it has been shown to antagonize cognitive function impairment induced by scopolamine in rats. Although physostigmine has shown promising results in improving cognitive functions in experiments, its clinical use in the treatment of AD is not very successful due to its short half life. Thus many new plants, along with already established traditional plants, are constantly being subjected to experimental studies in order to discover potent drugs suitable to provide therapeutic benefits in AD (Howes et al., 2003).

An important medicinal plant that shows benefits in neurological disorders and improves cognitive functions is *Galanthus nivalis*. *Galanthus nivalis* has been used traditionally in many countries for its beneficial effects in the treatment of cognitive disorders and neurological conditions. In current medical practice galantamine isolated from *Galanthus nivalis* and some other plants has been shown to have benefits in AD treatment (Heinrich and Teoh, 2004; Howes et al., 2003; Tewari et al., 2018). Also, galantamine has been prepared synthetically in laboratories. There is an interesting history behind the development of galantamine. In the early 1950s, a Russian pharmacologist in an unconfirmed report threw light on the use of Caucasian snowdrops (*Galanthus caucasicus*) by people living at the foot of the Ural mountains for treatment of poliomyelitis (Heinrich and Teoh, 2004). While working on galantamine isolated from *Galanthus woronowii*, Russian pharmacologists Maskovsky and Kruglikova-Lvova in 1951 demonstrated the ability of galantamine to antagonize the actions of curare. They also found acetylcholinesterase-inhibiting properties of galantamine. In 1952, for the first time, galantamine was isolated from *Galanthus woronowii*. During 1956–57 newer plant sources for isolation of galantamine emerged, including *Narcissus* species, *Galanthus nivalis*, and *Leucojum aestivum*. Isolation of galantamine from *Galanthus nivalis* was performed by D. Paskov (a Bulgarian chemist) and his team for the first time. Preclinical pharmacological studies on galantamine were carried out extensively in the 1950s to gain a better idea of the pharmacological profile of galantamine. "Nivalin" was the name given to the registered formulation of galantamine, which was marketed in Bulgaria. The anticholinesterase effect of galantamine was discovered in an anesthetized cat in 1960. During the 1980s, exploration of the therapeutic benefits of galantamine in AD began to take place with the help of preclinical studies. In the 1990s, various clinical studies helped in the development of galantamine as a medication for AD. Sanochemia Pharmazeutika was the first company to obtain a patent for the synthetic preparation of galantamine. Use of galantamine in the treatment of AD was approved by Asian countries, European countries, and the United States in the 2000s (Heinrich and Teoh, 2004; Howes et al., 2003; Tewari et al., 2018). AD is related to imbalances in cholinergic transmission. Overactivity of acetylcholinesterase, which degrades acetylcholine, is also found in AD patients. As mentioned earlier, acetylcholine is closely associated with cognitive function, thus decreases in the levels of acetylcholine lead to cognitive impairment in AD patients. Galantamine exerts a beneficial effect in AD by improving cholinergic transmission. The abovementioned action is produced due to the fact that galantamine selectively and reversibly inhibits the enzyme acetylcholinesterase and thus prevents degradation of acetylcholine. A study has demonstrated that galantamine inhibits acetylcholinesterase in the brain and thus increases acetylcholine levels in the synaptic cleft by inhibition of its degradation (Hussain et al., 2018).

Galantamine is more selective for acetylcholinesterase than for butylcholinesterase. Its bioavailability profile is also considered good. Galantamine is available in different countries for the treatment of AD. Stimulation of nicotinic receptors is another important function shown by galantamine, and this action may contribute to the improvement produced by galantamine in cholinergic functions and memory. Because of its additional actions on the nicotinic receptors it has a clear-cut advantage over other anticholinesterases (Howes et al., 2003). Researchers have shown that galantamine shows allosteric modulation of nicotinic cholinergic receptors, especially $\alpha7$ and $\alpha3\beta4$ on cholinergic neurons (Roy and Awasthi, 2017). By virtue of this action it increases the release of acetylcholine. Also, galantamine has demonstrated a positive stimulation in neurogenesis associated with the hippocampus by the a7 type of nicotinic receptors.

Aβ deposition and cytotoxicity are also inhibited by galantamine in many studies. Another important finding suggests that galantamine may protect neurons from oxidative damage by scavenging reactive oxidative species (Hussain et al., 2018). The detailed mechanism is presented in Fig. 3.2.9.2.

Many preclinical studies have established galantamine's efficacy in AD. NBM (nucleus basalis magnocellularis)-lesioned mice were used in a study to evaluate the special memory performance by using a swim maze test to investigate the effect of galantamine. It was found in this study that galantamine improved the performance of the mice in the swim maze test. In a different study deficiencies induced in learning and memory by administration of scopolamine to animals was reduced by galantamine (Girdhar et al., 2015). Studies have shown that galantamine possesses adverse effects including vomiting, nausea, anorexia, and diarrhea (Ng et al., 2015). Thus *Galanthus nivalis* is a medicinal plant that could prove to be of significant benefit in Alzheimer's disease treatment. Studies in the future may identify more phytochemicals in *Galanthus nivalis* in the treatment of AD.

Galanthus nivalis actions against other neurodegenerative diseases

Besides AD, *Galanthus nivalis* can be used in treatment of memory loss and dementia that are caused by a variety of neurological disorders (Zaganas and Schousboe, 2014). Mitochondrial dysfunction is also prevented by galantamine, which may prove helpful in dementia. This effect of galantamine is evident by the fact that Aβ25/35 or hydrogen peroxide-induced alterations in mitochondrial membrane potential and morphology are inhibited by galantamine. Also, due to its anticholinesterase and mitochondrial protective abilities, galantamine reduces the oxidative stress faced by cells, which results in a neuroprotective effect. Galantamine also inhibits the efflux transporter P-glycoprotein. P-glycoprotein is present in the vascular endothelium of the brain. It results in efflux of the drugs back to the bloodstream and prevents many drugs to enter the brain. By virtue of its inhibitory action on

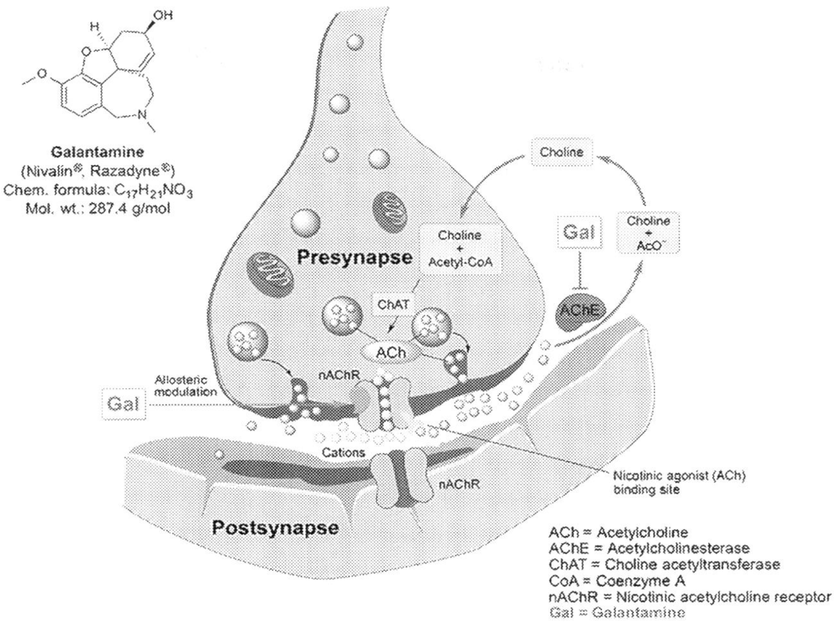

FIGURE 3.2.9.2 Chemical structure and targeted mechanisms of galantamine (Gal) against AD and dementia. The major biological effects of Gal lead to significant neuroprotection via dual AChE inhibition and allosteric stimulation of nAChRs. *(Tewari, D., Stankiewicz, A.M., Mocan, A., Sah, A.N., Tzvetkov, N.T., Huminiecki, L., Horbanczuk, J.O., Atanasov, A.G., 2018. Ethnopharmacological approaches for dementia therapy and significance of natural products and herbal drugs. Front. Aging Neurosci. 10 (3). https://doi.org/10.3389/fnagi.2018.00003. Copyright@2018 Tewari, Stankiewicz, Mocan, Sah, Tzvetkov, Huminiecki, Horbańczuk and Atanasov).*

P-glycoprotein, galantamine may be beneficial in enabling other drugs that are useful in dementia and AD to enter the brain along with it (Tewari et al., 2018).

Clinical studies on *Galanthus nivalis* with respect to Alzheimer's disease

Various clinical experiments by researchers have established the effect of *Galanthus nivalis* and galantamine in AD treatment and complications associated with AD treatment. Clinical studies in the 1990s on galantamine (8–32 mg/day) showed that administration of galantamine ameliorated the impaired day-to-day activities and cognitive functions of patients with AD (moderate or mild) over a 3–6-month time period. A dose of 24 mg/day of galantamine for a duration of 12 months was found to be effective in AD patients. In another study, the side-effect profiles of galantamine and donepezil were compared for a duration of 1 year. It was found that galantamine and donepezil produced a similar adverse effect profile (Ng et al., 2015). In a study carried out in AD patients it was found that after 2 years of treatment with galantamine and memantine, no beneficial effect was observed due to the addition of galantamine. Further investigations are required to assess the advantages of using a galantamine and memantine combination therapy in AD patients (Hager et al., 2016). In another experiment, a combination consisting of cilostazol and galantamine as well as monotherapy with the above individual drugs maintained or improved day-to-day activities, affective functions, and functions related to cognition in patients suffering from AD with asymptomatic lacunar infarction (Hishikawa et al., 2017). In a clinical study, AD patients that moved to galantamine therapy after donepezil therapy were observed. It was found that in comparison to donepezil, galantamine showed its action by two mechanisms. The dual mechanism by which galantamine acts is by enhancing the levels of acetylcholine as well as by modulation of nicotinic receptors within the frontal lobe. Galantamine treatment produced a reduction in apathy and executive dysfunctions in AD patients (Oka et al., 2016).

Another clinical study on AD patients showed that AD patients who were given a combination of galantamine and ambulatory cognitive rehabilitation showed better effects on affective and cognitive functions than the use of galantamine alone (Tokuchi et al., 2016). A clinical trial was performed on a population of elderly patients in Japan suffering from AD. This study was referred to as the "Okayama Galantamine Study." In this study the effect of galantamine on elderly AD patients was observed over a prolonged period of time. Comparison was done between the elderly patients in the experiment regarding the efficacy of galantamine with respect to their gender. It was observed that galantamine provided long-term effectiveness in elderly AD patients. Galantamine treatment maintained cognitive and affective activities and activities of daily living (ADL). Also, the effects of galantamine in AD were better for the male patients (Nakano et al., 2015). An experiment conducted on patients suffering from cognitive impairment found that patients who were treated with galantamine for 24 months showed a more significant decrease in brain atrophy than the placebo-treated group. However, no such decrease was observed with respect to hippocampus atrophy (Prins et al., 2014). The effects of galantamine alone and the combined effects of nimodipine and galantamine were evaluated in AD patients suffering from cerebrovascular disease. It was shown that galantamine improved the quality of life in AD patients with mixed dementia, but the combination of nimodipine with galantamine was not very advantageous. Since the number of patients used in the study was low, very definitive results could not be obtained by the study (Caramelli et al., 2014).

A study by Hager et al. (2014) showed that treatment with galantamine for a long period of time improved activities of daily living. Also, significant reductions in mortality and cognition decline were observed in patients suffering from AD with mild to moderate complications (Hager et al., 2014). Activities of daily living, cognition, and behavior were found to be improved in AD patients given treatment with galantamine for a duration of 12 months in a clinical study. A decline in all of the above parameters, except cognition, was observed after a follow-up of the patients for 3 years. Cognition was found to be higher in the galantamine treatment group in comparison to the untreated group, even after a 3-year follow-up of the patients (Richarz et al., 2014). In a 6-month study, the effects of galantamine were evaluated in AD patients. Galantamine (24 mg/day) was administered to these patients for a period of 6 months. Galantamine's efficacy and safety were proven in this experiment with respect to AD. Cognition was found to be significantly improved by galantamine at 6 months. Also, galantamine (24 mg/day) supported cognitive and daily functions for 12 months (Raskind et al., 2000). A clinical experiment at 43 centers (different countries) was conducted to study galantamine's efficacy and safety with respect to AD at 3 months using flexible dose escalation.

It was found that galantamine (24–32 mg/day) at 3 months showed a significantly better effect on cognition in comparison to the placebo-controlled group. Also, benefits concerning daily living were shown by the galantamine-treated group as compared to the placebo group. Galantamine's flexible dose escalation was tolerated well in the study (Rockwood et al., 2001). Also, in another clinical study slow dose escalation of galantamine was performed to showcase its tolerability and effectiveness in patients with AD showing mild to moderate symptomatology. The trial was placebo controlled, double blind, multicentric, and conducted for 5 months. The results of the study showed that galantamine in a dose of 16–24 mg/day showed significant benefits in behavioral and cognitive impairments associated with AD in comparison to the group administered the placebo formulation. It was observed that by slowly increasing the dose of galantamine, tolerability of galantamine was increased and the adverse effects of galantamine were decreased (Tariot et al., 2000). Galantamine was also evaluated in another study to highlight its tolerability and safety in AD by some researchers. In this study, AD patients whose symptoms ranged from mild to moderate were selected and the galantamine dose was increased over time. The results of the study at 6 months indicated that galantamine showed better effects than placebo by decreasing the decline in functional ability and cognitive functions. This study also established galantamine's effectiveness and tolerability profile in AD patients (Wilcock et al., 2000).

A study was also done to evaluate the effect of galantamine in improving the main symptoms associated with AD. For 3 months galantamine (18, 24, and 36 mg/day) was administered to 285 patients suffering from AD with mild to moderate symptoms. It was shown that the main symptoms of AD were improved significantly by galantamine with respect to the placebo group (Wilkinson and Murray, 2001).

Conclusion

Galanthus nivalis is a medicinal plant that has been used traditionally for various diseases over a very long period of time. It is distributed in various places around the world, especially in Europe. *G. nivalis* contains various active constituents, and galanthamine is the most important with respect to AD. *G. nivalis* has shown tremendous potential in treating AD. Galantamine is the main phytoconstituent that is held responsible for the activity of *G. nivalis* in AD. *G. nivalis* shows a beneficial effect in AD through its acetylcholinesterase inhibitory action. Also, it shows a modulatory effect on the cholinergic nicotinic receptors. Though *G. nivalis* is efficacious in AD, reliable data evaluating its therapeutic effects and adverse effects with respect to its use in AD remain to be obtained from dedicated and long-lasting clinical trials to showcase its true potential in the treatment of AD. Also, other phytoconstituents of *G. nivalis* might show beneficial effects in AD and their activity in AD needs to be evaluated by subjecting them to dedicated in vitro, preclinical, and clinical trials. Thus, with proper experimental studies *Galanthus nivalis* and its constituents might show tremendous potential in the treatment of AD.

References

Bastida, J., Lavilla, R., Viladomat, F., 2006. Chemical and biological aspects of Narcissus alkaloids. Alkaloids Chem. Biol. 63, 87–179.

Berkov, S., 2012. In: Codina, C. (Ed.), The Genus *Galanthus*: A Source of Bioactive Compounds. IntechOpen, Rijeka. https://doi.org/10.5772/28798. Ch. 11.

Bhushan, I., Kour, M., Kour, G., Gupta, S., Sharma, S., Yadav, A., 2018. Alzheimer's disease: Causes and treatment – A review. Ann. Biotechnol. 1 (1), 1002.

Bozoki, A.C., Korolev, I.O., Davis, N.C., Hoisington, L.A., Berger, K.L., 2012. Disruption of limbic white matter pathways in mild cognitive impairment and Alzheimer's disease: a DTI/FDG-PET study. Hum. Brain Mapp. 33, 1792–1802. https://doi.org/10.1002/hbm.21320.

Braak, H., Thal, D.R., Ghebremedhin, E., Del Tredici, K., 2011. Stages of the pathologic process in Alzheimer disease: age categories from 1 to 100 years. J. Neuropathol. Exp. Neurol. 70, 960–969. https://doi.org/10.1097/NEN.0b013e318232a379.

Budnikov, G., Kricsfalusy, V., 1994. Bioecological study of *Galanthus nivalis* L. in the East Carpathians. Thaiszia J. Bot. 4, 49–75.

Caramelli, P., Laks, J., Palmini, A.L.F., Nitrini, R., Chaves, M.L.F., Forlenza, O.V., Vale, F. de A.C. do, Barbosa, M.T., Bottino, C.M. de C., Machado, J.C., Charchat-Fichman, H., Lawson, F.L., 2014. Effects of galantamine and galantamine combined with nimodipine on cognitive speed and quality of life in mixed dementia: a 24-week, randomized, placebo-controlled exploratory trial (the REMIX study). Arq. Neuropsiquiatr. 72, 411–417.

Davis, A.P., ÖZHATAY, N., 2001. *Galanthus trojanus*: a new species of *Galanthus* (Amaryllidaceae) from north-western Turkey. Bot. J. Linn. Soc. 137, 409–412.

Davis, P.A., 2000. A Botanical Magazine Monograph; the Genus Galanthus.

Gilman, S., 2010. Oxford American Handbook of Neurology. Oxford University Press.

Girdhar, S., Girdhar, A., Verma, S.K., Lather, V., Pandita, D., 2015. Plant derived alkaloids in major neurodegenerative diseases: from animal models to clinical trials. J. Ayurvedic Herb. Med. 1, 91–100.

Hager, K., Baseman, A.S., Nye, J.S., Brashear, H.R., Han, J., Sano, M., Davis, B., Richards, H.M., 2016. Effect of concomitant use of memantine on mortality and efficacy outcomes of galantamine-treated patients with Alzheimer's disease: post-hoc analysis of a randomized placebo-controlled study. Alzheimer's Res. Ther. 8, 47. https://doi.org/10.1186/s13195-016-0214-x.

Hager, K., Baseman, A.S., Nye, J.S., Brashear, H.R., Han, J., Sano, M., Davis, B., Richards, H.M., 2014. Effects of galantamine in a 2-year, randomized, placebo-controlled study in Alzheimer's disease. Neuropsychiatric Dis. Treat. 10, 391–401. https://doi.org/10.2147/NDT.S57909.

Heinrich, M., Teoh, H.L., 2004. Galanthamine from snowdrop—the development of a modern drug against Alzheimer's disease from local Caucasian knowledge. J. Ethnopharmacol. 92, 147–162.

Hishikawa, N., Fukui, Y., Sato, K., Ohta, Y., Yamashita, T., Abe, K., 2017. Comprehensive effects of galantamine and cilostazol combination therapy on patients with Alzheimer's disease with asymptomatic lacunar infarction. Geriatr. Gerontol. Int. 17, 1384–1391. https://doi.org/10.1111/ggi.12870.

Holtzman, D.M., Morris, J.C., Goate, A.M., 2011. Alzheimer's disease: the challenge of the second century. Sci. Transl. Med. 3, 77sr1. https://doi.org/10.1126/scitranslmed.3002369.

Howes, M.R., Perry, N.S.L., Houghton, P.J., 2003. Plants with traditional uses and activities, relevant to the management of Alzheimer's disease and other cognitive disorders. Phyther. Res. 17, 1–18.

Hussain, G., Rasul, A., Anwar, H., Aziz, N., Razzaq, A., Wei, W., Ali, M., Li, J., Li, X., 2018. Role of plant derived alkaloids and their mechanism in neurodegenerative disorders. Int. J. Biol. Sci. 14, 341–357. https://doi.org/10.7150/ijbs.23247.

Jin, Z., 2013. Amaryllidaceae and sceletium alkaloids. Nat. Prod. Rep. 30, 849–868. https://doi.org/10.1039/c3np70005d.

Kaya, G.I., Sarıkaya, B., Onur, M.A., Somer, N.U., Viladomat, F., Codina, C., Bastida, J., Lauinger, I.L., Kaiser, M., Tasdemir, D., 2011. Antiprotozoal alkaloids from Galanthus trojanus. Phytochem. Lett. 4, 301–305. https://doi.org/10.1016/j.phytol.2011.05.008.

Kornienko, A., Evidente, A., 2008. Chemistry, biology, and medicinal potential of narciclasine and its congeners. Chem. Rev. 108, 1982–2014. https://doi.org/10.1021/cr078198u.

Korolev, I.O., 2014. Alzheimer's disease: a clinical and basic science review. Med. Student Res. J. 4, 24–33.

López, S., Bastida, J., Viladomat, F., Codina, C., 2002. Acetylcholinesterase inhibitory activity of some Amaryllidaceae alkaloids and Narcissus extracts. Life Sci. 71, 2521–2529. https://doi.org/10.1016/S0024-3205(02)02034-9.

McKhann, G.M., Knopman, D.S., Chertkow, H., Hyman, B.T., Jack Jr., C.R., Kawas, C.H., Klunk, W.E., Koroshetz, W.J., Manly, J.J., Mayeux, R., 2011. The diagnosis of dementia due to Alzheimer's disease: recommendations from the National Institute on Aging-Alzheimer's Association workgroups on diagnostic guidelines for Alzheimer's disease. Alzheimer's Dementia 7, 263–269.

McNulty, J., Nair, J.J., Codina, C., Bastida, J., Pandey, S., Gerasimoff, J., Griffin, C., 2007. Selective apoptosis-inducing activity of crinum-type Amaryllidaceae alkaloids. Phytochemistry 68, 1068–1074. https://doi.org/10.1016/j.phytochem.2007.01.006.

Nakano, Y., Matsuzono, K., Yamashita, T., Ohta, Y., Hishikawa, N., Sato, K., Deguchi, K., Abe, K., 2015. Long-term efficacy of galantamine in Alzheimer's disease: the okayama Galantamine Study (OGS). J. Alzheimers. Dis. 47, 609–617. https://doi.org/10.3233/JAD-150308.

Ng, Y.P., Or, T.C.T., Ip, N.Y., 2015. Plant alkaloids as drug leads for Alzheimer's disease. Neurochem. Int. 89, 260–270. https://doi.org/10.1016/j.neuint.2015.07.018.

Norfray, J.F., Provenzale, J.M., 2004. Alzheimer's disease: neuropathologic findings and recent advances in imaging. Am. J. Roentgenol. 182, 3–13. https://doi.org/10.2214/ajr.182.1.1820003.

Oka, M., Nakaaki, S., Negi, A., Miyata, J., Nakagawa, A., Hirono, N., Mimura, M., 2016. Predicting the neural effect of switching from donepezil to galantamine based on single-photon emission computed tomography findings in patients with Alzheimer's disease. Psychogeriatrics 16, 121–134. https://doi.org/10.1111/psyg.12132.

Plaitakis, A., Duvoisin, R.C., 1983. Homer's moly identified as Galanthus nivalis L.: physiologic antidote to stramonium poisoning. Clin. Neuropharmacol. 6, 1–5.

Plassman, B.L., Langa, K.M., Fisher, G.G., Heeringa, S.G., Weir, D.R., Ofstedal, M.B., Burke, J.R., Hurd, M.D., Potter, G.G., Rodgers, W.L., 2007. Prevalence of dementia in the United States: the aging, demographics, and memory study. Neuroepidemiology 29, 125–132.

Prins, N.D., van der Flier, W.A., Knol, D.L., Fox, N.C., Brashear, H.R., Nye, J.S., Barkhof, F., Scheltens, P., 2014. The effect of galantamine on brain atrophy rate in subjects with mild cognitive impairment is modified by apolipoprotein E genotype: post-hoc analysis of data from a randomized controlled trial. Alzheimer's Res. Ther. 6, 47. https://doi.org/10.1186/alzrt275.

Proskurina, N., Ordzhonikidze, S., 1953. Alkaloids of Galanthus woronovii. Structure of galanthine. Proc. Acad. Sci. SSSR 90, 565–567.

Querfurth, H.W., LaFerla, F.M., 2010. Alzheimer's disease. N. Engl. J. Med. 362, 329–344. https://doi.org/10.1056/NEJMra0909142.

Raskind, M.A., Peskind, E.R., Wessel, T., Yuan, W., Group, G.U.-S., 2000. Galantamine in AD A 6-month randomized, placebo-controlled trial with a 6-month extension. Neurology 54, 2261–2268.

Richarz, U., Gaudig, M., Rettig, K., Schauble, B., 2014. Galantamine treatment in outpatients with mild Alzheimer's disease. Acta Neurol. Scand. 129, 382–392. https://doi.org/10.1111/ane.12195.

Rockwood, K., Mintzer, J., Truyen, L., Wessel, T., Wilkinson, D., 2001. Effects of a flexible galantamine dose in Alzheimer's disease: a randomised, controlled trial. J. Neurol. Neurosurg. Psychiatr. 71, 589–595. https://doi.org/10.1136/jnnp.71.5.589.

Roy, S., Awasthi, H., 2017. Herbal medicines as neuroprotective agent: a mechanistic approach. Int. J. Pharm. Pharm. Sci. 9 (10), 1–7.

Salehi Sourmaghi, M.H., Azadi, B., Amin, G., Amini, M., Sharifzadeh, M., 2010. The first phytochemical report of Galanthus transcaucasicus Fomin. Daru 18, 124–127.

Shadlen, M.-F., Larson, E.B., 2010. Evaluation of Cognitive Impairment and Dementia. Waltham, MA UpToDate.

Silva, A.F.S., de Andrade, J.P., Machado, K.R.B., Rocha, A.B., Apel, M.A., Sobral, M.E.G., Henriques, A.T., Zuanazzi, J.A.S., 2008. Screening for cytotoxic activity of extracts and isolated alkaloids from bulbs of Hippeastrum vittatum. Phytomedicine 15, 882–885. https://doi.org/10.1016/j.phymed.2007.12.001.

Tariot, P.N., Solomon, P.R., Morris, J.C., Kershaw, P., Lilienfeld, S., Ding, C., 2000. A 5-month, randomized, placebo-controlled trial of galantamine in AD. The Galantamine USA-10 Study Group. Neurology 54, 2269–2276. https://doi.org/10.1212/wnl.54.12.2269.

Tewari, D., Stankiewicz, A.M., Mocan, A., Sah, A.N., Tzvetkov, N.T., Huminiecki, L., Horbanczuk, J.O., Atanasov, A.G., 2018. Ethnopharmacological approaches for dementia therapy and significance of natural products and herbal drugs. Front. Aging Neurosci. 10 https://doi.org/10.3389/fnagi.2018.00003.

Thomsen, T., Bickel, U., Fischer, J.P., Kewitz, H., 1990. Stereoselectivity of cholinesterase inhibition by galanthamine and tolerance in humans. Eur. J. Clin. Pharmacol. 39, 603—605.

Tokuchi, R., Hishikawa, N., Matsuzono, K., Takao, Y., Wakutani, Y., Sato, K., Kono, S., Ohta, Y., Deguchi, K., Yamashita, T., Abe, K., 2016. Cognitive and affective benefits of combination therapy with galantamine plus cognitive rehabilitation for Alzheimer's disease. Geriatr. Gerontol. Int. 16, 440—445. https://doi.org/10.1111/ggi.12488.

Unver, N., 2007. New skeletons and new concepts in Amaryllidaceae alkaloids. Phytochem. Rev. 6, 125—135.

Vera-Avila, H.R., Forbes, T.D.A., Randel, R.D., 1996. Plant phenolic amines: potential effects on sympathoadrenal medullary, hypothalamic-pituitary-adrenal, and hypothalamic-pituitary-gonadal function in ruminants. Domest. Anim. Endocrinol. 13, 285—296.

Wendelbo, P., 1971. Flora Iranica. Akad. Druck und Verlagsanatalt Graz 76, 1—100.

Wilcock, G.K., Lilienfeld, S., Gaens, E., 2000. Efficacy and safety of galantamine in patients with mild to moderate Alzheimer's disease: multicentre randomised controlled trial. Galantamine International-1 Study Group. BMJ 321, 1445—1449. https://doi.org/10.1136/bmj.321.7274.1445.

Wilkinson, D., Murray, J., 2001. Galantamine: a randomized, double-blind, dose comparison in patients with Alzheimer's disease. Int. J. Geriatr. Psychiatr. 16, 852—857.

Zaganas, I., Schousboe, A., 2014. A neurologist with a lifetime fascination for neurochemistry: preface for the special issue dedicated to. Dr. Andreas Plaitakis. Preface. Neurochem. Res. https://doi.org/10.1007/s11064-013-1211-0.

Chapter 3.2.10

Guggulu [*Commiphora wightii* (Arn.) Bhandari.]

Jyoti Upadhyay[1], Sumit Durgapal[2], Arvind Jantwal[2], Aadesh Kumar[2], Mahendra Rana[3], Nidhi Tiwari[2]

[1]*School of Health Sciences, University of Petroleum and Energy Studies, Dehradun, Uttarakhand, India;* [2]*Department of Pharmaceutical Sciences, Kumaun University, Nainital, Uttarakhand, India;* [3]*Department of Pharmaceutical Sciences, Center for Excellence in Medicinal Plants and Nanotechnology, Kumaun University, Nainital, Uttarakhand, India*

Introduction

Alzheimer's disease (AD) is a neurodegenerative disorder which is progressive and manifested by cognitive and memory impairment, loss of speech, and changes in personality, and is a major cause of hospital admissions (Wilson et al., 2007). Several reports have highlighted the importance of oxidative stress and inflammation of neurons in the pathogenesis of Alzheimer's disease. Key agents involved in neuroinflammation include β-amyloid, responsible for superoxide generation and α-carbon-centered radicals; COX generates both prostaglandins and free radicals; induction of nitric oxide synthase (iNOS) is responsible for generating reactive nitrogen species (RNS) and nitric oxide (NO) (Butterfield et al., 2007a,b; Mancuso et al., 2007a,b). All the above agents cause substantial destruction of the central nervous system, in particular the entorhinal cortex and hippocampus during the early phase of AD. This damage over several years spreads through the frontal, temporal, and parietal cortex (Thompson et al., 2007). Drug therapy used for AD includes acetylcholinesterase inhibitors, i.e., donepezil, galantamine, rivastigmine, and the NMDA channel opener memantine, which delay the onset and development of dementia (Farlow and Cummings, 2007). A recent therapeutic approach for the treatment of AD has been proposed which is based on the hypothesis that the

Naturally Occurring Chemicals against Alzheimer's Disease. https://doi.org/10.1016/B978-0-12-819212-2.00026-8

heat-shock family of proteins, i.e., Hsps, has a neuroprotective effect against free radical-induced oxidative damage that causes injuries and antioxidants regulate Hsps in many cells involving neurons. Some researchers have proposed that antioxidants from natural sources like curcumin, resveratrol, and carnitine counteract the oxidative damage in the brain that leads to AD (Pettegrew et al., 2000; Ramassamy, 2006). However, the pharmacokinetics and bioavailability of these antioxidants that are proposed must be evaluated.

Pathophysiology of Alzheimer's disease (AD)

The etiology of AD remains unclear, however certain predominant contributing factors responsible for AD are amyloid plaques and neurofibrillary tangles associated with the degeneration of neurons and synapse dysfunction, causing irreversible and progressive deterioration of memory, and affecting cognition, personality, and language. The neurofibrillary tangles are a collection of phosphorylated tau proteins produced in excess. Phosphate molecules are generally present in tau proteins and hyperphosphorylation of these proteins causes twisting of proteins around one another and the formation of insoluble tangles that disrupt neuronal transport. Amyloid plaques are formed by β-amyloid proteins derived from amyloid precursor protein (APP). Secretase enzymes α, β, and ϒ-secretase cleave the amyloid precursor protein into fragments which are soluble and then clear the fragments. When there is impairment in the cleavage of APP by these enzymes, insoluble amyloid plaques are formed and accumulate in the brain resulting in neurotoxicity and cell death (Nikl et al., 2019). Cholinergic dysfunction is another possible mechanism involved in the pathogenesis of AD, as it involves a decline in the activity of choline acetyl transferase enzyme activity. Also, loss of cholinergic neurons, muscarinic receptors, and nicotinic receptors is involved in AD pathogenesis. Free radical-induced oxidative stress is the major cause of Alzheimer's disease.

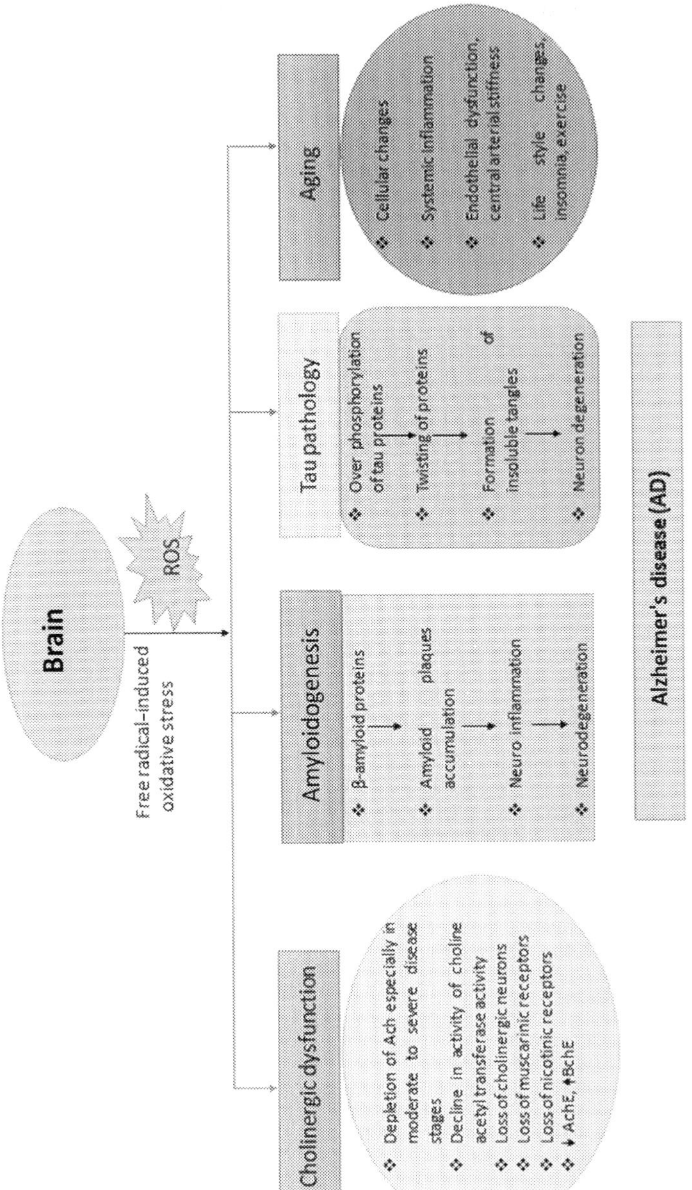

Pathophysiology of Alzheimer's disease.

Antioxidants and Alzheimer's disease

In AD, there is progressive disorientation and memory loss, and further progression to dementia. Most cases are sporadic, with more than 5%−10% of cases are familial (Mancuso et al., 2007a,b). In AD, gross analysis of the brain shows atrophy of the cortical cells with wide sulci and narrow gyri appearing in the frontal, temporal, and parietal lobes. Microscopic examination reveals the presence of neurofibrillary tangles, neurite plaques, β-amyloid peptide, angiopathy of amyloid, Hirano bodies, and granulovacuolar degeneration (Mott and Hulette, 2005). Some studies have reported that oxidative stress plays a major role in the pathogenesis of AD (Butterfield and Lauderback, 2002; Butterfield et al., 2001; Lauderback et al., 2001). β-Amyloid-associated generation of free radicals initiates lipid peroxidation, oxidation of proteins, reactive oxygen species formation, and calcium ion accumulation intracellularly as well as in the mitochondria, leading to death of neurons (Varadarajan et al., 2000). Antioxidants prevent or modulate the effect of β-amyloids. Studies related to AD show that the brains of patients with mild cognitive impairment have elevated levels of lipid peroxidation and increased protein oxidation when compared with aged control groups (Keller et al., 2005; Butterfield et al., 2006). Recent evidence shows that reactive nitrogen species (RNS) and nitric oxide (NO) may directly or indirectly be involved in death of neurons (Butterfield et al., 2007a,b). Also, other studies have suggested that both Ras p21 activator protein (or Ras GTPase activating protein) and Ras p21-dependent MAP-kinase pathways are intensely induced in Alzheimer's disease (Luth et al., 2000) (Fig. 3.2.10.1).

FIGURE 3.2.10.1 Guggulu plant. *Courtesy of Dr. Laxman Rautela.*

Guggulu (*Commiphora wightii* (Arn.) Bhandari)

Guggulu is an oleo-gum resin exudate obtained from the stem of *Commiphora wightii* (family Burseraceae). This plant is generally known as the guggulu tree and is found in seasonally dry tropical parts of India, Pakistan, and Bangladesh. It is bushy tree, small in size, with thorny branches. Guggulu is a yellowish gum resin in small ducts present in the bark (Fig. 3.2.10.1). The gum resin is collected by tapping the tree and making an incision on the bark. A guggulu tree yields approximately 250–500 g of dry resin during the collecting season from May to June (Ayurvedic Pharmacopoeia of India, 2007). In India, guggulu has been used in traditional medicine systems in the treatment of hypercholesterolemia, atherosclerosis, arthritis, gout, rheumatism, inflammation, and obesity. The synonyms of guggulu include gugar and Indian bdellium (Indian Pharmacopoeia, 1996). Guggulu was mentioned as early as 3000–10,000 years ago in the ancient Vedas—the Indian Holy Scriptures for the treatment of human diseases. The plant was first introduced to the scientific world by the Indian researcher G.V. Satyavati in 1966 (Satyavati, 1966). She studied the effect of guggulu in rabbits for the treatment of a condition known as coating and obstruction of channels resembling the condition of atherosclerosis. In 1986, guggulu was approved as a hypolipidemic drug and entered the Indian pharmaceutical market (Satyavati, 1988). In the 1990s guggulu was used in the treatment and management of several cardiovascular complications such as atherosclerosis, hypolipidemia, etc. The characteristic features of guggulu include that the guggulu tree is 4–6 feet tall, foliage-free and thorny, with ash-colored bark, which flakes off into papery flakes showing green-colored bark underneath that produces fragrant resin. The plant is also described as a shrub. The branches are spinescent and spirally ascending, and young parts of the plant are glandular and pubescent. One to three foliate compound leaves are present, whereas leaflets have irregularly toothed margins, are sessile to subsessile, ovate and rhomboid in shape, with alternate phyllotaxy. The flowers of the plant are small, unisexual, polygamous, and brown to pink in color. The calyx have glandular hairs and form a cylindrical cap. Eight to ten bilobed stigmas are present. The fruit of the *Commiphora* is a drupe, acuminate in shape, red in color, and when ripened the fruits splits into two halves showing brightly colored pseudo-arils. These pseudo-arils differ from species to species.

Chemistry

Guggulu contains steroids, terpenoids (di and tri), long-chain aliphatic tetrols and esters, lignans, ferulates, carbohydrates, and several inorganic ions and in addition a small quantity of sesamin and unknown constituents are also present. The oleo-gum resin also yields essential oil whose principal components are dimyrcene, myrcene, and polymyrcene (Bhati, 1950). Other components present in the oil are α-pinene, D-limonene, cineole, eugenol, linalool, bornyl acetate, α-terpineol, methylheptanone, geraniol, D-α-phellandrene, and some other unknown compounds. Cadinene, a bicyclic sesquiterpene has also been reported in gum resin of guggulu (Saxena and Sharma, 1998). Diterpenoids α-camphorene, cambrene-A, cemberene, and other cembrenoids are also found in guggulu. Mukulol, a cambrene alcohol, is also isolated from the aerial parts of guggulu oleo-gum resin. The main flavonoid constituent present in the flowers of guggulu was identified as quercetin (Sarup et al., 2015). The hypolipidemic activity of oleo-gum resin guggulu is associated with ketonic compounds, including a number of steroids including two isomers Z and E guggulsterone (*cis* and *trans*). Pharmacological investigations into these steroids have revealed that these isomers have profound hypolipidemic activity (Bajaj and Dev, 1982).

Antioxidant and antiinflammatory activity (preclinical and clinical studies)

Preclinical studies

Several scientific investigations have been carried out to determine the effects of medicinal herbs on Alzheimer's disease. Herbs showing potential antioxidant activities and antiinflammatory effects may have beneficial effects in Alzheimer's disease. Herbs with antiinflammatory potential may reduce inflammation of brain tissues. The active component Z-guggulsterone obtained from gum resin has been shown to enhance cognition and memory by activating the signaling pathway in the brain, i.e., brain-derived neurotrophic factor. Chen et al. conducted a study on mice to see the effect of Z-guggulsterone in scopolamine-induced memory impairment. Their study showed that pretreatment with z-guggulsterone reversed ascopolamine-induced increase in activity of enzyme acetylcholinesterase (AChE) and also reduced BDNF (brain-derived neurotrophic factor) protein expression, CREB (cAMP response element-binding protein), phosphorylation of protein kinase B (Akt) in the hippocampus and cortex region and also ERK1/2 (extracellular regulated kinase 1/2). These findings evaluated the memory-enhancing effect of Z-guggulsterone (Chen et al., 2016). Chronic stress impairs the functioning of CREB-BDNF signaling in the medial prefrontal cortex (mPFC) and hippocampus (Jiang et al., 2015). Huang et al. (2016) reported in their study that Z-guggulsterone causes a reduction in behavioral abnormalities caused by neuroinflammation in the tail suspension test (TST) and forced swimming test (FST). Liu et al. illustrated that Z-guggulsterone improves astrocyte-facilitated neuroinflammation after ischemia by restricting toll-like receptors. Toll-like receptors play a major role in stimulation and regulation of immune/inflammatory responses. Among them TLR 4 shows strong regulatory effects on inflammatory responses after ischemia by activation of transcriptional factors (Liu et al., 2018).

Free radicals cause oxidative stress, particularly in age-related disorders such as Alzheimer's disease. Oxidative damage occurs in the brain at an early stage in AD patients and its impact can be observed even before plaque pathology is initiated. Free radicals can cause neuronal death and neurodegeneration at the cellular and tissue levels (Evans et al., 1989). Several studies have investigated the association between amyloid precursor protein, cholesterol, and AD. Cholesterol is found to be an important modulator of the functional and physiochemical activity of the cell membrane that plays a major role in the regulation of neuronal plasticity. Guggulu is found to lower the cholesterol level, which contributes to its function of preventing AD development, as a low level of neuronal cholesterol causes inhibition of the Aβ-forming amyloid pathway, which is possible because of precursor protein removal from cholesterol (Butterfield and Boyd-Kimball, 2018). This association suggests that the lower cholesterol strategy of guggulu might influence the progression of AD.

It is suggested that excess production of nitric oxide causes oxidative damage to protein, DNA, and RNA, which are involved in the pathogenesis of several diseases including diabetes, cardiovascular disorders, arthritis, chronic inflammation, and neurodegenerative diseases. One study showed that E and Z-guggulsterone isomers show greater inhibitory activity against nitric oxide production caused by bacterial lipopolysaccharides (with IC50 values of 1.1 and 3.3 μM) in macrophages (Meselhy, 2003). These studies indicate that guggulsterone possesses therapeutic value in diseases related to oxidative stress like neurodegenerative diseases (e.g., AD). Some studies show the use of gugulipid as a positive control agent in evaluating the antioxidant, hypolipidemic, and cardioprotective activities of a number of synthetic compounds. Gugulipid was administered orally to rats at doses of 12.5−50 mg/kg for a duration of 30 days; it was observed that gugulipid decreases the total cholesterol level (i.e., 35%) and lipid peroxide level (i.e., 57%) in rats. It also significantly reduces the microsomal lipid peroxidation of liver (Batra et al., 2000). A mouse model experiment depicted the neuroprotective activity of guggulsterone. Mice were given streptozotocin (STZ) to induce neuronal damage and memory impairment. Parallel with this study, the glutathione enzyme level (GSH) in the brains of mice treated with gugulipid was also found to be significantly raised, suggesting oxidative stress inhibition in the brain by gugulipid (Saxena et al., 2007). Together, these studies reflect the antioxidant potential of guggulu and guggulsterone under several experimental conditions. Release of mediators causing inflammation and inflammatory signal pathway activation causes several diseases. Antiinflammatory agents have proved to be beneficial in delaying and preventing atherogenesis as the microenvironment in the atherosclerotic lesions is proinflammatory. Gugulipid is found to inhibit the inflammation caused by carrageenan in rat paw

(Duwiejua et al., 1993). Some research studies show the role of guggulsterone as an antiinflammatory agent in the treatment of inflammatory bowel syndrome induced by dextran sulfate sodium in mice (Cheon et al., 2006).

Clinical studies

The clinical trials related to the antiinflammatory and antioxidant activities of guggulu are very limited. A study showing cardioprotective effects of guggulu in combination with Ayurvedic Inularacemosa was conducted among 200 patients with ischemic heart disease and an abnormal electrocardiogram (ECG) (Singh et al., 1993; Miller, 1998). The treatment was given for 6 months and it was observed that the levels of triglycerides, total blood lipids, and cholesterol were reduced by 51%, 32%, and 39%, respectively, which is significantly consistent with the hypolipidemic effects of guggulu and inula. In 26% of the patients the normal electrocardiogram was restored and 56% of the patients showed improvement in the electrocardiogram at the end of the trials. These results indicate the cardioprotective effects of guggulu in ischemic patients, apparently through its antioxidant potential. Another clinical trial conducted in the United States investigated the median serum high-sensitivity reactive proteins hs-CRP level which was reduced by 29% in the subject group receiving a daily 2000 mg dose of gugulipid, while the hs-CRP level was found to be elevated in the group receiving a placebo drug (Szapary et al., 2003). The hs-CRP are acute-phase reactants produced in the liver in response to activation of cytokinins, which is an index of inflammation, and therefore this study investigated the antiinflammatory activity of guggulu. Another study examined the antiinflammatory activity in 30 patients suffering from arthritis of the knee. It was found that guggulu at a dose of 500 mg three times a day for the duration of 1 month improved the Western Ontario and McMaster Osteoarthritis Index (WOMAC) score and it continued to improve for 2 months (Singh et al., 2003).

Mechanism of action

Guggulsterone, a chemical constituent present in guggulu, shows its antioxidant effects by reversing both the production of enzyme xanthine oxidase and reduction of superoxide dismutase. Xanthine oxidase promotes the generation of reactive oxygen species, whereas superoxide dismutase is an antioxidant enzyme that catalyzes the reaction conversion of anion superoxide to hydrogen peroxide and oxygen. Nuclear factor NF-κB is a transcription factor thought to play a major role in regulation of various cellular functions like differentiation, proliferation, inflammation, immune responses, and apoptosis. Activation of this transcription factor can be achieved by proinflammatory molecule induction, such as tumor necrosis factor-α, interleukins, and PMA, i.e., phorbol myristate acetate. Also, NF-κB targets some genes which are responsible for inflammation like monocyte chemotactic proteins, interleukins, and motif ligands (C-X motif). This represents the regulatory role of NF-κB in inflammation. NF-κB, under resting conditions, is associated with the inhibitory unit in cytoplasm. When stimulated by several agents it is phosphorylated by I κB kinase enzyme (IKK) leading to nuclear translocation and stimulation of NF-κB target genes. Recent research investigated the inhibitory action of guggulsterone targeting NF-κB activation induced by several agents in various cell types. Such repression is mediated directly through inhibition of enzyme IKK by guggulsterone (Deng, 2007).

Safety issues

Guggulu has been used since ancient times for many health problems related to joints and bones, skin problems, and obesity. There are some side effects from the consumption of guggulu for a long period of time. Major side effects include headache, nausea, vomiting, hiccups, belching, loose stools, and an allergic reaction of the skin. Some major safety issues are discussed below (Richard, 2019).

1. Itching and skin rashes can be seen with those who have taken higher doses, such as 6000 mg per day (Richard, 2019).
2. Guggulu contains the plant phytosteroid guggulsterone, so it should be avoided by patients with hormone-sensitive conditions like breast cancer, ovarian cancer, uterine cancer, and prostate cancer.
3. According to preliminary in vitro studies, guggulsterones are found to decrease carcinogenic growth in head and neck cells from smokeless (chewing) tobacco (Macha, 2011).
4. Guggulu was found to be responsible for thyroid gland stimulation. Patients suffering from thyroid problems should consult with their doctors before using guggulu supplements(Tripathi et al., 1984).
5. For safety reasons, guggulu should not be used in pregnant and lactating women, or children (Richard, 2019).

Drug interactions

Guggulu is responsible for inhibiting blood clotting, and so should be avoided by those with bleeding disorders and who are using warfarin. Some other drugs that may have interactions with guggulu include the following:

1. Antifungal drugs like ketoconazole and itraconazole.
2. Antihistaminic drugs like fexofenadine.
3. An interaction was found with the β-blocker drug Propanolol and the calcium channel blocker drug diltiazem, used in the treatment of hypertension(Dalvi et al., 1994).

4. Sedatives drugs like triazolam and alprazolam could produce drug interactions with guggulu (Richard, 2019).
5. Hypolipidemic drugs like Lipitor atorvastatin and lovastatin may show drug interactions with guggulu (Richard, 2019).

Conclusion

Guggulu have been used in the treatment of cardiovascular diseases for many years, especially in the treatment of hyperlipidemia and atherosclerosis. Preclinical and clinical studies support the therapeutic claims of guggulu which were previously described in the Ayurvedic literature. Apart from clinical trials in cardiovascular diseases, clinical trials related to the antioxidant and anti-inflammatory activities of guggulu are very limited, although these activities have been performed in in vitro and in vivo preclinical studies. Oxidative stress and neuroinflammation are involved in the pathogenesis of Alzheimer's disease. The guggulu plant is has both antioxidant and antiinflammatory activities, and therefore it is used as a therapeutic agent in the treatment of Alzheimer's disease. Further studies are required to elaborate the possible mechanisms of action, bioavailability, and pharmacokinetics of guggulu in animal models as an antioxidant and antiinflammatory agent.

References

Bajaj, A.G., Dev, S., 1982. Chemistry of Ayurvedic crude drugs. Tetrahedron 38, 2949–2954.

Batra, S., Srivastava, S., Singh, K., Chander, R., Khanna, A.K., Bhaduri, A.P., 2000. Syntheses and biological evaluation of3-substituted amino-1-aryl-6-hydroxy-hex-2-ene-1-ones as antioxidant and hypolipidemic agents. Bioorg. Med. Chem. 8, 2195–2209.

Bhati, A., 1950. Essential oil from the resin of *Commiphora mukul*, Hook. Ex. stocks. J. Indian Chem. Soc. 27, 436–440.

Butterfield, D.A., Boyd-Kimball, D., 2018. Oxidative stress, amyloid-β peptide, and altered key molecular pathways in the pathogenesis and progression of Alzheimer's disease. J. Alzheimers Dis. 62, 1345–1367.

Butterfield, D.A., Drake, J., Pocernich, C., Castegna, A., 2001. Evidence of oxidative damage in Alzheimer's disease brain: central role for amyloid β -peptide. Trends Mol. Med. 7 (12), 548–554.

Butterfield, D.A., Lauderback, C.M., 2002. Lipid peroxidation and protein oxidation in Alzheimer's disease brain: potential causes and consequences involving amyloid β-peptide-associated free radical oxidative stress. Free Radic. Biol. Med. 32 (11), 1050–1060.

Butterfield, D.A., Poon, H.F., ST Clair, D., et al., 2006. Redox proteomics identification of oxidatively modified hippocampal proteins in mild cognitive impairment: insights into the development of Alzheimer's disease. Neurobiol. Dis. 22 (2), 223–232.

Butterfield, D.A., Reed, T.T., Perluigi, M., et al., 2007a. Elevated levels of 3-nitrotyrosine in brain from subjects with amnestic mild cognitive impairment: implications for the role of nitration in the progression of Alzheimer's disease. Brain Res. 1148, 243–248.

Butterfield, D.A., Reed, T., Newman, S.F., Sultana, R., 2007b. Roles of amyloid β -peptide-associated oxidative stress and brain protein modifications in the pathogenesis of Alzheimer's disease and mild cognitive impairment. Free Radic. Biol. Med. 43 (5), 658–677.

Cheon, J.H., Kim, J.S., Kim, J.M., Kim, N., Jung, H.C., Song, I.S., 2006. Plant sterol guggulsterone inhibits nuclear factor-kappaB signaling in intestinal epithelial cells by blocking IkappaB kinase and ameliorates acute murine colitis. Inflamm. Bowel Dis. 12, 1152–1161.

Chen, Z., Huang, C., Ding, W., December 2016. Z-guggulsterone improves the scopolamine-induced memory impairments through enhancement of the BDNF signal in C57BL/6J mice. Neurochem. Res. 41 (12), 3322–3332. https://doi.org/10.1007/s11064-016-2064-0.

Dalvi, S.S., et al., 1994. Effect of gugulipid on bioavailability of diltiazem and propranolol. J. Assoc. Physicians India.

Deng, R., 2007. Therapeutic effects of guggul and its constituent guggulsterone: Cardiovascular benefits. Cardiovas. Drug Rev. 25 (4), 375–390.

Duwiejua, M., Zeitlin, I.J., Waterman, P.G., Chapman, J., Mhango, G.J., Provan, G.J., 1993. Anti-inflammatory activity of resins from some species of the plant family Burseraceae. Planta Med. 59, 12–16.

Evans, P.H., Klinowski, J., Yano, E., Urano, N., 1989. Alzheimer's disease: a pathogenic role for aluminosilicate-induced phagocytic free radicals. Free Radic. Res. 6, 317–321.

Farlow, M.R., Cummings, J.L., 2007. Effective pharmacologic management of Alzheimer's disease. Am. J. Med. 120 (5), 388–397.

Huang, C., Wang, J., Lu, X., Hu, W., Wu, F., Jiang, B., Ling, Y., Yang, R., Zhang, W., 2016. Z-guggulsterone negatively controls microglia-mediated neuroinflammation via blocking IκB-α-NF-κB signals. Neurosci. Lett. 619, 34––42.

Indian Pharmacopoeia, 1996. The Controller of Publications. New Delhi, India.

Jiang, B., Huang, C., Chen, X.F., Tong, L.J., Zhang, W., 2015. Tetramethylpyrazine produces antidepressant-like effects in mice through promotion of BDNF signaling pathway. Int. J. Neuropsychopharmacol. 18.

Keller, J.N., Schmitt, F.A., Scheff, S.W., et al., 2005. Evidence of increased oxidative damage in subjects with mild cognitive impairment. Neurology 64 (7), 1152–1156.

Lauderback, C.M., Hackett, J.M., Huang, F.F., et al., 2001. The glial glutamate transporter, GLT-1, is oxidatively modified by 4-hydroxy-2-nonenal in the Alzheimer's disease brain: the role of Aβ1-42. J. Neurochem. 78 (2), 413–416.

Liu, T., Liu, M., Zhang, T., Liu, W., Xu, H., Mu, F., Ren, D., Jia, N., Li, Z., Ding, Y., Wen, A., Li, Y., 2018. Z-guggulsterone attenuates astrocytes mediated neuro-inflammation after ischemia by inhibiting toll like receptor 4 pathway. J. Neurochem. 147, 803–815.

Luth, H.J., Holzer, M., Gertz, H.J., Arendt, T., 2000. Aberrant expression of nNOS in pyramidal neurons in Alzheimer's disease is highly co-localized with p21ras and p16INK4a. Brain Res. 852 (1), 45–55. https://doi.org/10.1016/S0006-8993(99)02178-2.

Macha, M.A., et al., 2011. Guggulsterone targets smokeless tobacco induced P 13K/Akt pathway in ghead and neck cancer cells. PLoS One 6 (2), e14728.

Mancuso, C., Bates, T.E., Butterfield, D.A., Calafato, S., 2007a. Natural antioxidants in Alzheimer's disease. Expert Opin. Invest. Drugs 16 (12), 1921–1931.

Mancuso, C., Scapagnini, G., Curr, Ò.D., et al., 2007b. Mitochondrial dysfunction, free radical generation and cellular stress response in neurodegenerative disorders. Front. Biosci. 12, 1107–1123.

Meselhy, M.R., 2003. Inhibition of LPS-induced NO production by the oleo gum resin of *Commiphora wightii* and its constituents. Phytochemistry 62, 213–218.

Miller, A.L., 1998. Botanical influences on cardiovascular disease. Altern. Med. Rev. 3, 422–431.

Mott, R.T., Hulette, C.M., 2005. Neuropathology of Alzheimer's disease. Neuroimaging Clin. N. Am. 15 (4), 755–765.

Nikl, K., Castillo, S., Hoie, E., O Brien, K.K., 2019. Alzheimer's disease: current treatments and potential new agents. Neurology 44 (1), 20–23. https://www.uspharmacist.com/article/alzheimers-disease-current-treatments-and-potential-new-agents. (Accessed 10 July 2019).

Pettegrew, J.W., Levine, J., Mcclure, R.J., 2000. Acetyl-L-carnitine physical-chemical, metabolic, and therapeutic properties: relevance for its mode of action in Alzheimer's disease and geriatric depression. Mol. Psychiatry 5 (6), 616–632.

Ramassamy, C., 2006. Emerging role of polyphenolic compounds in the treatment of neurodegenerative diseases: a review of their intracellular targets. Eur. J. Pharmacol. 545 (1), 51–64.

Richard, N.F. (Ed.), 2019. Health Benefits of Guggul Ayurvedic remedy may aid in weight loss and cholesterol by Cathy Wong. Verywell Health. Updated on July, 11, 2020 at Verywell Health. https://www.verywellhealth.com/the-benefits-of-guggul-89567.

Sarup, P., Bala, S., Kamboj, S., 2015. Pharmacology and phytochemistry of oleo-gum resin of *Commiphora wightii* (guggulu). Scientifica 1–14.

Satyavati, G.V., 1966. Effect of an Indigenous Drug on Disorders of Lipid Metabolism with Special Reference to Atherosclerosis and Obesity (Medoroga) (M.D. thesis (Doctor of Ayurvedic Medicine)). Banaras Hindu University, Varanasi.

Satyavati, G.V., 1988. Gum Guggul (*Commiphora mukul*)–The success story of an ancient insight leading to a modern discovery. Indian J. Med. Res. 87, 327–335.

Saxena, G., Singh, S.P., Pal, R., Singh, S., Pratap, R., Nath, C., 2007. Gugulipid, an extract of *Commiphora wightii* with lipid-lowering properties, has protective effects against streptozotocin-induced memory deficits in mice. Pharmacol. Biochem. Behav. 86, 797–805.

Saxena, V.K., Sharma, R.N., 1998. Constituents of the essential oil from *Commiphora mukul* gum resin. J. Med. Aromat. Plant Sci. 20, 55–56.

Singh, B.B., Mishra, L.C., Vinjamury, S.P., Aquilina, N., Singh, V.J., Shepard, N., 2003. The effectiveness of *Commiphora mukul* for osteoarthritis of the knee: an outcome study. Altern. Ther. Health Med. 9, 74–79.

Singh, R.P., Singh, R., Ram, P., Batliwal, P.G., 1993. Use of Pushkar Guggul, an indeginousabti ischemic combination in the management of ischemic heart disease. Int. J. Pharmacogn. 31, 147–160.

Szapary, P.O., Wolfe, M.L., Bloedon, L.T., Cucchiara, A.J., DerMarderosian, A.H., Cirigliano, M.D., Rader, D.J., 2003. Guggul lipid for the treatment of hypercholesterolemia: a randomized controlled trial. J. Am. Med. Assoc. 290, 765–772.

The Ayurvedic Pharmacopoeia of India (Formulations), 2007. Department of Indian Systems of Medicine and Homeopathy, Ministry of Health and Family Welfare, Government of India, first ed. New Delhi, India.

Thompson, P.M., Hayashi, K.M., Dutton, R.A., et al., 2007. Tracking Alzheimer's disease. Ann. N.Y. Acad. Sci. 1097, 183–214.

Tripathi, Y.B., Malhotra, O.P., Tripathi, S.N., 1984. Thyroid stimulating action of Z guggulsterone obtained from *Commiphora mukul*. Planta Med. 50 (1), 78–80.

Varadarajan, S., Yatin, S., Aksenova, M., Butterfield, D.A., 2000. Review: Alzheimer's amyloid β-peptide-associated free radical oxidative stress and neurotoxicity. J. Struct. Biol. 130 (2–3), 184–208.

Wilson, R.S., Mccann, J.J., Li, Y., Aggarwal, N.T., Gilley, D.W., Evans, D.A., 2007. Nursing home placement, day care use, and cognitive decline in Alzheimer's disease. Am. J. Psychiatry 164 (6), 910–915.

Chapter 3.2.11

Lepidium meyenii

Amit Bahukhandi[1], Tanuj Joshi[2], Aadesh Kumar[2]
[1]*G.B. Pant National Institute of Himalayan Environment, Almora, Uttarakhand, India;*
[2]*Department of Pharmaceutical Sciences, Faculty of Technology, Kumaun University, Nainital, Uttarakhand, India*

Introduction

Lepidium meyenii a herbaceous plant, commonly known as maca, which belongs to the family Brassicae, and can be cultivated up to 4500 m above sea level (Quiros et al., 1996; Wang et al., 2007). The species has a typical structure and the leaves are approximately 12–20 cm in height, show dimorphism according to the reproductive stage, and are highly resistant to frost. The leaves are more prominent during the vegetative cycle and are continuously renewed from the center as the outer leaves die (Leon, 1964; Balick and Lee, 2002). The growth of the root starts 7 months after plantation and then initiates the development of an off-white, self-fertile flower on a central raceme. The species is planted in the months of September and October and are harvested in May to July (Wang et al., 2007). The plant species is a cash crop and a natural source of diverse bioactive constituents which have

Naturally Occurring Chemicals against Alzheimer's Disease. https://doi.org/10.1016/B978-0-12-819212-2.00027-X

capabilities to prevent or reduce several diseases. It is consumed both fresh and in dried form, and is used in traditional medicine systems (Quiros et al., 1996; Balick and Lee, 2002). These substances have shown antioxidant properties and a small portion is capable of reducing degenerative and neurodegenerative diseases (Bhatt et al., 2013; Bahukhandi et al., 2018). Today, the demands for potential plant species is increasing constantly, and they are important for the food industry and preparation of products for health promotion, and their nutritional/antioxidant properties also are gaining popularity all over the world. Giving consideration to their potential role, their cultivation is widespread both locally and globally, with their increased utilization in fresh or product forms. Today, maca has attracted a great deal of interest from consumers and researchers based on its nutritional and biochemical attributes.

Chemical composition of maca

L. meyenii is used as a food supplement and natural source of starch, polyphenolics, dietary fiber, alkaloids, protein, macamides, etc. (Korkmaz, 2018; Ribeiro and Carvalho, 2019; Wang and Zhu, 2019). The antioxidant effect depends on its various bioactive components (i.e., phenols, glucosinolates, alkamides, and polysaccharides) and a positive relationship with antioxidant activity was reported (Gan et al., 2017; Korkmaz, 2018). Lee and Chang (2019) found 25 chemical constituents from leaf and root portions of the species and categorized them as phenolics, flavonoids, saponins, steroids, amines, and alkybenzenes. They concluded that maca is a good source of gingerol (phenol), ergosterol peroxide (steroid), tanshinone I, panaxytriol, rotundifolioside (saponins), macaene, macamides, campesterol, sitosterol, and glucosinolates (Fig. 3.2.11.1) (Lee and Chang, 2019). Similarly, leaves of maca were reported to be natural sources of polysaccharides (MLP-1; molecular weight 42,756 Da and MLP-2; molecular weight 93,541 Da), and MLP-1 was recommended as a potential antioxidant agent (Caicai et al., 2018) and free amino acid, minerals, protein (10%–18%), and carbohydrate (59%–76%) were reported in the root (Zha et al., 2014). Researchers suggested that the presence of active constituents in plants act as antimicrobial, anticancer, antifatigue, etc. agents (Campos et al., 2013; Lee and Chang, 2019).

Several pieces of research on animal and human models have suggested that consumption of maca can enhance sexual performance, stimulate spermatogenesis, increase energy and growth rates, and reduce stress (Leon, 1964; Quiros et al., 1996; Balick and Lee, 2002; Gonzales et al., 2002; Valentova et al., 2006; Ye et al., 2019). Some reports have indicated that maca has the power to increase sexual performance in higher altitude regions in sheep and guinea pigs (Sandoval et al., 2002; Zheng et al., 2000). The fresh roots of maca contain 80% water and its dehydrated powder is the source of micronutrients,

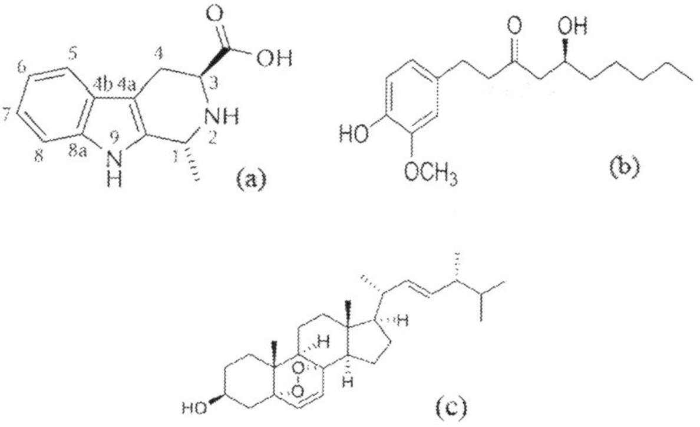

FIGURE 3.2.11.1 Chemical structure of (A) (1R,3S)-1-methyletrahydro-carboline-3-carboylic acid; (B) gingerol, and (C) ergosterol peroide in maca.

energy, carbohydrate, proteins, fibers, amino acids, etc. (Guo et al., 2016). Numerous factors such as soil composition, maca ecotype, harvesting time, extraction process, UV radiation, harsh environmental conditions, etc. influence the concentration and composition of antioxidant compounds and regulate the metabolic process in plants (Campos et al., 2013; Reyes et al., 2019).

Mechanism of action and pharmacological actions of *Lepidium meyenii*

Maca improves memory by increasing levels of acetylcholine (Wang et al., 2006). Animal studies have shown that it can increase memory retention and learning abilities in ovariectomized mice by its acetylcholinesterase-inhibitory and antioxidant activities (Rubio et al., 2011). In addition it improves fertility and increases sex drive. Researchers have performed several in vitro and in vivo studies for determination of the effects of maca on Alzheimer's disease and found it to be effective in decreasing complications associated with Alzheimer's disease in humans. The effect of maca on Alzheimer's disease is probably due to its immunostimulant action (Agarwal et al., 2013). A study was carried out for 35 days in albino mice in which scopolamine was administered at a dose of 1 mg/kg to induce learning and memory deficits. It was observed that *L. meyenii* aqueous and hydroalcoholic extract in different doses ameliorated learning, memory deficits, spatial learning, and memory impairments. A 1.5-fold increase in acetylcholinesterase (AChE) activity was produced due to scopolamine. Maca administration decreased brain AChE activity. In one study, mice was treated with ethanol and subjected to the Morris maze test. Administration of *L. meyenii* reversed the harmful effects of alcohol on mice in the Morris water maze test (Jivad and Rabiei, 2014).

Researchers investigating the neuroprotective effect of *L. meyenii* also have found a decrease in autophagy signaling and mitochondrial function which may be involved in age-related cognitive decline. Administration of *L. meyenii* to middle-aged mice led to cognitive function improvement which was associated with changes in autophagy signaling in mouse cortex and amelioration of mitochondrial activity. In this study administration of *L. meyenii* powder by gavage for a duration of 5 weeks in mice showed positive effects on motor coordination, endurance capacity, and cognitive functions. Also, improvements in respiratory function of the mitochondria and upregulation of proteins associated with autophagy in the cortex were observed. Improvements in mitochondrial function, cognitive function, and enhancement of autophagy-related proteins indicated that maca has the ability to reduce the effects of Alzheimer's disease (Guo et al., 2016). In one study, macamides (M 18:1, M 18:2, and M18:3) isolated from *L. meyenii* and ethanolic extracts of *L. meyenii* were evaluated against neurotoxicity induced in pheochromocytoma cells of rats by corticosterone (Yu et al., 2019). Corticosterone showed various effects, such as increased release of lactic acid dehydrogenase (LDH), increased intracellular levels of reactive oxygen species, decreased viability of cells, and decreased rates of matrix metalloproteinase (MMP). Macamides and extract of *L. meyenii* led to amelioration of neurotoxicity induced by corticosterone and the results reported an increase in viability of cells and a decrease in lactic acid dehydrogenase. In addition, certain other actions were also shown by macamides (M 18:1, M1 8:2, and M 18:3). M 18:2 and M 18:3 led to a decrease in Bcl-2-associated X protein (BAX)/B-cell lymphoma (Bcl)-2 ratios along with inhibitory effects on MMP reduction. The intracellular levels of reactive oxygen species (ROS) were decreased by M 18:1 without affecting any other factor. M 18:3 improved the expression levels of Bcl-2 and produced restraint in the expression level of important proapoptotic proteins like cytochrome C, Bax, caspase-3, etc. Similarly, M 18:3 prevented mitochondrial apoptosis induced by corticosterone (CORT) and increased the phosphorylation of AKT or protein kinase B. These effects of macamides may be beneficial for the treatment of Alzheimer's disease (Yu et al., 2019). Rubio and coworkers conducted a step-down avoidance test and Morris maze test for evaluation of learning and memory in ovariectomized mice. Step-down latency was increased in the group administered black maca as compared to the ovariectomized mice which did not receive treatment. Levels of malonaldehyde and acetylcholinesterase were decreased, whereas no effect was observed in the levels of monoamine oxidase in the brains of mice with administration of *L. meyenii*. Thus the study pointed out that *L. meyenii* showed beneficial effects in memory impairment in ovariectomized mice. The beneficial effects were probably due to acetylcholinesterase inhibitory activity or antioxidant activity of *L. meyenii* (Rubio et al., 2011).

Clinical studies of *Lepidium meyenii*

Maca has been cultivated by the natives of the Peruvian Central Andes for a very long time and increased research to determine its beneficial effects on human health is still required. Researchers conducted a survey on two populations living in the Peruvian Central Andes, where one of these populations consumed maca and the other did not. In the study a survey was conducted regarding consumption of maca, health status, sociodemographic aspects, and fractures occurring in males and females in the 35–75 years age group. In a subsample of the population, hemoglobin values, hepatic function, and kidney function were also evaluated. It was observed that 80% of the participants belonged to the region that consumed maca and of these 85% consumed it for nutritional benefits. Maca was helpful in reducing the rate of fractures and in treating the symptoms of chronic mountain sickness. It also proved helpful in lowering systolic blood pressure and body mass index. Those who consumed maca showed a normal lipid, hepatic, renal, and carbohydrate profile. Traditionally, many people believe that maca can improve mental performance in children and reduce anxiety and depression. Thus maca can be developed into a safe formulation that is useful in cognitive dysfunction associated with Alzheimer's disease (Zheng et al., 2000; Rubio et al., 2011; Gonzales, 2012; Yu et al., 2019).

Toxicity profile of *Lepidium meyenii*

The consumption of maca around the world has been taking place for centuries (Dini et al., 1994; Campos et al., 2013; Zha et al., 2014; Guo et al., 2016). The aqueous and methanolic extracts of maca have not demonstrated any signs of toxicity in in vitro studies. Likewise, in a study on preimplanted mouse embryos, aqueous extract of maca (1 g/kg body weight) did not show any toxic effects. Dried hypocotyls of yellow, black, and red maca have shown no toxicity in rats at a specific dose. When maca was administered to rats (1 g/kg body weight) there were no adverse effects observed and the histopathological conditions of their livers were similar to those of the control group. Maca administration for 90 days (0.6 g per day) in patients with metabolic syndrome led to a moderate increase in diastolic arterial pressure and AST levels. However, consumption of maca at a dose <1 g/kg is considered safe in humans (Gonzales, 2012; Agarwal et al., 2013; Yu et al., 2019).

Conclusion

Maca is a natural source of macro- and micronutrients and bioactive constituents. It contains starch, dietary fiber, protein, carbohydrates, amino acids, minerals, polyphenols, and phytosterols. Their regular consumption in the diet is helpful for strengthening and development of the body, and enhancement of energy, stamina, and endurance. Maca also helps to improve reproductive health, and has antioxidant, anticancer, hepatoprotective, and immunomodulator activities, which indicate its position as a superfood to reduce aging. This species improves memory by increasing the level of acetylcholine. Several research studies performed on in vitro and in vivo models have depicted its effectiveness in decreasing the effects of Alzheimer's disease due to its natural immune stimulant and hypothyroidism action. Researchers have also focused on investigating the neuroprotective effect of *L. meyenii* and found a decrease in autophagy signaling and mitochondrial function, which may be involved in age-related cognitive decline. The species can be utilized in the preparation of various useful products which also could improve the economic condition of the local Himalayan population.

References

Agarwal, P., Fatima, A., Alok, S., Singh, P.P., 2013. Herbal remedies for neurodegenerative disorder (Alzheimer's disease): a review. Int. J. Pharm. Sci. Res. 4, 3328–3340.

Bahukhandi, A., Sekar, K.C., Barola, A., Bisht, M., Mehta, P., 2018. Total phenolic content and antioxidant activity of *Meconopsis aculeata* Royle: a high value medicinal herb of Himalaya. Proc. Natl. Acad. Sci. India B Biol. Sci. 1–8.

Balick, M.J., Lee, R., 2002. Maca: from traditional food crop to energy and libido stimulant. Altern. Ther. Health Med. 8 (2), 96–98.

Bhatt, I.D., Rawat, S., Rawal, R.S., 2013. Antioxidants in medicinal plants. In: Biotechnology for Medicinal Plants. Springer, Berlin, Heidelberg, pp. 295–326.

Caicai, K., Limin, H., Liming, Z., Zhiqiang, Z., Yongwu, Y., 2018. Isolation, purification and antioxidant activity of polysaccharides from the leaves of maca (*Lepidium Meyenii*). Int. J. Biol. Macromol. 107, 2611–2619.

Campos, D., Chirinos, R., Barreto, O., Noratto, G., Pedreschi, R., 2013. Optimized methodology for the simultaneous extraction of glucosinolates, phenolic compounds and antioxidant capacity from maca (*Lepidium meyenii*). Ind. Crop. Prod. 49, 747–754.

Dini, A., Migliuolo, G., Rastrelli, L., Saturnino, P., Schettino, O., 1994. Chemical composition of *Lepidium meyenii*. Food Chem. 49 (4), 347–349.

Gan, J., Feng, Y., He, Z., Li, X., Zhang, H., 2017. Correlations between antioxidant activity and alkaloids and phenols of maca (*Lepidium meyenii*). J. Food Qual. 1–10.

Gonzales, G.F., Cordova, A., Vega, K., Chung, A., Villena, A., Gonez, C., Castillo, S., 2002. Effect of *Lepidium meyenii* (maca) on sexual desire and its absent relationship with serum testosterone levels in adult healthy men. Andrologia 34 (6), 367–372.

Gonzales, G.F., 2012. Ethnobiology and ethnopharmacology of *Lepidium meyenii* (maca), a plant from the Peruvian highlands. Evid. Based Complement. Altern. Med.

Guo, S.S., Gao, X.F., Gu, Y.R., 2016. Preservation of cognitive function by *Lepidium meyenii* (maca) is associated with improvement of mitochondrial activity and upregulation of autophagy-related proteins in middle-aged mouse cortex. Evid. Based Complement. Altern. Med. 1–9.

Jivad, N., Rabiei, Z., 2014. A review study on medicinal plants used in the treatment of learning and memory impairments. Asian Pac. J. Trop. Biomed. 4 (10), 780–789.

Korkmaz, S., 2018. Antioxidants in maca (*Lepidium meyenii*) as a supplement in nutrition. Antioxid. Foods Appl. 138–154.

Lee, Y.K., Chang, Y.H., 2019. Physicochemical and antioxidant properties of methanol extract from maca (*Lepidium meyenii* Walp.) leaves and roots. Food Sci. Technol.

Leon, J., 1964. The "maca" (*Lepidium meyenii*), a little known food plant of Peru. Econ. Bot. 18 (2), 122–127.

Quiros, C.F., Epperson, A., Hu, J., Holle, M., 1996. Physiological studies and determination of chromosome number in maca, *Lepidium meyenii* (Brassicaceae). Econ. Bot. 50 (2), 216–223.

Reyes, T.H., Scartazza, A., Pompeiano, A., Guglielminetti, L., 2019. Physiological responses of *Lepidium meyenii* plants to ultraviolet-B radiation challenge. BMC Plant Biol. 19 (1), 186.

Ribeiro, P.R., Carvalho, F.V., 2019. Structural diversity, biosynthetic aspects, and LC-HRMS data compilation for the identification of bioactive compounds of *Lepidium meyenii*. Food Res. Int. 108615.

Rubio, J., Qiong, W., Liu, X., Jiang, Z., Dang, H., Chen, S., Gonzales, F., 2011. Aqueous extract of black maca (*Lipidium Meyenii*) on memory impairment induced by ovariectomy in mice. Evid. Based Complement. Altern. Med. 1–5.

Sandoval, M., Okuhama, N.N., Angeles, F.M., Melchor, V.V., Condezo, L.A., Lao, J., Miller, M.J., 2002. Antioxidant activity of the cruciferous vegetable maca (*Lepidium meyenii*). Food Chem. 79 (2), 207–213.

Valentova, K., Buckiova, D., Kren, V., Peknicova, J., Ulrichova, J., Simanek, V., 2006. The *in vitro* biological activity of *Lepidium meyenii* extracts. Cell Biol. Toxicol. 22 (2), 91–99.

Wang, R., Yan, H., Tang, X., 2006. Progress in studies of huoerzine A, a natural cholinesterase inhibitor from Chinese herbal medicine. Acta Phamacol. Sin. 27 (1), 1–26.

Wang, S., Zhu, F., 2019. Chemical composition and health effects of maca (*Lepidium meyenii*). Food Chem. 288, 422–443.

Wang, Y., Wang, Y., McNeil, B., Harvey, L.M., 2007. Maca: an Andean crop with multi-pharmacological functions. Food Res. Int. 40 (7), 783–792.

Ye, Y.Q., Ma, Z.H., Yang, Q.F., Sun, Y.Q., Zhang, R.Q., Wu, R.F., Ren, X., Mu, L.J., Jiang, Z.Y., Zhou, M., 2019. Isolation and synthesis of a new benzylated alkamide from the roots of *Lepidium meyenii*. Nat. Prod. Res. 33 (19), 2731–2737.

Yu, Z., Jin, W., Cui, Y., Ao, M., Liu, H., Xu, H., Yu, L., 2019. Protective effects of macamides from *Lepidium meyenii* Walp. against corticosterone-induced neurotoxicity in PC12 cells. Royal Soc. Chem. Adv. 9 (40), 23096–23108.

Zha, S., Zhao, Q., Chen, J., Wang, L., Zhang, G., Zhang, H., Zhao, B., 2014. Extraction, purification and antioxidant activities of the polysaccharides from maca (*Lepidium meyenii*). Carbohydr. Polym. 111, 584–587.

Zheng, B.L., He, K., Kim, C.H., Rogers, L., Shao, Y.U., Huang, Z.Y., Lu, Y., Yan, S.J., Qien, L.C., Zheng, Q.Y., 2000. Effect of a lipidic extract from *Lepidium meyenii* on sexual behavior in mice and rats. Urology 55 (4), 598–602.

Chapter 3.2.12

Acorus calamus

Ajay Singh Bisht[1], Amit Bahukhandi[2], Mahendra Rana[3], Amita Joshi Rana[3], Aadesh Kumar[3]

[1]*Himalayan Institute of Pharmacy and Research, Dehradun, Uttarakhand, India;* [2]*G.B. Pant National Institute of Himalayan Environment Almora, Uttarakhand, India;* [3]*Department of Pharmaceutical Sciences, Kumaun University, Nainital, Uttarakhand, India*

Introduction

In India, the plant *Acorus calamus* Linn., generally known as sweet flag or "Bach," is a significant medicinal and aromatic herb with wide uses in almost all herbal-based systems (Mehrotra et al., 2003; Bahukhandi et al., 2013; Koeduka et al., 2019). Based on its morphological similarities with other members within the family, sweet flag has traditionally been placed in the family Araceae. It is a perennial herb with long, erect, narrow, aromatic leaves ascending from a spreading, buried rhizome. *Acorus* is approximately 6 feet long with sword-shaped leaves. It flourishes with small yellow/green flowers and has spreading rhizomes (Fig. 3.2.12.1) (Balakumbahan et al., 2010). It is a peculiar but widespread, semiaquatic plant found in amphibious habitats of moderate to submoderate temperature zones. On the ground it can be found as numerous plants above ground in a population probably arising from a single plant connected by an extensive underground rhizome. The color of the rhizome is a whitish pink internally, and it smells pleasantly aromatic and citrus, producing a bitter taste. The bloom consists of a leaf-like covering and nail-like spadix, originating from the center of the covering, heavily capped with yellow and green blossoms. On the basis of genome differences, the species has been redefined principally. *Acorus calamus* var. *americanus* (Raf.) Wulff contains two members which are distributed from North America to Siberia. The arid triploid *A. calamus* var. *calamus* (*vulgaris* L.), is scattered throughout Europe, temperate India, and the Himalayan region, and the tetraploid variety, *A. calamus* var. *angustatus* Bess., is found in eastern and tropical southern Asia including Japan and Taiwan (Raina et al., 2003; Balakumbahan et al., 2010; Bahukhandi et al., 2013).

Naturally Occurring Chemicals against Alzheimer's Disease. https://doi.org/10.1016/B978-0-12-819212-2.00028-1

FIGURE 3.2.12.1 Actual image of aerial part of *Acorus calamus* of Himalayan region of Uttarakhand.

The species is found growing plentifully in the wild, ascending to 2200 m in the Himalayas (Raina et al., 2003). *Acorus calamus* incorporates four cytotypes: diploid (2× = 24), triploid (3× = 36), tetraploid (4× = 48), and hexaploid (6× = 72) and is thought to have originated in India, spreading along commercial routes, where it was admired for its rhizome and volatile oil. The dried unpeeled rhizomes are steam distilled for extraction of volatile oil. The oil yield ranges from 0.85% to 2.5% depending on the type of calamus rhizomes collected from different sources. The most distinctive element of calamus oil is β-asarone. European and North American calamus are characterized by the presence of 3%–19% β-asarone in the rhizome oil and 31%–44% β-asarone in the leaf top oils (Raina et al., 2003; Mehrotra et al., 2003; Ahlawat et al., 2010). The Indian, Indonesian, and Taiwanese *A. calamus* contain up to 96% of β-asarone (Rost and Bos, 1979) in the rhizome oils and 60%–70% β-ararone in their leaf top oils, while the tetraploids of Japan and far-eastern Russian (East Siberia) are enriched by the presence of 10%–40% β-asarone in the rhizome and 20%–50% β-asarone in the leaf top oils. The essential oil is yellow to yellowish-brown in color, and is a somewhat syrupy liquid, possessing a burning aromatic flavor with a peculiar, warm, and slightly camphoraceous odor. Calamus oil consists of calamol, calameone, asarone, eugenol, methyl-eugenol, pinene, camphere, cineole, camphor, asaronaldehyde, calamenol, fatty acids, acorone, and isoacorone. The volatile oil of rhizomes is used in flavorings, particularly in liquors and perfumery. Calamus is also employed for medicinal preparations. The leaf oil is used for making aromatic vinegar. Dried and powered roots are used to make flavoring snuffs. It is also employed for scenting beer and to give it a clear appearance. The roots of these plants are sliced and boiled in maple syrup and when dried a delicious candid sweet dish is formed. In kwashiorkor disease of children, small pieces of roots are tied around the neck (Shreelaxmi et al., 2018).

Acorus calamus extracts have antirheumatic and pain-relieving activity. The extract of calamus is used in the dried form and as an oily substance in the treatment of upper respiratory tract problems such as cough, asthma, and bronchitis. In Ayurvedic medicine, the stem is thought to possess antispasmodic, carminative, and anthelmintic properties and is used for the treatment of epilepsy, chronic diarrhea, dysentery, bronchial catarrh, and abdominal tumors (CSIR, 1985). Calamus extract is also used for the treatment of kidney and liver problems, rheumatism, and eczema. Traditionally, calamus has been used in Indian and Chinese prescriptions for the treatment of memory disorder, grasping capacity, and antiaging activity (Mukherjee et al., 2007). The essential oil is used in the preparation of perfumes, flavors, and medicines. The essential oil is reported to be effective in bronchitis, heart, lungs, pain in the liver, kidney, etc. (Bown, 1995).

Phytochemical investigation of *Acorus calamus* L.

Phytochemical investigation of this species showed that it contains an essential oil or volatile, which is commonly known as "calamus oil," composed mainly of some major compounds like α- and β-pinene, α-terpineol, β-caryophyllene, linalool, α-bisabolol, α-asarone, β-asarone, etc. (Balakumbahan et al., 2010; Deepalakshmi et al., 2016). The rhizome (root) and leaves of *A. calamus* produce glycoside (Huxley, 1992), oxalic acid (Chopra et al., 1986), and volatile oil.

The chemical investigations of different parts of the plant, viz., leaves, underground part, and essential oil indicate the presence of different components, such as asarone (α and β), caryophylene, isoasarone, methyl isoeugenol, and safrol (Han-Zhang et al., 1998; Venskutonsis and Dagilyte, 2003; Zahin et al., 2009) Table 3.2.12.1 represents principal constituents of *A.Calamus* with their structure.

α- and β-asarone have been identified as the principal chemical constituents in rhizomes, roots, leaves, and essential oils that are responsible for almost all of the plant's pharmacological indications. Composition of the essential oil and especially β-asarone content (shown in Table 3.2.12.1) is claimed to be dependent on the ploidy level of the taxon with a high range, i.e., 70%–96% in tetraploids, relatively low around 5%–19% in triploids, and zero in diploids (Todorova and Ognyanov 1995; Subramanian et al., 2008). *A. calamus* found in India, Indonesia, and Taiwan resembles the "Indian type," which consists of up to 96% β-asarone (Rost and Bos, 1979; Zaiba et al., 1999) in the rhizome oils and 60%–70% β-asarone in the oils from aerial parts,

TABLE 3.2.12.1 Principal constituents of *A. calamus* with their structure.

Name of active constituents	Quantities present (%)	Structure
β-Asarone	42.5–78.4	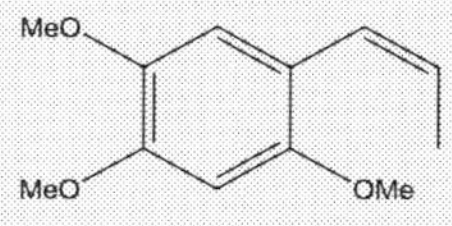
Isoeugenol	2.3–25.0	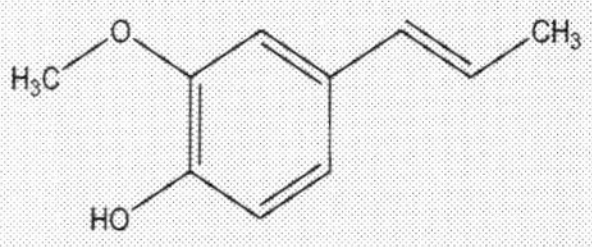
Calamenene	3.8–5.0	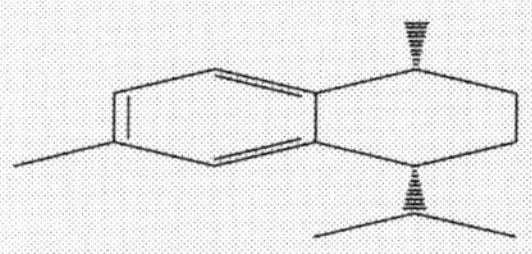
Calamene	3.8%	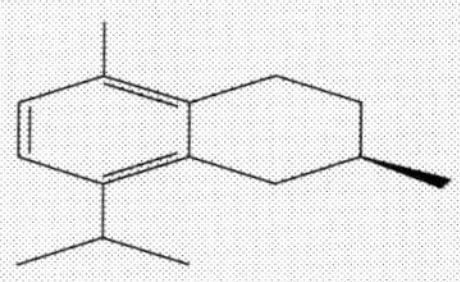
Methyl isoeugenol	2.8	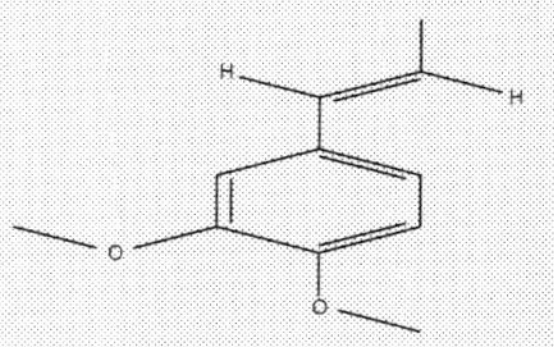
Methyleugenol	2.0	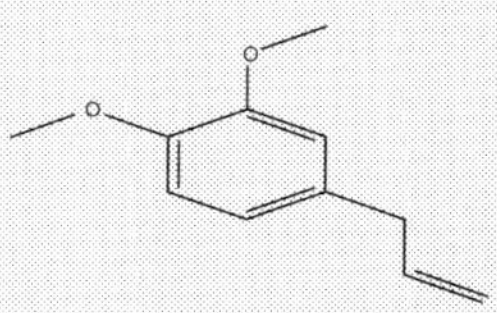

TABLE 3.2.12.1 Principal constituents of *A. calamus* with their structure.—cont'd

Name of active constituents	Quantities present (%)	Structure
α-Asarone	1.3–6.8	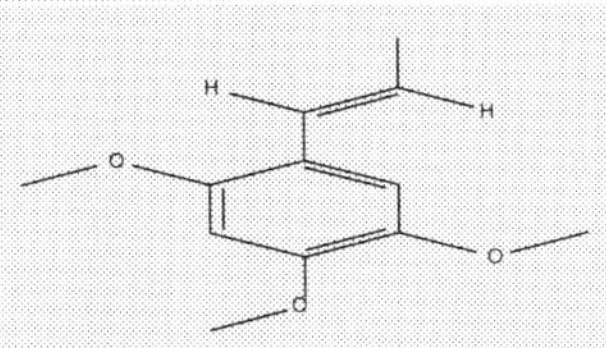

while 10%–40% β-asarone in rhizomes and 20%–50% β-asarone in the aerial part oils are contained by the tetraploids from Japan and far-east Russia (East Siberia). The larger amount of β-asarone in the rhizome oils and appreciable amounts of β-farnesene in the leaf top oils was found from tetraploid *A. calamus* from Thailand, Singapore, and Vietnam, which differs from the eastern Asiatic type. Gas chromatography (GC) and gas chromatography–mass spectroscopy (GC-MS) of essential oils obtained from leaves of *A. calamus* growing wild in 19 different locations was reported to show a total of 84 chemical constituents representing at least 86% of the essential oil. Reports showed the dominating presence of phenolic compounds, i.e., (Z)-asarone (15.7%–25.5%) and (Z)-methyl isoeugenol (2.0%–4.9%) in essential oils. Other major components were identified, viz., (E)-caryophyllene, a-humulene, germacrene, linalool, camphor, and isoborneol (Radusiene et al., 2007). Gyawali and SuKim (2009) studied the rhizomes of Nepalese *A. calamus* L for volatile organic compounds (VOCs) which was confined with the help of a simultaneous distillation-extraction (SDE) process and evaluated by GC-MS and reported a total of 50 VOCs so far related to chemical classes of aldehyde, alcohol, ester, furan, hydrocarbon, ketone, and N-containing miscellaneous. The yield from rhizomes of volatile compounds obtained by *A. calamus* was 7493.59 mg/kg. Raina et al. (2003) analyzed the rhizomes and leaf oils of *Acorus calamus* L. from the lower altitudes of the Himalaya region by GC and GC-MS methods and reported the identification of 29 and 30 constituents consisting of 99.7% of each of the β-asarone (83.2%) and α-asarone (9.7%) that were principal constituents in the rhizome oil, while β-asarone (85.6%) and linalool (4.7%) were reported as major components in the leaf oil from the rhizome and leaf oils, respectively. In China also, detailed study of leaf oil composition by the GC-MS method was carried out by Li and Jiang (1993) and the major constituents of the oils are given in Table 3.2.12.2.

TABLE 3.2.12.2 Major chemical constituents found in leaf oil of *A.calamus*.

Name of constituents	Quantity (%)	Name of constituents	Quantity (%)
(1,10)- (Z)-methyl isoeugenol	36.4	δ-cadinene	4.1
aristolen-2-one	6.6	isoacolamone	3.6
β-caryophyllene	4.1	epi-α-muurolol	3.2
acoragerm crone	4.1	shyobunone	3.2
(E)-methyl isoeugenol	3.4	α-muurolol	2.6

Pharmacology of *Acorus calamus* L.

The aromatic oil obtained by ethanolic extraction of the underground part is used by pharmaceutical industries (Bertea et al., 2005; Kumar et al., 2015: Olas and Brys, 2018). The pharmacological studies of *A. calamus* showed that the rhizome and its constituents, especially α- and β-asarone, have a broad range of pharmacological activities, i.e., antifungal (Lee et al., 2004), antibacterial (McGaw et al., 2002; Phongpaichit et al., 2005), allopathic (Nawamaki and Kuroyanagi, 1996), anticellular, and immunosuppressive (Mehrotra et al., 2003). A study by Phongpaichit et al. (2005) acknowledged the strong antimicrobial (antifungal and antiyeast) activity of the crude methanol extracts of *A. calamus* rhizome. In this study it was found that the methanolic extract of the rhizome showed strong activity against filamentous fungi, *Trichophyton rubrum*, *Microsporum gypseum*, and *Penicilium marneffei*. It is further well known that α- and β-asarones found in the whole plant of *A. calamus* are responsible for all types of antimicrobial activities of *A. calamus* (McGaw et al., 2002). Lee et al. (2004) investigated the fungicidal property of *A. gramineus* rhizome (methanol extract) against phytopathogenic fungi apart from human pathogenic microorganisms and this activity is associated with the existence of α-asarone and asaronaldehyde. Sulaiman et al. (2008) reported that the raw hexane extract of *A. calamus* showed a larvicidal activity against fourth-instar Ae. Shukla et al. (2002) indicated that ethanolic extract of *Acorus calamus* blocked acrylamide-induced hindlimb paralysis, depreciated GSH and GST, and raised dopamine receptors in the corpus striatum.

Essential oil of *Acorus calamus* was also found to acquire an antigonadal property in insects (Mathur and Saxena, 1975; Koul et al., 1977). Aqil et al. (2008) reported the antimutagenic potential of the plant; however they also found that the total phenolic content of the extract did not correlate with the antimutagenic activity in *A. calamus*. Flavonoides and phenolics are the most likely among the methanol extracts for producing the antimutagenic effect and countering the oxidative damage (Edenharder and Grunhage, 2003). The ethanolic extract of calamus produces powerful antisecretory, antiulcer, and cytoprotective activity in rats tested by giving an oral dose of 500 mg/kg, subjected to pyloric ligation, indometacin, reserpine, and cysteamine administration (Rafatullah et al., 1994). These results advocate the use of calamus in the treatment of gastropathy, which is adopted as a traditional medicine. Other than its medicinal uses, *Acorus* oil is used as a flavoring in the production of alcoholic beverages, liquors, vermouths, bitters, and in small amounts it can be used in foods such as frozen desserts, yogurts, bakery items, snacks, and confectionary (McGaw et al., 2002). Besides its various uses, the β-asarone content of *Acorus calamus* has been reported as toxic and unsafe on account of its tumor-producing activity (Lander and Schreier, 1990)—screening of different accessions for their asarone content offers a solution to this problem (Mazza, 1985; Lander and Schreier, 1990).

Cognitive impairment

Basically, cognition is the mental action and process of acquiring knowledge and understanding by thought, experience, and senses, including awareness, perception, reasoning, and judgment. Cognitive behavior can be categorized into memory, attention, creativity, and intelligence as functions. It is intuitive in nature and influenced by various aspect including aging, hypertension, stress, and numerous pathological indications such as dementia related to Parkinson's disease (PD), **Alzheimer's disease (AD)**, schizophrenia, cancer, and HIV (Ringman and Cummings, 2006; Lanni et al., 2008). Amplification of the basic functions of the mind with improvement of the internal and external message transformation systems is known as cognitive enhancement. The concept of cognitive enhancement, including learning and memory capacity, could be achievable for people with normal age-related decline and in healthy people, although the effects of these cognitive enhancers are not significant.

Numerous AChEIs, such as rivastigmine, donepezil, and galanthamine, are used for the management of mild to moderate AD. All discussed compounds have already been proven effective in healthy aged people for increasing learning and memory (Lanni et al., 2008; Narashi et al., 2004).

Cholinergic system and cognitive impairment: mechanism of action

The possible mechanism of action would be based on the cholinergic hypothesis of AD in which there are presynaptic losses in the brain of Alzheimer's disease patients and the performance of ACh in animal and human nature is established; this hypothesis concluded that cholinomimetic drugs would enhance cognitive function (Paul et al., 1999).

It has long been recognized that the cholinergic system is critically involved in the control of cognition (Everitt and Robbins, 1997). Agents that block muscarinic cholinergic receptors (mAChRs) disrupt cognitive functions and cause a transient loss of short-term memory (Drachman and Leavitt, 1974; Roldan et al., 1997; Coyle et al., 1983). Drugs that potentiate central cholinergic functions enhance short-term memory and alleviate memory deficiencies caused by muscarinic blockers.

Clinical trials and safety issues related to *A. calamus* in AD

Acorus calamus is a traditionally well-known medicinal plant used in the treatment of various ailments. Shah et al. (2012) studied the rational toxicity of an ethanolic extract of *Acorus calamus* Linn. rhizomes in Wistar rats. For an acute toxicity study, ethanolic extract was given to female Wistar rats by the oral route at doses of 175, 550, 1750, and 5000 mg/kg body weight with respect to OECD 425. Animals were checked regularly during the first 24 h after administration of the extract, and on a daily basis thereafter for 14 days. Also in a chronic toxicity study, the ethanolic extract of *Acorus calamus* was given orally at doses of 0, 200, 400, and 600 mg/kg body weight daily for 90 days to Wistar rats. The results were observed on the basis of clinical signs including total body weight, food intake, organ weight, blood profile, clinical biochemistry, as well as histology. There was no mortality, abdominal breathing, piloerection, and tremor-like clinical signs were detected for 30 min in rats dosed with 1750 and 5000 mg/kg body weight of alcoholic extract. There was no statistically significant change in body weight and feed consumption. No change was observed in hematological and biochemical parameters. Pathologically, neither gross abnormalities nor histopathological changes were observed. It was concluded that the ethanolic extract of *A. calamus* does not appear to have any toxicity on acute or chronic administration in Wistar rats (Shah et al., 2012).

Morales et al. performed an acute toxicity study on the chief constituent, α-asarone, at different doses (150, 200, 250, 300, and 350 mg/kg b.w.; i.p.) in male BALB/c mice for 14 days. Most recurrent clinical signs like ptosis, ataxia, piloerection, and dyspnea were observed among the animals (Morales-Ramiez and Madrigal-Bujaidar, 1992). Liu et al. studied the effect of β-asarone at different doses (0.5, 0.75, 1, 1.25, 1.5, 1.75, and 2 g/kg b.w.; i.v.) on BALB/c mice to discover the preliminary toxicity profile. They found that the animals that received β-asarone did not show any behavioral changes within 24 h of dosing. In a long-term toxicity study for 90 days at dosages of 10, 20, and 50 mg/kg/d, the results showed a dose-dependent toxicity with an increase in the number of white blood cells (WBCs), and red blood cell (RBC) count was decreased at 10 mg/kg dose. Serum K^+ levels also decreased at a dose of 20 mg/kg. A subacute toxicity study was performed on Sprague-Dawley rats treated with β-asarone (100 mg/kg/d b.w.; i.p.) displayed characteristic weight loss and decreased food consumption. Furthermore, the weights of the heart and thymus were reduced and the adrenal glands weight was increased. Additionally, no significant changes were found in hematological and biochemical parameters inducing hepatotoxicity (European-Commission 2002). In another study either β-asarone or calamus oil (400, 800, or 2000 mg/kg) was given orally as part of the diet for a 2-year feeding study in which Osborne-Mendel found that the rats developed an increased incidence of leiomyosarcomas in the small intestine. By microscopic analysis cardiac atrophy, fat infiltration, and fibrosis in the heart were also found in these animals. Further, studies were carried out with calamus oil obtained from Jammu; rats fed with the Jammu oil of calamus (0, 50, 100, or 5000 mg/kg) for 2 years developed early mortality, and liver and heart lesions at a dose of 5000 mg/kg and Cl^- increased in 50 mg/kg treated mice (Liu, 2013).

Role of *Acorus calamus* in AD

On the basis of experiments as well as clinical evidence, the central cholinergic system is considered as the most important neurotransmitter involved in the regulation of cognitive functions (Vanderwolf, 1988; Blockland, 1995). In the pharmacological data, there are thousands of pieces of evidence which prove the involvement of both muscarinic and nicotinic acetylcholine receptors in the encoding of new memories (Erberk and Rezaki, 2007). By the Ellman colorimetric method it was previously confirmed that *Acorus calamus* roots showed prominent inhibitory action on AchE. Powerful enzyme inhibition was seen at a 200 μg/mL dose of extracts from *Acorus calamus* (Oh et al., 2004). The results of many studies have revealed that extract of *Acorus calamus* leaves in acetone and methanol have a promising effect on acetylcholine feedback at lower doses, which shows an AChE-inhibiting effect at 250−1000 μg/mL. These results were established to be rational with the earlier findings on anticholinesterase activities with methanol extracts of *Acorus calamus* rhizomes (Oh et al., 2004; Ahmed and Urooj, 2010).

Conclusion

An improvement in cognition, including learning and consciousness, is possible today for people who are suffering with normal age-related deterioration and in healthy people, although so far the effects of these cognition boosters are modest. Various clinical and experimental studies have depicted that the central cholinergic system is responsible for the most important aspects of neurotransmission in the regulation of cognitive function. The role of both muscarinic and nicotinic acetylcholine receptors in encoding new memories was proved by the pharmacological data and thousands of pieces of evidence from earlier studies. The strong inhibitory effect on AchE was shown by methanolic extracts of *Acorus calamus* roots. Compelling enzyme inhibition was recognized at 200 μg/mL for extracts of *Acorus calamus*. It was stated from the results of various studies that extract of *Acorus calamus* leaves in acetone and methanol apply a promising effect on acetylcholine feedback at lower doses, which shows an AChE-inhibiting effect at 250–1000 μg/mL. It has been discovered that most of the acetyl cholinesterase inhibitors are known to contain nitrogen and rich alkaloid content, which is responsible for the higher activities.

References

Ahlawat, A., Katoch, M., Ram, G., Ahuja, A., 2010. Genetic diversity in *Acorus calamus* L. as revealed by RAPD markers and its relationship with β-asarone content and ploidy level. Sci. Hortic. 124, 294–297.

Ahmed, F, Urooj, A., 2010. Anticholinesterase activities of cold and hot aqueous extracts of F. racemosa stem bark. Pharmacogn. Mag. 6 (22), 142–144.

Aqil, F., Zahin, M., Ahmad, I., 2008. Antimutagenic activity of methanolic extracts of four Ayurvedic medicinal plants. Indian J. Exp. Biol. 46, 668–672.

Bahukhandi, A., Rawat, S., Bhatt, I.D., Rawal, R.S., 2013. Influence of solvent types and source of collection on total phenolic content and antioxidant activities of *Acorus calamus* L. Natl. Acad. Sci. Lett. 36, 93–99.

Balakumbahan, R., Rajamani, K., Kumanan, K., 2010. *Acorus calamus*: an overview. J. Med. Plants Res. 4, 2740–2745.

Bertea, C.M., Azzolin, C.M.M., Bossi, S., Doglia, G., Maffei, M.E., 2005. Identification of an EcoR1 restriction site for rapid and precise determination of β-asarone-free *Acorus calamus* cytotypes. Phytochemistry 66, 507–514.

Blockland, A., 1995. Acetylcholine: a neurotransmitter for learning and memory. Brain Res. Brain Res. Rev. 21 (3), 285–300.

Bown, D., 1995. Encyclopedia of Herbs and Their Uses. Dorling Kindersley, London. ISBN- 0-7513-020-31.

Chopra, R.N., Nayar, S.L., Chopra, I.C., 1986. Glossary of Indian Medicinal Plants (Including the Supplement). Council of Scientific and Industrial Research, New Delhi.

Coyle, J.T., Price, D.L., Delong, M.R., 1983. Alzheimer's disease: a disorder of cortical cholinergic innervation. Science 11 (219), 1184–1190.

CSIR, 1985. The Wealth of India-Dictionary of Raw Materials and Industrial Products, vol. 1(A). CSIR, New Delhi, India, p. 63.

Deepalakshmi, P.D., Odgerel, K., Thirugnanasambantham, P., Yungeree, O., Khorolragchaa, A., Senthil, K., 2016. Metabolite profiling of *in vitro* cultured and field grown rhizomes of *Acorus calamus* from Mongolia using GC–MS. Chromatographia 79, 1359–1371.

Drachman, D.A., Leavitt, J., 1974. Human memory and the cholinergic system. Arch. Neurol. 30, 113–121.

Edenharder, R., Grunhage, D., 2003. Free radical scavenging ability of flavanoids as mechanism of protection against mutagenicity induced by test butyl hydroxide or cumen hydroperoxide in *Salmonella typhimurium* TA102. Mutat. Res. Genet. Toxicol. Environ. Mutagen 540, 1–18.

Erberk, O.N., Rezaki, M, 2007. Prefrontal cortex: Implications for memory functions and dementia. Turkish J. Psychiatry 18, 262–269.

European-Commission, 2002. Opinion of the Scientific Committee on Food on the Presence of β-asarone in Flavourings and Other Food Ingredients with Flavouring Properties, Scientific

Committee on Food, pp. 1–15. https://ec.europa.eu/food/sites/food/files/safety/docs/sci-com_scf_out111_en.pdf.

Everitt, B.J., Robbins, T.W., 1997. Central cholinergic systems and cognition. Annu. Rev. Psychol. 48, 649–684.

Gyawali, R., SuKim, K., 2009. Volatile organic compounds of medicinal values from Nepalese *Acorus calamus* L. Kathmandu Univ. J. Sci. Eng. Technol. 5, 51–65.

Han-Zhang, W.A.N.G., You-Gen, C.H.E.N., Cui-Sheng, F.A.N., 1998. Review of studies on chemical constituents and pharmacology of genus *Acorus* in China. Plant Divers. 20, 1–3.

Huxley, A., 1992. The New RHS Dictionary of Gardening. ISBN-0-333- 47494-5.

Koeduka, T., Hatada, M., Suzuki, H., Suzuki, S., Matsui, K., 2019. Molecular cloning and functional characterization of an O-methyltransferase catalyzing 4′-O-methylation of resveratrol in *Acorus calamus*. J. Biosci. Bioeng. 127, 539–543.

Koul, O., Saxena, B.P., Tikku, K., 1977. Mode of action of *Acorus calamus* L oil vapours on adult male sterility in red cotton bugs. Experientia 33, 29–31.

Kumar, A., Sharma, S., Verma, G., 2015. Insecticidal and genotoxic potential of *Acorus calamus* rhizome extract against *Drosophila melanogaster*. Asian J. Pharm. Clin. Res. 8, 113–116.

Lander, V., Schreier, P., 1990. Acorenone and c-asarone: indicators of the origin of calamus oils (*Acorus calamus* L.). Flavour Fragrance J. 5, 75–79.

Lanni, C., Lenzken, S.C., Pascale, A., 2008. Pharmacol. Res. 57, 196–213.

Lee, J.Y., Lee, J.Y., Yun, B.S., Hwang, B.K., 2004. Antifungal activity of β-asarone from rhizomes of *Acorus gramineus*. J. Agric. Food Chem. 52, 776–780.

Li, M.X., Jiang, Z.R., 1993. The volatile constituents of the essential oil and their distribution in *Acorus calamus* L. Chinese Trad. Herb. Drug 24, 459–461.

Liu, L., 2013. β-Asarone induces senescence in colorectal cancer cells by inducing lamin B1 expression. Phytomedicine 20 (6), 512–520.

Mathur, A.C., Saxena, B.R., 1975. Induction of sterility in male house flies by vapours of *Acorus calamus* L. oil. Naturwissenschaften 62, 576–577.

Mazza, G., 1985. Gas chromatographic and mass spectrometric studies of the constituents of the rhizome of calamus. The volatile constituents of the essential oil. J. Chromatogr. A 328, 179–194.

McGaw, L.J., Jager, A.K., Van Staden, J., 2002. Isolation of ß-asarone, an antibacterial and anthelmintic compound, from *Acorus calamus* in South Africa. South Afr. J. Bot. 68, 31–35.

Mehrotra, S., Mishra, K.P., Maurya, R., Srimal, R.C., Yadav, V.S., Pandey, R., Singh, V.K., 2003. Anticellular and immunosuppressive properties of ethanolic extract of *Acorus calamus* rhizome. Int. Immunopharm. 3, 53–61.

Morales-Ramiez, P., Madrigal-Bujaidar, E, et al., 1992. Sister-chromatid exchange induction produced by *in-vivo* and *in-vitro* exposure to alpha-asarone. Mutat. Res. 279 (4), 269–273.

Mukherjee, P.K., Kumar, V., Mal, M., Houghton, P.J., 2007. *Acorus calamus*: scientific validation of Ayurvedic tradition from natural resources. Pharmaceut. Biol. 45, 651–666.

Narashi, T., Moriguchi, S., Zhao, X., et al., 2004. Mechanism of action of cognitive enhancers on Neuroreceptors. Biol. Pharm. Bull. 27 (11), 1701–1706.

Nawamaki, K., Kuroyanagi, M., 1996. Sesquiterpenoides from *Acorus calamus* as germination inhibitors. Phytochemistry 43, 1175–1182.

Oh, M.H., Houghton, P.J., Whang, W.K., Cho, J.H., 2004. Screening of Korean herbal medicines used to improve cognitive function for anti-cholinesterase activity. Phytomedicine 11 (6), 544–548.

Olas, B., Brys, M., 2018. Is it safe to use Acorus calamus as a source of promising bioactive compounds in prevention and treatment of cardiovascular diseases? Chem. Biol. Interact. 281, 32–36.

Paul, T., Francis, A.M.P., Snape, M., Wilcock, G.K., 1999. The cholinergic hypothesis of Alzheimer's disease: a review of progress. J. Neurol. Neurosurg. Psychiatry 66, 137–147.

Phongpaichit, S., Pujenjob, N., Rukachaisirikul, V., Ongsakul, M., 2005. Antimicrobial activities of the crude methanol extract of *Acorus calamus* Linn. Songklanakarin J. Sci. Technol. 27, 517–523.

Radusiene, J., Judzentiene, A., Peciulyte, D., Janulis, V., 2007. Essential oil composition and antimicrobial assay of *Acorus calamus* leaves from different wild populations. Plant Genet. Resour. 5, 37–44.

Rafatullah, S., Tariq, M., Mossa, J.S., Al-Yahya, M.A., Al-Said, M.S., Ageel, A.M., 1994. Antisecretagogue, anti-ulcer and cytoprotective properties of *Acorus calamus* in rats. Fitotherpia 1, 19–23.

Raina, V.K., Srivastava, S.K., Syamasunder, K.V., 2003. Essential oil composition of *Acorus calamus* L. from the lower region of the Himalaya. Flavour Fragrance J. 18, 18–20.

Ringman, J.M., Cummings, J.L., 2006. Current and emerging Oharmacological treatment options for dementia. Behav. Neurol. 17 (1), 5–16.

Roldan, G., Bolanos-Badillo, E., Gonzalez-Sanchez, H., Quirarte, G.L., Prado-Alcala, R.A., 1997. Selective M1 muscarinic receptor antagonists disrupt memory consolidation of inhibitory avoidance in rats. Neurosci. Lett. 230, 93–96.

Rost, L.C.M., Bos, R., 1979. Biosystematic investigations with *Acorus*. Communication. Constituents of essential oils. Planta Med. 26, 350–361.

Shah, P., Ghag, S., Deshmukh, M., Kulkarni, P., Joshi, Y., Vyas, S., Bhavin, Shah, D.R., 2012. Toxicity study of ethanolic extract of Acorus calamus rhizome. Int. J. Green Pharm. 6, 29–35. https://doi.org/10.4103/0973-8258.97119.

Shreelaxmi, S.H., Ramachandra, C.T., Roopa, R.S., Hanchinal, S.G., 2018. Antimicrobial activity of supercritical fluid extracted *Acorus calamus* oil against different microbes. J. Pharmacogn. Phytochem. 7, 2836–2840.

Shukla, P.K., Khanna, V.K., Ali, M.M., Maurya, R.R., Handa, S.S., Srimal, R.C., 2002. Photother. Res. 16, 256–260.

Subramanian, R.B., Oza, V.P., Parmar, P., Mehta, S.R., 2008. Rapid determination of β-asarone-free *Acorus calamus* cytotypes by HPTLC. Curr. Trends Biotechnol. Pharm. 2, 506–513.

Sulaiman, S., Kamarudin, A., D.S.F., Othman, H., 2008. Evaluation of bifenthrin and *Acorus calamus* linn. extract against *Aedes aegypti* L. and *Aedes albopictus* (Skuse). J. Arthropod-Borne Dis. 2, 7–11.

Todorova, M.N., Ognyanov, I.V., 1995. Chemical composition of essential oil from Mongolian *Acorus calamus* L. rhizomes. J. Essent. Oil Res. 7, 191–193.

Vanderwolf, C.H., 1988. Cerebral activity and behavior: control by central cholinergic and serotonergic system. Int. Rev. Neurobiol. 30, 225–340.

Venskutonsis, P.R., Dagilyte, A., 2003. Composition of essential oil of sweet flag (*Acorus calamus* L.) leaves at different growing phases. J. Essent. Oil Res. 15, 313–318.

Zahin, M., Aqil, F., Ahmad, I., 2009. The *in vitro* antioxidant activity and total phenolic content of four Indian medicinal plants. Int. J. Pharm. Pharmaceut. Sci. 1, 88–95.

Zaiba, I.A., Beg, A.Z., Mahmood, Z., 1999. Antimicrobial potency of selected medicinal plants with special interest in activity against phytopathogenic fungi. Indian Vet. Med. J. 23, 299–306.

Further reading

ElayaRaja, A., Vijayalakshmi, M., Devalara, G., 2009. *Acorus calamus* linn.: chemistry and biology. Res. J. Pharm. Technol. 2 (2), 256–261.

Farnsworth, N.R., Soejarto, D.D., 1991. Global importance of medicinal plants. In: Akerele, O., Heywood, V., Synge, H. (Eds.), The Conservation of Medicinal Plants, vol. 26. Cambridge University Press, Cambridge, UK, pp. 25–51.

Ghosh, M., 2006. Antifungal properties of haem peroxidase from *Acorus calamus*. Ann. Bot. 98, 1145–1153.

Hamilton, A.C., 2004. Medicinal plants, conservation and livelihoods. Biodivers. Conserv. 13, 1477–1517.

Kala, C.P., Dhyani, P.P., Sajwan, B.S., 2006. Developing the medicinal plants sector in northern India: challenges and opportunities. J. Ethnobiol. Ethnomed. 2, 1–15.

Kamboj, V.P., 2000. Herbal medicine. Curr. Sci. 78, 35–39.

Laird, S.A., Ten Kate, K., 2002. Linking biodiversity prospecting and forest conservation. In: Pagiola, S., Bishop, J., Landell-Mills, N. (Eds.), Selling Forest Environmental Services-Market-Based Mechanisms for Conservation and Development, pp. 151–172.

Motley, T.J., 1994. The ethnobotany of sweet flag, *Acorus calamus* (Araceae). Econ. Bot. 48, 397–412.

Nalamwar, V.P., Khadabadi, S.S., Aswar, P.B., Kosalge, S.B., Rajurkar, R.M., 2009. In vitro licicidal activity of different extracts of *Acorus calamus* Linn. (Araceae) Rhizome. Int. J. Pharm. Tech. Res. 1, 96–100.

Narmatha, K., Arul, P., Jeyamurugan, S., Srinivasan, R., Vignesh, T., 2018. Standardisation of rhizome extract of *Acorus calamus* linn. Asian J. Res. Chem. 11, 423–426.

Paithankar, V.V., Belsare, S.L., Charde, R.M., Vyas, J.V., 2011. *Acorus calamus*: an overview. Int. J. Biomed. Res. 2 (10), 518–529.

Satheesh Kumar, N., Mukherjee, P.K., Bhadra, S., Saha, B.P., 2010. Acetyl cholinesterase enzyme inhibitory potential of standardized extract of *Trigonella foenum graecum* L and its constituents. Phytomedicine 17, 292–295.

Surveswaran, S., Cai, Y.Z., Corke, H., Sun, M., 2007. Systematic evaluation of natural phenolic antioxidant from 133 Indian medicinal plants. Food Chem. 102, 938–953.

Wojdylo, A., Oszmianski, J., Czemerys, R., 2007. Antioxidant activity and phenolic compounds in 32 selected herbs. Food Chem. 105, 940–949.

Chapter 3.2.13

Tinospora cordifolia

Osama M. Ahmed
Physiology Division, Zoology Department, Faculty of Science, Beni-Suef University, Salah Salem St., Beni-Suef, Egypt

Plants and their extracts

Tinospora cordifolia

Classification, description, and distribution

Tinospora cordifolia, commonly known as guduchi, belongs to kingdom Plantae, class Magnoliopsida, order Ranunculales, family Menispermaceae, genus *Tinospora*, and species *cordifolia* (Fig. 3.2.13.1) (Spandana et al., 2013; Reddy and Rajasekhar, 2015).

T. cordifolia is a perennial, climbing, deciduous, fleshy, and robust shrub with succulent stem and papery bark (Fig. 3.2.13.1). The leaf of the plant is heart-shaped, membranous, alternate, long petiolate, cordate, exstipulate, and glabrous with multicoated reticulate venation. The plant stems are fibrous and the cross-section in such stems has radially located V-shaped wood bundles with canals, separated with medullary rays. The bark is gray to white in color. The plant's flower is in an axillary position, 2- to 9-cm-long raceme on leaflet branches, small, unisexual, and yellow in its color. The female is usually solitary and male flowers are clustered. The plant seed is curved. The fruit is succulent and single-seeded and red when ripe. The flower grows during the summer, while fruit develops during the winter (Reddy and Rajasekhar, 2015; Sharma et al., 2019; Shetty and Singh, 2010; Upadhyay et al., 2010; Tiwari et al., 2018).

It is widely distributed in various countries including China, Vietnam, India, Myanmar, Philippines, Sri Lanka, Thailand, Malaysia, Borneo, Indonesia, Burma, Bangladesh, Ceylon, South Africa, West Africa, and North Africa (Pendse et al., 1981; Jain et al., 2010; Mia et al., 2009; Singh et al., 2003; Mutalik and Mutalik, 2011; Joshi and Kaur, 2016).

Naturally Occurring Chemicals against Alzheimer's Disease. https://doi.org/10.1016/B978-0-12-819212-2.00029-3

FIGURE 3.2.13.1 Pictures of various parts of *Tinospora cordifolia*: (A) stems; (B) roots; (C) leaves; (D) flowers; (E) fruits (Sharma et al., 2019). *(Under permission from Elsevier; License Number: 4751540556057 at January 17, 2020).*

Chemical constituents

As reported by different previous publications, the chemical components found in *T. cordifolia* belong to various classes of compounds such as alkaloids, glycosides, diterpenoid lactones, aliphatic compounds, steroids, aliphatic compounds, sesquiterpenoids, and polysaccharides (Tiwari et al., 2018; Singh et al., 2003).

Palmatine and barberine are major alkaloids found in the stem of *T. cordifolia*. Also, glucosides such as furanoid diterpene glucosides, tinocordisides, 18-norclerodane glucosides, tinocordifoliosides, tinocordifolins, sesquiterpenes tinocordiosides, cordiosides, cordifoliosides (A and B), syringin, syringin apiosyl glycosides, cordifolisides (A, B, C, D and E), tinosponone, and palmatosides (C and F) are important constituents in the stem (Spandana et al., 2013; Pendse et al., 1981; Rana, 2014). Steroids such as giloinsterol, hydroxyl ecdysone, ecdysterone, δ-sitosterol, β-sitosterol, and makisterone A were also reported to be found in the stem (Spandana et al., 2013; Rana, 2014; Sanklala et al., 2012). Aliphatic compounds such as octacosanol, heptacosanol, and nonacosan-15-one are important constituents

(Rana, 2014). Furthermore, the stem contains immunologically active substances (1,4)-alpha-D-glucan and arabinogalactan (Chintalwar et al., 1999).

Tinosporin, tembetarine, magnoflorine, choline, jatrorrhizine, isocolumbin, tetrahydropalmatine, and palmatine are major alkaloids in the root of *T. cordifolia* (Spandana et al., 2013; Sanklala et al., 2012).

Furanolactone, clerodane derivatives, [(5R,10R)-4R-8R dihydroxy-2S-3R:15,16-diepoxycleroda-13 (16), 14-dieno-17,12S:18,1S-dilactone], tinosporon, β-sitosterol, tinosporides, jateorine, columbin, δ-sitosterol, 20-β-hydroxy ecdysone, octacosanol, tinosporidine, heptacosanol, miscellaneous nonacosan-15-one3, (α,4-di hydroxy-3-methoxy-benzyl)-4-(4-hydroxy-3-methoxy-benzyl)-tetrahydrofuran, cordifol, cordifelone, N-transferuloyl tyramine as diacetate, giloinin, giloin, and tinosporic acid were reported in whole plant (Spandana et al., 2013; Antul et al., 2019). The β-sitosterol, β-hydroxy ecdysone, and δ-sitosterol were stated to be found in the *T. cordifolia* aerial parts (Sanklala et al., 2012).

T. cordifolia contains high fiber (15.9%), sufficient carbohydrates (61.66%), sufficient proteins (4.5%−11.2%), and low fats (3.1%). The herb's nutritive energy calories are 292.54/100 g. It also contains enough iron (0.28%), enough calcium (0.131%), high chromium (0.006%), and high potassium (0.845%), all of which are important in different regulatory functions (Mutalik and Mutalik, 2011; Nile and Khobragade, 2009).

Medicinal uses and biologic activities

T. cordifolia has been used in Ayurvedic preparations throughout the centuries worldwide for the treatment of various ailments like general debility, fever, secondary syphilis, dyspepsia, gonorrhea, urinary diseases, chronic diarrhea, impotency, cardiac disease, skin diseases, gout, viral hepatitis, jaundice, dysentery, helminthiasis, diabetes, leprosy rheumatoid arthritis, and many more diseases (Reddy and Rajasekhar, 2015; Sharma et al., 2019; Tiwari et al., 2018; Joshi and Kaur, 2016).

The important medicinal properties of *T. cordifolia* are antiinflammatory, antiarthritic, antioxidant, antiperiodic, antispasmodic, antiallergic, antidiabetic, antimicrobial, antistress, antileprotic, antimalarial, hepatoprotective, antineoplastic, immunomodulatory, anticancer, antitoxic, antihyperglycemic, antihyperlipidemic, antituberculosis, antiosteoporotic, antiangiogenic, antimalarial, and cognition (learning and memory) activities (Reddy and Rajasekhar, 2015; Sharma et al., 2019; Singh et al., 2003; Saha and Gosh, 2012; Pandey et al., 2012; Dwivedi, 2016; Agarwal et al., 2019).

T. cordifolia *against Alzheimer disease and other related memory disorders*

In traditional medicine, *T. cordifolia* is commonly used as a memory and learning enhancer (Tillotson et al., 2001; Roy, 2018). It was reported by many publications that it has memory enhancement properties in normal as well as in memory-deficient animals (Roy, 2018; Singhal et al., 2012; Gupta et al., 2013; Dileepkumar et al., 2014; Mishra et al., 2016; Akram and Nawaz, 2017; Roy et al., 2017; Sharma and Kaur, 2018). However, clinical studies in human beings that confirm these findings are very scarce.

Preclinical studies revealed that the treatment with *T. cordifolia* ameliorates cognitive functions in rat models of sleep deprivation (Mishra et al., 2016) and drug-induced amnesia (Gupta et al., 2013). These improvements in cognitive functions were reported to be owing to the suppressive effects on inflammation and cell death (Mishra et al., 2016; Sharma and Kaur, 2018).

It was found by Agarwal et al. in 2002 that hydroalcoholic extracts of *T. cordifolia* improved the cognitive behavior in normal rats as manifested by the passive avoidance task as well as by the Hebb William maze (Agarwal et al., 2002). The same authors also revealed that *T. cordifolia* may have a potential therapeutic use in neurodegenerative diseases affecting cerebral neurons and immunosuppression-induced memory.

In 2014, Malve et al. noticed that *T. cordifolia* and its combination with *Phyllanthus emblica* and *Ocimum sanctum* improved learning and memory performance in rats with diazepam-, scopolamine-, and cyclosporine-induced memory deteriorations and amnesia (Malve et al., 2014). The time taken to trace food and number of errors was significantly reduced. The authors also stated that these three medicinal plants by virtue of their immunomodulatory effects alleviate learning and memory. Furthermore, they reported that the beneficial effect of the plant drugs against scopolamine-induced amnesia and memory impairment is suggestive of effects of these plants on the cholinergic system to increase acetylcholine synthesis. In support to this later proposed mechanism of action of *T. cordifolia*, Tripathi (2013) found that citicholine, a compound derived from choline and cytidine, used as memory and cognitive enhancer, produced short-term improvement in behavior and memory in cerebrovascular disorders.

Jyothi et al., in 2016, demonstrated that the oral supplementation of *T. cordifolia* alcoholic extract to albino mice with alprazolam-induced amnesia has enhanced cognition as indicated by cognition and memory tests, elevated plus maze, and step-down type passive avoidance task; the effects were comparable to those of the standard drug, piracetam (Jyothi et al., 2016).

Roy (2018) reported that *T. cordifolia* root aqueous extract produced verbal learning and logical memory enhancement in normal and memory-lacking animals. The author attributed this enhancement to immune stimulation and increasing acetylcholine synthesis. The choline content of *T. cordifolia* (Spandana et al., 2013; Sanklala et al., 2012) may lead to the increase in synthesis of acetylcholine, which in turn improves cognitive function (Roy, 2018; Lannert and Hoyer, 1998).

In clinical studies, *T. cordifolia* is applied in the form of a polyherbal formulation for general amendment of memory function. The formulation also contains *Convolvulus pluricaulis*, *Bacopa monnieri*, *Withania somnifera*, *Centella asiatica*, and *Celastrus paniculata* (Mutalik and Mutalik, 2011).

In a 21-day randomized double-blind placebo controlled study, *T. cordifolia* root aqueous extract improved logical memory (Bairy et al., 2004).

The formulation, composed of alcoholic extract/juice of roots of *Withania somnifera*, leaves of *Centella asiatica*, seeds of *Mucuna pruriens*, stems of *T. cordifolia*, and rhizomes of *Curcuma longa*, mixed in ranging ratios 1:1:1:1:2 and 1:0.5:1:1:2 by weights, can alleviate memory as a brain tonic in presenile and senile dementia (Mutalik and Mutalik, 2011). Furthermore, the improvement effects of this formulation were proven by applying elevated plus maze behavioral test and step-down test, which are predictable and reliable procedures for studying the cognition in Alzheimer syndrome as well as the effect of drug responses to senile dementia in intracerebroventricular streptozotocin-treated rats (Palpu et al., 2008). As reported by Mutalik and Mutalik (2011), this polyherbal formulation is used in humans as soft gelatin capsules or as an emulsion, and it has been found to amend memory and the general health.

Modern pharmacologic substances used to manage and treat Alzheimer disease are agents that concentrate acetylcholine (by inhibiting its breakdown or by increasing its synthesis) or agents that protect against neurodegeneration (Roy, 2018; Harvey et al., 2009). *T. cordifolia* has this dual effect as it has been reported to produce protective effects against neuronal damage and contains an alkaloid, choline, which is essential for acetylcholine synthesis. In addition, the antioxidant property and the immunomodulatory activity of *T. cordifolia* may be involved in the protection against Alzheimer disease and in the enhancement of memory (Malve et al., 2014; Palpu et al., 2008).

Side effects/adverse effects

Few side effects of *T. cordifolia* were demonstrated in limited clinical trials (Mukherjee et al., 2010; Purandare and Supe, 2007). *Tinospora* produced hepatotoxicity in a 49-year-old man within 4 weeks of taking 10 pellets/day (Langrand et al., 2014). Signs and functional laboratory biomarker abnormalities normalized within 2 months of stopping *T. cordifolia* administration (Langrand et al., 2014).

At Ayurvedic therapeutic doses, no toxicity has been elucidated (Panchabhai et al., 2008). No deleterious effects were found in healthy volunteers given 0.5 g/day for 21 days (Karkal and Bairy, 2007). Moreover, no deleterious side effects were noticed when *T. cordifolia* stem extract was supplemented up to the biggest oral dose of 1600 mg/kg body weight (b. w.) to rabbits (Jagetia et al., 1998; Ikram et al., 1987) and at a dose of 1000 mg/kg b. w. of the extract of the whole plant in rats (Rao et al., 2005). However, 40% of deaths were attained in mice after administration of 0.5 mg/kg b. w. of *Tinospora* stem extract (Goel et al., 2004). Genotoxicity tests in rats indicated that no clastogenicity or DNA damage was found when *T. cordifolia* was given at doses up to 250 mg/kg b. w. for 7 days, and no mutagenic effects were attained in *Salmonella typhimurium* (Chandrasekaran et al., 2009).

Conclusion and future directions

In conclusion, *T. cordifolia* supplementation alone or in formulation improves cognitive and memory functions in healthy humans and in conditions of presenile and senile dementia and memory deficits as well as in experimental animals with drug-induced amnesia and memory impairments. These ameliorative effects may be owing to the suppressive effects on inflammation and cell death in addition to enhancement of the cholinergic system to increase the acetylcholine biosynthesis. At therapeutic doses, *T. cordifolia* has no side effects.

Further studies are required to assess effects of *T. cordifolia* extracts and constituents on Alzheimer disease and memory disorders especially at clinical levels and to elucidate the physiologic and molecular mechanisms of their actions. Thus, this study calls for collaboration between pharmacognosists, physiologists, biochemists, immunologists, molecular biologists, and clinicians to scrutinize the potential effects and to investigate various aspects of the mechanisms of action of *T. cordifolia* in Alzheimer disease.

References

Agarwal, A., Malini, S., Bairy, K.L., Rao, S.M., 2002. Effect of *Tinospora cardifolia* on learning and memory in normal and memory deficit rats. Indian J. Pharm. 34, 339–349.

Agarwal, S., Ramamurthy, P.H., Fernandes, B., Rath, A., Sidhu, P., 2019. Assessment of antimicrobial activity of different concentrations of *Tinospora cordifolia* against *Streptococcus mutans*: an *in vitro* study. Dent. Res. J. 16 (1), 24–28.

Akram, M., Nawaz, A., 2017. Effects of medicinal plants on Alzheimer's disease and memory deficits. Neural Regen. Res. 12 (4), 660–670.

Antul, K., Amandeep, P., Gurwinder, S., Anuj, C., 2019. Review on pharmacological profile of medicinal vine: *Tinospora cordifolia*. CJAST 35 (5), 1–11.

Bairy, K.L., Rao, Y., Kumar, K.B., 2004. Efficacy of *Tinospora cordifolia* on learning and memory in healthy volunteers: a double blind, randomised, placebo controlled study. Iran. J. Pharmacol. Ther. 3, 57–60.

Chandrasekaran, C.V., Mathuram, L.N., Daivasigamani, P., Bhatnagar, U., 2009. *Tinospora cordifolia*, a safety evaluation. Toxicol. In Vitro 23 (7), 1220–1226.

Chintalwar, G., Jain, A., Sipahimalani, A., Banerji, A., Sumariwalla, P., Ramakrishnan, R., et al., 1999. An immunologically active arabinogalactan from *Tinospora cordifolia*. Phytochemistry 52 (6), 1089–1093.

Dileepkumar, K.J., Shreevathsa, S., Bharathi, H., Shivappa, P., 2014. Alzheimer's disease: an Ayurvedic perspective. J. Ayurveda Holist. Med. 2 (9), 36–41.

Dwivedi, S.K., 2016. Enespa, *Tinospora cordifolia* with reference to biological and microbial properties. Int. J. Curr. Microbiol. App. Sci. 5 (6), 446–465.

Goel, H.C., Prasad, J., Singh, S., et al., 2004. Radioprotective potential of an herbal extract of *Tinospora cordifolia*. J. Radiat. Res. 451 (1), 61–68.

Gupta, A., Raj, H., Karchuli, M.S., et al., 2013. Comparative evaluation of ethanolic extracts of *Bacopa monnieri*, *Evolvulus alsinoides*, *Tinospora cordifolia* and their combinations on cognitive functions in rats. Curr. Aging Sci. 6, 239–243.

Harvey, R.A., Champe, P.C., Finkel, R., Clark, M.A., Cubeddu, L.X., 2009. Lippincott's Illustrated Reviews: Pharmacology, fourth ed. Lippincott Williams & Wilkins, Philadelphia Baltimore.

Ikram, M., Khattak, S.G., Gilani, S.N., 1987. Antipyretic studies on some indigenous Pakistani medicinal plants: II. J. Ethnopharmacol. 19 (2), 185–192.

Jagetia, G., Nayak, V., Vidyasagar, M.S., 1998. Evaluation of the antineoplastic activity of guduchi (*Tinospora cordifolia*) in cultured HeLa cells. Cancer Lett. 127 (1–2), 71–82.

Jain, S., Sherlekar, B., Barik, R., 2010. Evaluation of antioxidant potential of *Tinospora cordifolia* and *Tinospora sinensis*. Int. J. Pharm. Sci. Res. 1 (11), 122–128.

Joshi, G., Kaur, R., 2016. *Tinospora cordifolia*: a phytopharmacological review. Int. J. Pharm. Sci. Res. 7 (3), 890–897.

Jyothi, C.H., Shashikala, G., Vidya, H.K., Shashikala, G.H., 2016. Evaluation of effect of alcoholic extract of *Tinospora cordifolia* on learning and memory in alprazolam induced amnesia in albino mice. Int. J. Basic Clin. Pharmacol. 5, 2159–2163.

Karkal, Y.R., Bairy, L.K., 2007. Safety of aqueous extract of *Tinospora cordifolia* (Tc) in healthy volunteers: a double blind randomised placebo controlled study. Iran. J. Pharmacol. Ther. 6 (1), 59–61.

Langrand, J., Regnault, H., Cachet, X., et al., 2014. Toxic hepatitis induced by a herbal medicine: *Tinospora crispa*. Phytomedicine 21 (8–9), 1120–1123.

Lannert, H., Hoyer, S., 1998. Intracerebroventricular administration of streptozotocin causes long-term diminutions in learning and memory abilities and in cerebral energy metabolism in adult rats. Behav. Neurosci. 112, 1199–1208.

Malve, H.O., Raut, S.B., Marathe, P.A., Rege, N.N., 2014. Effect of combination of *Phyllanthus emblica*, *Tinospora cordifolia*, and *Ocimum sanctum* on spatial learning and memory in rats. J. Ayurveda Integr. Med. 5 (4), 209–215.

Mia, M.M.K., Kadir, M.F., Hossan, M.S., Rahmatullah, M., 2009. Medicinal plants of the Garo tribe inhabiting the Madhupur forest region of Bangladesh. Am.-Eurasian J. Sustain. Agric. (AEJSA) 3 (2), 165–171.

Mishra, R., Manchanda, S., Gupta, M., et al., 2016. Tinospora cordifolia ameliorates anxiety-like behavior and improves cognitive functions in acute sleep deprived rats. Sci. Rep. 6, 25564.

Mukherjee, R., De, U.K., Ram, G.C., 2010. Evaluation of mammary gland immunity and therapeutic potential of *Tinospora cordifolia* against bovine subclinical mastitis. Trop. Anim. Health Prod. 42 (4), 645–651.

Mutalik, M., Mutalik, M., 2011. *Tinospora cordifolia*: role in depression, cognition, and memory. Aust. J. Med. Herbalism 23 (4), 168–173.

Nile, S.H., Khobragade, C.N.N., 2009. Determination of nutritive value and mineral elements of some important medicinal plants from Western part of India. J. Med. Plants 8 (5), 79–88.

Palpu, P., Rao, C.V., Kishore, K., Gupta, Y.K., Kartik, R., Govindrajan, R., 2008. Herbal Formulation as Memory Enhancer in Alzheimer Condition. Council of Scientific and Industrial Research, United States. Patent 7429397. http://www.freepatentsonline.com/7429397.html. (Accessed 31 May 2011).

Panchabhai, T.S., Kulkarni, U.P., Rege, N.N., 2008. Validation of therapeutic claims of *Tinospora cordifolia*: a review. Phytother. Res. 22 (4), 425–444.

Pandey, M., Chikara, S.K., Vyas, M.K., Sharma, R., 2012. *Tinospora cordifolia*: a climbing shrub in health care management. Int. J. Pharm. Bio. Sci. 3 (4), 612–628.

Pendse, V.K., Mahavir, M.M., Khanna, K.C., Somani, S.K., 1981. Anti-inflammatory and related activity of *Tinospora cordifolia* (Neemgiloe). Indian Drugs 19, 14–71.

Purandare, H., Supe, A., 2007. Immunomodulatory role of *Tinospora cordifolia* as an adjuvant in surgical treatment of diabetic foot ulcers: a prospective randomized controlled study. Indian J. Med. Sci. 61 (6), 347–355.

Rana, M.K., 2014. Vegetables and Their Allied as Protective Food. Scientific Publishers, Jodhpur, India, p. 276. www.scientificpub.com.

Rao, P.R., Kumar, V.K., Viswanath, R.K., Subbaraju, G.V., 2005. Cardioprotective activity of alcoholic extract of *Tinospora cordifolia* in ischemia-reperfusion induced myocardial infarction in rats. Biol. Pharm. Bull. 28 (12), 2319–2322.

Reddy, N.M., Rajasekhar, R.N., 2015. *Tinospora cordifolia* chemical constituents and medicinal properties: a review. Sch. Acad. J. Pharm. 4 (8), 364–369.

Roy, A., 2018. Role of medicinal plants against Alzheimer's disease. Int. J. Complement. Alternat. Med. 11 (4), 205–208.

Roy, A., Kundu, K., Saxena, G., Bharadvaja, N., 2017. Estimation of asiaticoside by using RP–HPLC and FAME analysis of medicinally important plant *Centella asiatica*. J. Plant Biochem. Physiol. 5, 198.

Saha, S., Gosh, S., 2012. *Tinospora cordifolia*: one plant, many roles. Anc. Sci. Life. 31 (4), 151–159.

Sanklala, L.N., Saini, R.K., Saini, B.S., 2012. A review on chemical and biological properties of *Tinospora cordifolia*. Int. J. Med. Arom. Plants 2 (2), 340–344.

Sharma, A., Kaur, G., 2018. Tinospora cordifolia as a potential neuroregenerative candidate against glutamate induced excitotoxicity: an in vitro perspective. BMC Complement. Alternat. Med. 18, 268.

Sharma, P., Dwivedee, B.P., Bisht, D., Dash, A.K., Kumar, D., 2019. The chemical constituents and diverse pharmacological importance of *Tinospora cordifolia*. Heliyon 5, e02437, 1-8.

Shetty, B.V., Singh, V., 2010. Flora of Rajasthan, first ed., vol. 1. Merrut Publishers and Distributors, Merrut. 756-100.

Singh, J., Sinha, K., Sharma, A., Mishra, N.P., Khanuja, S.P., 2003. Traditional uses of *Tinospora cordifolia* (guduchi). J. Med. Aromat. Plant Sci. 25, 748–751.

Singhal, A.K., Naithani, V., Bangar, O.P., 2012. Medicinal plants with a potential to treat Alzheimer and associated symptoms. Int. J. Nutr. Pharmacol. Neurol. Dis. 2, 84–91.

Spandana, U., Ali, S.L., Nirmala, T., Santhi, M., Sipai Babu, S.D., 2013. A review on *Tinospora cordifolia*. Int. J. Curr. Pharm. Rev. Res. 4 (2), 61–68.

Tillotson, A.K., Tillotson, N.H., Abel Jr., R., 2001. The One Earth Herbal Sourcebook: Everything You Need to Know about Chinese, Western, and Ayurvedic Herbal Treatments. Twin Streams, Kesington Publishing Corp., New York NY.

Tiwari, P., Nayak, P., Prusty, S.K., Sahu, P.K., 2018. Phytochemistry and pharmacology of *Tinospora cordifolia*: a review. Sys. Rev. Pharm. 9 (1), 70–78.

Tripathi, K.D., 2013. CNS Stimulants and Cognitive Enhancers. Essentials of Medical Pharmacology, seventh ed. Jaypee Brothers Medical Publishers (P) Ltd, New Delhi, pp. 486–491.

Upadhyay, A.K., Kumar, K., Kumar, A., Mishra, H.S., 2010. *Tinospora cordifolia* (Willd.) Hook. F. And Thoms. (Guduchi)-validation of the Ayurvedic pharmacology through experimental and clinical studies. Int. J. Ayurveda Res. 1, 112–121.

Chapter 3.2.14

Magnolia officinalis Rehder & E.H.Wilson

Ipek Süntar[1], Gülsüm Bosdancı[1,2]
[1]*Department of Pharmacognosy, Faculty of Pharmacy, Gazi University, Etiler, Ankara, Turkey;*
[2]*Department of Pharmacognosy, Faculty of Pharmacy, Selcuk University, Campus-Konya, Turkey*

Alzheimer disease and treatment approaches

Alzheimer disease (AD) is a neurologic disorder that causes cognitive decline and memory loss due to neurofibrillary tangles and senile plaques in the brain. It is one of the most common cases of dementia (Tiraboschi et al., 2004; Blennow et al., 2006; Ferri et al., 2009). Postmortem investigations on the brain of AD patients have revealed several pathologic abnormalities such as microglial activation, deep synapse loss, and inflammation (Pratico and Trojanowski, 2000). Senile plaques are agglomerated indissoluble extracellular amyloid beta-peptide (*Aβ*). *Aβ* is a hydrophobic polypeptide that is generated from amyloid precursor protein (*APP*) by β- or γ-secretase (De Strooper and Annaert, 2000). *Aβ* accumulation causes loss of neurons and, therefore, progressive degeneration in the brain (Golde et al., 2006). β-secretase reduces the reaction of *Aβ* production (Haass and Selkoe, 2007). Therefore, β-secretase cleaving enzyme I (*BACE1*) is regarded as the main target for the treatment and/or prevention of AD (Yan et al., 1999; Lin et al., 2000). *BACE1* is influenced and orchestrated by various factors including oxidative stress and hypoxia (Cole and Vassar, 2007). Among the several different factors involved, oxidative stress has a vital role in the AD's pathophysiology. It occurs as a result of the imbalance in the oxidant and antioxidant status (Manoharan et al., 2016). Previous research indicated that oxidative stress increases in the brains of AD patients (Aksenov et al., 2000; Gibson and Huang, 2005; Zhang et al., 2007).

In folk medicine, a great number of medicinal plants have been utilized for the treatment of neurodegenerative disorders. Scientific research has been carried out to reveal beneficial effects of natural products against neurodegenerative diseases by using in vivo models. In terms of neuroprotective action of the medicinal plants in AD, *Centella asiatica* (L.) Urb. (Apiaceae), *Alpinia*

Naturally Occurring Chemicals against Alzheimer's Disease. https://doi.org/10.1016/B978-0-12-819212-2.00030-X

galanga Willd. (Zingiberaceae), *Juglans regia* L. (Juglandaceae), *Aloe arborescens* Mill. (Alloaceae), *Capparis spinosa* L. (Capparaceae), *Abelmoschus esculentus* Moench (Malvaceae), *Vanda roxburghii* R. Br. (Orchidaceae), and *Carica papaya* L. (Caricaceae) were reported to have promising effects (Dhanasekaran et al., 2009; Hanish Singh et al., 2011; Barbagallo et al., 2015; Clementi et al., 2015; Turgut et al., 2015; Uddin et al., 2015; Manoharan et al., 2016). In Korean, Chinese, and Japanese traditional medicine, *Magnolia officinalis* Rehder & E.H.Wilson (Magnoliaceae) has been used as a medicinal herb for a long time (Iwasaki et al., 2000; Luo et al., 2000). Upon the folkloric use of the *Magnolia* species, various experimental studies have been carried out on this genus. In this chapter, we aim to present previous findings on the potential effect of *M. officinalis* and its effective secondary metabolites particularly on AD (Table 3.2.14.1).

Magnolia officinalis Rehder & E.H.Wilson (Magnoliaceae)

M. officinalis is a deciduous tree growing wild in China. It grows to 20 m height and is commonly recognized as magnolia bark or houpu magnolia. The plant has broad ovate leaves (20–40 cm long, and 11–20 cm broad) and white fragrant flowers (10–15 cm wide, 9–12 tepals). There are two varieties, namely, *Magnolia officinalis* var. *officinalis* and *Magnolia officinalis* var. *biloba* (Rivers, 2015).

The root barks and stems of *Magnolia* species are utilized for the treatment of several disorders such as headache, fever, allergic reactions, anxiety, stroke, neurosis, gastrointestinal tract diseases, and cancer in eastern folk medicine (Watanabe et al., 1983; Song et al., 1989; Kuribara et al., 1998). Owing to several bioactivities of *M. officinalis*, the medicinal products containing its extracts are used by many people worldwide as dietary supplements. Previous studies have revealed that *Magnolia* bark extract was shown to display protective effects against anxiety, stress (Weeks, 2009), depression (Xu et al., 2008), Alzheimer disease, and stroke (Lee et al., 2010, 2011a).

In previous studies, it was reported that *M. officinalis* has several bioactive compounds including phenylpropanoids, lignans, flavonoids, coumarins, terpenoids, and alkaloids (Ito et al., 1982; Tachikawa et al., 2000). *Magnolia* extracts were shown to have lignanoid type compounds including magnolol, honokiol, and obovatol. These individual compounds were also reported to possess antioxidant (Fujita and Taira, 1994; Chen et al., 2007; Dikalov et al., 2008), antiinflammatory (Lin et al., 2007; Munroe et al., 2007; Zhou et al., 2008), and neuroprotective activities (Lin et al., 2006). Particularly, 4-*O*-methylhonokiol exerted memory-enhancing activity (Lee et al., 2009, 2010; Choi et al., 2011; Lee et al., 2011b). It was reported that the neuroprotective and antineuroinflammatory activities of magnolol, honokiol, and obovatol from *Magnolia* sp. were attributed to their antioxidant and redox regulatory

TABLE 3.2.14.1 Experimental studies on *M. officinalis* and its bioactive compounds on memory enhancement.

Extract/ compound	Experimental model	Mechanism	References
***M. officinalis* ethanol extract**	***In vivo:*** AD (C57BL/6 mice) model LPS-induced memory impairment mouse (ICR mice) model Passive avoidance test Probe test Scopolamine-induced mouse (ICR mice) model Water maze test	***In vivo:*** ↓AChE activitiy ↓Apoptotic cell death ↓*Aβ* accumulation ↓*APP*, C99, and *BACE1* ↓β-secretase activity ↓*GFAB* and *Iba1* level ↓i-NOS and COX-2 expressions Ameliorated learning and memory impairment	Lee et al. (2012a) Lee et al. (2012b) Lee et al. (2013)
Magnolol	***In vitro:*** Rat pheochromacytoma cell (P12 cell) ***In vivo:*** Body weight, brain weight, and survival rate Location and object novelty recognition test Passive avoidance test SAMP8/TaSlc mouse model	***In vitro:*** ↓*Aβ*-induced cell death ↓*Aβ*-induced increase in ROS accumulation ↓Caspase-3 activity ↓Intacelular calcium level ***In vivo:*** ↓Loss of ChAT-positive cell in the forebrain ↑p-Akt levels in the forebrain ↑Spatial memory performance and visual recognition Prevented age-relating learning and memory impairment	Matsui et al. (2009) Hoi et al. (2010)
Honokiol	***In vitro:*** Rat pheochromacytoma cell (P12 cell) ***In vivo:*** Body weight, brain weight, and survival rate Location and object novelty recognition test Passive avoidance test Probe test	***In vitro:*** ↓*Aβ*-induced cell death ↓*Aβ*-induced increase in ROS accumulation ↓Caspase-3 activity ↓Intacelular calcium level ***In vivo:*** ↓AChE activity ↑Level of p-ERK1/2 ↓Loss of ChAT-positive cell in the forebrain ↑p-Akt levels in the forebrain ↓PGE2 production and	Matsui et al. (2009) Hoi et al. (2010) Xian et al. (2015) Talarek et al. (2017)

Continued

TABLE 3.2.14.1 Experimental studies on *M. officinalis* and its bioactive compounds on memory enhancement.—cont'd

Extract/ compound	Experimental model	Mechanism	References
	SAMP8/TaSlc mouse model Scopolamine induced by ICR mouse model Water maze test	mRNA expression of COX-2 ↑Protein and mRNA level of IL-10 ↓Protein and mRNA level of L-1β ↑Spatial memory performance and visual recognition Ameliorated learning and memory impairment Prevented age-relating learning and memory impairment	
4-***O***-**methylhonokiol**	***In vitro:*** Human P12 cell culture LPS-treated astrocytes and microglial BV-2 cells culture ***In vivo:*** $A\beta_{1-42}$ infused mouse model LPS-induced memory impairment mouse (ICR mice) model Passive avoidance test Presenilin 2 mutant mouse model Probe test Scopolamine-induced mouse (ICR mice) model Water maze test	***In vitro:*** ↓*APP*, C99 and *BACE1* ↓$A\beta$ fibrillogenesis ↓$A\beta_{1-42}$ -induced neurotoxicity and ↑cell viability ↓β-secretase and γ-secretase activity ↓iNOS, COX-2 and NF-κB ↓ROS ***In vivo:*** ↓$A\beta_{1\text{-}42}$ accumulation ↓$A\beta$-induced cell death ↓AChE activitiy ↓Amyloidogenesis ↓Astrocyte activation ↓Apoptotic neuronal cell death ↓Caspase-3 activity ↓ERK activation ↑Glutathione level ↓*GFAB* and *Iba1* level ↓iNOS and COX-2 expression ↓Lipid peroxidation and carbonyl protein level ↑Memory improving	Lee et al. (2009) Lee et al. (2010) Lee et al. (2011a) Lee et al. (2011b) Lee et al. (2011c)

TABLE 3.2.14.1 Experimental studies on *M. officinalis* and its bioactive compounds on memory enhancement.—cont'd

Extract/ compound	Experimental model	Mechanism	References
		↓NO, NF-κB, TNF-α, IL-1β ↓Phosphorylated p38 MAP kinase ↑Spatial memory function and memory impairment	

AChE, Acetylcholinesterase; *AD*, Alzheimer disease; *APP*, Amyloid precursor protein; *BACE1*, Beta-secretase enzyme 1; *ChAT-positive*, Choline acetyltransferase positive; *COX-2*, Cyclooxygenase-2; *ERK*, Extracellular regulated kinases; *GFAB*, Glial fibrillary acidic protein; *Iba1*, Ionized calcium binding adaptor molecule 1; *ICR*, Imprinting control region; *IL-10*, Interleukin-10; *IL-1β*, Interleukin -beta; *i-NOS*, Inducible-nitric oxide synthase; *LPS*, Lipopolysaccharide; *NF-κB*, Nuclear factor-kappa B; *NO*, Nitric oxide; *p38 MAP kinase*, P38 mitogen-activated protein kinase; *p-Akt*, Phosphor-Akt; *p-ERK1/2*, Phosphor-extracellular regulated kinases 1/2; *PGE2*, Prostaglandin E2; *ROS*, Reactive oxygen species; *TNF-α*, Tumor necrosis factor alpha.

effects (Lin et al., 2006; Cui et al., 2007; Hoi et al., 2010; Ock et al., 2010). Moreover, magnolol and honokiol treatment was demonstrated to increase the release of hippocampal acetylcholine in in vivo studies (Hou et al., 2000) and to inhibit age-related memory loss in senescence-accelerated-prone 8 (SAMP8) mice (Matsui et al., 2009). In an in vivo study, Lee et al. (2012a) demonstrated that oral application of *M. officinalis* ethanol extract (at 2.5, 5, or 10 mg/kg doses) improved the $A\beta_{1-42}$ (0.5 μg/mouse, i.c.v.)-induced memory impairments, inhibited $A\beta_{1-42}$ accumulation, apoptosis-related neuronal cell death, and expression of *BACE1* in human *APP* 695-expressing Tg2576 mice (Lee et al., 2012a).

Neuroprotective effects of *M. officinalis* and its secondary metabolites

A transgenic AD model is used to assess neuroprotective effects of the extracts. This model is induced by overexpression of *APP*, presenilin 1 or 2 genes, and is apparent with age-related memory impairment and $A\beta$ plaque accumulation (Hsiao, 1998; Janus and Westaway, 2001; Duff and Suleman, 2004). In a study by Lee et al. (2012a), promising data were obtained for *M. officinalis* ethanol extract, which was reported to have magnolol (12.9%), honokiol (16.5%), 4-*O*-methylhonokiol (16.6%), and other compounds (42%−45%). *M. officinalis* ethanol extract (10 mg/kg) was given to mice for 3 months prior to memory impairment model induction. Water maze, probe, and passive avoidance step-through tests were employed in Tg2576 mice to evaluate the activity of *M. officinalis* extract. Swimming speed and pattern, escape latency, and distance of mice were observed and compared with non-treated Tg2576. In the end of the experiment, oral treatment of *M. officinalis* ethanol extract for 3 months remarkably improved memory function in the AD mice model. Escape latency time of the treated group was recorded as shorter than that of the animals in the nontreated group, and escape distance was also inhibited by the treatment. On the other hand, no remarkable difference was detected on the average speed between the test and control groups. In the same study, the probe test was performed on the animals after 24 h of water maze test. In this test, the mice were permitted to swim freely, and percentage time in target area transitions as well as the swimming pattern of mice were recorded. The results have shown that time spent in the target quadrant by the animals treated with ethanol extract of *M. officinalis* was notably enhanced compared to the nontreated group animals. At 24 h after the probe test, the effect of *M. officinalis* extract on the learning and memory capacity was investigated by passive avoidance test, another broadly accepted method. In this assay, the mouse was placed in the illuminated compartment with the dark compartment facing the back. By entering the dark compartment, the mouse received an electric shock (1 mA, 3 s). After a day, the animal was placed in the illuminated compartment and the latency time was measured. Tg2576 mice treated with *M. officinalis* ethanol extract remarkably enhanced the step-through latency when compared to the animals of the nontreated group. In immunohistochemical and quantification analysis, administration of *M. officinalis* ethanol extract significantly decreased the amount of *Aβ* accumulated in the brain. In parallel to those results, activities of β-secretase and *BACE1* and amounts of *APP* and C99 levels were also reduced in the hippocampus and cortex of transgenic AD model Tg2576 mice with the administration of *M. officinalis* ethanol extract. Briefly, *M. officinalis* ethanol extract improved memory decline in Tg2576 mice and decreased *Aβ* accumulation by inhibiting β-secretase effect and *BACE1* expression (Lee et al., 2012a).

In another study by the same research group, the effects of the *M. officinalis* extract and a commercially available dietary supplement (*Magnolia* Extract, Health Freedom Nutrition LLC, USA) were comparatively evaluated by using the same in vivo models. Even though both extracts displayed protective effect against *Aβ* accumulation and memorial dysfunction, the activity of prepared ethanol extract was found to be higher than that of the commercial product, which was suggested to be as a result of cultivating area and manufacturing methods (Lee et al., 2012b).

Another study of Lee et al. (2013) investigated memory enhancing and antineuroinflammatory and antiamyloidogenic activities of the ethanol extract *M. officinalis* in the lipopolysaccharide (LPS)-induced AD model in mice. Previous reports have indicated that intraperitoneal (i.p.) administration of LPS results in potent and persistent increase of IL-1β, IL-6, and TNF-α, leading to neuroinflammation and causing a gradual decrease in dopaminergic neurons in a significant amount of substantia nigra. These inflammatory components also induce amyloidogenesis by upregulating β-secretase activity leading to memory impairment and *Aβ* accumulation (Vassar, 2001; Sastre et al., 2003; Qin et al., 2007; Lee et al., 2008). In light of this previous knowledge on the LPS-induced AD, Lee et al. (2013) aimed to investigate the probable mechanisms of *M. officinalis* in an AD model in mice (Lee et al., 2013).

For this purpose, ethanol extract of *M. officinalis* (10 mg/kg) was given to mice in drinking water for 4 weeks, prior to inducing memory impairment. LPS (250 mg/kg) or saline (as control) was injected i.p. once a day throughout 7 days. Passive avoidance and Morris water maze and tests were performed on mice after 15 training periods (three times in a day during 5 days). It was observed that *M. officinalis* ethanol extract notably improved the LPS-induced memory impairment in mice on the escape latencies when compared to the control group that received no treatment. More to the point, *M. officinalis* ethanol extract suppressed amyloidogenesis through antineuroinflammatory action in a systemic model and provided memory enhancement (Lee et al., 2013).

Bioactive constituents of the bark of *M. officinalis* are lignanoid compounds, namely honokiol, 4-*O*-methylhonokiol, and magnolol (Fig. 3.2.14.1). A previous study aimed to examine 7-day oral application of 5 and 10 mg/kg doses of *M. officinalis* ethanol extract and 0.75 and 1.5 mg/kg doses of 4-*O*-methylhonokiol in mice. Scopolamine at a dose of 1 mg/kg was intraperitoneally injected to induce memory impairment. Step-down avoidance and Morris water maze tests were used to assess learning and memory. Ethanolic extract of *M. officinalis* and compound 4-*O*-methylhonokiol dose-dependently prevented memory decline. The ethanolic extract and its component dose-dependently decreased the enhanced acetylcholinesterase (AChE) activity (Lee et al., 2009).

In another study, possible activities of *M. officinalis* ethanol extract and 4-*O*-methylhonokiol on memory impairment was investigated. Five-week pretreatment with *M. officinalis* ethanol extract at doses of 2.5, 5, and 10 mg/kg and 4-*O*-methylhonokiol at a dose of 1 mg/kg was reported to suppress the intraventricular injection of $A\beta_{1-42}$ (0.5 μg/mouse, i.c.v.)-induced memory dysfunction. Moreover, expression of β-secretase and the apoptotic cell death was prevented by 4-*O*-methylhonokiol. It also inhibited $A\beta$ fibrilization, H_2O_2, and neurotoxicity in cultured PC12 cells and neurons (Lee et al., 2010). In another report, orally administrated 4-*O*-methylhonokiol dose-dependently

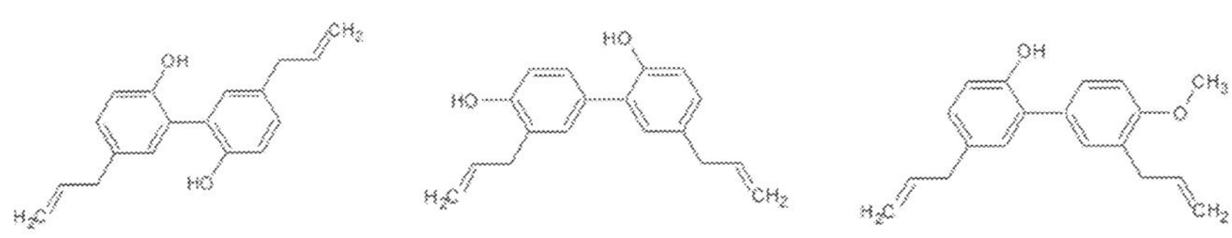

1 2 3

FIGURE 3.2.14.1 Compounds isolated from *M. officinalis*: (**1**) magnolol; (**2**) honokiol; (**3**) 4-*O*-methylhonokiol.

improved the memory impairment induced by LPS. Furthermore, LPS-induced inflammatory protein expression, cyclooxygenase-2 (COX-2), inducible nitric oxide synthase (iNOS), and astrocytes activation in the brain were prevented by 4-*O*-methylhonokiol. The results were also supported by in vitro study (Lee et al., 2012c).

Other studies on AD were carried out on the isolated compounds of *M. officinalis*, especially on magnolol and honokiol. A study by Matsui et al. (2009) was conducted to assess the preventive effect of honokiol and magnolol on age-associated learning and memory dysfunction by location and object novelty recognition and passive avoidance tests in senescence-accelerated mice. Thus, over 14 days, honokiol (0.1, 1 mg/kg) or magnolol (1, 10 mg/kg) were orally given daily to the animals. Notable impairment of learning was detected in 4- and 6-month-old SAMP8 mice. On the other hand, immunohistochemical analysis on choline acetyltransferase (ChAT) revealed a remarkable cholinergic deficit at 6-month-old mice. This age-associated learning and memory dysfunction and cholinergic deficits were found to be prevented by either honokiol (1 mg/kg) or magnolol (10 mg/kg) pretreatment. Furthermore, honokiol or magnolol treatment enhanced phosphorylation of Akt in the forebrain in 2-month-old mice. The outcome of this study suggested that honokiol and magnolol can prevent age-associated learning and memory dysfunction by preserving forebrain cholinergic neurons (Matsui et al., 2009).

In another study by Hoi et al. (2010), magnolol and honokiol markedly reduced $A\beta$-induced cell death. The neuroprotective effects of these two compounds were attributed to their possible mediatory actions on the production of ROS, suppression capacities on intracellular calcium increase, and inhibitory effects on caspase-3 activity (Hoi et al., 2010).

Lee et al. (2011b) reported that 3 weeks pretreatment of 4-*O*-methylhonokiol at doses of 0.2, 0.5, and 1.0 mg/kg prevented neuronal cell death and improved $A\beta_{1-42}$-induced memory dysfunction in a dose-dependent manner. Furthermore, 4-*O*-methylhonokiol prevented the astrocytes activation and p38 mitogenic activated protein (MAP) kinase as well as decreased $A\beta_{1-42}$ infusion-induced oxidative damages in mice brains (Lee et al., 2011b). In another study by the same research group, 3 months of oral treatment of 4-*O*-methylhonokiol at a dose of 1.0 mg/kg was demonstrated to prevent PS2 mutation-induced memory dysfunction and neuronal cell death along with a decline in $A\beta_{1-42}$ accumulation. 4-*O*-Methylhonokiol inhibited extracellular-signal-regulated kinase (ERK) and β-secretase activation, TNF-α, IL-1β production, nitric oxide, and ROS, as well as recovered glutathione, so it prevented oxidative protein and lipid damage (Lee et al., 2011c).

Xian et al. (2015) investigated the in vivo beneficial effects of honokiol in reversing the scopolamine-induced memory and learning dysfunction. During 21 days, mice were intraperitoneally injected honokiol at doses of 10 and 20 mg/kg/day. The results of Morris water maze test revealed a remarkable improvement activity of honokiol in spatial memory and learning function in mice. Moreover, honokiol remarkably reduced mRNA and protein levels of interleukin (IL)-1β and acetylcholinesterase activity, notably enhanced mRNA and protein levels of IL-10, and acetylcholine level, and markedly suppressed the production of prostaglandin E 2 (PGE2) and mRNA expression of cyclooxygenase-2 (COX-2) in the brain. Mechanistic studies demonstrated that honokiol significantly reversed extracellular regulated kinases 1/2 (ERK1/2) changes and phosphorylated Akt (Xian et al., 2015).

Safety of *M. officinalis* extract and its principle components

According to the outcome obtained from in vitro and in vivo genotoxicity experiments, *M. officinalis* bark extract was reported to have neither mutagenic nor genotoxic effect. In a subchronic study conducted according to the Organisation for Economic Co-operation and Development guidelines, it was stated that daily adverse effect level for *M. officinalis* bark extract is 240 mg/kg. Food safety authorities assessed honokiol and magnolol as the principle ingredients of *M. officinalis* and considered them safe (Sarrica et al., 2018).

Conclusion and future prospects

For several years, *M. officinalis* has been utilized in folk medicine in China, Japan, and Korea due to its therapeutic effects for human health. Currently, it is used as a supplement or tea to contribute to gastrointestinal, cardiovascular, and central nervous system conditions. *M. officinalis* extract displays antioxidant and antineuroinflammatory effects and, thus, inhibits age-related memory impairment (Lin et al., 2006; Cui et al., 2007; Hoi et al., 2010; Ock et al., 2010). Small lignan type compounds honokiol, magnolol, and 4-*O*-methylhonokiol isolated from *M. officinalis* were reported to have capacity to interact with proteins of cell membrane, displaying a wide range of pharmacologic effects. In preclinical studies, these compounds were also reported to possess several effects including antimicrobial, antioxidant, antiinflammatory, antidiabetic, antineurodegenerative, antidepressant, and anticarcinogenic (Chen et al., 2011; Talarek et al., 2017). According to the outcome of in vivo studies regarding the neuroprotective action, honokiol, magnolol, and 4-*O*-methylhonokiol could be considered a preventive agent against AD development and progression via reduction of inflammation, oxidative stress, neuronal cell death, and amyloidogenesis (Lee et al., 2011b, 2012c). Nowadays, safe and effective neuroprotective agents are extremely in need. According to the scientific experimental reports, *M. officinalis* and its secondary metabolites may be regarded as essential candidates in the search for effective neuroprotective agents to be clinically used for the management of AD. However clinical studies should be carried out before the application to human patients.

References

Aksenov, M., Aksenova, M., Butterfield, D.A., Markesbery, W.R., 2000. Oxidative modification of creatine kinase BB in Alzheimer's disease brain. J. Neurochem. 74 (6), 2520–2527.

Barbagallo, M., Marotta, F., Dominguez, L.J., 2015. Oxidative stress in patients with Alzheimer's disease: effect of extracts of fermented papaya powder. Mediat. Inflamm. 2015, 624801.

Blennow, K., de Leon, M.J., Zetterberg, H., 2006. Alzheimer's disease. Lancet 368, 387–403.

Chen, C.L., Chang, P.L., Lee, S.S., Peng, F.C., Kuo, C.H., Chang, H.T., 2007. Analysis of magnolol and honokiol in biological fluids by capillary zone electrophoresis. J. Chromatogr. A 1142, 240–244.

Chena, Y.H., Huang, P.H., Lin, F.Y., Chen, W.C., Chen, Y.L., Yin, W.H., Mana, K.M., Liu, P.L., 2011. Magnolol: a multifunctional compound isolated from the Chinese medicinal plant *Magnolia officinalis*. Eur. J. Integr. Med. 3, 317–324.

Choi, I.S., Lee, Y.J., Choi, D.Y., Lee, Y.K., Lee, Y.H., Kim, K.H., Kim, Y.H., Jeon, Y.H., Kim, E.H., Han, S.B., Jung, J.K., Yun, Y.P., Oh, K.W., Hwang, D.Y., Hong, J.T., 2011. 4-*O*-Methylhonokiol attenuated memory impairment through modulation of oxidative damage of enzymes involving amyloid-beta generation and accumulation in a mouse model of Alzheimer's disease. J. Alzheimers Dis. 27 (1), 127–141.

Clementi, M.E., Tringali, G., Triggiani, D., Giardina, B., 2015. *Aloe arborescens* extract protects IMR-32 cells against Alzheimer amyloid beta peptide via inhibition of radical peroxide production. Nat. Prod. Commun. 10 (11), 1993–1995.

Cole, S.L., Vassar, R., 2007. The Alzheimer's disease beta-secretase enzyme, BACE1. Mol. Neurodegener. 15 (2), 22.

Cui, H.S., Huang, L.S., Sok, D.E., Shin, J., Kwon, B.M., Youn, U.J., Bae, K., 2007. Protective action of honokiol, administered orally, against oxidative stress in brain of mice challenged with NMDA. Phytomedicine 14, 696–700.

De Strooper, B., Annaert, W., 2000. Proteolytic processing and cell biological functions of the amyloid precursor protein. J. Cell Sci. 113 (11), 1857–1870.

Dhanasekaran, M., Holcomb, L.A., Hitt, A.R., Tharakan, B., Porter, J.W., Young, K.A., Manyam, B.V., 2009. *Centella asiatica* extract selectively decreases amyloid β levels in hippocampus of Alzheimer's disease animal model. Phytother. Res. 23 (1), 14–19.

Dikalov, S., Losik, T., Arbiser, J.L., 2008. Honokiol is a potent scavenger of superoxide and peroxyl radicals. Biochem. Pharmacol. 76, 589–596.

Duff, K., Suleman, F., 2004. Transgenic mouse models of Alzheimer's disease: how useful have they been for therapeutic development. Brief. Funct. Genomic. Proteomic. 3, 47–59.

Ferri, C.P., Sousa, R., Albanese, E., Ribeiro, W.S., Honyashiki, M., 2009. World Alzheimer Report 2009-Executive Summary. Alzheimer's Disease International, London.

Fujita, S., Taira, J., 1994. Biphenyl compounds are hydroxyl radical scavengers: their effective inhibition for UV-induced mutation in *Salmonella typhimurium* TA102. Free Radic. Biol. Med. 17, 273–277.

Gibson, G.E., Huang, H.M., 2005. Oxidative stress in Alzheimer's disease. Neurobiol. Aging 26, 575–578.

Golde, T.E., Dickson, D., Hutton, M., 2006. Filling the gaps in the abeta cascade hypothesis of Alzheimer's disease. Curr. Alzheimer Res. 3, 421–430.

Haass, C., Selkoe, D.J., 2007. Soluble protein oligomers in neurodegeneration: lessons from the Alzheimer's amyloid beta-peptide. Nat. Rev. Mol. Cell Biol. 8, 101–112.

Hanish Singh, J.C., Alagarsamy, V., Sathesh Kumar, S., Narsimha Reddy, Y., 2011. Neurotransmitter metabolic enzymes and antioxidant status on Alzheimer's disease induced mice treated with *Alpinia galanga (L.) Willd.* Phytother. Res. 25 (7), 1061–1067.

Hoi, C.P., Ho, Y.P., Baum, L., Chow, A.H., 2010. Neuroprotective effect of honokiol and magnolol, compounds from *Magnolia officinalis*, on beta-amyloid-induced toxicity in PC12 cells. Phytother. Res. 24, 1538–1542.

Hou, Y.C., Chao, P.D., Chen, S.Y., 2000. Honokiol and magnolol increased hippocampal acetylcholine release in freely-moving rats. Am. J. Chin. Med. 28 (3–4), 379–384.

Hsiao, K., 1998. Transgenic mice expressing Alzheimer amyloid precursor proteins. Exp. Gerontol. 33, 883–889.

Ito, K., Iida, T., Ichino, K., Tsunezuka, M., Hattori, M., Namba, T., 1982. Obovatol and obovatal, novel biphenyl ether lignans from the leaves of *Magnolia obovata Thunb.* Chem. Pharm. Bull. (Tokyo) 30, 3347–3353.

Iwasaki, K., Wang, Q., Seki, H., Satoh, K., Takeda, A., Arai, H., Sasaki, H., 2000. The effects of the traditional Chinese medicine, "Banxia Houpo Tang (Hange- Koboku To)" on the swallowing reflex in Parkinson's disease. Phytomedicine 7, 259–263.

Janus, C., Westaway, D., 2001. Transgenic mouse models of Alzheimer's disease. Physiol. Behav. 73, 873–886.

Kuribara, H., Stavinoha, W.B., Maruyama, Y., 1998. Behavioural pharmacological characteristics of honokiol, an anxiolytic agent present in extracts of *Magnolia* bark, evaluated by an elevated plus-maze test in mice. J. Pharm. Pharmacol. 50, 819–826.

Lee, J.W., Lee, Y.K., Lee, B.J., Nam, S.Y., Lee, S.I., Kim, Y.H., Kim, K.H., Oh, K.W., Hong, J.T., 2010. Inhibitory effect of ethanol extract of Magnolia officinalis and 4-*O*-methylhonokiol on memory impairment and neuronal toxicity induced by beta-amyloid. Pharmacol. Biochem. Behav. 95, 31–40.

Lee, J.W., Lee, Y.K., Yuk, D.Y., Choi, D.Y., Ban, S.B., Oh, K.W., Hong, J.T., 2008. Neuroinflammation induced by lipopolysaccharide causes cognitive impairment through enhancement of beta-amyloid generation. J. Neuroinflammation 5, 37.

Lee, Y.J., Choi, D.Y., Choi, I.S., Kim, K.H., Kim, Y.H., Kim, H.M., Lee, K., Cho, W.G., Jung, J.K., Han, S.B., Han, J.Y., Nam, S.Y., Yun, Y.W., Jeong, J.H., Oh, K.W., Hong, J.T., 2012c. Inhibitory effect of 4-*O* methylhonokiol on lipopolysaccharide-induced neuroinflammation, amyloidogenesis and memory impairment via inhibition of nuclear factor-κB *in vitro* and *in vivo* models. J. Neuroinflammation 9, 35.

Lee, Y.J., Choi, D.Y., Han, S.B., Kim, Y.H., Kim, K.H., Hwang, B.Y., Kang, J.K., Lee, B.J., Oh, K.W., Hong, J.T., 2012a. Inhibitory effect of ethanol extract of *Magnolia officinalis* on memory impairment and amyloidogenesis in a transgenic mouse model of Alzheimer's disease via regulating β-secretase activity. Phytother. Res. 26, 1884–1892.

Lee, Y.J., Choi, D.Y., Han, S.B., Kim, Y.H., Kim, K.H., Seong, Y.H., Oh, K.W., Hong, J.T., 2012b. A comparison between extract products of *Magnolia officinalis* on memory impairment and amyloidogenesis in a transgenic mouse model of Alzheimer's disease. Biomol. Ter. 20 (3), 332–339.

Lee, Y.J., Choi, D.Y., Yun, Y.P., Han, S.B., Kim, H.M., Lee, K., Choi, S.H., Yang, M.P., Jeon, H.S., Jeong, J.H., Oh, K.W., Hong, J.T., 2013. Ethanol extract of *Magnolia officinalis* prevents lipopolysaccharide-induced memory deficiency via its antineuroinflammatory and anti-amyloidogenic effects. Phytother. Res. 27, 438–447.

Lee, Y.J., Lee, Y.M., Lee, K., Jung, J.K., Han, S.B., Hong, J.T., 2011a. Therapeutic applications of compounds in the *Magnolia* family. Pharmacol. Ther. 130, 157–176.

Lee, Y.K., Choi, I.S., Ban, J.O., Lee, H.J., Lee, U.S., Han, S.B., Jung, K.J., Kim, Y.H., Kim, K.H., Oh, K.W., Hong, J.T., 2011b. 4-*O*-Methylhonokiol attenuated beta-amyloid-induced memory impairment through reduction of oxidative damages via inactivation of p38 MAP kinase. J. Nutr. Biochem. 22, 476–486.

Lee, Y.J., Choi, I.S., Park, M.H., Lee, Y.M., Song, J.K., Kim, Y.H., Kim, K.H., Hwang, D.Y., Jeong, J.H., Yun, Y.P., Oh, K.W., Jung, J.K., Han, S.B., Hong, J.T., 2011c. 4-*O*-Methylhonokiol attenuates memory impairment in presenilin 2 mutant mice through reduction of oxidative damage and inactivation of astrocytes and the ERK pathway. Free Radic. Biol. Med. 50, 66–77.

Lee, Y.K., Yuk, D.Y., Kim, T.I., Kim, Y.H., Kim, K.T., Kim, K.H., Lee, B.J., Nam, S.Y., Hong, J.T., 2009. Protective effect of the ethanol extract of *Magnolia officinalis* and 4-*O*-methylhonokiol on scopolamine-induced memory impairment and the inhibition of acetylcholinesterase activity. J. Nat. Med. 63, 274–282.

Lin, X., Koelsch, G., Wu, S., Downs, D., Dashti, A., Tang, J., 2000. Human aspartic protease memapsin 2 cleaves the beta-secretase site of beta-amyloid precursor protein. Proc. Natl. Acad. Sci. U.S.A. 97, 1456–1460.

Lin, Y.R., Chen, H.H., Ko, C.H., Chan, M.H., 2006. Neuroprotective activity of honokiol and magnolol in cerebellar granule cell damage. Eur. J. Pharmacol. 537, 64–69.

Lin, Y.R., Chen, H.H., Ko, C.H., Chan, M.H., 2007. Effects of honokiol and magnolol on acute and inflammatory pain models in mice. Life Sci. 81, 1071–1078.

Luo, L., Nong Wang, J., Kong, L.D., Jiang, Q.G., Tan, R.X., 2000. Antidepressant effects of Banxia Houpu decoction, a traditional Chinese medicinal empirical formula. J. Ethnopharmacol. 73, 277–281.

Manoharan, S., Guillemin, G.J., Abiramasundari, R.S., Essa, M.M., Akbar, M., Akbar, M.D., 2016. The role of reactive oxygen species in the pathogenesis of Alzheimer's disease, Parkinson's disease, and Huntington's disease: a mini review. Oxid. Med. Cell Longev. 2016 (15), 8590578. Article ID 8590578.

Matsui, N., Takahashi, K., Takeichi, M., Kuroshita, T., Noguchi, K., Yamazaki, K., Tagashira, H., Tsutsui, K., Okada, H., Kido, Y., Yasui, Y., Fukuishi, N., Fukuyama, Y., Akagi, M., 2009. Magnolol and honokiol prevent learning and memory impairment and cholinergic deficit in SAMP8 mice. Brain Res. 1305, 108–117.

Munroe, M.E., Arbiser, J.L., Bishop, G.A., 2007. Honokiol, a natural plant product, inhibits inflammatory signals and alleviates inflammatory arthritis. J. Immunol. 179, 753–763.

Ock, J., Han, H.S., Hong, S.H., Lee, S.Y., Han, Y.M., Kwon, B.M., Suk, K., 2010. Obovatol attenuates microglia-mediated neuroinflammation by modulating redox regulation. Br. J. Pharmacol. 159, 1646–1662.

Pratico, D., Trojanowski, J.Q., 2000. Inflammatory hypotheses: novel mechanisms of Alzheimer's neurodegeneration and new therapeutic targets? Neurobiol. Aging 21, 441–445.

Qin, L., Wu, X., Block, M.L., Liu, Y., Breese, G.R., Hong, J.S., Knapp, D.J., Crews, F.T., 2007. Systemic LPS causes chronic neuroinflammation and progressive neurodegeneration. Glia 55, 453–462.

Rivers, M.C., 2015. *Magnolia officinalis*. The IUCN Red List of Threatened Species 2015, e.T34963A2857694. https://doi.org/10.2305/IUCN.UK.2015-4. RLTS.T34963A2857694.

Sarrica, A., Kirika, N., Romeo, M., Salmona, M., Diomede, L., 2018. Safety and toxicology of magnolol and honokiol. Planta Med. 84 (16), 1151–1164.

Sastre, M., Dewachter, I., Landreth, G.E., Willson, T.M., Klockgether, T., van Leuven, F., Heneka, M.T., 2003. Nonsteroidal anti-inflammatory drugs and peroxisome proliferator-activated receptor-c agonists modulate immunostimulated processing of amyloid precursor protein through regulation of beta-secretase. J. Neurosci. 23, 9796–9804.

Song, W.Z., Cui, J.F., Zhang, G.D., 1989. Studies on the medicinal plants of Magnoliaceae tu-houpo of Manglietia. Yao Xue Xue Bao 24, 295–299.

Tachikawa, E., Takahashi, M., Kashimoto, T., 2000. Effects of extract and ingredients isolated from *Magnolia obovata* Thunberg on catecholamine secretion from bovine adrenal chromaffin cells. Biochem. Pharmacol. 60, 433–440.

Talarek, S., Listos, J., Barreca, D., Tellone, E., Sureda, A., Nabavi, S.F., Braidy, N., Nabavi, S.M., 2017. Neuroprotective effects of honokiol: from chemistry to medicine. Biofactors 43 (6), 760–769.

Tiraboschi, P., Hansen, L.A., Thal, L.J., Corey-Bloom, J., 2004. The importance of neuritic plaques and tangles to the development and evolution of AD. Neurology 62, 1984–1989.

Turgut, N.H., Kara, H., Arslanbaş, E., Mert, D.G., Tepe, B., Güngör, H., 2015. Effect of *Capparis spinosa L.* on cognitive impairment induced by D-galactose in mice via inhibition of oxidative stress. Turk. J. Med. Sci. 45 (5), 1127–1136.

Uddin, M.N., Afrin, R., Uddin, M.J., Alam, A.H., Rahman, A.A., Sadik, G., 2015. *Vanda roxburghii* chloroform extract as a potential source of polyphenols with antioxidant and cholinesterase inhibitory activities: identification of a strong phenolic antioxidant. BMC Complement. Altern. Med. 15 (1), 195.

Vassar, R., 2001. The beta-secretase, BACE: a prime drug target for Alzheimer's disease. J. Mol. Neurosci. 17, 157–170.

Watanabe, K., Watanabe, H., Goto, Y., Yamaguchi, M., Yamamoto, N., Hagino, K., 1983. Pharmacological properties of magnolol and honokiol extracted from *Magnolia officinalis*: central depressant effects. Planta Med. 49, 103–108.

Weeks, B.S., 2009. Formulations of dietary supplements and herbal extracts for relaxation and anxiolytic action: Relarian. Med. Sci. Monit. 15, 256–262.

Xian, Y.F., Ip, S.P., Quing, Q.M., Su, Z.R., Chen, J.N., Xiao, P.L., 2015. Honokiol improves learning and memory impairments induced by skopolamine in mice. Eur. J. Pharmacol. 760, 88–95.

Xu, Q., Yi, L.T., Pan, Y., Wang, X., Li, Y.C., Li, J.M., Wang, C.P., Kong, L.D., 2008. Antidepressant-like effects of the mixture of honokiol and magnolol from the barks of *Magnolia officinalis* in stressed rodents. Prog. Neuropsychopharmacol. Biol. Psychiatry 32, 715–725.

Yan, R., Bienkowski, M.J., Shuck, M.E., Miao, H., Tory, M.C., Pauley, A.M., Brashier, J.R., Stratman, N.C., Mathews, W.R., Buhl, A.E., Carter, D.B., Tomasselli, A.G., Parodi, L.A., Heinrikson, R.L., Gurney, M.E., 1999. Membrane-anchored aspartyl protease with Alzheimer's disease beta-secretase activity. Nature 402, 533–537.

Zhang, X., Zhou, K., Wang, R., Cui, J., Lipton, S.A., Liao, F.F., Xu, H., Zhang, Y.W., 2007. Hypoxia-inducible factor 1alpha (HIF-1alpha)-mediated hypoxia increases BACE1 expression and beta-amyloid generation. J. Biol. Chem. 282, 10873–10880.

Zhou, H.Y., Shin, E.M., Guo, L.Y., Youn, U.J., Bae, K., Kang, S.S., Zou, L.B., Kim, Y.S., 2008. Anti-inflammatory activity of 4-methoxyhonokiol is a function of the inhibition of iNOS and COX-2 expression in RAW 264.7 macrophages via NF-kappa B, JNK and p38 MAPK inactivation. Eur. J. Pharmacol. 586, 340–349.

Chapter 3.2.15

Collinsonia canadensis L.

Mohamad Fawzi Mahomoodally[1], Preethisha Devi Dursun[1], Katharigatta N. Venugopala[2]
[1]*Department of Health Sciences, Faculty of Medicine and Health Sciences, University of Mauritius, Réduit, Mauritius;* [2]*Department of Biotechnology and Food Technology, Durban University of Technology, Durban, South Africa*

Introduction

One of the most common forms of dementia is Alzheimer's disease (AD), which is a neurodegenerative disease by nature. The disease is progressive and affects mostly groups of people who are 65 years and greater AD seems to be described by disorientation, decreased speech function, gait irregularities, progressive loss of memory, increased apathy, and deterioration of nearly all intellectual functions. The key characteristic of AD is loss of memory. The latter is believed to be due to a shortage in the nerve transmitter acetylcholine and free radicals that are produced during the initiation and progression stages of AD. This form of dementia is not curable, but it is managed to some extent with medication. Acetylcholinesterase is an enzyme that splits or breaks down the transmitter substance, thus making it unavailable. Hence, drugs that inhibit the activity of the enzyme were devised, consequently aiding in delaying the progression and development of AD. Natural antioxidants, such as β-carotene, vitamin C, and vitamin E, also aid in scavenging those free radicals (Singhal et al., 2012).

Many herbs have been studied in relation to AD for their antioxidant and antiinflammatory properties. Because acetylcholine has a major role in cognitive function and reasoning, any compound enhancing the cholinergic system of the brain could be valuable in managing AD (Singhal et al., 2012).

Botanical description

Collinsonia canadensis L., best known under the common name of richweed or stone root, is a perennial plant reported to be native to eastern North America;

Naturally Occurring Chemicals against Alzheimer's Disease. https://doi.org/10.1016/B978-0-12-819212-2.00031-1

it is geographically distributed from Quebec to Florida and furthermore to west Missouri and mainly to east of the Mississippi River (Kaur and Sekhon, 2015). The plant has common names such as horse balm, broadleaf collinsonia, horseweed, ox balm, richweed, knobroot, stone root (derived from its dense, hard roots), heal-all, and hard hack. It is named plants usually found in woodlands, and the latter prefer sandy loam soils for better growth (Kaur and Mukhtar, 2018).

C. canadensis can be described as an upright herb that forms part of the mint family (Kaur and Mukhtar, 2018). It has ovoid-shaped, large, simple leaves and light-yellow fragrant flowers, and when the leaves of the latter are crushed, a lemon-like distinct smell can be noticed. The fruit can be described as a nutlet bearing four seeds.

Traditional uses of *Collinsonia canadensis*

C. canadensis has its own traditional uses in its place of origin, mainly among Appalachians. The flowers and leaves of the latter were frequently used as fragrant deodorants, and when they are applied to wounds, sprains, bruises, and contusions, healing is reported. Other reported traditional curative uses are for laryngitis, phthisis, and chronic bronchitis. *C. canadensis* has been documented to alleviate irritation and cough (Scudder, 1870).

Furthermore, *C. canadensis* was found to strengthen the function of the kidneys, hence aiding in urinary tract infections and hemorrhoids as well. Another reported benefit of the latter includes improving blood circulation around the body and fortifying the heart, when taken as a decoction. According to the British Herbal Medicine Association (1983), the preparation and dosage of richweed is 1–4 g to be taken three times a day. In the United States, *C. canadensis* is regulated as a food supplement.

Collinsonia canadensis and Alzheimer's disease

The root of *C. canadensis* has been found to decrease lipopolysaccharide (LPS)-stimulated inflammatory response in murine macrophages (Hanlon et al., 2009). Because AD is closely linked to inflammation, the root extract could be considered promising in future research. Interestingly, the phytochemical analysis showed the presence of rosmarinic acid as a major component in *Collinsonia* shoots and the roots at 20.1 and 15.8 mg/g, respectively. Other phenolic compounds were observed in the crude methanol extract of the *Collinsonia* root (Hanlon et al., 2009). The combination of all three major phenolic compounds equimolar to 5 mg/mL of *Collinsonia* extract treatment showed significant inhibition of LPS-stimulated nitrite production. This initial characterization of a methanol extract from *Collinsonia* demonstrated high concentrations of phenolic compounds with antiinflammatory activity in macrophage cells (Hanlon et al., 2009).

C. canadensis has been found to prevent the breakdown of acetylcholine via inhibition of acetylcholinesterase, a key clinical enzyme involved in neurodegenerative diseases such as AD. The extract has been found to contain carvacrol and thymol as main components, which are used in AD. In addition, *C. canadensis* constitutes of several compounds, such as resins, tannins, alkaloids, rosmarinic acid, mucilage, and flavonoids (Kaur and Mukhtar, 2018), which are known bioactive compounds with antioxidant, antiinflammatory, and neuroprotective properties. However, the main chemical components of this particular plant are thymol and carvacrol, and both are isomers of each other (Jukic et al., 2007). These two compounds are of primary interest owing to their positive effects in AD. The blood–brain barrier does not allow harmful compounds to reach the brain, and helpful medicines can undergo the same fate. However, the compounds carvacrol and thymol get through the gap, which gives an idea of their capacity to be efficient in the management of AD. Moreover, they are known to have properties that enhance the cholinergic system of the body.

Thymol is chemically known as 2-isopropyl-5-methylphenol. It is a monoterpene phenol found in the diet and forms part of other plants, such as *Thymus vulgaris* and *Nigella sativa* seed, among others. Thymol has antiinflammatory, antifungal, antibacterial, larvicidal, antioxidant, anticonvulsant, analgesic, and wound-healing properties. In relation to AD, thymol was shown to have an acetylcholinesterase inhibitory property. The thymol in question was, however, not from the *C. canadensis* herb, but nevertheless the properties found were still anti-AD. According to Jan et al. (2017), one mechanism of action of AD is the deposit of amyloid beta (Aβ) proteins in the central nervous system. Consequentially, amyloid plaques are formed, area-specific neuronal loss occurs, and there are neurofibrillary tangles and synaptic changes in the brain. Thymol, in the proportions of 0.5–2.0 mg/kg, has been shown to inhibit cognitive impairments caused by the heightened level of Aβ or cholinergic hypofunction in Aβ (Singhal et al., 2012).

Furthermore, a study carried out by Asadbegi et al. (2018) investigated the effects of thymol on impairments induced by Aβ in hippocampal synaptic plasticity in rats fed a fat-rich diet. It was stated that obesity and high-fat diets were also contributors of AD. Reports showed a decrease in Aβ-induced impairments in rats fed high-fats diets when they were given thymol. The latter was shown to increase total antioxidant capacity and glutathione levels in the rats; hence, it had therapeutic effects against these risk factors for AD (Asadbegi et al., 2018).

Jukic et al. (2007) carried out an in vitro study in which they discovered that carvacrol exerted 10 times stronger acetylcholinesterase inhibitory activity than thymol. Carvacrol, a molecule with the structure 1, 5-isopropyl-2-methylphenol, is present in essential oils such as thyme and oregano (Aydın et al., 2013).

No studies have yet been discovered in which carvacrol found in *C. canadensis* L. was tested against AD. Aluminum chloride was reported to be one cause of AD. Rats were administered 100 mg/kg body weight of aluminum chloride, and the activity levels of acetylcholinesterase and expressions of amyloid precursor protein as well as reactive oxygen species were significantly elevated (Rather et al., 2018). In another study, AD was induced by aluminum chloride in albino rats. The disease was identified by the presence of a significant increase in neurotransmitters that were accompanied by an increased amount of oxidative stress and cholinesterase levels, and an important decrease in superoxide dismutase and glutathione peroxidase. Then, the diseased rats were treated with carvacrol oil and nanoemulsion. A reverse in the antioxidant status could be observed, in which the biomarkers associated with AD improved (Medhat et al., 2019).

On a final note, because both carvacrol and thymol have a great capacity to enhance the cholinergic system of the brain and are the main compounds in the plant *C. canadensis*, the latter might be extremely important in managing AD.

References

Asadbegi, M., Komaki, A., Salehi, I., Yaghmaei, P., Ebrahim-Habibi, A., Shahidi, S., Sarihi, A., Soleimani Asl, S., Golipoor, Z., 2018. Effects of thymol on amyloid-β-induced impairments in hippocampal synaptic plasticity in rats fed a high-fat diet. Brain Res. Bull. 137, 338–350.

Aydın, E., Türkez, H., Keleş, M., 2013. The effect of carvacrol on healthy neurons and N2a cancer cells: some biochemical, anticancerogenicity and genotoxicity studies. Cytotechnology 66 (1), 149–157.

British Herbal Medicine Association, 1983. https://bhma.info/product/british-herbal-pharmacopoeia-1983.

Hanlon, P.R., Williams, M., Barnes, D., 2009. A methanol extract from *Collinsonia canadensis* root decreases lipopolysaccharide-induced inflammatory response in murine macrophages. FASEB J. 23 (1_Suppl.), 716.3-716.3.

Jan et al., 2017. https://www.ncbi.nlm.nih.gov/pmc/articles/PMC5671974/pdf/fnagi-09-00356.pdf.

Jukic, M., Politeo, O., Maksimovic, M., Milos, M., Milos, M., 2007. In vitro acetylcholinesterase inhibitory properties of thymol, carvacrol and their derivatives thymoquinone and thymohydroquinone. Phytother. Res. 21 (3), 259–261.

Kaur, M., Mukhtar, H.M., 2018. Phytochemical screening and TLC fingerprinting of various extracts of roots and rhizomes of *Collinsonia canadensis*. Indo Am. J. Pharm. Sci. 05 (01), 355–358.

Kaur, M., Sekhon, B.S., 2015. Pharmacognostical standardization of roots and rhizomes of collinsonia canadensis. Int. J. Med. Pharm. Sci. 5 (1), 49–52.

Medhat, D., El-mezayen, H., El-Naggar, M., Farrag, A., Abdelgawad, M., Hussein, J., Kamal, M., 2019. Evaluation of urinary 8-hydroxy-2-deoxyguanosine level in experimental Alzheimer's disease: impact of carvacrol nanoparticles. Mol. Biol. Rep. https://doi.org/10.1007/s11033-019-04907-3.

Rather, M.A., Thenmozhi, A.J., Manivasagam, T., Bharathi, M.D., Essa, M., Guillemin, G., 2018. Neuroprotective role of Asiatic acid in aluminium chloride induced rat model of Alzheimer's disease. Front. Biosci. 10 (1), 262–275. https://doi.org/10.2741/s514.

Scudder, J.M., 1870. Specific Medications and Specific Medicines. https://www.henriettes-herb.com/eclectic/spec-med/index.html.

Singhal, A., Bangar, O., Naithani, V., 2012. Medicinal plants with a potential to treat Alzheimer and associated symptoms. Int. J. Nutr. Pharmacol. Neurol. Dis. 2 (2), 84. https://doi.org/10.4103/2231-0738.95927.

Chapter 3.2.16

Bertholletia excelsa

Arti Bisht[1], Sushil Kumar Singh[2], Rahul Kaldate[2]
[1]*G.B. Pant National Institute of Himalayan Environment, Almora, Uttarakhand, India;*
[2]*Department of Agriculture Biotechnology, Assam Agricultural University, Jorhat, Assam, India*

Introduction

The forests provide timber and other beneficial forest products. Traditionally products other than timber have represented an important source of livelihood for forest-dependent communities. It is a good earning resource for many people worldwide (Ticktin, 2004). Examples include sources used as food such as herbs, spices, edible nuts, edible mushrooms, fruits; fibers utilized for making furniture, clothing, or utensils; and aromatic plants used for medicinal, cultural, or cosmetic purpose (Chandel et al., 2018).

Among others, *Bertholletia excelsa* (Family Lecythidaceae) is extensively used as a nontimber forest product, an important resource for local populations and drawn from the natural forests present in Amazon basins (Rockwell et al., 2015). *B. excelsa,* well-known for producing the Brazil nut, is a tropical tree, prevalent in the Amazonian region, with a maximum height of 150 feet (wikipedia.org, 2009). It is normally grow on undernourished, well-drained oxisol and ultisol soils (Peres and Baider, 1997). Morphologic characteristic of the long-lived Brazil nut tree are that it has large, uneven, and rough leaves up to 60 cm in length and 7 cm in width. The Brazil nuts are covered with a hard shell and are morphologically a round shape with a diameter of around 7 cm. The age of Brazil nut trees has been estimated to be 427 years (Brienen and Zuidema, 2006), which is in the category of pioneer tree. It is first to colonize in the forest, and its growth and development depends on forest clearings (Mori and Prance, 1990).

Different parts of *B. excelsa* such as timber, bark, seeds, and fruits are mostly used (Almeida, 1963). Yang (2009) reported that Brazil nut seeds contain a reliable source of nutrients vital for proper functioning of growth, reproduction, and metabolism of cells, such as fiber, protein, thiamine, selenium, magnesium, and phosphorus. They also include 65% oil, 17% protein, 3% ash, 0.9% crude fiber, 4% water, and 10.1% carbohydrates. They also contain vitamins (vitamin B_6, niacin, and vitamin E) and micronutrients

Naturally Occurring Chemicals against Alzheimer's Disease. https://doi.org/10.1016/B978-0-12-819212-2.00032-3

(potassium, calcium, iron, zinc, and copper). They have high concentrations of sulfur-containing amino acid methionine and polyunsaturated fatty acid (omega 6), causing rancidity of the nut, which is relatively inadequate in the Amazon diet and often found in meat and beans (Maldonado et al., 2020).

It is well-known for its antioxidant and anticarcinogenic properties due to the presence of selenium, which helps to prevent neurodegenerative diseases (Fillion, 2011). Consumption of too many seeds with high selenium may result in nausea and hair and fingernail loss (Kerdell-Vegas, 1966). At present, the Brazil nut has much economic value due to extensive use in the cosmetics industry for development of daily used products such as oils, perfumes, soaps, and other derivatives (Barbosa and Moret, 2015).

Distribution

B. excelsa is distributed in Amazonian regions such as Brazil, Peru, Colombia, Venezuela, Ecuador, and Bolivia (Mori and Prance, 1990). The tree grows mainly in the Amazon's rainforest (yearly rainfall ranges between 1445 and 3399 mm) of South America, 400 m above sea level, with an annual mean temperature range between 23 and 28°C (Almeida, 1963).

Chemical constituents

Brazil nuts have an average of 15% saturated fatty acids (SFAs), 25% monounsaturated fatty acids (MUFAs), and 21% polyunsaturated fatty acids (PUFAs). The main phenolic compounds found in Brazil nut are gallic acid, ellagic acid, vanillic acid, protocatechuic acid, and catechin (John and Shahidi, 2010). Sterols and tocopherols were found in a significant amount in these nuts. The chemical structure of some these bioactive compounds is given in Fig. 3.2.16.1. The major source of selenium, providing 8–83 g of selenium per nut, has already been established in Brazil nut.

Mode of action of Brazilian nuts (*B. excelsa*) against Alzheimer disease

Oxidative stress is recognized as a key player for the development of neurodegenerative disorders, such as Alzheimer disease (AD), which ultimately break the cell metabolism pathways such as DNA repair mechanism of selenium catalyzed by selenoprotein enzyme thioredoxin reductase (Trnd) and mitochondrial dysfunction (Arner and Holmgren, 2000). The brain requires a higher level of oxygen, and its function may be impaired through the development of reactive oxygen species (Selkoe, 2003; Kozlowski et al., 2012; Dominy et al., 2019). For the proper functioning of the central nervous system, selenium has a vital role, and its paucity due to inadequate food habit has been associated with cognitive decline of dementia, causing AD, as described in Fig. 3.2.16.2 (Berr et al., 2000; Cardoso et al., 2010).

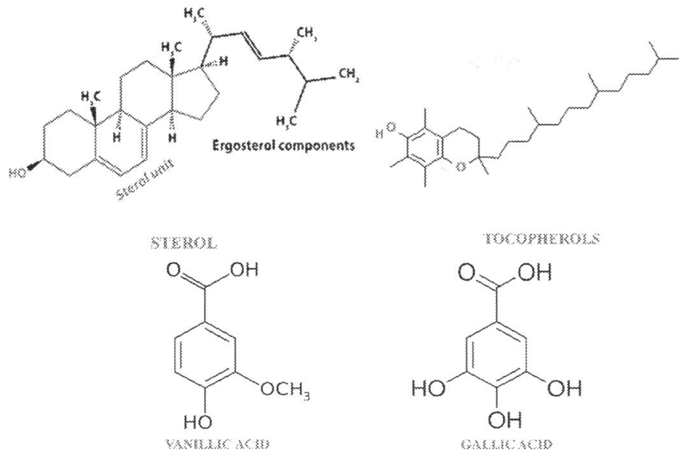

FIGURE 3.2.16.1 Chemical structure of sterol, tocopherols, gallic acid, and vanillic acid.

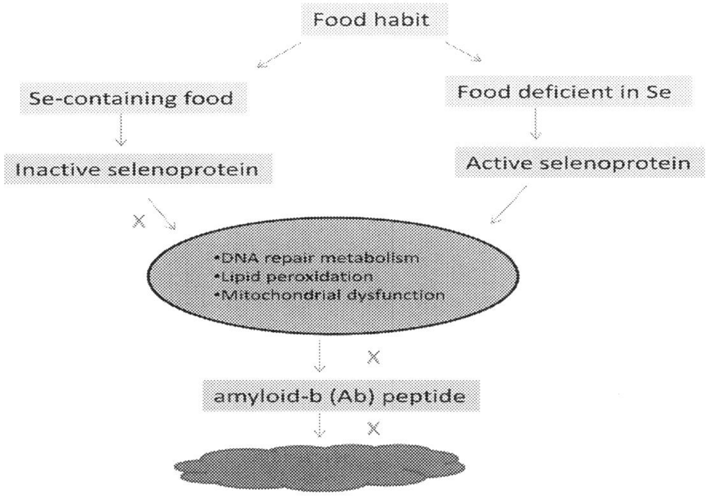

FIGURE 3.2.16.2 The selenium-containing diet influences the antioxidant activity of selenium-containing selenoproteins, causing Alzheimer disease.

Selenium may also have role at the molecular level, especially in degeneration of neurons, causing AD, embracing a cascade of signal transduction, redox pathways, assembly of the cytoskeleton, and pathology of the spinal cord (Loef et al., 2011; Cardoso et al., 2015). The high selenium content in Brazil nuts increases the intake of selenium in our diet mainly as selenomethionine and regulates the activity of selenoprotein (Weeks et al., 2012; Zhang et al., 2013). Selenium contains antioxidant properties of enzymes such as selenoprotein P and glutathione peroxidase (GPx) that play protective roles against oxidative stress responsible for AD.

Humans have 25 different selenoproteins that contain selenocysteine (Sec), a 21st amino acid that has a strong propriety of reduction, essential for the functioning of antioxidant selenoproteins. The Single Nucleotide Polymorphism (SNP) genotyping and gene expression studies suggest that both proteins have a regulatory role in AD. The presence of SNPs in genes encoding selenoproteins may manipulate dietary intake of selenium, thus the genetic variation that influences the selenoprotein homeostasis in our body.

SELENOP (Selenoprotein P) affects the accumulation of mass of protein amyloid-β and hyperphosphorylated tau responsible for death of neurons through apoptosis and breakdown of synaptic transmission. It also interacts with redox-active metals required for proper functioning of the brain, such as Cu, Fe, and Hg and thus regulates the metal-induced Ab aggregation and neurotoxicity in AD. It has an antioxidant enzyme activity that may act as signaling function via neuronal ApoER2 (Solovyev et al., 2018). The gene-encoding selenoproteins P SNPs, rs7579 (G/A substitution), and rs3877899 were investigated to play a vital role at the protein level (Solovyev et al., 2018).

The role of SNPs linked to selenium deficiency in selenoprotein genes may affect the protective shield against oxidative stress played by selenium-containing proteins and act as stress biomarkers in AD patients (Cardoso et al., 2015; Shinkai et al., 2006). The activity of antioxidant is affected by the presence of SNP Pro198Leu in selenoprotein glutathione peroxidase *GPX1 rs1050450 gene* (Jablonska et al., 2009). The allelic variant was also found to differentiate between GPx activity and selenium levels of erythrocytes (Cardoso et al., 2012; Paz-y-Miño et al., 2016). Another vital protein playing a major role in AD is selenoprotein M (SelM). It reduces the intracellular level of reactive oxygen species and inhibits the Ab aggregations depending on the presence of CXXU redox-active motif in selenoprotein. This motif functions as a metal regulator by binding with metal ions zinc and copper, which modulate Ab aggregation and neurotoxicity in AD (Du et al., 2013).

Clinical studies

The effect of Brazil nut on AD has not been studied much in live models. Some reports of studies have been found based on selenium compound that plays an important role in AD (Chang et al., 1995; Kannamkumarath et al., 2002; Vonderheide et al., 2002; Thomson et al., 2008; Dumont et al., 2006; Moodley et al., 2007; Cominetti et al., 2012). Considering the US recommended daily allowance of selenium, one Brazil nut supplies 160% selenium (Yang, 2009). Selenium helps to enhance antioxidant activities, functioning with selenoproteins like GSH-Px. Selenium associates with amino acids to assemble selenoproteins (Papp et al., 2010). The antioxidant property of this crucial micronutrient reduces the danger of cancer and neurodegenerative diseases (Meplan and Hesketh, 2014; Cardoso et al., 2015).

Many studies have reported that a hallmark of degenerative diseases is the age-related reduction in amount of antioxidants causing an oxidative anxiety, which would have a prime role in pathogenesis of multiple degenerative disorders such as AD.

The glutathione peroxidase family consists of four diverse selenoenzymes, which work as antioxidant enzymes (Benton, 2002). In AD, these enzymes protect from reactive oxygen species responsible for cell collapse (Filipcik et al., 2006). It has been found that dietary intake of selenium changes the function of glutathione peroxidase action (Arthur, 2001; Cominetti et al., 2012; Reddy et al., 2017).

Oxidative stress is a primary dangerous factor in aging (Finkel and Holbrook, 2000), which triggers the beginning and progression of AD (Markesbery, 1997; Pratico, 2008; Zhao and Zhao, 2013). Thomson et al. (2008) assessed the value of Brazil nuts on enhancing selenium status in 59 New Zealand adults compared with selenomethionine. After consumption of two Brazil nuts, 100 μg of selenomethionine as a source of selenium, and placebo each day for 12 weeks, the observed percentage enhancement of selenium in the blood plasma was 64.2%, 61.0%, and 7.6%; glutathione peroxidase by 8.3%, 3.4%, and 1.2%; while gross blood glutathione peroxidase was 13.2%, 5.3%, and 1.9% in the Brazil nut, selenomethionine, and placebo groups, respectively.

Interestingly, Cardoso et al. (2010) showed significantly lower selenium levels in blood plasma, nails, and erythrocytes of 28 AD patients in contrast to 29 controls based on evaluation of 3-day dietary food record of selenium intake. This study suggested the strong correlation of AD with selenium deficiency in humans. In 2010, Stockler-Pinto reported that the eating of a single brazil nut of nearly 5 g/day for 3 months enhanced the amount of selenium and glutathione peroxidase activity along with simultaneously upgrading the antioxidant ability (Stockler-Pinto et al., 2010). Cominetti et al. (2011) also reported that the eating of a single Brazil nut enhanced the amount of selenium and glutathione peroxidase action.

In a different study, Cardoso et al. (2016) reported that the consumption of a single Brazil nut with 288.75 μg/day intake up to 6 months restores selenium deficiency and has an encouraging effect on cognitive function with mild cognitive impairment in elderly patients.

Reddy et al. (2017) performed a meta-analysis by systematically reviewing the studies in available literature associated with selenium level in AD. This analysis demonstrated that amount of plasma selenium is considerably lower in AD patients compared with controls, revealing the involvement of antioxidant enzyme GPx in decreased selenium level in AD patients. In another meta-analysis study, 14 independent studies with 40 separate observations on selenium level of brain tissue show significant decline of selenium concentration in brains of AD patients in contrast to controls (non-AD person) and support evidence of selenium's role as an antioxidant and involvement with oxidative stress in AD development (Varikasuvu et al., 2019).

Adverse effect

It was reported that consumption of a huge amount of *Lecythis ollaria* nuts with high concentration of selenium may consequently cause nausea and the loss of fingernail and hair growth (Kerdell-Vegas, 1966). It has been found that a considerable amount of metal toxicants [barium (Ba), strontium (Sr)] and small amount of cancer-causing element [radium (Ra)] accumulate in Brazil nuts (Goncalves et al., 2009; Lemire et al., 2010; Parekh et al., 2008).

The amount of strontium present in Brazil nuts is between 38.7 and 184 μg/g, whereas a single nut can supply 193.5−920 μg (Lemire et al., 2010; Welna and Szymczycha-Madeja, 2014). It has been reported that strontium is not toxic for adults, but it may affect children by impairing the mineralization of the emergent bones (Agency for Toxic Substances & Disease Registry, 2004).

Conclusion

B. excels (Brazil nut) is one of the most potent sources of unsaturated fatty acids, protein, fiber, micronutrients, vitamins, and photochemicals. Clinical evidence suggest that it contains selenium, which helps in retarding the aging process, stimulates the immune system, and cures congenital heart disease, some types of cancer, and AD. The health benefits of Brazil nuts are mainly because of the occurrence of enormous level of phytonutrients responsible for antioxidant, cholesterol-reducing, and antiproliferative properties. Many forest inhabitants collect and sell Brazil nuts for their living. The high market demand and extreme expenditure of timber used by local inhabitants creates ruthless deforestation, causing fragmentation of natural habitats of species. As these species are overexploited for the active constituents, the species are placed under threatened category in International Union for Conservation of Nature, so species emerge as a conservation priority. There is an urgent need to develop strategies for conservation and effectively harnessing the potential of these species. There is a requirement of large-scale plantation or micropropagation of Brazil nut to meet the demands of the herbal and cosmetic industries.

References

Almeida, C. P. de, 1963. Brazil nut: its exportation and importance in the Amazon economy, vol. 19, pp. 1−86. Editions S. I. A. Brazilian Studies.

Arnér, E.S., Holmgren, A., 2000. Physiological functions of thioredoxin and thioredoxin reductase. Eur. J. Biochem. 267 (20), 6102−6109.

Arthur, J.R., 2001. The glutathione peroxidases. Cell. Mol. Life Sci. 57 (13−14), 1825−1835.

Barbosa, M.A.M., Moret, A.S., 2015. Production and marketing of Brazil nuts: economy and financial availability (subsistence of families residing in extractive reserves). Environ. Manag. Sustain. Mag. 4 (2), 413−428.

Benton, D., 2002. Selenium intake, mood and other aspects of psychological functioning. Nutr. Neurosci. 5 (6), 363−374.

Berr, C., Balansard, B., Arnaud, J., Roussel, A.M., Alpérovitch, A., EVA Study Group, 2000. Cognitive decline is associated with systemic oxidative stress: the EVA study. J. Am. Geriatr. Soc. 48 (10), 1285–1291.

Brienen, R.J.W., Zuidema, P.A., 2006. Lifetime growth patterns and ages of bolivian rain forest trees obtained by tree ring analysis. J. Ecol. 94 (2), 481–493.

Cardoso, B.R., Apolinário, D., da Silva Bandeira, V., Busse, A.L., Magaldi, R.M., Jacob-Filho, W., Cozzolino, S.M.F., 2016. Effects of Brazil nut consumption on selenium status and cognitive performance in older adults with mild cognitive impairment: a randomized controlled pilot trial. Eur. J. Nutr. 55 (1), 107–116.

Cardoso, B.R., Ong, T.P., Jacob-Filho, W., Jaluul, O., Freitas, M.I.D.Á., Cozzolino, S.M.F., 2010. Nutritional status of selenium in Alzheimer's disease patients. Br. J. Nutr. 103 (6), 803–806.

Cardoso, B.R., Ong, T.P., Jacob-Filho, W., Jaluul, O., Freitas, M.I.D.Á., Cominetti, C., Cozzolino, S.M.F., 2012. Glutathione peroxidase 1 pro198leu polymorphism in Brazilian Alzheimer's disease patients: relations to the enzyme activity and to selenium status. Lifestyle Genom. 5 (2), 72–80.

Cardoso, B.R., Roberts, B.R., Bush, A.I., Hare, D.J., 2015. Selenium, selenoproteins and neurodegenerative diseases. Metallomics 7 (8), 1213–1228.

Chandel, P.K., Prajapati, R.K., Dhurwe, R.K., 2018. Documentation of traditional collection methods of different NTFPs in Dhamtari forest area. J. Pharmacogn. Phytochem. 7 (1), 1531–1536.

Chang, J.C., Gutenmann, W.H., Reid, C.M., Lisk, D.J., 1995. Selenium content of Brazil nuts from two geographic locations in Brazil. Chemosphere 30 (4), 801–802.

Cominetti, C., de Bortoli, M.C., Garrido Jr., A.B., Cozzolino, S.M., 2012. Brazilian nut consumption improves selenium status and glutathione peroxidase activity and reduces atherogenic risk in obese women. Nutr. Res. 32 (6), 403–407.

Cominetti, C., de Bortoli, M.C., Purgatto, E., Ong, T.P., Moreno, F.S., Garrido Jr., A.B., Cozzolino, S.M.F., 2011. Associations between glutathione peroxidase-1 Pro198Leu polymorphism, selenium status, and DNA damage levels in obese women after consumption of Brazil nuts. Nutrition 27 (9), 891–896.

Dominy, S.S., Lynch, C., Ermini, F., Benedyk, M., Marczyk, A., Konradi, A., et al., 2019. Porphyromonas gingivalis in Alzheimer's disease brains: evidence for disease causation and treatment with small-molecule inhibitors. Sci. Adv. 5 (1), eaau3333.

Du, X., Li, H., Wang, Z., Qiu, S., Liu, Q., Ni, J., 2013. Selenoprotein P and selenoprotein M block Zn^{2+}-mediated Aβ 42 aggregation and toxicity. Metallomics 5 (7), 861–870.

Dumont, E., De Pauw, L., Vanhaecke, F., Cornelis, R., 2006. Speciation of Se in Bertholletia excelsa (Brazil nut): a hard nut to crack? Food Chem. 95 (4), 684–692.

Filipcik, P., Cente, M., Ferencik, M., Hulin, I., Novak, M., 2006. The role of oxidative stress in the pathogenesis of Alzheimer's disease. Bratisl. Lek. Listy 107 (9–10), 384–394.

Fillion, M., 2011. Risks and Benefits of Eating of Local Communities in Amazonia Brazilian: The Effects of Mercury, Lead, Selenium and Omega-3 Fatty Acids on Top Visual Functions (Ph.D. thesis). University of Quebec in Montreal, Montreal, Canada.

Finkel, T., Holbrook, N.J., 2000. Oxidants, oxidative stress and the biology of ageing. Nature 408 (6809), 239.

Gonçalves, A.M., Fernandes, K.G., Ramos, L.A., Cavalheiro, É.T., Nóbrega, J.A., 2009. Determination and fractionation of barium in Brazil nuts. J. Braz. Chem. Soc. 20 (4), 760–769.

Jablonska, E., Gromadzinska, J., Reszka, E., Wasowicz, W., Sobala, W., Szeszenia-Dabrowska, N., Boffetta, P., 2009. Association between GPx1 Pro198Leu polymorphism, GPx1 activity and plasma selenium concentration in humans. Eur. J. Nutr. 48 (6), 383–386.

John, J.A., Shahidi, F., 2010. Phenolic compounds and antioxidant activity of Brazil nut (*Bertholletia excelsa*). J. Funct. Foods 2 (3), 196–209.

Kannamkumarath, S.S., Wrobel, K., Wrobel, K., Vonderheide, A., Caruso, J.A., 2002. HPLC–ICP–MS determination of selenium distribution and speciation in different types of nut. Anal. Bioanal. Chem. 373 (6), 454–460.

Kerdel-Vegas, F., 1966. The depilatory and cytotoxic action of "coco de mono"(*Lecythis ollaria*) and its relationship to chronic seleniosis. Econ. Bot. 20 (2), 187–195.

Kozlowski, H., Luczkowski, M., Remelli, M., Valensin, D., 2012. Copper, zinc and iron in neurodegenerative diseases (Alzheimer's, Parkinson's and prion diseases). Coord. Chem. Rev. 256 (19–20), 2129–2141.

Lemire, M., Fillion, M., Barbosa Jr., F., Guimarães, J.R.D., Mergler, D., 2010. Elevated levels of selenium in the typical diet of Amazonian riverside populations. Sci. Total Environ. 408 (19), 4076–4084.

Loef, M., Schrauzer, G.N., Walach, H., 2011. Selenium and Alzheimer's disease: a systematic review. J. Alzheim. Dis. 26 (1), 81–104.

Maldonado, S.A.S., Fernandez, I.M., Aleman, R.S., Fuentes, J.A.M., Ferreira, M.I.D.C.C., 2020. Determination of total phenolic compounds, antioxidant activity and nutrients in Brazil nuts (Bertholletia excelsa HBK). J. Med. Plant Res. 14 (8), 373–376.

Markesbery, W.R., 1997. Oxidative stress hypothesis in Alzheimer's disease. Free Radic. Biol. Med. 23 (1), 134–147.

Méplan, C., Hesketh, J., 2014. Selenium and cancer: a story that should not be forgotten-insights from genomics. In: Advances in Nutrition and Cancer. Springer, Berlin, Heidelberg, pp. 145–166.

Moodley, R., Kindness, A., Jonnalagadda, S.B., 2007. Elemental composition and chemical characteristics of five edible nuts (almond, Brazil, pecan, macadamia and walnut) consumed in Southern Africa. J. Environ. Sci. Health B 42 (5), 585–591.

Mori, S.A., Prance, G.T., 1990. Taxonomy, ecology, and economic botany of the Brazil nut (*Bertholletia excelsa* Humb. & Bonpl.: Lecythidaceae). Adv. Econ. Bot. 8, 130–150.

Papp, L.V., Holmgren, A., Khanna, K.K., 2010. Selenium and selenoproteins in health and disease. Antioxid. Redox Signal. 12 (7), 793–795.

Parekh, P.P., Khan, A.R., Torres, M.A., Kitto, M.E., 2008. Concentrations of selenium, barium, and radium in Brazil nuts. J. Food Compos. Anal. 21 (4), 332–335.

Paz-y-Miño, C., Sacoto, M.J.G., Leone, P.E., 2016. Genetics and genomic medicine in Ecuador. Mol. Genet. Genom. Med. 4 (1), 9.

Peres, C.A., Baider, C., 1997. Seed dispersal, spatial distribution and population structure of Brazil nut trees (*Bertholletia excelsa*) in southeastern Amazonia. J. Trop. Ecol. 13 (4), 595–616.

Pratico, D., 2008. Oxidative stress hypothesis in Alzheimer's disease: a reappraisal. Trends Pharmacol. Sci. 29 (12), 609–615.

Reddy, V.S., Bukke, S., Dutt, N., Rana, P., Pandey, A.K., 2017. A systematic review and meta-analysis of the circulatory, erythrocellular and CSF selenium levels in Alzheimer's disease: a metal meta-analysis (AMMA study-I). J. Trace Elem. Med. Biol. 42, 68–75.

Rockwell, C.A., Guariguata, M.R., Menton, M., Quispe, E.A., Quaedvlieg, J., Warren-Thomas, E., et al., 2015. Nut production in *Bertholletia excelsa* across a logged forest mosaic: implications for multiple forest use. PLoS One 10 (8), e0135464. J046BETORoA02.lrcex au 2iTovP2ft-cr1oili0Pnkeso-cr0h8wLawkgd39 lit.

Selkoe, D.J., 2003. Folding proteins in fatal ways. Nature 426 (6968), 900.

Shinkai, T., Müller, D.J., De Luca, V., Shaikh, S., Matsumoto, C., Hwang, R., Hori, H., 2006. Genetic association analysis of the glutathione peroxidase (GPX1) gene polymorphism (Pro197Leu) with tardive dyskinesia. Psychiatry Res. 141 (2), 123–128.

Solovyev, N., Drobyshev, E., Bjørklund, G., Dubrovskii, Y., Lysiuk, R., Rayman, M.P., 2018. Selenium, selenoprotein P, and Alzheimer's disease: is there a link? Free Radic. Biol. Med. 127, 124–133.

Stockler-Pinto, M.B., Mafra, D., Farage, N.E., Boaventura, G.T., Cozzolino, S.M.F., 2010. Effect of Brazil nut supplementation on the blood levels of selenium and glutathione peroxidase in hemodialysis patients. Nutrition 26 (11–12), 1065–1069.

Thomson, C.D., Chisholm, A., McLachlan, S.K., Campbell, J.M., 2008. Brazil nuts: an effective way to improve selenium status. Am. J. Clin. Nutr. 87 (2), 379–384.

Ticktin, T., 2004. The ecological implications of harvesting non-timber forest products. J. Appl. Ecol. 41 (1), 11–21.

Varikasuvu, S.R., Prasad, S., Kothapalli, J., Manne, M., 2019. Brain selenium in Alzheimer's disease (BRAIN SEAD Study): a systematic review and meta-analysis. Biol. Trace Elem. Res. 189 (2), 361–369.

Vonderheide, A.P., Wrobel, K., Kannamkumarath, S.S., B'Hymer, C., Montes-Bayón, M., Ponce de León, C., Caruso, J.A., 2002. Characterization of selenium species in Brazil nuts by HPLC– ICP-MS and ES-MS. J. Agric. Food Chem. 50 (20), 5722–5728.

Weeks, B.S., Hanna, M.S., Cooperstein, D., 2012. Dietary selenium and selenoprotein function. Med. Sci. Monit. 18 (8), RA127.

Welna, M., Szymczycha-Madeja, A., 2014. Improvement of a sample preparation procedure for multi-elemental determination in Brazil nuts by ICP-OES. Food Addit. Contam. 31 (4), 658–665.

Yang, J., 2009. Brazil nuts and associated health benefits: a review. LWT-Food Sci. Technol. 42 (10), 1573–1580.

Zhang, Q., Chen, L., Guo, K., Zheng, L., Liu, B., Yu, W., Guo, C., Liu, Z., Chen, Y., Tang, Z., 2013. Effects of different selenium levels on gene expression of a subset of selenoproteins and antioxidative capacity in mice. Biol. Trace Elem. Res. 154, 255–261.

Zhao, Y., Zhao, B., 2013. Oxidative stress and the pathogenesis of Alzheimer's disease. Oxid. Med. Cell. Longev. 2013.

Chapter 3.2.17

Urtica diocia

Sumit Durgapal[1], Arvind Jantwal[1], Jyoti Upadhyay[2], Mahendra Rana[3], Aadesh Kumar[1], Tanuj Joshi[1], Amita Joshi Rana[1]

[1]*Department of Pharmaceutical Sciences, Kumaun University, Nainital, Uttarakhand, India;* [2]*School of Health Sciences, University of Petroleum and Energy Studies, Dehradun, Uttarakhand, India;* [3]*Department of Pharmaceutical Sciences, Kumaun University, Nainital, Uttarakhand, India*

Introduction

The 21st century has witnessed most of use of herbs, herbal drugs, and herbal-based formulations. The importance of herbs in the manufacturing of pharmaceuticals is well established as presently more than 80% of the global market has been captured by herbal formulations. Despite the availability of a number of synthetic drugs and synthetic drug–based formulations for almost all the known disorders and diseases in the market, the development of herbal formulations for the same again speaks about the immense potential, importance, and day-by-day increasing promise and development of herbs in the effective management of a global scenario of various life-threatening diseases. Today, most researchers have sifted their interest toward naturally occurring plant-based drugs because of their immense potential, maximum benefits, and no or little toxicity aspects. The concept of nanotechnology for herbal drugs has recently gained much attention from present era scientists and researchers to get maximum advantages by the development of nano dosage forms from a therapeutic point of view as drugs administered in nano dosage forms are known to maintain optimum concentration of drug in plasma for longer periods of time without much effort, thus increasing the bioavailability of drug to the maximum extent compared to conventional dosage forms. Another great advantage of development of herbal drugs in a nano dosage form is easy targeting of drugs to their respective site of action to reduce toxic effects.

Earth bestows us with an abundance of medicinal plants that are known to be highly efficient if formulated properly in a suitable dosage form with the principles of biopharmaceutical design of dosage forms. One such plant that is

Naturally Occurring Chemicals against Alzheimer's Disease. https://doi.org/10.1016/B978-0-12-819212-2.00033-5

highly effective, possess great potential, and is well known for the treatment of some severe and life-threatening disease but neglected and underrated for a long time is *Urtica dioica*. There are number of literatures and evidences that support the folk and traditional use of this plant for a number of diseases because of its high nutritional value. It was also very common as one of the magical remedies among the tribes of Native America. Some of the great physicians and writers of all time such as Hippocrates, Dioscorides, and Paracelsus also mentioned this plant, and it has potential in treating numerous diseases and health-related problems such as cancerous ulcers, burns, menstrual problems, asthma, pneumonia, facial ring worms, etc.

U. dioica, a perennial herb belongs to the family Urticaceae (Ahmed and Parsuraman, 2014; Upton, 2013), is one of the most common plants among ancient herbal and traditional medicinal practitioners and is known for number of its therapeutic activities since time immemorial. Originally, it was a native of Europe and Eurasia, now growing wild in much of the temperate zone of the world, such as temperate Asia and western North Africa. Due to its unique climatic adaptability (Di Virgilio et al., 2015), it is now found commonly in almost all regions of the world including North America and New Zealand. Genus *Urtica* of this remedial plant is coined from the word "uro" from the Greek period and signifies burning. Similarly, the species name *dioica* is obtained from "oikia" meaning bipod or two houses, which signifies the dioecious nature of the plant in which male and female parts required for the reproduction exist in separate plants (Upton, 2013; Lahigi et al., 2011). It is commonly known by a number of names like stinging nettle (Blumenthal et al., 2000; Marzell, 1979), greater nettle (Marzell, 1979), common nettle (Marzell, 1979; Taxon, 1990; Applequist, 2006), giant nettle, European nettle (Marzell, 1979), or simply nettle (Blumenthal et al., 2000; Applequist, 2006) and *Urtica urens* (burning nettle, lesser nettle, or dwarf nettle). It is interesting to know that the common name nettle was derived from the Anglo-Saxon word that means "needle" (Kregiel et al., 2018). In India, it is widely distributed in almost all the high-altitude, hilly terrains. It is one of the most ubiquitous and commonly seen plants in Kumaun and Garhwal regions of Uttarakhand, called shisuun or kandali in local dialect, and used in cuisine for a number of traditional occasions. Some of the other known names of this plant in different languages in India are *vrishchhiyaa-shaaka* in Sanskrit, bichu butti in Hindi and Punjabi, and anjuraa in Unani. *U. dioica* is a flowering plant with 500 different species spread worldwide among 46 genuses (Kregiel et al., 2018). The most important, well-known, and extensively studied species of this genus are *U. dioica* L. and the small nettle *U. urens* L. The plant shown below in Fig. 3.2.17.1 is Urtica dioica in its natural habitat and is used effectively in various traditional medicines.

FIGURE 3.2.17.1 *Urtica dioica* plant in natural habitat.

Nomenclature of plant

Scientific name
Urtica dioica L, *Urtica urens* L, *Urtica galeopsifolia, Urtica major Kanitz*

International names

English:	California nettle; stinging nettle, dwarf stinging nettle
French:	Grande ortie; ortie; ortie dioique
Russian:	Krapiva
German:	Brennessel, grosse brennessel
Italian:	Ortica
Arabic:	Qurrays
Spanish:	Chichicaste; ortiga; ortiga mayor

Local names

Hindi and Punjabi:	Bichu butti
Sanskrit:	Vrishchhiyaa-shaaka
Netherlands:	Grote brandnetel

Plant description

U. dioica is originally a native of cold places but now grows throughout the world as a perennial herbaceous flowering and fruiting plant with 100–200 cm of height. It blooms and bears fruits every year in summer from July to September (Baumgardner, 2016) Generally, it grows naturally in nitrogen-rich, moist soil, particularly in neglected places like dump yards, sideways of pedestrian tracks, rail tracks, waterways, forest lands, etc. Only after 1940, when it was realized as an industrial crop due to its immense potential in obtaining natural fibers as well as active herbal constituents, has it gained much attention for commercialization and scientific work (Di Virgilio et al., 2015). It can be identified with the characteristic presence of sharp, hairy fuzz of stinging trichomes all over the plant (Lahigi et al., 2011). Because of the presence of these needle-like piercing trichomes throughout the plant, it results in the aggravation of contact urticarial dermatitis in some persons when these plants come in close contact with the skin because of the introduction of inflammatory chemicals like histamine, acetylcholine, serotonin, and formic acid (Bourgeois et al., 2016; Kregiel et al., 2018). When it comes to rhizomes and stolen, these are also found in abundance in this plant and generally appear bright yellowish in color (Joshi et al., 2014). Leaves of this plant can be identified as soft and greenish bright in color, possessing a characteristic feature of strong serrated boundaries and showing the presence of venation in the lower leaf surface. The presence of striking hairs throughout the leaf and stem of this plant that contains various inflammatory chemicals has been reported and gives the plant a very distinctive appearance.

Chemical constituents and mechanism of action of *U. dioica*

This plant is one of the most common and highly efficient plants of ancient times and has been known as a miraculous plant for the effective treatment of many severe ailments. There are a number of evidences of traditional and folk use of this plant (Cummings and Olsen, 2011). Various experimental studies conducted for this plant revealed the presence of a wide variety of chemicals that support the use of this plant in number of ways. Studies have shown the presence of flavanoids, sterols, terpenoids, isolectins, polysaccharides, volatile compounds, fatty acids, vitamins, proteins, and minerals as the main chemical constituents (De Vico et al., 2018). The most common and well-known mechanism of action of this plant actually comes from its stinging effect due to the abundance of chemicals like formic acid, histamine, acetylcholine, serotonin, and low quantity of leukotrienes in the ubiquitous hair present in the entire plant, for which it is also known as stinging nettle (Kregiel et al., 2018). Excessive burning, irritation, and rashes followed by urticarial dermatitis are the common symptoms with this plant when it comes in contact with bare skin. For this reason, it is quite common among laymen for its antiinflammatory activity. However, experimental studies claim the inhibition and decrease in the binding efficiency of nuclear factor kappa B (NF-kB) to DNA and further the inhibition of proinflammatory cytokines due to *U. dioica* is the possible mechanism rendering its antiinflammatory action (Ahmed and Parsuraman, 2014). Scientifically, these are characterized as biochemical and mechanical irritation actions of the plant (Cummings and Olsen, 2011). Both the leaves and roots of this plant are of utmost importance from a therapeutic point of view and are used for different indications. Some common uses of leaves of this plant includes the treatment of lower urinary tract infections, rheumatoid arthritis, and allergic rhinitis such as hay fever, while the root of this plant has been found to have special indication in the treatment of prostate hyperplasia. The plausible mechanism is the inhibition of binding of sex hormone binding globulin with specific receptors in prosthetic membrane due to the presence of lignans in the root extract. Aromatase gene expression inhibition with the administration of plant extract could also be the mechanism of action. A change in the conversion profile of testosterone into estrogen due to the inhibition of androstenedione also supports the effectiveness of this plant in prostate hyperplasia (De Vico et al., 2018). Recent studies suggested the therapeutic efficacy and use of this plant in a chronic neurodegenerative disorder called Alzheimer disease (AD), which shows the symptoms of memory loss, cognitive impairment, and drastic changes in behavioral patterns. Oxidative stress (OS) and inflammatory cascades along with other mechanisms such as senile plagues, neurofibrillary tangles, and decrease in neurotransmitter acetylcholine are the major underlying causes of this disease (Dastmalchi et al., 2007). These symptoms of AD make a right platform for the use of *U. dioica* for its treatment and management as evidences have shown that this possesses high antioxidant and antiinflammatory activity. A number of experimental studies support the significance and beneficial role of this herbal remedy for the management of AD. Thus, to know how it is highly effective in AD, the thorough understanding of the role of OS and inflammatory cascade is highly required.

Oxidative stress and Alzheimer disease

OS is a burden due to the imbalance between the natural antioxidant of the body and concentration of free radicals like reactive oxygen species (ROS) in the body. ROS is defined as the species that are either free radicals or species that possess immense potential to generate free radicals. Several such kinds of species that play a profound role in the prevalence of OS and leave body highly vulnerable for various OS-induced diseases are superoxide ($O_2{}^{*-}$) and nitric oxide (NO*) radicals (Yoshikawa and Naito, 2002). As oxygen is required for normal functioning of the body and its various mechanisms, similarly, an optimum concentration of ROS is required for a number of normal physiologic processes of the body for its survival such as an effective body defense mechanism to keep infections and other life-challenging inflammatory responses at bay, gene expressions, and growth of cells. As long as the concentration of ROS remains in its steady state because of the maintenance of equilibrium between rate of production and consumption of ROS and its elimination by the normal antioxidant system, the body will behave normally. Problems occur when this equilibrium is no longer maintained, and the body under such a condition results in a state called OS (Kohen and Nyska, 2002). Through extensive research studies, it has been found that 2% of the total amount of oxygen that is required for the fulfillment of normal physiologic needs of the body gets converted to O_2*^- via processes that occur in mitochondria. But there occurs a drastic change in this normal percentage of ROS during any kind of infection, disease, or exposure to ionizing radiations, UV radiation, or several pollutants. Studies showed a marked increase in the ROS percentage during exercise as well because at the time of exercise the overall amount of oxygen consumption gets tremendously increased to supply oxygen to each tissue of the body under activity. The other free radical species, i.e., NO·, which is equally responsible for OS, is a neurotransmitter produced by nitric oxide synthase enzyme. Further, these radicals undergo a number of transformation reactions that increase their oxidizing potential many fold by generating oxidizing radicals such as hydroxyl radical (˙OH), alkoxy radicals (RO·), peroxyl radicals (ROO·), and singlet oxygen ($^1O_2$), and these are the main players of OS that damage essential biomolecules such as DNA, lipids, and proteins. This lays a solid foundation for the pathogenesis of various disorders like AD and many more oxidative-induced diseases (Kunwar and Priyadarsini, 2011). With the advancements in various scientific tools, techniques, and increased understanding of physiology of the body, now it has become a well-established fact that OS, along with some other parameters, is the main underlying cause in the commencement of AD (Perry et al., 2002). In spite of a very small size of brain in comparison to the entire body the overall consumption of oxygen of the brain is almost one-fourth of the total oxygen supplied by the respiratory system because it has to perform complex physiologic mechanisms and processes to control each and every single organ of the body. Thus to cater the needs of each organ, it always requires a high amount of oxygen, so it is more at risk of OS than any other organ (Chen and Zhong, 2014). Being a functional unit of the brain, neurons are always at higher risk of damage due to the OS than any other normal brain cells, and the excessive damage of lipids, proteins, and DNA of neurons is the main cause behind the episodes of AD. Patients suffering from AD show the tangible accumulation of amyloid β (Aβ) peptide, a kind of neurotoxic oligomer, hyperphosphorylation of τ protein, and dementia. These amyloid β and τ proteins act like prominent targets and biomarkers for the effective treatment of AD because under the influence of excessive OS, these are known to form a platform for the neurodegeneration and result in a number of conditions such as neuronal damage, neuronal loss, severe inflammation of neurons, neurotransmitter imbalance, impaired synaptic plasticity, cholinergic denervation, substantial synaptic loss, and certain dendritic alterations. Moreover, ROS-generated mitochondrial impairment again plays an important role in neurodegeneration and results in the pathogenesis of AD (Huang et al., 2016). This explains the specific and prominent role of OS in the development of AD. Thus, close association between OS and prevalence of AD supports, strengthens, and forms a great platform for the exploration of plants rich in antioxidants for the treatment of AD.

Inflammatory cascades and AD

Extensive research studies in the patients of AD revealed the presence of activated microglial cell inflammatory cytokines in the brain, which are the characteristic features of this disease. Findings of clinical studies done in this field so far also have claimed the presence of inflammatory markers specifically in the brain of persons suffering from AD. Therefore, treatment using these targets with the help of antiinflammatory agents possesses enough potential for the management of AD (Dastmalchi et al., 2007).

As it is a well-known and established fact that plants possess immense potential to treat diseases because of their capability to give maximum therapeutic benefits with no or little toxicity aspects, *U. dioica* is one such ordinary plant with extraordinary benefits that has been reported for its excellent antioxidant activity since the time of Hippocrates. Despite such great potential, it has been one of the most neglected plants for so long and requires extensive research studies to be utilized for the benefits of mankind. The antioxidant and antiinflammatory properties of this plant make it highly beneficial and a magical remedy to be used in AD patients, as the main underlying cause of this disease is OS because of excessive free radical generation when the equilibrium between production of free radical and natural antioxidant system of the body gets disturbed. It has also been found through research studies that the antioxidant is a key role player in the reduction of dementia (Shahat et al., 2015). The importance of antioxidants in AD can also be determined by the fact that most physicians of the present era who are dealing with the treatment of AD are now prescribing antioxidants due to their effectiveness with the drug regimen of AD. The other reason of pathogenesis of AD is the inflammatory cascade, which can be best treated by *U. dioica* because of its great antiinflammatory potential. *U. dioica* is known to contain a mineral called boron that raises the levels of estrogen hormone found to be involved greatly in short-term memory, and it also elevates mood in patients suffering from AD (Singhal et al., 2012). Numbers of research studies have been carried out to date by a number of researchers working in this arena to know the significant role of antioxidant and antiinflammatory potential of

U. dioica in the management of AD. Ghasemi et al. carried out one of the research studies to determine the advantageous effect of *U. dioica* on acetylcholinestrase activity and OS in brain using a scopolamine-induced memory impairment rat model, and with the result of the studies, it was concluded that *U. dioica* is significant enough to recover the effects of scopolamine to a great extent due to its antioxidant potential (Ghasemi et al., 2019). Bazazzadegan et al. studied the neuroprotective effect of *U. dioica* in the hippocampus region of a streptozotocin rat model in sporadic AD by selecting the AD-related genes: *Daxx*, *Nfκβ*, and *Vegf*. Expression levels of these three genes were evaluated with the help of qPCR to know the effect of herbal extract and to establish the molecular mechanism of AD. Further studies related to behavioral, learning, and memory levels were also carried out. Results of the study revealed that herbal extract showed significant effect on gene expressions and memory levels, but no significant effect was observed in behavioral level (Bazazzadegan et al., 2017). Daneshmand et al. determined the neuroprotective role of *U. dioica* along with some other plants for sporadic AD on rat models. Studies were conducted by dividing 40 adult male Wistar rats into five animal groups respectively, and it was determined that the herbal extract of plants possesses enough potential to treat the disorder by improving learning and memory in rat models, as significant changes in the expression of *Syp* and *Psen1* genes, which are key players in the neuronal physiology and pathogenesis, were observed in the herbal-treated group (Daneshmand et al., 2016). Toldy et al. conducted one of their research works to determine the effect of stinging nettle leaf on OS in brains of rats models. Three animal groups were made in which rats were divided as per the treatment they received. One group of animals were subjected to swimming training for forced swim test, a second group of animals were given dried stinging nettle leaf in a concentration of 1% w/w, and third group received the combination of both. OS during the study was measured with the help of electron spin resonance and carbonylated proteins concentration during study. Results of their study showed a marked decrease in OS created by ROS particularly in the cerebellum and frontal lobe of rats who received nettle leaves. Increase in DNA binding of activator protein-1 was also found with stinging nettle leaves. Researchers with this study concluded that stinging nettle is a potent antioxidant as it showed marked decrease in free radical concentration (Toldy et al., 2005). The same group of researchers in their other study determined the effect of stinging nettle leaves administered with exercise regime in brain lesions created by N-methyl-D-aspartate and memory in rat animal models and found that the rats subjected to nettle supplementation showed remarkable recovery from the brain lesion, which was tested by using passive avoidance test and open field test (Toldy et al., 2009).

Toxicologic studies

Various toxicologic studies have been done to know the toxic profile of this plant, and these studies clearly suggested that there occurs little or no toxicity with the use of this plant. Dar et al. conducted the toxicity of extract of *U. dioica* by using standard method called *Artemia salina*. Outcome of the study revealed that both the aqueous extract and herbal formulation coded as HF2 of hexane extract of *U. dioica* (*HEUD*) possess extremely higher margins of safety with $LC_{50} > 1000$ μg/mL when used against *A. Salina* larvae. Acute toxicity test in Wistar rats was also conducted by Dar et al., and results showed no mortality after 24 h of administration of aqueous and hexane extract of *U. dioica*, revealing enough safety potential of this herbal drug (Dar et al., 2013). Tekin M et al. conducted an acute toxicity study in rat models in a dose ranging from 0.2 to 12.8 mL/kg and observed no death within the 72 h of study even in the highest dose, i.e., 12.8 mL/kg. With these studies, the authors suggested the safe profile of *U. dioica* (Tekin et al., 2009). Bathe et al. found the signs and symptoms of distress, ataxia, muscle weakness, and urticaria with an instant stinging nettle rash causing apparent neurologic disorders in three horses used as models in the study (Bathe, 1994). Lasheras et al. conducted the acute and chronic toxicity study of this plant and determined LD_{50} using the technique of Reed Muench-Pizzi. Results of the study clearly revealed that the aqueous extract of this plant had 3625 mg/kg of intraperitoneal LD_{50} and doses larger than 750 mg/kg were found to be highly significant for muscle tone loss, incidence of hypothermia, and decrease in spontaneous activity (Lasheras et al., 1986). Starkenstein et al. conducted a research study to determine the toxic profile of this plant using rabbit models. In this study, they administered 50 mL of 50% ethanolic extract of stinging nettle given orally to rabbits of about 2 kg body weight for 10 days, and diarrhea was occasionally observed. Good tolerance with 5–20 mL of subcutaneous injection was observed at the time of administration, but later on, within 24–36 h, death of animals was observed. Same incidence of death after 8 days with a loss of 40% of body weight of animals was observed with chronic subcutaneous administration of 5–10 mL of extract after its initial bolos dose of 15–20 mL. With this study, the authors determined the intravenous lethal dose of extract is 1.5 mL in fivefold concentrations, and boiling of extract decreases its toxicity. Increase in respiration and central excitatory behavior were the main symptoms shown by the animals under study (Starkenstein and Wasserstrom, 1933).

Conclusion

U. dioica is an herb that has immense potential to treat a number of various disorders and has been used as a folk medicine since the time of Hippocrates. There has been a long history of use of this herb in a number of indications due to its diverse chemical composition and characteristics. Despite of all these advantages, this plant is one of the most neglected plants and has not been utilized much to obtain its maximum potential. Great antioxidant and antiinflammatory potential of this plant makes it highly efficient and a choice of plant for the effective management and treatment of AD, as OS along with the inflammatory cascade are the main underlying causes for the pathogenesis of AD. A number of research studies have been conducted to explore the antioxidant and antiinflammatory effect of UD by using a number of different animal models, and all these experimental findings gave strong evidence regarding the effectiveness of UD in the treatment of AD.

Summary

U. dioica is a perennial herb belongs to the family Urticaceae that has a long history of medicinal uses. Its traditional and folk use has been well documented and highly evident since the time of Hippocrates. Basically, it is a native of Europe and Eurasia, but because of its unique climatic adaptability, now, it grows wild in almost all regions worldwide. It is known by a number of common names like stinging nettle, common nettle, giant nettle, and some of the other known names of this plant in different languages in India are *vrishchhiyaa-shaaka* in Sanskrit, bichu butti in Hindi and Punjabi, and anjuraa in Unani. It has a unique presence of trichomes throughout the plant, which are like sharp piercing needles that aggravate contact urticarial dermatitis when it comes in contact with the skin due to the introduction of inflammatory chemicals like histamine, acetylcholine, serotonin, and formic acid. Presence of these sharp hairs gives it a characteristic identity feature. This plant is known to contain a wide variety of chemicals such as flavanoids, sterols, terpenoids, isolectins, polysaccharides, volatile compounds, fatty acids, vitamins, proteins, and minerals as its main chemical constituents, so it is used for the effective treatment of number of diseases and disorders. An abundance of chemicals like formic acid, histamine, acetylcholine, serotonin, and a low quantity of leukotrienes in the ubiquitous hair present in the entire plant gives its stinging effect, and for this reason, it is also known as stinging nettle. Both leaves and roots of this plant possess high therapeutic value, for which it is used for a number of indications such as lower urinary tract infections, rheumatoid arthritis, and allergic rhinitis. Recent studies have suggested the use and effectiveness of this plant in AD because of its great antioxidant and antiinflammatory potential, as the main underlying causes in the pathogenesis of AD are OS and inflammatory cascades along with other mechanisms. A number of research studies conducted globally support the beneficial role of *U. dioica* in AD. From a toxicity point of view, this plant has been found to be safe enough with very few incidences of side effects. Thus, it has been concluded with all the research findings that *U. dioica* is highly effective in the treatment and management of AD.

References

Ahmed, K.M., Parsuraman, S., 2014. *Urtica dioica* L.,(Urticaceae): a stinging nettle. Sys. Rev. Pharm. 5 (1), 6.

Applequist, W., 2006. The Identification of Medicinal Plants: A Handbook of the Morphology of Botanicals in Commerce. Missouri Botanical Garden Press, St. Louis, MO. Austin, TX: American Botanical Council.

Bathe, A.P., 1994. An unusual manifestation of nettle rash in three horses. Vet. Rec. 134, 11–12.

Baumgardner, D.J., 2016. Stinging nettle: the bad, the good, the unknown. J. Patient Cent. Res. Rev. 3 (1), 48–53.

Bazazzadegan, N., Shasaltaneh, M.D., Saliminejad, K., Kamali, K., Banan, M., Khorshid, H.R., September 2017. Effects of herbal compound (IMOD) on behavior and expression of Alzheimer's disease related genes in streptozotocin-rat model of sporadic Alzheimer's disease. Adv. Pharm. Bull. 7 (3), 491.

Blumenthal, M., Goldberg, A., Brinckmann, J. (Eds.), 2000. Herbal Medicine: Expanded Commission E Monographs. Integrative Medicine Communications, Newton, MA. Austin, TX: American Botanical Council.

Bourgeois, C., Leclerc, É.A., Corbin, C., Doussot, J., Serrano, V., Vanier, J.R., Seigneuret, J.M., Auguin, D., Pichon, C., Lainé, É., Hano, C., September 1, 2016. Nettle (*Urtica dioica* L.) as a source of antioxidant and anti-aging phytochemicals for cosmetic applications. Compt. Rendus Chem. 19 (9), 1090–1100.

Chen, Z., Zhong, C., April 1, 2014. Oxidative stress in Alzheimer's disease. Neurosci. Bull. 30 (2), 271–281.

Cummings, A.J., Olsen, M., June 1, 2011. Mechanism of action of stinging nettles. Wilderness Environ. Med. 22 (2), 136–139.

Daneshmand, P., Saliminejad, K., Shasaltaneh, M.D., Kamali, K., Riazi, G.H., Nazari, R., Azimzadeh, P., Khorshid, H.R., July 2016. Neuroprotective effects of herbal extract (*Rosa canina*, *Tanacetum vulgare* and *Urtica dioica*) on rat model of sporadic Alzheimer's disease. Avicenna J. Med. Biotechnol. 8 (3), 120.

Dar, S.A., Ganai, F.A., Yousuf, A.R., Balkhi, M.U., Bhat, T.M., Sharma, P., February 1, 2013. Pharmacological and toxicological evaluation of *Urtica dioica*. Pharm. Biol. 51 (2), 170–180.

Dastmalchi, K., Dorman, H.D., Vuorela, H., Hiltunen, R., 2007. Plants as potential sources for drug development against Alzheimer's disease. Int. J. Biomed. Pharm. Sci. 1 (2), 83–104.

De Vico, G., Guida, V., Carella, F., March 26, 2018. *Urtica dioica* (Stinging Nettle): a neglected plant with emerging growth promoter/immunostimulant properties for farmed fish. Front. Physiol. 9, 285.

Di Virgilio, N., Papazoglou, E.G., Jankauskiene, Z., Di Lonardo, S., Praczyk, M., Wielgusz, K., June 1, 2015. The potential of stinging nettle (*Urtica dioica* L.) as a crop with multiple uses. Ind. Crop. Prod. 68, 42–49.

Ghasemi, S., Moradzadeh, M., Hosseini, M., Beheshti, F., Sadeghnia, H.R., March 4, 2019. Beneficial effects of *Urtica dioica* on scopolamine-induced memory impairment in rats: protection against acetylcholinesterase activity and neuronal oxidative damage. Drug Chem. Toxicol. 42 (2), 167–175.

Huang, W.J., Zhang, X.I., Chen, W.W., May 1, 2016. Role of oxidative stress in Alzheimer's disease. Biomed. Rep. 4 (5), 519–522.

Joshi, B.C., Mukhija, M., Kalia, A.N., 2014. Pharmacognostical review of *Urtica dioica* L. Int. J. Green Pharm. 8 (4).

Kohen, R., Nyska, A., October 2002. Oxidation of biological systems: oxidative stress phenomena, antioxidants, redox reactions, and methods for their quantification. Toxicol. Pathol. 30 (6), 620–650.

Kregiel, D., Pawlikowska, E., Antolak, H., July 2018. Urtica spp.: ordinary plants with extraordinary properties. Molecules 23 (7), 1664.

Kunwar, A., Priyadarsini, K.I., July 1, 2011. Free radicals, oxidative stress and importance of antioxidants in human health. J. Med. Allied Sci. 1 (2), 53–60.

Lahigi, S.H., Amini, K., Moradi, P., Asaadi, K., 2011. Investigating the chemical composition of different parts extracts of bipod nettle *Urtica dioica* L. in Tonekabon region. Physiology 2 (1), 339–342.

Lasheras, B., Turillas, P., Cenarruzabeitia, E., 1986. Étude pharmacologique préliminaire de *Prunus spinosa* L. Amelanchier ovalis Medikus, *Juniperus communis* L. et *Urtica dioica* L. Plantes Méd. Phytothér. 20, 219–226.

Marzell, H., 1979. Wörterbuch der deutschen Pflanzennamen, Band 4. Franz Steiner Verlag, Wiesbaden, Germany, pp. 914–927. Stuttgart, Germany: S. Hirzel Verlag.

Perry, G., Cash, A.D., Smith, M.A., 2002. Alzheimer disease and oxidative stress. BioMed Res. Int. 2 (3), 120–123.

Shahat, A.A., Ibrahim, A.Y., Ezzeldin, E., Alsaid, M.S., April 4, 2015. Acetylcholinesterase inhibition and antioxidant activity of some medicinal plants for treating neuro degenarative disease. Afr. J. Tradit. Complement. Altern. Med. 12 (3), 97–103.

Singhal, A.K., Naithani, V., Bangar, O.P., May 1, 2012. Medicinal plants with a potential to treat Alzheimer and associated symptoms. Int. J. Nutr. Pharmacol. Neurol. Dis. 2 (2), 84.

Starkenstein, E., Wasserstrom, T., 1933. Pharmakologische undchemische Untersuchungen über die wirksamen Bestandteile der Urtica dioica und Urtica urens, vol. 27. Dtsch Univ Prague, Prague, pp. 137–148.

Taxon: *Urtica dioica* L. US National Plant Germplasm System Website. Blackwell WH. Poisonous and Medicinal Plants, 1990. Prentice-Hall, Inc., Edgewood Cliffs, NJ. Available from: https://npgsweb.ars-grin.gov/gringlobal/taxonomydetail.aspx?id=40944. (Accessed 29 February 2016).

Tekin, M., Özbek, H., Him, A., 2009. Investigation of acute toxicity, anti-inflammatory and analgesic effect of *Urtica dioica* L. Pharmacologyonline 1, 1210–1215.

Toldy, A., Stadler, K., Sasvári, M., Jakus, J., Jung, K.J., Chung, H.Y., Berkes, I., Nyakas, C., Radák, Z., May 30, 2005. The effect of exercise and nettle supplementation on oxidative stress markers in the rat brain. Brain Res. Bull. 65 (6), 487–493.

Toldy, A., Atalay, M., Stadler, K., Sasvári, M., Jakus, J., Jung, K.J., Chung, H.Y., Nyakas, C., Radák, Z., December 1, 2009. The beneficial effects of nettle supplementation and exercise on brain lesion and memory in rat. J. Nutr. Biochem. 20 (12), 974–981.

Upton, R., March 1, 2013. Stinging nettles leaf (*Urtica dioica* L.): extraordinary vegetable medicine. J. Herb. Med. 3 (1), 9–38.

Yoshikawa, T., Naito, Y., 2002. What is oxidative stress? Jpn. Med. Assoc. J. 45 (7), 271–276.

Chapter 3.2.18

Withania somnifera

Vaibhav Rathi[1], Ashwani K. Dhingra[2], Bhawna Chopra[2]
[1]*School of Health Sciences, Quantum University, Roorkee, Uttarakhand, India;* [2]*Guru Gobind Singh College of Pharmacy, Yamuna Nagar, Haryana, India*

Introduction

Withania somnifera is an important medicinal plant of the Indian subcontinent. It is a woody shrub commonly known as "Indian ginseng" or "winter cherry." It is also known as ashwagandha in Sanskrit and asgand in Urdu (Dhuley, 1998; Ziauddin et al., 1996). It was widely used alone or in combination with other herbs to treat numerous biological problems in humans. It possesses a wide spectrum of pharmacological properties, such as antimicrobial, antiinflammatory, antistress, antitumor, neuroprotective, cardioprotective, and many more for use in the treatment of biological approaches. In addition to this, it was able to reduce the level of reactive oxygen species, modulate mitochondrial function, regulate the process of apoptosis, and enhance endothelial function. It was documented in research to treat various clinical conditions related to the nervous system (John Dar et al., 2012). Molecularly, ashwagandha root was useful in Alzheimer's disease (AD) by inhibiting nuclear factor-kB activation, blocking the production of amyloid beta (Aβ), restoring synaptic function, and improving antioxidant effects through the migration of Nrf2 (Sandhir and Sood, 2017). It was well-documented that many herbal formulations are available in the market, such as ashwagandharishta, and stresswin. Thus, the plant and its steroidal components mitigate pathophysiological aspects of the disease; still, further studies are needed to prove the safety and efficacy of this compound in humans.

Pharmacokinetics

Two major constituents—withaferin A and withanolide A—have been identified after oral administration of the aqueous extract of the plant in mice using multiple reaction monitoring. About a 1000-mg/kg extract dose with an equivalent dose of 0.485 mg/kg of withaferin A and 0.4785 mg/kg of withanolide A demonstrated similar pharmacokinetic patterns for both of these with a mean plasma concentration of 16.69 ± 4.02 and 26.59 ± 4.47 ng/mL

Naturally Occurring Chemicals against Alzheimer's Disease. https://doi.org/10.1016/B978-0-12-819212-2.00034-7

for withaferin A and withanolide A with a time taken to reach the maximum concentration of 10 and 20 min, respectively, indicating a rapid absorption process. Thus, oral bioavailability was found to be 1.44 times greater for withaferin A compared with withanolide A (Patil et al., 2013).

In addition, the study performed by Thaiparambil shows that withaferin A has a peak concentration up to 2 L M in plasma with a half-life of 1.36 h followed by a single dose in 7- to 8-week-old female BALB/C mice, whereas clearance from plasma is at rapid (Thaiparambil et al., 2011).

It was also concluded that a single dose of 500 mg/kg in six healthy buffalo claves resulted in a mean peak plasma concentration 248.16 $\pm$ 16.12 Lg/mL at 0.75 h. The mean half-life of *W. somnifera* was 0.92 $\pm$ 0.032 h with total body clearance from 2.26 to 3.09 L/kg/h with a mean value of 2.78 $\pm$ 0.12 L/kg/h (Dahikar et al., 2012).

In the case of albino rabbits with a weight range of 1.5–1.8 kg, both sexes absorbed a single oral dose of 0.42 g/kg when they fasted overnight before administration (Sumanth and Nedunuri, 2014).

Ashwagandha in Alzheimer's disease

The aqueous extract of ashwagandha, improves cognitive and psychomotor performance in healthy human participants (Pingali et al., 2014). The root extract was known to reverse behavioral deficits and pathological conditions as well as Aβ clearance in AD by upregulation of the lipoprotein receptor protein in the liver (Sehgal et al., 2012).

In addition, it also shows protective effects against hydrogen peroxide and Aβ(1–42)–induced cytotoxicity in differentiated PC12 cells (Kumar et al., 2010). One pilot study revealed that the root extract ameliorates deficiencies in cognition and memory loss in adults with mild cognitive impairment (Choudhary et al., 2017).

Research has concluded that the root extract contains an active component, withanone, which shows inhibition of Aβ-42. It also intensifies the activity of acetylcholine (Ach), glutathione, and secretase enzyme (beta and gamma) and thus helps in improve the level of the proinflammatory cytokine (Pandeya et al., 2018).

Its constituents, withanamides A and C, bind to the Aβ(25–35) and prevent the formation of fibrils, thus protecting cells from Aβ toxicity (Jayaprakasam et al., 2010). Studies also prove that withanoside IV and sominone attenuate Aβ(25–35)-induced neurodegeneration, thus improving memory deficits in mice and preventing the loss of axons, dendrites, and synapses (Kuboyama et al., 2006).

It was also found to restore cellular morphology in the Aβ-treated SK-N-MC cell line by enhancing cell viability and PPAR-c level (Kurapati et al., 2013). It also abolished the inhibition of acetylcholinesterase (AchE) activity and cognitive impairment caused by subchronic exposure to propoxur to rats (Yadav et al., 2010).

It also ameliorated oxidative damage induced by streptozocin in cognitive impairment (Ahmed et al., 2013) and thus led to the inhibition of Ach activity (Kurapati et al., 2014).

Withanolide A and the root extract of the plant mitigate memory loss and hippocampal neurodegeneration in male Sprague Dawley rats by attenuating the glutathione level via activation of glutathione biosynthesis in hippocampal cells induced by hypobaric hypoxia. All of these were facilitated by the NrF-2 pathway and NO in a corticosterone-dependent manner and AchE activity (Baitharu et al., 2013, 2014).

Oral administration of *W. somnifera* in mice reduces neurite atrophy, modulates synaptic integrity, and improves memory. The plant's active constituents (withanolide A, withanolide IV, and withanolide VI), preserved the axons and dendrites (Kuboyama et al., 2002).

A water extract of the ashwagandha has been employed in treating the deficiency of cognitive and motor coordination associated with systemic inflammation in rats by regulating the protein expression involved in synaptic plasticity and neuronal cell survival (Gupta and Kaur, 2019). In addition, withanamides and a water extract of the plant rescued PC-12 cells against H_2O_2- and Aβ(1−42)-induced toxicity (Kumar et al., 2010).

The root part of the plant contains glycowithanolides and sitoindosides VII−X. Therefore, the root extract significantly reverses ibotenic acid−induced cognitive impairments in animal models of AD (Bhattacharya et al., 1995).

Studies report that it protects SK-N-SH neuron−like cells against Aβ- and acrolein-induced toxicity. It exhibits effective Ach inhibitory activity and demonstrates a decrease in intracellular reactive oxygen species levels generated by Aβ- and acrolein-induced toxicity (Singh and Ramassamy, 2017).

It also enhances memory consolidation caused by to chronic electroconvulsive shock in mice (Dhuley, 1998).

Oral administration of withanoside IV yields sominone as its active metabolite, which reduces memory deficits by improving dendritic, axonal, and synaptic connections in mice subjected to Aβ-induced neurodegeneration (Uddin et al., 2019).

It was also evident from a computational study that *W. somnifera* has a substantial neuroprotective ability because of the presence of withanolide A. It enhances the central cholinergic function by inhibiting AchE in the treatment of AD. To elucidate the associated binding or ligand interactions involved in cholinesterase inhibition, a docking simulation predicted the residues Thr78, Trp81, Ser120, and His442 for human Ach esterase, which were found to be active sites critical for activity. It also provides evidence for withanolide A as a valuable ligand molecule in the treatment and prevention of AD (Grover et al., 2012; Zahiruddin et al., 2020).

Withaferin A reduces the level of secreted Aβ at lower concentrations (0.5−2 μM) without inducing cytotoxicity in SH-SY5Y cells (Tiwari et al., 2018).

Withanolides have been used in the treatment of the AD (Khan et al., 2016). Withanolide A was found to be a potent inhibitor of AchE activity and reduces the production or level of Aβ protein (Mahrous et al., 2017).

In addition, *W. somnifera* was found to be a potent inhibitor of glycogen synthase kinase-3β, a target involved in AD (Joshi et al., 2018).

Toxicological studies

It can be used in both males and females of all age groups and even during pregnancy without causing side effects (Sharma et al., 1985). As observed in rats, hydroalcoholic root extract with a maximum dose of 2000 mg/kg for 14 days did not result in a toxicity profile or any significant change in body weight, organ weight, or hematobiochemical parameters (Prabu et al., 2013). It does not show toxicity even in the fetus of pregnant rats or any significant change in growth, body weight, or other parameters (Prabu and Panchapakesan, 2015).

A study performed on Swiss albino and Wistar rats indicated that intraperitoneal injections of 1100 mg/kg did not result in death within 24 hours but caused a small increase in the mortality rate with a median lethal dose of 1260 mg/kg of body weight. However, no changes were observed in peripheral blood constituents, but some significant changes in a reduction in the spleen, thymus, and adrenal gland weight were optimized (Tiwari et al., 2014; Sharada et al., 1993).

Aphale and coworkers evaluated the combination of ashwagandha and ginseng in rats. The study was conducted for 90 days in three doses. Upon evaluation, it was observed that liver and body weight increased, but there was an improvement in hematological parameters. Histopathological studies indicate that the brain, liver, kidney, heart, spleen, testis, and ovaries were normal and did not show toxic effects (Aphale et al., 1998). Another study concluded that ashwagandharishta did not show significant changes in kidney functions in both male and female rats at low, medium, and high doses, but serum creatinine levels were found to increase at the medium dose level in males, whereas no changes were observed in the case of females (Rahman et al., 2019).

References

Ahmed, M.E., Javed, H., Khan, M.M., Vaibhav, K., Ahmad, A., et al., 2013. Attenuation of oxidative damage-associated cognitive decline by *Withania somnifera* in rat model of streptozotocininduced cognitive impairment. Protoplasma 250, 1067−1078.

Aphale, A.A., Chhibba, A.D., Kumbhakarna, N.R., Mateenuddin, M., Dahat, S.H., April 1998. Subacute toxicity study of the combination of ginseng (*Panax ginseng*) and ashwagandha (*Withania somnifera*) in rats: a safety assessment. Indian J. Physiol. Pharmacol. 42 (2), 299−302.

Baitharu, I., Jain, V., Deep, S.N., Hota, K.B., Hota, S.K., Prasad, D., Ilavazhagan, G., 2013. *Withania somnifera* root extract ameliorates hypobaric hypoxia induced memory impairment in rats. J. Ethnopharmacol. 145 (2), 431−441.

Baitharu, I., Jain, V., Deep, S.N., Shroff, S., Sahu, J.K., Naik, P.K., Ilavazhagan, G., 2014. Withanolide A prevents neurodegeneration by modulating hippocampal glutathione biosynthesis during hypoxia. PloS One 9 (10) e105311.

Bhattacharya, S.K., Kumar, A., Ghosal, S., 1995. Effects of glycowithanolides from *Withania somnifera* on an animal model of Alzheimer's disease and perturbed central cholinergic markers of cognition in rats. Phytother Res. 9 (2), 110−113.

Choudhary, D., Bhattacharyya, S., Bose, S., 2017. Efficacy and safety of Ashwagandha (*Withania somnifera* (L.) Dunal) root extract in improving memory and cognitive functions. J. Diet. Suppl. 14 (6), 599−612.

Dahikar, P.R., Kumar, N., Sahni, Y., 2012. Pharmacokinetics of *Withania somnifera* (ashwagandha) in healthy buffalo calves. Buffalo Bull 31, 219.

Dhuley, J.N., 1998. Effect of ashwagandha on lipid peroxidation in stress-induced animals. J. Ethnopharmacol. 60 (2), 173−178.

Grover, A., Shandilya, A., Agrawal, V., Bisaria, V.S., Sundar, D., 2012. Computational evidence to inhibition of human acetyl cholinesterase by withanolide A for Alzheimer treatment. J. Biomol. Struct. Dynam. 29 (4), 651−662.

Gupta, M., Kaur, G., 2019. *Withania somnifera* (L.) Dunal ameliorates neurodegeneration and cognitive impairments associated with systemic inflammation. BMC Complement. Alternat. Med. 19 (1), 1−18.

Jayaprakasam, B., Padmanabhan, K., Nair, M.G., 2010. Withanamides in *Withania somnifera* fruit protect PC-12 cells from beta-amyloid responsible for Alzheimer's disease. Phytother Res. 24, 859−863.

John Dar, N., Hamid, A., Ahmad, M., 2012. Pharmacologic overview of *Withania somnifera*, the Indian ginseng. Cell. Mol. Life Sci. https://doi.org/10.1007/s00018-015-2012-1.

Joshi, A., Sharma, A., Kumar, R., 2018. Docking of GSK-3 with novel inhibitors, a targetβprotein involved in Alzheimer's disease. Biosci. Biotech. Res. Comm. 11 (2), 277−284.

Khan, S.A., Khan, S.B., Shah, Z., Asiri, A.M., 2016. Withanolides: biologically active constituents in the treatment of Alzheimer's disease. Med. Chem. 12 (3), 238−256.

Kuboyama, T., Tohda, C., Zhao, J., Nakamura, N., Hattori, M., Komatsu, K., 2002. Axon- or dendritepredominant outgrowth induced by constituents from Ashwagandha. Neuroreport 13 (14), 1715−1720.

Kuboyama, T., Tohda, C., Komatsu, K., 2006. Withanoside IV and its active metabolite, sominone, attenuate Abeta(25−35)-induced neurodegeneration. Eur. J. Neurosci. 23, 1417−1426.

Kumar, S., Seal, C.J., Howes, M.J., Kite, G.C., Okello, E.J., 2010. In vitro protective effects of *Withania somnifera* (L.) dunal root extract against hydrogen peroxide and beta-amyloid(1-42)-induced cytotoxicity in differentiated PC12 cells. Phytother Res. 24 (10), 1567−1574.

Kurapati, K.R., Atluri, V.S., Samikkannu, T., Nair, M.P., 2013. Ashwagandha (*Withania somnifera*) reverses beta-amyloid1-42 induced toxicity in human neuronal cells: implications in HIV associated neurocognitive disorders (HAND). PloS One 8 e77624.

Kurapati, K.R., Samikkannu, T., Atluri, V.S., Kaftanovskaya, E., Yndart, A., et al., 2014. beta-Amyloid1-42, HIV-1Ba-L (clade B) infection and drugs of abuse induced degeneration in human neuronal cells and protective effects of ashwagandha (*Withania somnifera*) and its constituent Withanolide A. PloS One 9 e112818.

Mahrous, R.S., Ghareeb, D.A., Fathy, H.M., Abu EL-Khair, R.M., Omar, A.A., 2017. The protective effect of Egyptian *Withania somnifera* against. Alzheimer's. Med. Aromat Plants (Los Angeles). 6 (2), 1–6.

Pandeya, A., Bani, S., Dutt, P., Kumar Satti, N., Avtar Suri, K., Nabi Qazi, G., 2018. Multifunctional Neuroprotective Effect of Withanone, a Compound from Withania Somnifera Roots in Alleviating Cognitive Dysfunction. Cytokine, p. 221.

Patil, D., Gautam, M., Mishra, S., Karupothula, S., Gairola, S., et al., 2013. Determination of withaferin A and withanolide A in mice plasma using high-performance liquid chromatography-tandem mass spectrometry: application to pharmacokinetics after oral administration of *Withania somnifera* aqueous extract. J. Pharmaceut. Biomed. Anal. 80, 203–212.

Pingali, U., Pilli, R., Fatima, N., 2014. Effect of standardized aqueous extract of *Withania somnifera* on tests of cognitive and psychomotor performance in healthy human participants. Pharmacogn. Res. 6, 12–18.

Prabu, P.C., Panchapakesan, S., 2015. Prenatal developmental toxicity evaluation of *Withania somnifera* root extract in Wistar rats. Drug Chem. Toxicol. 38, 50–56.

Prabu, P.C., Panchapakesan, S., Raj, C.D., 2013. Acute and subacute oral toxicity assessment of the hydroalcoholic extract of *Withania somnifera* roots in Wistar rats. Phytother Res. 27, 1169–1178.

Rahman, T., Rakib Hasan, M., Choudhuri, M.S.K., 2019. Effect of Ashwagandharista (*Withania somnifera*) on the kidney functions of male and female rats. J. Biol. Sci. 8 (1), 1–7 (June).

Sandhir, R., Sood, A., 2017. Europrotective potential of *Withania somnifera* (ashwagandha) in neurological conditions. In: Kaul, S., Wadhwa, R. (Eds.), In Science of Ashwagandha: Preventive and Therapeutic Potentials. Springer International Publishing, Cham, Germany, pp. 373–387.

Sehgal, N., Gupta, A., Valli, R.K., Joshi, S.D., Mills, J.T., et al., 2012. *Withania somnifera* reverses Alzheimer's disease pathology by enhancing low-density lipoprotein receptor-related protein in liver. Proc. Natl. Acad. Sci. U. S. A. 109, 3510–3515.

Sharada, A., Solomon, F.E., Devi, P.U., 1993. Toxicity of *Withania somnifera* root extract in rats and mice. Pharm. Biol. 31, 205–212.

Sharma, S., Dahanukar, S., Karandikar, S., 1985. Effects of longterm administration of the roots of ashwagandha and shatavari in rats. Indian Drugs 22, 133.

Singh, M., Ramassamy, C., 2017. In vitro screening of neuroprotective activity of Indian medicinal plant *Withania somnifera*. J. Nutrition. Sci. 6. https://doi.org/10.1017/jns.2017.48.

Sumanth, M., Nedunuri, S., 2014. Comparison of bioavailability and bioequivalence of herbal anxiolytic drugs with marketed drug alprazolam. World J. Pharmaceut. Res. 3, 1358–1366.

Thaiparambil, J.T., Bender, L., Ganesh, T., Kline, E., Patel, P., et al., 2011. Withaferin A inhibits breast cancer invasion and metastasis at sub-cytotoxic doses by inducing vimentin disassembly and serine 56 phosphorylation. Int. J. Canc. 129, 2744–2755.

Tiwari, R., Chakraborty, S., Saminathan, M., Dhama, K., Singh, S.V., 2014. Ashwagandha (*Withania somnifera*): role in safeguarding health, immunomodulatory effects, combating infections and therapeutic applications: a review. J. Biol. Sci. 14 (2), 77–94.

Tiwari, S., Atluri, V.S.R., Yndart Arias, A., Jayant, R.D., Kaushik, A., Geiger, J., Nair, M.N., 2018. Withaferin A suppresses beta amyloid in APP expressing cells: studies for tat and cocaine associated neurological dysfunctions. Front. Aging Neurosci. 10, 291. https://doi.org/10.3389/fnagi.2018.00291.

Uddin, M.S., Al Mamun, A., Kabir, M.T., Jakaria, M., Mathew, B., Barreto, G.E., Ashraf, G.M., 2019. Nootropic and anti-Alzheimer's actions of medicinal plants: molecular insight into therapeutic potential to alleviate Alzheimer's neuropathology. Mol. Neurobiol. 56 (7), 4925–4944.

Yadav, C.S., Kumar, V., Suke, S.G., Ahmed, R.S., Mediratta, P.K., et al., 2010. Propoxur-induced acetylcholine esterase inhibition and impairment of cognitive function: attenuation by *Withania somnifera*. Indian J. Biochem. Biophys. 47, 117–120.

Zahiruddin, S., Basist1, P., Parveen, A., Parveen, R., Khan, W., Gaurav, Ahmad, S., 2020. Ashwagandha in brain disorders: a review of recent developments. J. Ethnopharmacol. 257, 112876.

Ziauddin, M., Phansalkar, N., Patki, P., Diwanay, S., Patwardhan, B., 1996. Studies on the immunomodulatory effects of Ashwagandha. J. Ethnopharmacol. 50, 69–76.

Chapter 3.2.19

Convolvulus prostratus

Deepak Kumar Semwal[1], Ankit Kumar[2], Ruchi Badoni Semwal[3], Harish Chandra Andola[4]

[1]*Department of Phytochemistry, Faculty of Biomedical Sciences, Uttarakhand Ayurved University, Dehradun, Uttarakhand, India;* [2]*Research and Development Centre, Faculty of Biomedical Sciences, Uttarakhand Ayurved University, Dehradun, Uttarakhand, India;* [3]*Department of Chemistry, Pt. Lalit Mohan Sharma Government Postgraduate College, Rishikesh, Uttarakhand, India;* [4]*School of Environment & Natural Resources, Doon University, Dehradun, Uttarakhand, India*

Introduction

Globally, neurologic and mental disorders are severe health challenges. They are particularly common in developing countries, in which the cultural aspects and poor or limited healthcare facilities make it more important to use traditional medicines. Neurological disorders are caused by a dysfunction in part of the nervous system or brain, resulting in physical and/or psychological symptoms. They are generally classified according to the location, cause, and type of dysfunction, and usually described as central and/or peripheral nervous system disorders. They are congenital, acquired, and idiopathic (unknown cause). Globally, neurological disorders are the main cause of mortality and constitute about 12% of total deaths. Some examples of neurological disorders are Alzheimer's and other dementias, multiple sclerosis, Parkinson's disease, cerebrovascular disease, migraine, neuroinfections, neurological injuries, nutritional deficiencies, neuropathies, tetanus, poliomyelitis, meningitis, Japanese encephalitis, epilepsy, etc. In lower-/middle-income countries, neurological disorders constitute about 16.8% of total deaths, while they account for 13.2% of total deaths in economically strong countries. According to an estimation from 2005, among the neurologic disorders, Alzheimer's and other dementias caused about 2.84%, 0.46%, 0.41%, and 0.34% of the total deaths in high, upper-middle, low, and lower-middle income countries, respectively, in 2005 (WHO, 2006). Alzheimer's disease (AD) was first identified by a German pathologist and psychiatrist Alois Alzheimer in 1906 in a 50-year-old woman (Berchtold and Cotman, 1998). AD causes brain cells to degenerate; it starts slowly but the condition worsens with time. This disease is the cause of 60%–70% cases of dementia in which a patient is unable to function

Naturally Occurring Chemicals against Alzheimer's Disease. https://doi.org/10.1016/B978-0-12-819212-2.00035-9

independently due to a continuous decline in thinking and social skills (WHO, 2017; Burns and Iliffe, 2009). In 2015, about 30 million people were reported as having Alzheimer's disease worldwide (WHO, 2017; Querfurth and LaFerla, 2010). In 2005, a panel of Alzheimer's Disease International (ADI) estimated 24.3 million dementia patients globally, with 4.6 million new cases annually and that this number would double every 20 years to 81 million by 2040. In developed countries, it is estimated that the number of AD patients will increase by up to 100% by the year 2040, however, in China, India, and other South-East Asian countries, this increase may be as much as 300% (WHO, 2006). AD is counted among the most financially costly diseases, mainly in developed countries (Bonin-Guillaume et al., 2005; Meek et al., 1998). Although currently available drugs for AD can temporarily improve symptoms, there is no permanent treatment available (WHO, 2017). Besides the genetic differences in 1%−5% patients, the actual cause of AD remains unknown. Hypothetically, there may be many causes of AD, such as a reduction in the synthesis of the neurotransmitter acetylcholine (Mathew and Subramanian, 2014), extracellular amyloid β (Aβ) deposition (Hardy and Allsop, 1991), the formation of intraneuronal neurofibrillary tangles of hyperphosphorylated τ protein (Kizhakke et al., 2019), poor functioning of the blood−brain barrier (Deane and Zlokovic, 2007), smoking (Cataldo et al., 2010), air pollution, oxidative stress (Moulton and Yang, 2012; Xu et al., 2014), dysfunction of oligodendrocytes (Bartzokis, 2011), gum disease (Miklossy, 2011), and fungal infection (Pisa et al., 2015).

The genus *Convolvulus* is a major group of the Convolvulaceae family of angiosperms or flowering plants; it has up to 72 species that are distributed worldwide. *Convolvulus prostratus* Forssk. is one of the highly valuable medicinal plant used in the treatment of human ailments. It is commonly known as prostrate bindweed and aloe weed. *Convolvulus pluricaulis* Choisy and *Convolvulus pluricaulis* var. *macra* Clarke are synonyms of *Convolvulus prostratus* Forssk (The Plant List, 2013). It is a prostrate, spreading, perennial, wild herb. The stems are 10−40 cm long, prostrate or ascending, and densely velvety with appressed to spreading hairs. The leaves are 0.8−3 cm long and 1.5−6 mm broad, nearly stalkless, linear to oblong, lance-shaped or inverted-lance shaped, and wedge-shaped at the base. Flowers are develop in 1−3 flowered cymes on stalks of up to 3 cm long. Bracts are 3−7 mm long and linear to lance-shaped. The flowers are pale pink or white, the style is 2−4 mm long, while stigma-lobes are 3−5 mm long. The capsule is 3−4 mm in diameter and round in shape. Seeds are 2−2.5 mm long, numbering 2−4, and are dark brown (Flowers of India, 2019). The flowering season is September to October. It is distributed in Egypt, Qatar, Oman, Saudi Arabia, the Sinai Peninsula, Yemen, United Arab Emirates, Pakistan, Nepal, India, south Algeria, Libya, Somalia, Morocco, Mali, Sudan, and the Cape Verde Islands (Catalogue of Life, 2019). All parts of the plant are known to have therapeutic benefits and have long been used in indigenous medicine, mainly in the treatment of epilepsy, hepatic disorder, CNS-related problems, cardiac diseases, respiratory disorder (chronic bronchitis and asthma), fever, skin treatment, hair loss, infectious diseases, and also as a nervine tonic and brain tonic to improve memory (Agarwa et al., 2014).

In India, it is found throughout the country and is known as Shankhpushpi, which is one of the most popular herbs in Ayurveda. According to Ayurveda, it is a laxative, brain tonic, aphrodisiac, and cures psychological diseases. It is astringent in taste, hot in potency, acts as a tissue vitalizer, boosts memory power, promotes luster and vigor, and is an appetizer. It alleviates disease of three Dosas, epilepsy, evil spirits, poverty, skin disease, worm infestation, and poisons (Sitsram, 2015). It is cold in potency and Rasa is bitter. It improves the intellect and quality of the voice. It controls the affections of evil spirits/planets and bestows power to subdue others (Sankhyadhar, 2012). Several bioactive compounds (Fig. 3.2.19.1) have been reported in *Convolvulus prostratus* such as shankhpushphin, convolvine, convolamine, convolidine, convoline, phylabine, confoline, sterol I—II, β-sitosterol, stigmasterol, scopoletin, ayapanin (herniarin), scopolin, caffeic acid, ferulic acid, lupeol, 20-oxodotriacontanol, tetra-triacontanoic acid, 29-oxodotriacontano, volatile oil, etc. (Basu and Dandiya, 1948; Bihaqi et al., 2009; Malik et al., 2016; Irshad and Khatoon, 2018). Some other studies indicate the existence of several other compounds, namely 1,2-benzenedicarboxylic acid, 10-bromodecanoic acid, 1-octadecanesulphonyl chloride, 2-butanone, 2-pentanol, ascorbic acid, cinnamic acid, cyclononasiloxane, cyclononasiloxane, octadecamethyl, decanoic acid, eicosane, heneicosane, nonacosane, octatriacontyl, pentafluropropionate, pentanoic acid, phthalic acid, pyrimidine, silane, squalene, sulfurous acid pentadecyl 2-propyl ester, tridecane, and vitamin E in *C. prostrates* (Rachitha et al., 2018).

Several other species are also known as Shankhpusphi in India, and are used in the place of *C. prostrates* and are commercially available, they are *Clitoria ternatea* L., *Evolvulus alsinoides* (L.) L., *Tephrosia purpurea* (L.) Pers., and *Canscora alata* (Roth) Wall. (syn. *Canscora decussata* (Roxb.) Schult. & Schult.f.), etc. (Sethiya et al., 2009; Irshad and Khatoon, 2018). It is an important ingredient of several herbal formulations such as Remem (Zydus Industries, India), Abana (The Himalaya Drug and Co, India), Tirukati,

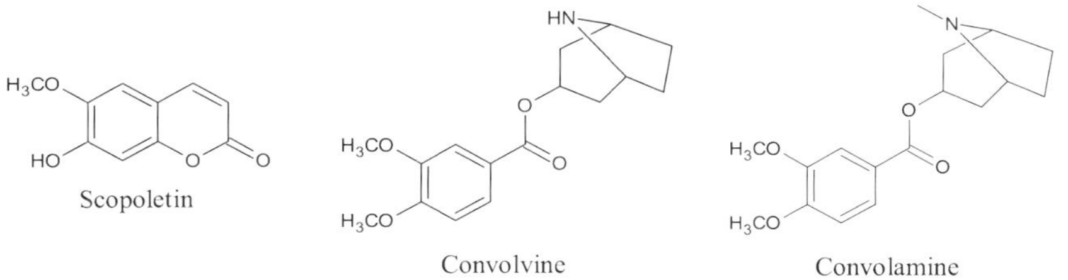

FIGURE 3.2.19.1 Structure of selected bioactive compounds of *C. prostrates*.

Ayumemo (Welexlabs, India) (Agarwa et al., 2014), Brahmi Ghrta, Agastya-haritaki Rasayana, Brahma, Rasayana, Manasmitra Vataka, Gorocanadi Vataka, Brahmi Vati (API, 1999), Dabur Shankhpushpi, Unjha Shankhpushpi, Baidyanath Shankhpushpi, etc.

Pharmacological activities of *C. prostratus*

Anti-Alzheimer's activity

Effect on amyloid β (Aβ) inhibition

The in vitro inhibitory activity of alcoholic extract of leaves on Aβ (amyloid peptide) production and amyloid precursor protein modulation was screened in mouse neuroblastoma cells expressing Swedish (N2a-Swe) APP. MTT assay was also done to determine the toxic concentration of the extract. The alcoholic extract showed greater inhibitory action on Aβ generation; therefore, it was chosen for further examination of APP modulation. The extract was found to be inactive against the multiplication rate as it did not affect cell viability at concentrations of 4, 12, and 36 μg/mL. The extract showed no effect except reducing Aβ production without amyloid precursor protein modulation (Liu et al., 2012). A methanol extract of the whole plant was tested for β-amyloid (2.5 μM)-induced neuroprotection on a neuroblastoma Neuro-2a cell line. The extract (10−200 μM) increased cell viability and the highest concentration (200 μM) showed better effect against neurotoxicity induced by β-amyloid in the brain cell line. The extract also possessed antioxidant activity, acetylcholinesterase, and lipoxygenase enzyme inhibition (Sethiya et al. (2019)).

Effect on acetylcholinesterase and lipoxygenase inhibition

The methanol extracts of the whole plant (0.1 mg/mL) were tested for AChE inhibitory activity using Ellman's microplate colorimetric method. The extract showed 40.6% ± 5.4% inhibition of acetylcholinesterase with an IC_{50} value of 234 μg/mL. The IC_{50} for physostigmine was 0.075 μg/mL (Mathew and Subramanian, 2014). The methanol extracts of the whole plant (50, 100, 150, 200, and 250 μg/mL) were tested for their acetylcholinesterase inhibition against β-amyloid ($A\beta_{1-40}$ 2.5 μM) induced neurotoxicity on a neuroblastoma cell line (Neuro-2a). The IC_{50} values of extract for acetylcholinesterase inhibition were 107.99 and 4.68 μg/mL for galantamine, used as a standard in microplate assay (Sethiya et al., 2019). The study also found that the extract showed lipooxygenase (LOX) inhibitory activity against β-amyloid-induced neurotoxicity on Neuro-2a cells with an IC_{50} value of 81.76 μg/mL. The IC_{50} value for rutin (positive control) was recorded as 4.21 ± 0.17 μg/mL.

Effect on neuroprotection

The methanol, ethanol, and water extracts of the whole plant (1.5−50 μg/mL) and quercetin (standard drug) displayed cytoprotective activity when tested against cytotoxicity induced by hydrogen peroxide (H_2O_2) in human IMR32 neuroblastoma cell lines. Among the tested extracts, methanol extract significantly decreased H_2O_2-induced cell death. A noteworthy reduction in neurofilament (NF-200), heat shock protein (HSP70), and mortalin expression have been recorded in the cultures treated with a combination of methanolic extract and H_2O_2. The levels of first-line defense antioxidants viz. SOD, CAT, GPX, and GSH were increased, whereas lipid peroxidation was decreased with both quercetin and the methanol extract. The study suggested the protective effect of *Convolvulus prostratus* was caused by induction of antioxidant machinery of the cells (Dhuna et al., 2012). The ethanolic leaf extract was also studied for the same effect in SHSY5Y cells. The pretreatment of ethanol extract (50 μg/mL) showed 50% cell survival against 100 μM H_2O_2 challenge for 24 h and it also diminished the lactate dehydrogenase leakage. The pretreatment with ethanol extract improved and regulated the antioxidant and apoptosis markers (SOD, CAT, caspase-3, and p53 inhibited the generation of reactive oxygen species and mitochondrial membrane depolarization) (Rachitha et al., 2018).

The aqueous root extract was evaluated for neuroprotective effects against neurotoxicity induced by aluminum chloride in rat cerebral cortex. The extract potential to inhibit the toxicity was compared with standard drug rivastigmine tartrate (1 mg/kg). The elevated enzymatic activity of AChE (~60%) was reduced when the aqueous extract was administered daily at a dose of 150 mg/kg with aluminum chloride (50 mg/kg) for 3 months and this dose also inhibited the decline in Na^+/K^+ ATPase activity which resulted from the aluminum intake. The extract of root preserved the levels of mRNA in muscarinic receptor 1, choline acetyltransferase (ChAT), and nerve growth factor (NGF)-tyrosine kinase A receptor. The extract also improved the upregulated expression of cyclin-dependent kinase 5 (CDK5) induced by aluminum (Bihaqi et al., 2009). The neuroprotective effect of aqueous roots extract (150 mg/kg) against scopolamine (1 mg/kg)-induced neurotoxicity was screened in the cerebral cortex of male Wistar rats. As a standard drug, rivastigmine tartrate (1 mg/kg, p.o.) was used. Pretreatment with aqueous root extracts significantly decreased the scopolamine-induced increase in the transfer latency in the elevated plus maze, while in the Morris water maze, the extract administration ameliorated the loss of spatial memory induced by scopolamine. Significant inhibition was observed by the extract in the activity of acetylcholinesterase within the cortex (~46%) and hippocampus (~56%). Administration of rivastigmine tartrate also inhibited elevated acetylcholinesterase activity by ~24% in the cerebral cortex and ~30% in the hippocampus. The extract elevated the diminished activities of glutathione reductase (GR), glutathione (GSH) and superoxide dismutase (SOD) within the cortex and hippocampus induced by scopolamine (Bihaqi et al., 2011). In a neuroprotective study, the treatment with aqueous root extract (150 mg/kg) to scopolamine (2 mg/kg, i.p.) treated rats decreased the increase in tau protein and mRNA levels, and AβPP (β-amyloid precursor protein) levels followed by a decrease in amyloid β levels. The results were compared with rivastigmine tartrate (1 mg/kg p.o.), which was used as a positive control (Bihaqi et al., 2012). The neuroprotective action of aqueous extract (root) against human microtubule-associated protein tau (hMAPτ) induced neurotoxicity in an AD *Drosophila* model was studied. A separate study by Kizhakke et al. (2019) also found a neuroprotective effect of aqueous root extract against hMAPτ-induced neurotoxicity in the *Drosophila melanogaster* (Oregon K) strain. The extract was found to control hMAPτ-induced early death. The extract also increased the lifespan and reduced the level of τ protein in tauopathy *Drosophila*. The extract treatment also enhanced the antioxidant enzyme action, ameliorated the oxidative stress induced by τ, and restored the depleted acetylcholinesterase action.

Effect on cognitive function

The effect of ethanol extract of aerial parts, its ethyl acetate, and aqueous fractions (100 and 200 mg/kg p.o.) were evaluated for nootropic activity. The administration of the extract and fractions for 7 days potentially increased the number of avoidance responses in the training trials and retention trials. A dose of 100 mg/kg of extract and fractions was found to be statistically significant, while the results were increased with a higher dose, i.e., 200 mg/kg. The step-down latency in the passive avoidance test was significantly increased by extract and fractions. The ethanol extract, ethyl acetate, and aqueous fraction increased the inflexion ratio to 13.14, 13.00, and 14.68 at the dose of 100 mg/kg, and 17.42, 14.29, and 15.66 at the dose of 200 mg/kg, respectively, as compared to the control group (3.69). The results were compared with the standard drug piracetam (100 mg/kg p.o.). Treatment with the extracts for 15 days showed better retention and recovery than the vehicle-treated animals in a dose-dependent manner against scopolamine-produced amnesia. The decrease in the number of avoidance responses was observed after administration of scopolamine butylbromide (0.3 mg/kg i.p.) on days 9–15 30 min before the daily dosing of the extract and fractions. The animals treated with 100 mg/kg extract as well as fractions took 5–7 days, while animals treated with 200 mg/kg of the same extracts as well as fractions took 3–4 days only to get to the point of reversal, signifying better retention and recovery. In an active avoidance test, the animals treated with extract and fractions showed a remarkable increase in the percentage ARs as compared to the vehicle. These findings suggested the extract and fraction from aerial parts of *C. prostratus* enhance learning and memory retention (Nahata et al., 2008).

The effects of pretrial and posttrial administration of ethanol extract (whole plant) on learning and memory in young and aged mice were evaluated by an elevated plus maze test and by measuring acetylcholinesterase activity. In pretrial administration of the extract, the effect was measured in acute, subchronic, and chronic studies. The animals were treated with distilled water orally (control), piracetam 10 mg/kg, i.p. (standard), and ethanol extract at doses of 100 and 200 mg/kg, p.o. (test). In an acute study in young mice, transfer latency (TL) was noted 60 min after drug administration and again after 24 h. A dose-dependent improvement of memory was observed in extract-treated animals as compared to the vehicle-treated control group when tested on the second day. The extract (200 mg/kg, p.o) showed significantly higher percent retentions (81.42%) than piracetam (48.7%). In a subchronic study in young mice, TL was noted 60 min after drug administration on the third day and again after 24 h in all groups. Multiple treatments with the extract for 3 days showed a significant dose-dependent increase in percentage retentions as compared to the control group. The same effect was observed in piracetam-treated animals. The extract showed a more prominent effect of 84.2% as compared to piracetam 75.9%. In a chronic study in aged mice, TL was noted 60 min after drug administration on the seventh day and again after 24 h. Significant lower percentage retention (3.7%) in aged mice was seen as compared to young mice (18.3%). Mice aged 18–20 months exhibited higher transfer latency (TL) values on the first and second days as compared to young mice, showing impairment in learning and memory. Pretreatment with extract for 7 days improved memory as a significant increase in percent retention as compared to control aged mice. Significantly higher retention (55.3%) was observed at a dose of 200 mg/kg as compared with piracetam (which was 42.84%). To evaluate the effect of posttrial administration of the extract on memory the same treatment was given to animals after training them on an elevated plus-maze. During acquisition trials, latency times were similar across all experimental groups. However, posttrial administration of extract revealed a significant decrease in latency times during retention trials. The percent retention of memory after 24 h in the control group was significantly less than that of the extract-treated groups during these trials. The effect on acetylcholinesterase activity was also screened. Hippocampal regions associated with the learning and memory functions exhibited an increase in acetylcholinesterase activity in a dose-dependent manner in Cornu Ammonis 3 (CA3) area with extract treatment. The principal mechanism of these extract actions may be due to their antioxidant, neuroprotective, and cholinergic properties (Sharma et al., 2010).

The nootropic activity of aqueous-methanol extract (whole plant) at doses of 50, 100, 200, and 400 mg/kg for 30 days was studied against amnesia induced by scopolamine hydrobromide (3 mg/kg, p.o.). The activity was tested using elevated plus-maze (EPM) and step-down models. As a standard drug, piracetam (100 mg/kg) was administered orally. In the EPM test, a decrease in the transfer latency indicated the memory-enhancing action of the tested drug. In the EPM model, the extract showed memory-enhancing action at all tested doses, and at the dose of 100 mg/kg the activity was maximum. When compared to animal training-day latency, the extract showed a 70% reduction in transfer latency. In the step-down passive avoidance model, the extract (100 and 200 mg/kg) profoundly improved the step-down latency. The results at 100 mg/kg were similar to piracetam and were significantly changed from vehicle-treated groups. Moreover, CPE also exhibited significant action at the higher doses (200 and 400 mg/kg), in comparison to controls (Malik et al., 2011). Three compounds, scopoletin, scopolin, and ayapanin (coumarins), isolated from chloroform and ethyl acetate fractions at doses of 2.5, 5, 10, and 15 mg/kg, p.o. were tested for memory-enhancing activity against amnesia induced by scopolamine. As a standard drug, donepezil (1 mg/kg, p.o.) was used. Compounds of scopoletin and scopolin significantly and dose-dependently diminished the amnesic action induced by scopolamine in both elevated plus maze and step-down paradigms. At the dose of 15 mg/kg, the effect of both compounds was similar to that of donepezil. Ayapanin was found to be inactive at all tested doses. However, no significant activity on locomotor action was shown by any of the tested compounds. Both scopoletin and scopolin exhibited significant reversal of increased acetylcholinesterase activity induced by scopolamine, at doses of 10 and 15 mg/kg the activity was comparable to that of donepezil (Malik et al., 2016). The nootropic action of methanol extract (whole plant) at a dose of 400 mg/kg, p.o. was tested against scopolamine (0.3 mg/kg, i.p.) induced memory retrieval. The administration of extract decreased the number of avoidance responses of scopolamine-induced amnesia when compared to piracetam (100 mg/kg p.o.) (Sethiya et al., 2019).

Activity against other CNS-related disorders

Depression

The consecutive 1-week treatment of methanolic extract prepared from the aerial parts decreased immobility time and increased the sucrose preference index and the number of rearings in chronic mild stress rats at oral doses of 50 and 100 mg/kg. The results were found to be comparable to those of standard drug fluoxetine at a dose of 10 mg/kg. The extract also reversed the increased elevated levels of IL-1β, IL-6, TNF-α, alanine transaminase, and aspartate transaminase in rats. In addition, this extract also reestablished the noradrenaline and serotonin levels in the hippocampus and prefrontal cortex of experimental rats. This study suggested that the exerted antidepressant-like effect of methanol extract could be facilitated by antiinflammatory action, reestablishing liver biomarkers or mono-aminergic responses in stressed animals (Gupta and Fernes, 2019). An aqueous-methanol extract of the whole plant at a dose of 400 mg/kg showed an antidepressant effect by decreasing the immobility period by 37% in experimental animals. The extract improved the immobility time to 156, 167, and 191 s at doses of 100, 200, and 400 mg/kg, respectively. A known antidepressant drug, imipramine (12.5 mg/kg), was used as a positive control (Malik et al., 2011). A chloroform fraction of the ethanolic extract of whole plant showed a significant reduction in the immobility time of mice at oral doses of 50 and 100 mg/kg without showing any effect on locomotor activity. This activity was found to be comparable to that of the standard drugs fluoxetine (20 mg/kg) and imipramine (15 mg/kg). The study concluded that the antidepressant-like effect of the chloroform fraction in mice was through the interaction with adrenergic, dopaminergic, and serotonergic systems (Dhingra and Valecha, 2007).

The chloroform, ethanol, and aqueous extracts of the aerial parts were evaluated for their neuropharmacological activity in mice. The extracts (500 mg/kg) were given orally 30 min before thiopental sodium (40 mg/kg) administration. A dose of 1 mg/kg of diazepam was used as a standard. The study showed that only aqueous and ethanol extracts had promising results in general behavior models, while the chloroform extract was less active (Siddiqui et al., 2014).

Stress

The aqueous extract was evaluated for antistress potential, and the stressed animals were post-treated with 100, 150, and 200 mg/kg b.w. of the aqueous extract which significantly restored altered serum cortisol, lipid peroxidation levels, and body weight, compared with the stress control group. The stress was experimentally induced using a cold water forced swimming stress model. The significant dose-dependent antistress activity was observed in this study. This effect was due to its neuroprotective and antioxidant potential (Yuvaraj et al., 2018). The oral administration of methanol extract of aerial parts at a dose of 100 mg/kg showed a potent anxiolytic effect in conditioned avoidance response (CAR) induction of experimental stress test, when the extract was given 45 min before the test. The extract significantly lowered the stress-induced epinephrine level and potentiated more prolonged sleeping time (Sethiya et al., 2009).

Anxiety

The ethyl acetate fraction of ethanolic extract from the aerial parts was tested for central nervous system activity in rat and mouse models. The extract at a dose of 100 mg/kg showed an anxiolytic effect and increased the open-field exploratory behavior. At a dose of 200 mg/kg, ethyl acetate fraction potentially decreased the neuromuscular coordination. Diazepam (1 mg/kg i.p.) was taken as a reference drug. The aqueous fraction was also used in this study but it was not found to be effective at doses of 100 and 200 mg/kg (Nahata et al., 2009).

The ethanol and chloroform extracts of the aerial parts at doses of 500 mg/kg showed anxiolytic-like effects in mice by increasing the total time spent and the open arms in the elevated plus-maze models. The results were compared with an anxiolytic drug, diazepam (1 mg/kg) (Siddiqui et al., 2014). A similar study by Malik et al. (2011) studied the aqueous-methanol extract of the whole plant (50, 100, 200, and 400 mg/kg) to evaluate the anxiolytic effect. The extract exhibited significant anxiolytic action at all dose levels. The effect of extract at the dose of 100 mg/kg was comparable to the standard drug, diazepam (2 mg/kg). The treatment of animals with the extract showed an average of 9.4 entries in the open arms.

Huntington's disease

An ethyl acetate fraction obtained from the methanolic extract of the whole plant showed neuroprotective activity by attenuating body weight loss, enhancing locomotor activity, grip strength, and gait abnormalities in nitropropionic acid-induced Huntington's disease in rats at a dose of 20 mg/kg. The extract also reduced the elevated malondialdehyde and nitrite levels, and reestablished the superoxide dismutase and decreased the glutathione enzyme action in the striatum and cortex. This study revealed that the extract showed a neuroprotective effect through accelerating brain antioxidant defense mechanisms in nitropropionic acid-treated rats (Kaur et al., 2016).

The hydromethanol extract of whole plant (200 mg/kg) and its ethyl acetate (15 and 30 mg/kg), butanol (25 and 50 mg/kg), and aqueous (50 and 100 mg/kg) fractions were significantly in attenuated 3-nitropropionic acid-induced reduction in locomotor activity, memory, grip strength, oxidative defense, and body weight of mice in 15 days. This study suggested that *Convolvulus prostratus* has a protective effect against 3-nitropropionic acid-induced neurotoxicity and can be developed as a drug to treat Huntington's disease (Malik et al., 2015).

Epilepsy (seizure disorder)

The anticonvulsant action of chloroform, ethanol, and aqueous extracts of the aerial part was evaluated against pentylenetetrazol-induced seizures in mice. The extracts were orally administered at a dose of 500 mg/kg body weight, 30 min before the subcutaneous injection of pentylenetetrazol (80 mg/kg). Diazepam, at a dose of 2.0 mg/kg i.p., was used as a standard drug. The ethanol, chloroform, and aqueous extract showed highly efficacious protection against pentylenetetrazol-induced clonic convulsions (453.5, 673.8, and 544.1 s) while the absence of any convulsion was observed in the diazepam-treated groups (Siddiqui et al., 2014). The methanol extract of the whole plant was evaluated for anticonvulsant activity in a seizure model. The extract at doses of 500 and 1000 mg/kg reduced the mean recovery time from convulsion but did not abolish the hind limb extension. The results were compared with phenytoin which was used as a standard drug (Verma et al., 2012).

CNS stimulation

Aqueous-methanol extract of the whole plant (100, 200, 400, and 600 mg/kg) was studied for CNS-depressant activity by measuring the locomotor activity using an actophotometer. As a standard drug, diazepam was administered orally at a dose of 10 mg/kg. The extract gave a significant CNS-depressant activity in a dose-dependent manner. A higher dose (400 and 600 mg/kg) gave an average of 27% and 45% decrease, respectively, in locomotor activity (measured as a decrease in the number of counts), whereas diazepam showed a 52% reduction in activity (Malik et al., 2011). Similarly, chloroform, ethanol, and aqueous extracts of aerial parts were also evaluated for locomotor activity using the same mouse model. As compared with the positive control group (amphetamine, 2 mg/kg), ethanol and aqueous extracts (500 mg/kg) showed a significant reduction in motor activity (Siddiqui et al., 2014). The effect of water-soluble alcoholic extractives of different parts of the plant on the potentiation of barbiturate hypnosis was tested in albino rats (300 mg/kg, i.p.). The maximum barbiturate hypnosis potentiation (sleeping time) was shown by the leaf and flower extractives. A comparison of the activity of plant parts when collected in different seasons (rainy, winter, spring, and summer) was also carried out. The activity was found to be higher when the plant was collected in the spring season while it was lower in the rainy season (Mudgal, 1975).

Alcohol addiction

Shankhpushpi Churna, a marketed formulation (100 and 200 mg/kg) gave dose-dependent activity against acute alcohol withdrawal anxiety in Swiss albino mice screened by an elevated plus-maze test. Diazepam at 1 mg/kg was used as a standard drug. Both Shankhpushpi (200 mg/kg) and diazepam (1 mg/kg) were shown to have comparable anxiolytic potential. However, pretreatment with $GABA_A$ antagonist prevented Shankhpushpi-mediated reversal of withdrawal anxiety. Treatment with $GABA_B$ agonist and Shankhpushpi did not give any significant change in withdrawal anxiety compared with Shankhpushpi-treated animals. These results suggested that Shankhpushpi may reverse ethanol withdrawal anxiety in a $GABA_A$-dependent manner.

On chronic ethanol consumption (21 days), the effect of Shankhpushpi was evaluated using a two-bottle choice protocol of voluntary drinking. The effect was also evaluated on cortico-hippocampal GABA levels. After 30 days of alcohol treatment, animals showed a significant increase in cortico-hippocampal GABA as compared with the control group. Shankhpushpi treatment at a dose of 200 mg/kg for 10 days showed a significant increase in GABA as compared with alcohol-consuming animals. This significance was further increased in the diazepam-treated group as compared with the alcohol group. Animals treated with Shankhpushpi showed a significant decrease in ethanol and water intake as compared with the control group after day 24 or 4 days post-Shankhpushpi therapy. This was comparable with diazepam-treated animals, which also presented a significant decrease in ethanol intake. However, administration with $GABA_A$ blocker to animals followed by

Shankhpushpi failed to show a decrease in ethanol and an increase in water intake till day 30. These findings also suggested that Shankhpushpi prevented chronic ethanol intake and showed antiaddictive potential, which may be mediated by $GABA_A$ receptors (Heba et al., 2017).

Effect on oxidation and oxidative stress

It is well-established that antioxidants are extensively used as health-promoting agents which can protect from many age-related ailments. According to Kamat et al. (2008), there is enough evidence that antioxidants are able to prevent CNS-related diseases. The total phenolic content in methanol leaves extract in 50, 100, and 200 μg/mL was determined to be 0.2092, 0.2380, and 0.3608 mg GAE/g, respectively. The radical-scavenging activity of the extract was recorded to be 52.56% and for standard (butylated hydroxytoluene) to be 93.48% at the highest concentration of 100 μg/mL. The IC_{50} values of 90.56 and 29.02 μg/mL were calculated for the extract and standard, respectively (Balaji et al., 2014). The methanol extract of the whole plant was evaluated for antioxidant activity by DPPH assay and the IC_{50} value of the extract was observed at 41.00 μg/mL as compared to 2.03 μg/mL of ascorbic acid which was used as a standard (Verma et al., 2012). In another study, the significant antioxidant effect was observed with ethanol extract of aerial parts, its ethyl acetate and aqueous fractions, in comparison of ascorbic acid. In a dose-dependent manner, the best superoxide radical-scavenging action was displayed by the ethanol extract followed by the ethyl acetate and aqueous fractions (Nahata et al., 2009). The methanol extracts of the whole plant (0.1 mg/mL) showed 21.9% inhibition of DPPH with an IC_{50} 275 μg/mL. As a positive control, gallic acid and ascorbic acid were used, which showed IC_{50} values of 1 and 2.5 μg/mL, respectively (Mathew and Subramanian, 2014). Four subfractions (FrA, FrB, FrC, and FrD) collected from chloroform and ethyl acetate fractions of methanol extract of the whole plant were evaluated for antioxidant potential. The IC_{50} were 61, 51, 100, and 91 μg/mL, respectively, while the IC_{50} for ascorbic acid was recorded as 40 μg/mL. As a pure compound, scopoletin was obtained from FrB on further purification (Kaur et al., 2016). The extract showed IC_{50} values of 25.46, 26.94, and 22.11 μg/mL in DPPH, FRAP assay, and phosphomolybdenum complex method, respectively. As a standard, vitamin E displayed IC_{50} values of 13.62, 12.39, and 4.41 μg/mL in the respective methods (Sethiya et al., 2019).

Clinical evidence to support CNS-related activities of *C. prostrates*

Sankhapuspi has been used as a brain tonic mainly for improving memory for the past several years. Today, a number of formulations are available in the market which contain *Convolvulus prostrates* as a major component. Its potency against CNS-related disorders has been proven clinically. The effect of oral administration of Sankhapuspi Rasayana (3 g three times a day for 6 weeks) in patients with anxiety disorders was evaluated in a clinical study and the effect was compared with the Jaladhara treatment (30 min daily for 6 weeks). Jaladhara treatment provided 67.49% relief on the Hamilton scale, whereas Sankhapuspi Rasayana provided 72.06% relief. The study also proved that 12.5% patients gained marked improvement and 87.5% patients showed moderate improvement in the Jaladhara group, and in the Sankhapuspi Rasayana group, 47.7% patients showed marked improvement and 58.3% patients showed moderate improvement. Therefore, the overall effect of Sankhapuspi Rasayana for management was better than Jaladhara treatment, as it provided a comparative improvement in anxiety, irritability, inability to relax, lack of concentration, disturbed sleep, loss of memory, palpitations and headache, dryness of mouth, upset stomach, and restlessness. Sankhapuspi Rasayana was made from Sankhapuspi Churna processed with 7 Bhavana of Sankhapuspi Kwath (Dass, 2012). Dandekar et al. (1992) observed an encouraging improvement of seizure control and reduction in plasma phenytoin levels in patients who were continuously taking Sankhapuspi for a long time.

The antihypertensive action of *Convolvulus prostrates* was determined in a randomized, single-blind, active-controlled clinical study. In this study, the aqueous extract of *Convolvulus prostrates* at a dose of 3 g (in a gelatin capsule) was orally administered (twice daily) in combination with Gokhru (*Tribulus terrestris*) 7 g (once a day) for 28 days. The combination of benzthiazide (25 mg) and triaterene (50 mg) was given as a control. After 28 days of therapy, significant activity was noticed in both the experimental and control groups (Rizwan and Khan, 2014).

Toxicity studies

The continuous use of the *Convolvulus pluricaulis* plant for the treatment of many ailments for hundreds of years has confirmed its clinical safety. There are many preclinical studies available that support its nontoxic nature within a dose limit. Its aqueous leaf extract was found to be safe up to a dose of 2000 mg/kg in an acute oral toxicity study in rats. Histopathology showed that the extract is also safe for the brain, heart, and liver (Ravichandra et al., 2013). Similarly, its ethanolic and aqueous extracts at a dose of 5000 mg/kg did not show any toxicity or behavioral changes in rats during an acute oral toxicity study (Agarwal et al., 2014).

Conclusion

Convolvulus pluricaulis, popularly known as Sankhpuspi in Ayurveda, has been used as a brain and nervine tonic for the past several years, mainly in India and other South Asian countries. Its remarkable effect on CNS-related activities makes it an important plant for the treatment of Alzheimer's disease. There are many preclinical studies available which support its role against AD, however, clinical trials, particularly for AD, are not available. Hence, there is a need for further preclinical and clinical studies to enable its use for the treatment of AD.

References

Agarwa, P., Sharma, B., Fatima, A., Jain, S.K., 2014. An update on Ayurvedic herb *Convolvulus pluricaulis* Choisy. Asian Pac. J. Trop. Biomed. 4 (3), 245–252.

Balaji, K., Hean, K.C., Ravich&ran, K., Shikarwar, M., 2014. In-vitro evaluation of antioxidant activity & total phenolic content of methanolic extract of *Convolvulus pluricaulis*. Res. J. Pharm. Biol. Chem. Sci. 5 (6), 959–964.

Bartzokis, G., 2011. Alzheimer's disease as homeostatic responses to age-related myelin breakdown. Neurobiol. Aging 32 (8), 1341–1371.

Basu, N.K., Dandiya, P.C., 1948. Chemical investigation of *Convolvulus pluricaulis* Chois. J. Am. Pharm. Assoc. 37 (1), 27–28.

Berchtold, N.C., Cotman, C.W., 1998. Evolution in the conceptualization of dementia and Alzheimer's disease: Greco-Roman period to the 1960s. Neurobiol. Aging 19 (3), 173–189.

Bihaqi, S.W., Sharma, M., Singh, A.P., Tiwari, M., 2009. Neuroprotective role of *Convolvulus pluricaulis* on aluminium induced neurotoxicity in rat brain. J. Ethnopharmacol. 124 (3), 409–415.

Bihaqi, S.W., Singh, A.P., Tiwari, M., 2011. *In vivo* investigation of the neuroprotective property of *Convolvulus pluricaulis* in scopolamine-induced cognitive impairments in Wistar rats. Indian J. Pharmacol. 43 (5), 520–525.

Bihaqi, S.W., Singh, A.P., Tiwari, M., 2012. Supplementation of *Convolvulus pluricaulis* attenuates scopolamine-induced increased tau & amyloid precursor protein (AβPP) expression in rat brain. Indian J. Pharmacol. 44 (5), 593.

Bonin-Guillaume, S., Zekry, D., Giacobini, E., Gold, G., Michel, J.P., 2005. The economical impact of dementia. Presse Med. 34 (1), 35–41.

Burns, A., Iliffe, S., 2009. Alzheimer's disease. BMJ 338, b158.

Cataldo, J.K., Prochaska, J.J., Glantz, S.A., 2010. Cigarette smoking is a risk factor for Alzheimer's disease: an analysis controlling for tobacco industry affiliation. J. Alzheimers Dis. 19 (2), 465–480.

Catalogue of Life, 2019. Annual Checklist. Retrieved from: http://www.catalogueoflife.org/col/details/species/id/e9548700372fcb059b5012a680b6774f/synonym/1bb8ca4da2bd2439d17488dc7e291839.

Dandekar, U.P., Chandra, R.S., Sharma, A.V., Gokhale, P.C., 1992. Analysis of a clinically important interaction between phenytoin and shankhapushpi, an Ayurvedic preparation. J. Ethnopharmacol. 35, 285–288.

Dass, R.K., 2012. A clinical study to compare the role of Jaladhara & Sankhapuspi rasayan in the management of chittodvega (Anxiety disorders). Int. J. Res. Ayurveda Pharm. 3 (6), 872–875.

Deane, R., Zlokovic, B.V., 2007. Role of the blood–brain barrier in the pathogenesis of Alzheimer's disease. Curr. Alzheimer Res. 4 (2), 191–197.

Dhingra, D., Valecha, R., 2007. Screening for antidepressant-like activity of *convolvulus pluricaulis* choisy in mice. Pharmacologyonline 1, 262–278.

Dhuna, K., Dhuna, V., Bhatia, G., Singh, J., Kamboj, S.S., 2012. Neuroprotective effect of *Convolvulus pluricaulis* methanol extract on hydrogen peroxide induced oxidative stress in human IMR32 neuroblastoma cell line. Br. Biotechnol. J. 2 (4), 192–210.

Flowers of India, 2019. Retrieved from: http://www.flowersofindia.net.

Gupta, G.L., Fernes, J., 2019. Protective effect of *Convolvulus pluricaulis* against neuroinflammation associated depressive behavior induced by chronic unpredictable mild stress in rat. Biomed. Pharmacother. 109, 1698–1708.

Hardy, J., Allsop, D., 1991. Amyloid deposition as the central event in the aetiology of Alzheimer's disease. Trends Pharmacol. Sci. 12, 383–388.

Heba, M., Faraz, S., Banerjee, S., 2017. Effect of Shankhpushpi on alcohol addiction in mice. Phcog. Mag. 13 (49), S148–S153.

Irshad, S., Khatoon, S., 2018. Development of a Validated High-Performance Thin-Layer Chromatography Method for the Simultaneous Estimation of Caffeic Acid, Ferulic Acid, β-Sitosterol, and Lupeol in Convolvulus pluricaulis Choisy and Its Adulterants/Substitutes. JPC-J Planar Chromat 31, 429–436. https://doi.org/10.1556/1006.2018.31.6.2.

Kamat, C.D., Gadal, S., Mhatre, M., Williamson, K.S., Pye, Q.N., Hensley, K., 2008. Antioxidants in central nervous system diseases: preclinical promise and translational challenges. J. Alzheimers Dis. 15 (3), 473–493.

Kaur, M., Prakash, A., Kalia, A.N., 2016. Neuroprotective potential of antioxidant potent fractions from *Convolvulus pluricaulis* Chois in 3-nitropropionic acid challenged rats. Nutr. Neurosci. 19 (2), 70–78.

Kizhakke, P.,A., Olakkaran, S., Antony, A., Tilagul, K.,S., Hunasanahally, P.,G., 2019. *Convolvulus pluricaulis* (Shankhapushpi) ameliorates human microtubule-associated protein tau (hMAPτ) induced neurotoxicity in Alzheimer's disease Drosophila model. J. Chem. Neuroanat. 95, 115–122.

Liu, L.F., Durairajan, S.S.K., Lu, J.-H., Koo, I., Li, M., 2012. In vitro screening on amyloid precursor protein modulation of plants used in Ayurvedic and Traditional Chinese medicine for memory improvement. J. Ethnopharmacol. 141 (2), 754–760.

Malik, J., Karan, M., Vasisht, K., 2011. Nootropic, anxiolytic & CNS-depressant studies on different plant sources of Shankhpushpi. Pharm. Biol. 49 (12), 1234–1242.

Malik, J., Choudhary, S., Kumar, P., 2015. Protective effect of *Convolvulus pluricaulis* standardized extract and its fractions against 3-nitropropionic acid-induced neurotoxicity in rats. Pharm. Biol. 53 (10), 1448–1457.

Malik, J., Karan, M., Vasisht, K., 2016. Attenuating effect of bioactive coumarins from *Convolvulus pluricaulis* on scopolamine-induced amnesia in mice. Nat. Prod. Res. 30 (5), 578–582.

Mathew, M., Subramanian, S., 2014. In vitro screening for anti-cholinesterase & antioxidant activity of methanolic extracts of Ayurvedic medicinal plants used for cognitive disorders. PLoS One 9 (1), e86804.

Meek, P.D., McKeithan, E.K., Schumock, G.T., 1998. Economic considerations in Alzheimer's disease. Pharmacotherapy 18 (2P2), 68–73.

Miklossy, J., 2011. Alzheimer's disease—a neurospirochetosis. Analysis of the evidence following Koch's and Hill's criteria. J. Neuroinflammation 8 (1), 90.

Moulton, P.V., Yang, W., 2012. Air pollution, oxidative stress, and Alzheimer's disease. J. Environ. Public Health 2012.

Mudgal, V., 1975. Studies on medicinal properties of *Convolvulus pluricaulis* and *Boerhaavia diffusa*. Planta Med. 28 (1), 62–68.

Nahata, A., Patil, U.K., Dixit, V.K., 2008. Effect of *Convulvulus pluricaulis* Choisy. on learning behaviour and memory enhancement activity in rodents. Nat. Prod. Res. 22 (16), 1472–1482.

Nahata, A., Patil, U.K., Dixit, V.K., 2009. Anxiolytic activity of *Evolvulus alsinoides* and *Convulvulus pluricaulis* in rodents. Pharmaceut. Biol. 47 (5), 444–451.

Pisa, D., Alonso, R., Rábano, A., Rodal, I., Carrasco, L., 2015. Different brain regions are infected with fungi in Alzheimer's disease. Sci. Rep. 5, 15015.

Querfurth, H.W., LaFerla, F.M., 2010. Mechanisms of disease. N. Engl. J. Med. 362 (4), 329–344.

Rachitha, P., Krupashree, K., Jayashree, G.V., Kandikattu, H.K., Amruta, N., Gopalan, N., Khanum, F., 2018. Chemical composition, antioxidant potential, macromolecule damage & neuroprotective activity of *Convolvulus pluricaulis*. J. Tradit. Complement. Med. 8 (4), 483–496.

Ravichandra, V.D., Ramesh, C., Sridhar, K.A., 2013. Hepatoprotective potentials of aqueous extract of *Convolvulus pluricaulis* against thioacetamide induced liver damage in rats. Biomed. Aging Pathol. 3 (3), 31–135.

Rizwan, M., Khan, A.A., 2014. Assessment of efficacy of Sankhahuli (*Convolvulus pluricaulis* Chois.) & Gokhru (*Tribulus terrestris* L.) in the management of hypertension. Indian J. Tradit. Know. 13 (2), 313–318.

Sankhyadhar, S.C., 2012. Raj Nighantu, Shri Narhari Pandit. Chaukhambha Orientalia, Varanasi.

Sethiya, N.K., Thakore, S.G., Mishra, S.H., 2009. Comparative evaluation on commercial sources of indigenous medicine Shankhpushpi for anti-stress potential: a preliminary study. Pharmacologyonline 2, 460–467.

Sethiya, N.K., Nahata, A., Singh, P.K., Mishra, S.H., 2019. Neuropharmacological evaluation on four traditional herbs used as nervine tonic and commonly available as Shankhpushpi in India. J. Ayurveda Integr. Med. 10 (1), 25–31.

Sharma, K., Bhatnagar, M., Kulkarni, S.K., 2010. Effect of *Convolvulus pluricaulis* Choisy. and *Asparagus racemosus* willd on learning & memory in young and old mice: a comparative evaluation. Indian J. Exp. Biol. 48 (5), 479–485.

Siddiqui, N.A., Ahmad, N., Musthaq, N., Chattopadhyaya, I., Kumria, R., Gupta, S., 2014. Neuropharmacological profile of extracts of aerial parts of *Convolvulus pluricaulis* choisy in mice model. Open Neurol. J. 8 (1), 11–14.

Sitsram, B., 2015. Bhavaprakasha Nighantu: Bhavaprakasha of Bhavamishra. Chaukhambha Orientalia, Varanasi.

The Ayurvedic Pharmacopoeia of India, 1999. Part-I, vol. II, p. 156.

The Plant List, 2013. Version 1.1. Published on the Internet. http://www.theplantlist.org/. (Accessed 1 January 2020).

Verma, S., Sinha, R., Kumar, P., Amin, F., Jain, & J., Tanwar, S., 2012. Study of *Convolvulus pluricaulis* for antioxidant & anticonvulsant activity. Cent. Nerv. Syst. Agents Med. Chem. 12 (1), 55–59.

World Health Organization, 2006. Neurological Disorders Public Health Challenges.

World Health Organization, 2017. Dementia Fact Sheet.

Xu, H., Finkelstein, D.I., Adlard, P.A., 2014. Interactions of metals and Apolipoprotein E in Alzheimer's disease. Front. Aging Neurosci. 6, 121.

Yuvaraj, B.K., Saraswathi, P., Vijayaraghavan, R., Mohanraj, K.G., Vishnu, P.V., 2018. Anti-stress potential of *Convolvulus pluricaulis* choisy in chronic cold swimming stress rat model. Int. J. Res. Pharm. Sci. 9 (2), 349–352.

Chapter 3.2.20

Celastrus paniculatus

Harikesh Maurya[1], Rajewshwar K.K. Arya[2], Tarun Belwal[3], Mahendra Rana[2], Aadesh Kumar[2]

[1]*M.G.B. Rajat College of Pharmacy, Ambedkar Nagar, Uttar Pradesh, India;* [2]*Department of Pharmaceutical Sciences, Center for Excellence in Medicinal Plants and Nanotechnology, Kumaun University, Nainital, Uttarakhand, India;* [3]*College of Biosystems Engineering and Food Science, Zhejiang University, China*

Introduction

Alzheimer's disease is a neurodegenerative and age-associated progressive disorder characterized by severe memory loss, cognitive dysfunction, unusual behaviors, and deficits in personal activities (Alam and Haque, 2011). The neurotransmitters, acetylcholine transformed in the brain principally by acetylcholinesterase (AChE) and subsequently by butyrylcholinesterase (BuChE), are thought to play a pathological role in Alzheimer's disease (Hebert et al., 1995) and are distinguished by neurofibrillary tangles and neuritic plaques (Beard et al., 1995). The quantity of acetylcholine (Ach) elevated through AChE has been established as an efficient treatment approach in Alzheimer's disease (Arnold and Kumar, 1993). In view of this, the development of new AChE inhibitors that are nontoxic and more effective will be indispensable for the management of those with worsening conditions. These clarifications suggest that the existing medication and the inaccessible phyto-active complexes may be potential directions to the expansion of new improved pharmaceuticals (Huang et al., 2013).

Celastrus paniculatus Wild (Celastraceae) is also called Malkangni in Hindi, Jyotishmati in Sanskrit, and black oil plant in English, and is an important herb, traditionally used for the improvement of memory and cognitive functions (Malik et al., 2017). It has been distributed all over India at altitudes $\geq$2000 m in the subtropical Himalayan region and grows wild in Indonesia, Laos, Maharashtra, Orissa, and the Andaman Nicobar group of islands (Sharada et al., 2003). The plant grows as a large woody climber (gynodioecious category), with a yellow corky bark, and climbs up to over 10 m supported by poles or a nearby tree. The stem is approximately 23 cm in diameter, and is covered with lenticels, alternate, oblong, and broadly elliptic

Naturally Occurring Chemicals against Alzheimer's Disease. https://doi.org/10.1016/B978-0-12-819212-2.00036-0

FIGURE 3.2.20.1 *Celastrus paniculatus* Wild. (A) The plant with young fruits. (B) Flowering branchlets. (C) Close-up of flowers. (D) Inflorescence with bracts. (E) Infructescence. (F) Close-up of a capsule. (G) Capsule with seeds (John et al., 2017). *Source: https://www.researchgate.net/publication/318882459/figure/fig2/AS:525142481739778@1502215132942/Celastrus-paniculatus-Willd-ssp-angladeanus-SJ-Britto-B-Mani-S-Thomas-A-The.png.*

leaves (Suttee et al., 2013). Malkangni are available in all seasons through yellowish-green flowers, unisexual at the terminal with pendulous panicles. The fruits are well encapsulated, fully developed globules, three-valved, three-celled, with three to six seeds enclosed in the completely red arillus, that are ovoid and brown (Harish et al., 2008; Singh et al., 1996). The image of seeds (capsule with seed), leaves, stings, flower, fruit (close-up of capsule) could be seen clearly in the Fig. 3.2.20.1 (John et al., 2017).

Botanical description (Bhanumathy et al., 2010a; John et al., 2017)

Botanical name: *Celastrus paniculatus* belonging to the family Celastraceae
English: Black seed-oil tree, intellect tree, climbing-staff tree
Hindi: Malkangni, Kondgaidh, Malkagani, Sankhu
Sanskrit: Jyotishmati, Svarnalota, Jyotishka, Katabhi

Plant taxonomy(*Celastrus paniculatus* Wild, 1875; Premila, 2006)

Kingdom: Plantae
Subkingdom: Viridiplantae

Infrakingdom: Streptophyta
Superdivision: Embryophyta
Division: Tracheophyta
Subdivision: Spermatophyta
Class: Magnoliopsida
Superorder: Rosanae
Order: Celastrales
Family: Celastraceae
Genus: Celastrus
Species: Paniculatus

Medicinal parts

The root, leaves, bark, seeds, and oil are clinically useful in Ayurvedic remedies to alleviate several human ailments, i.e., skin diseases, leprosy, asthma, depression, fever, paralysis, and arthritis (Warintorn et al., 2018; Lal and Singh, 2010). The seed of the plant is reported to have sedative, anxiolytic, and anticonvulsant activities, while the root and bark extracts are considered useful in the treatment of brain tumors and malaria, as a brain tonic, useful in snake bite, and taken internally as an abortifacient (Deodhar and Shinde, 2015). The leaves are widely used for combating addiction, while its juice is given in the treatment of opium addiction. The alcohol extract of the seed exerted free-radical scavengers and antioxidant properties in individual ailments (Ramammoorthy et al., 2007).

The characteristic properties of *Celastrus paniculatus* seeds are acridity, and using thermogenics it is clinically used as a cerebral circulatory stimulant, promoting intellect, treating hepatitis, cardiotonic, laxative, emetic, expectorant, aphrodisiac, diuretic, diaphoretic, febrifugic, moisturizer, antileprosy, skin diseases, paralytic, antimigraine, antiarthralgia, asthmatic, leucodermic, antiulcer, cardiac debility, nephropathic, dysmenorrheal properties, inducing menstruation, and sometime causing abortion (Sharma et al., 2013; Shivwanshi and Gaikwad, 2017). While the taste of the seed oil is unpleasant, it is useful in abdominal disorders, beriberi, blistering, and intellect promotion while also reducing the rate of bowel movement although it had no affect on gastrointestinal transit (Palle et al., 2018). Consumption for 2 weeks is sufficient to produce a notable effect on memory, eyesight, cognition, and mental alertness (Humber, 2002).

Physical properties (Deodhar and Shinde, 2014)

Taste: Pungent, bitter
Quality: Acrid, unctuous (oily)
Potency: Hot (warm effect on the body)
Therapeutic effect: Nootropic (brain tonic)

Beneficial for organs: Nerves, joints, and bones

Chemical constituents

The phytochemical analysis of leaf and seed extract of *Celastrus paniculatus* has illustrated different constituents, i.e., alkaloids, glycosides, proteins, amino acids, phenolic compounds, tannins, fixed oil, carbohydrates, flavonoids, and saponins in aqueous extract, while the aqueous and ethanol extracts contain sterols and triterpenoids (Deodhar and Shinde, 2014; Venkataramaiah and Wudayagiri, 2013). The seed extract possesses around 30% oil and contains sterols, alkaloids like celapanine, celapanigine, celapagine, celastrine, and paniculatine that are rich in oleic acid (55%) together with linoleic acid (16%), palmitic acid (20%), and stearic acid (4%) (Parimala et al., 2009; Zohera et al., 2010), and also contains sesquiterpenes like dipalmitoyl glycerol (Debnath et al., 2012). (The chemical structure of active components is given in Table 3.2.20.1.) The sesquiterpene alkaloids result from a new sesquiterpene

TABLE 3.2.20.1 Chemical structures of the constituents present in *Celastrus paniculatus*.

Name of chemical constituent	Chemical structure
Celapanine	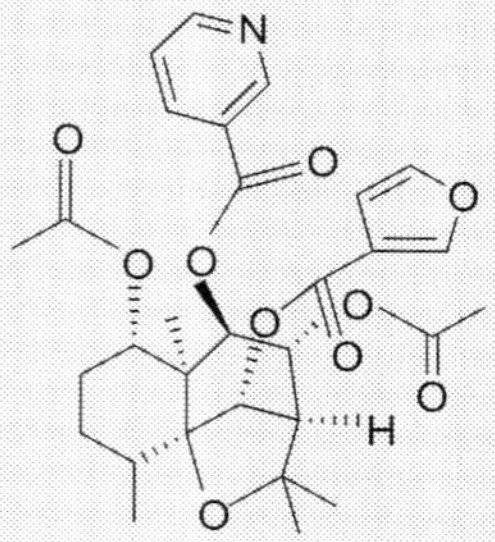
Paniculatine	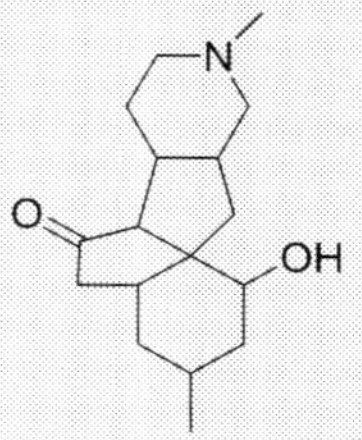

TABLE 3.2.20.1 Chemical structures of the constituents present in *Celastrus paniculatus.*—cont'd

Name of chemical constituent	Chemical structure
Oleic acid	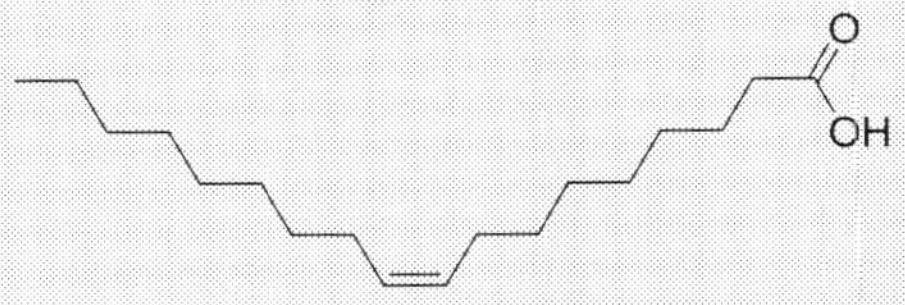 oleic acid
Linoleic acid	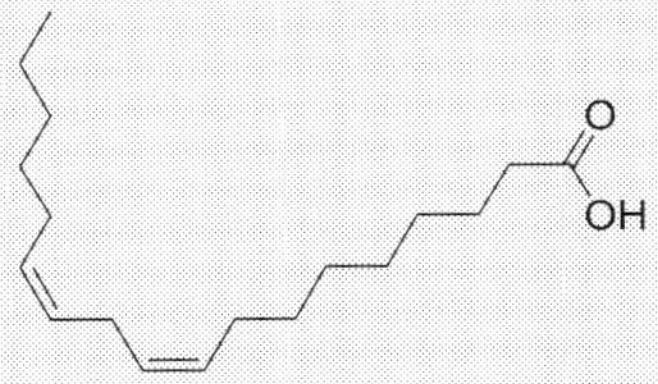 linoleic acid
Palmitic acid	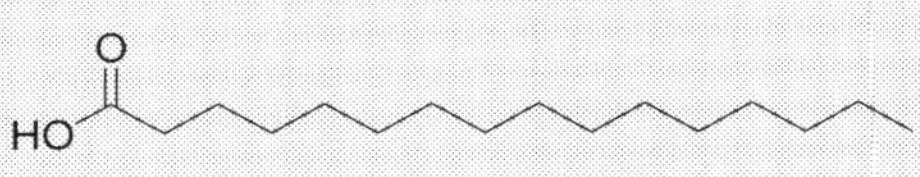 palmitic acid
Stearic acid	 stearic acid

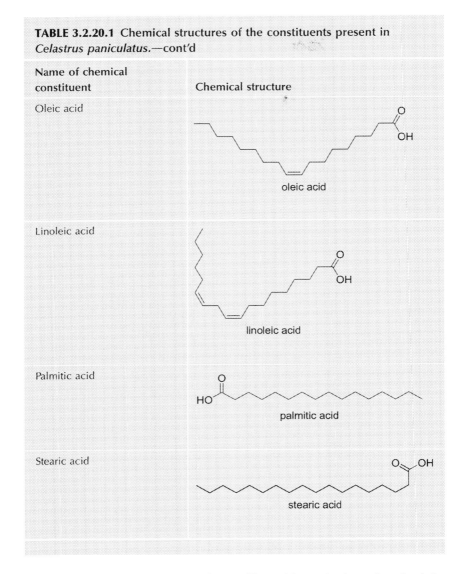

tetra-ol (celapanol) that alternately esterifies with acetic, benzoic, nicotinic, and β-furoic acids (Kalaskar et al., 2012; Wagner and Heckel, 1975).

Therapeutic effects

In the Ayurveda system *Celastrus paniculatus* is recognized as effective an nervine tonic, sedative, antidepressant and produces a significant effect in the

TABLE 3.2.20.2 Bioactive components and therapeutic benefits of different parts of *Celastrus paniculatus*.

Parts	Bio-active components	Therapeutic effects
Leaves	Saponin	Antifungal, antimicrobial, and antitussive
Root and bark	β-sitosterol, pristimerin, terpenes, Zeylasteral, celastrol, zeylasterone	Used in the treatment of malaria, arthritis, sciatica, and bradycardia
Seed	Alkaloids like celastrine and paniculatin, fatty acids, acetic acids, benzoic acids, sterol, and tetracasanol	Sedative and antidepressant actions, febrifugal, insomnia, tranquilizer, anticognitive, and diaphoretic properties Used to treat Alzheimer's disease

chronic debilitating diseases of the nervous system. The *Celastrus paniculatus* plant was traditionally used as a powerful brain tonic, to increase mental activity and suppress symptoms like those of Alzheimer's disease (Bhagy et al., 2016). It radically improves brain power as well as retention and recall of memories in Alzheimer's patients (Howes and Houghton, 2003). It facilitates the overcoming of physical weakness and mental confusion and also relieves asthmatic symptoms, and reduces headaches and pain in arthritis (The component present in different parts of *C. paniculatus* and their therapeutic effects were concisely given in Table 3.2.20.2) (Arora and Rai, 2012; Gamlath et al., 1990).

Pharmacological activities

Around two-thirds of Alzheimer's patients suffering from dementia, which includes short-term memory loss, difficulty with language, disorientation, mood swings, loss of motivation, and behavioral disturbances. The most effective medicines used to treat mild-to-moderate Alzheimer's disease are memantine, tacrine, rivastigmine, galantamine, and donepezil (Crews and Masliah, 2010).

The pharmacology of Alzheimer's disease may be divided into two classes:

- NMDA receptor antagonists;
- Acetylcholinesterase inhibitors.

NMDA receptors: Memantine is a noncompetitive antagonist and glutamate is an important excitatory neurotransmitter in the brain. By inhibiting NMDA receptors, memantine inhibits the excitatory effects of its ligand, glutamate. Memantine is used to treat moderate-to-severe symptoms of Alzheimer's disease (Shankar et al., 2007).

Acetylcholinesterase inhibitors: Tacrine, rivastigmine, galantamine, and donepezil inhibit the breakdown of the enzyme acetylcholinesterase, while extra acetylcholine is presented to conduct neuronal activity. A number of possibilities are under investigation, including the formation of pore-like structures with channel activity (Shankar et al., 2008), variation in glutamate receptors and excitotoxicity (Li et al., 2009), circuitry hyperexcitability; mitochondrial dysfunction (Nakamura and Lipton, 2010), lysosomal failure, and alterations in signaling pathways interrelated to synaptic plasticity, neuronal cell death, and neurogenesis (Palop et al., 2007). The seed oil of *Celastrus paniculatus* shows significant pharmacological activities such as nootropic, antiarthritic, hypolipidemic, antioxidant, central nervous system stimulant, antifertility, analgesic, antiinflammatory effects and a positive effect on cardiovascular disorders (Godkar et al., 2003).

Benefits and uses

Celastrus paniculatus seeds are used in powder form and taken with milk. The beneficial effects on the following organs, brain, nerves, joints, and bones are given as:

Intellect improvement: The seeds improve concentration, alertness, and cognitive function in the brain by acting on the acetylcholine level (Godkar et al., 2003).

Loss of memory: Generally it is used as a memory booster (brain tonic) and is effective in forgetfulness (memory disorder) at a dose of 5−15 drops seed oil or 1 g seed powder with milk (Godkar et al., 2004).

Dementia: It prevents the progress of cell damage in the brain due to its potent antioxidant effect. It was reported that the seed oil increases the glutathione and catalase levels but decreases the level of malonaldehyde in the brain, which may be responsible for its cognitive-enhancing and neuroprotective effects (Da Rocha et al., 2011).

Nootropic: It acts as a nervine stimulant that induces alertness, improves concentration, ability to think, reasoning, reduces nerve cell death, and tackles stress disorders (Kumar and Gupta, 2002).

Insomnia: It produces an antistress and calming effect that helps to induce good sleep.

Anti-Alzheimer: *Celastrus paniculatus* seed extract and its organic fraction shows moderate anticholinesterase activity and is effectively used in the treatment of Alzheimer's disease (Bhanumathy et al., 2010b).

TABLE 3.2.20.3 General dosage of *Celastrus paniculatus* (seed and oil) in individual categories of users (Singh, 2016).

S.N.	Individual category	Seed powder	Seed oil
1	Children	10 mg/kg body weight	1–5 drops
2	Adults	500 mg–2 g	5–15 drops
3	Pregnancy	CONTRAINDICATION	
4	Maximum dosage	4 g/day in divided dose	45 drops/day in divided dose
5	Administration	Twice a day 2 h after food with butter or milk	

Dosage and administration

The dose of seeds and oil of *Celastrus paniculatus* are discussed in Table 3.2.20.3.

Generally, a low dose of *Celastrus paniculatus* (seed powder and oil) is well tolerated and likely safe in all categories of patients. Inappropriate dosages or wrong uses may cause a mild side effect such as restlessness, giddiness, heat sensation, burning sensation, and excessive sweating (Singh, 2016).

Safety issue

Yogesh A. Kulkarni et al. (2015) evaluated a toxicity study on *Celastrus paniculatus* seed extract with an initial dose of 300 mg/kg carried out as per of the Organization for Economic Cooperation and Development (OECD) guidelines No. 423. It was reported that the extract did not show any toxic signs or symptoms during or after the study and it was reported that doses of 300, 2000, and 5000 mg/kg body weight were well tolerated after a single oral administration. Thus the maximum accepted dose (5000 mg/kg) of *Celastrus paniculatus* seed extract has been considered as safe and confirmed nontoxic on single administration (Kulkarni et al., 2015).

Conclusion

Alzheimer's disease is generally known to be caused by a cholinergic insufficiency, principally in part of the cerebral cortex of the human brain. It is assumed that this is as a result of enhancing cholinergic levels, which play a significant role in improvement of mild-to-moderate Alzheimer's disease (Nakamura and Lipton, 2009).

Celastrus paniculatus is a versatile plant with excellent potency to treat several abnormal health conditions. The seeds, fruits, leaves, and root of this plant possess several chemical constituents such as alkaloids, glycosides, protein, amino acid, phenol, tannins, fixed oil, carbohydrates, flavonoids, saponin, sterols and triterpinoids. Active constituents of *C. paniculatus* show excellent effect on CNS antioxidant action, nootropic, and anti-Alzheimer actions. The extract protects neuronal cells against H_2O_2-induced toxicity due to their antioxidant properties and good ability to protect neuronal cells against glutamate-induced toxicity by modulating function of glutamate receptor. The phytoconstituents of this plant possess potent free radical-scavenging properties, reducing lipid peroxidation, and giving a better ability to induce the antioxidant enzyme catalase.

The phytochemical properties of seed extracts were reported the presence of oleic acid, palmitic acid, linoleic acid, stearic acid with few novel compounds such as benzendiol, cinnamic acid, benzoquinone, butylated hydroxytoluene and eudalene. The pharmacological activities have been reported as nootropic, anti-Alzheimer, memory enhancer, antioxidant, nervine-tonic, antiarthritic, hypolipidemic, antifertility, and analgesic.

Celastrus paniculatus seed oil has been traditionally used to improve learning and memory, while its antioxidant properties decrease lipid peroxidation and free radicals. The mechanism by which *Celastrus paniculatus* seed oil enhances cognition may be attributed to its antioxidant properties. Finally, it was concluded that the pharmacological evidence suggests that *Celastrus paniculatus* seeds are a promising supplement and may protect the brain and improve cognition in Alzheimer's patients.

References

Alam, B., Haque, E., 2011. Anti-Alzheimer and antioxidant activity of *Celastrus paniculatus* seed. Iran. J. Pharm. Sci. Winter 7, 49–56.

Arnold, S.E., Kumar, A., 1993. Reversible dementias. Med. Clin. N. Am. 77, 215–225.

Arora, N., Rai, S.P., 2012. *Celastrus paniculatus*, an endangered Indian medicinal plant with miraculous cognitive and other therapeutic properties: an overview. Int. J. Pharm. Biol. Sci. 3, 290–303.

Beard, C.M., Kokmen, E., Kurland, L.T., 1995. Prevalence of dementia is changing over time in Rochester, Minnesota. Neurology 45, 75–79.

Bhagy, V., Christofer, T., Shankaranarayana, R.B., 2016. Neuroprotective effect of *Celastrus paniculatus* on chronic stress-induced cognitive impairment. Indian J. Pharmacol. 48, 687–693.

Bhanumathy, M., Chandrasekar, S.B., Chandur, U., Somasundaram, T., 2010a. Phytopharmacology of *Celastrus paniculatus*: an overview. Int. J. Pharm. Sci. Drug Res. 2, 176–181.

Bhanumathy, M., Harish, M.S., Shivaprasad, H.N., Sushma, G., 2010b. Nootropic activity of *Celastrus paniculatus* seed. Pharm. Biol. 48, 324–327.

Celastrus paniculatus Willd, 1875. Sp. Pl. 1: 1125. 1797. Hook. f., Fl. Brit., India, p. 617.

Crews, L., Masliah, E., 2010. Molecular mechanisms of neurodegeneration in Alzheimer's disease. Hum. Mol. Genet. 19 (R1), R12–R20.

Da Rocha, M.D., Viegas, F.P., Campos, H.C., Nicastro, P.C., Fossaluzza, P.C., Fraga, C.A., Barreiro, E.J., Viegas, C.J., 2011. The role of natural products in the discovery of new drug candidates for the treatment of neurodegenerative disorders II: Alzheimer's disease. CNS Neurol. Disord. - Drug Targets 10, 251–270.

Debnath, M., Biswas, M., Nishteswar, K., 2012. Evaluation of analgesic activity of different leaf extracts of *Celastrus paniculatus* Willd. J. Adv. Pharm. Educ. Res. 2, 68–73.

Deodhar, K.A., Shinde, N.W., 2014. Phytochemical constituents of leaves of *Celastrus paniculatus* Willd: endangered medicinal plant. Int. J. Pharmacogn. Phytochem. Res. 6, 792–794.

Deodhar, K.A., Shinde, N.W., 2015. *Celastrus paniculatus* traditional uses and ethnobotanical study. Indian J. Adv. Plant Res. 2, 18–21.

Gamlath, Celastrus, B., Gunatilaka, A.L., Tezuka, Y., Kikuchi, T., Balasubramaniam, S., 1990. Quinone-methide phenolic and related triterpenoids of plants of Celastraceae further evidence for the structure of Celastranhydride. Phytochemistry 29, 3189–3192.

Godkar, P., Gordon, R.K., Ravindran, A., Doctor, B.P., 2003. Celastrus paniculatus seed water soluble extracts protect cultured rat forebrain neuronal cells from hydrogen peroxide-induced oxidative injury. Fitoterapia 74, 658–669.

Godkar, P.B., Gordon, R.K., Ravindran, A., Doctor, B.P., 2004. Celastrus paniculatus seed water soluble extracts protect against glutamate toxicity in neuronal cultures from rat forebrain. J. Ethnopharmacol. 93, 213–219.

Harish, B.G., Krishna, V., Santosh Kumar, H.S., Khadeer Ahamed, B.M., Sharath, R., Kumar Swamy, H.M., 2008. Wound healing activity and docking of glycogen-synthase-kinase-3-beta-protein with isolated triterpenoid lupeol in rats. Phytomedicine 15, 763–767.

Hebert, L.E., Scherr, P.A., Beckeff, L.A., 1995. Age-specific incidence of Alzheimer's disease in a community population. J. Am. Med. Assoc. 273, 1354–1359.

Howes, M.J., Houghton, P.J., 2003. Plants used in Chinese and Indian traditional medicine for improvement of memory and cognitive function. Pharmacol. Biochem. Behav. 75, 513–527.

Huang, L., Su, T., Li, X., 2013. Natural products as sources of new lead compounds for the treatment of Alzheimer's disease. Curr. Top. Med. Chem. 13, 1864–1878.

Humber, J.M., 2002. The role of complementary and alternative medicine accommodating pluralism. J. Am. Med. Assoc. 288, 1655–1656.

John, B.S., Mani, B., Thomas, S., Prabhu, S., 2017. A new subspecies of *Celastrus* (Celastraceae) from the Palni hills of South India. Taiwania 62, 311–314.

Kalaskar, M.G., Saner, S.Y., Pawar, M.V., Rokade, D.L., Surana, S.J., 2012. Pharmacognostical investigation and physicochemical analysis of *Celastrus paniculatus* Willd. leaves. Asian Pac. J. Trop. Biomed. 2, 1232–1236.

Kulkarni, Y.A., Agarwal, S., Garud, M.S., 2015. Effect of Jyotishmati (*Celastrus paniculatus*) seeds in animal models of pain and inflammation. J. Ayurveda Integr. Med. 56, 82–88.

Kumar, M.H., Gupta, Y.K., 2002. Antioxidant property of *Celastrus paniculatus* Willd.: a possible mechanism in enhancing cognition. Phytomedicine 9, 302–311.

Lal, D., Singh, N., 2010. Mass multiplication of *Celastrus paniculatus* an important medicinal plant under in vitro conditions using nodal segments. J. Am. Sci. 6, 55–61.

Li, S., Hong, S., Shepardson, N.E., Walsh, D.M., Shankar, G.M., Selkoe, D., 2009. Soluble oligomers of amyloid Beta protein facilitate hippocampal long-term depression by disrupting neuronal glutamate uptake. Neuron 62, 788–801.

Malik, J., Karan, M., Dogra, R., 2017. Ameliorating effect of *Celastrus paniculatus* standardized extract and its fractions on 3-nitropropionic acid induced neuronal damage in rats: possible antioxidant mechanism. Pharm. Biol. 55, 980–990.

Nakamura, T., Lipton, S.A., 2009. Cell death: protein misfolding and neurodegenerative diseases. Apoptosis 14, 455–468.

Nakamura, T., Lipton, S.A., 2010. Redox regulation of mitochondrial fission, protein misfolding, synaptic damage, and neuronal cell death: potential implications for Alzheimer's and Parkinson's diseases. Apoptosis 15 (11), 1354–1363.

Palle, S., Kanakalatha, A., Kavitha, C.N., 2018. Gastroprotective and antiulcer effects of *Celastrus paniculatus* seed oil against several gastric ulcer models in rats. J. Diet. Suppl. 15 (4), 373–385.

Palop, J.J., Chin, J., Roberson, E.D., Wang, J., Thwin, M.T., Bien-Ly, N., Yoo, J., Ho, K.O., Yu, G.Q., Kreitzer, A., 2007. Aberrant excitatory neuronal activity and compensatory remodeling of inhibitory hippocampal circuits in mouse models of Alzheimer's disease. Neuron 55, 697–711.

Parimala, S., Shashidhar, G.H., Sridevi, C.H., Jyothi, V., Suthakaran, R., 2009. Anti-inflammatory activity of *Celastrus paniculatus* seeds. Int. J. Pharm. Tech. Res. 1, 1326–1329.

Premila, M.S., 2006. Ayurvedic Herbs: A Clinical Guide to the Healing Plants of Traditional Indian Medicine. Haworth Press, New York.

Ramammoorthy, R., Elavarasan, P.K., Suresh, S., Bhojraj, S., 2007. Evaluation of anxiolytic potential of Celastrus oil in rat models of behaviour. Fitoterapia 78, 120–124.

Shankar, G.M., Bloodgood, B.L., Townsend, M., Walsh, D.M., Selkoe, D.J., Sabatini, B.L., 2007. Natural oligomers of the Alzheimer amyloid-beta protein induce reversible synapse loss by modulating an NMDA-type glutamate receptor-dependent signaling pathway. J. Neurosci. 27, 2866–2875.

Shankar, G.M., Li, S., Mehta, T.H., Garcia-Munoz, A., Shepardson, N.E., Smith, I., Brett, F.M., Farrell, M.A., Rowan, M.J., Lemere, C.A., 2008. Amyloid-beta protein dimers isolated directly from Alzheimer's brains impair synaptic plasticity and memory. Nat. Med. 14, 837–842.

Sharada, M., Ahuja, A., Kaul, M.K., 2003. Regeneration of plantlets *via* Callus cultures in *Celastrus paniculatus* Willd. A rare endangered medicinal plant. J. Plant Biochem. Biotechnol. 12, 65–69.

Sharma, M., Sahu, S., Khemani, N., Kaur, R., 2013. Ayurvedic medicinal plants as psychotherapeutic agents. Int. J. Appl. Biol. Pharm. Technol. 4, 214–218.

Shivwanshi, R., Gaikwad, P.D., 2017. Identification of best culture media from medicinal plant *Celastrus paniculatus* (Malkangni). Agri Update 12, 1595–1598.

Singh, U., Wadhwani, A.M., Johri, B.M., 1996. Dictionary of Economical Plants of India. Pbl. Indian Council of Agricultural Research, New Delhi, India, p. 46.

Singh, J., 2016. *Celastrus paniculatus* (Jyotishmati Malkangani). Medicinal Plants. Access from: https://www.ayurtimes.com/celastrus-paniculatus-jyotishmati-malkangani/.

Suttee, A., Bhandari, A., Singh, B., Sharma, A., 2013. Pharmacognostical and phytochemical evaluation of *Celastrus paniculata*. Int. J. Pharmacol. Phytochem. Res. 4, 227–333.

Venkataramaiah, C.H., Wudayagiri, R., 2013. Phytochemical screening of bioactive compounds presents in the seed of *Celastrus paniculatus*: role in traditional medicine. Indo Am. J. Pharm. Res. 3, 9104–9111.

Wagner, H., Heckel, E., 1975. Struktur und stereochemie eines sesquiterpenesters und dreier sesquiterpen alkaloide von *Celastrus paniculatus* Willd. Tetrahedron 31, 1949–1956.

Warintorn, R., Jakkapan, S., Chiranan, K., Krot, L., Pensak, J., 2018. Skin penetration and stability enhancement of *Celastrus paniculatus* seed oil by 2-Hydroxypropyl-β-cyclodextrin inclusion complex for cosmeceutical applications. Sci. Pharm. 86, 33.

Zohera, F., Habib, M., Imam, M., Mazumder, M., Rana, M., 2010. Comparative antioxidant potential of different extracts of *Celastrus paniculatus* Willd. seed. Stamford J. Pharm. Sci. 3 (1), 68–74.

Chapter 3.2.21

Uncaria rhynchophylla (Miq.) Jacks

Devina Lobine, Mohamad Fawzi Mahomoodally
Department of Health Sciences, Faculty of Medicine and Health Sciences, University of Mauritius, Réduit, Mauritius

Abbreviations

6-OHDA 6-hydroxydopamine
AChE acetylcholinesterase
AD Alzheimer's disease
PD Parkinson's disease
UR *Uncaria rhynchophylla* (Miq.) Jacks

Introduction

A member of the Rubiaceae family, the *Uncaria* genus comprises of 34 species worldwide, which are mainly distributed in the tropical areas of Southeast Asia, Africa, and South America. It is documented that several *Uncaria* species are represented in China, with 14 endemic to various provinces. Several members of the *Uncaria* genus have a long history of use in the traditional medicine systems for treating and/or managing various ailments (Ndagijimana et al., 2013; Guo et al., 2018).

Uncaria rhynchophylla (Miq.) Jacks, also known as Gou-teng or cat's claw, is an essential component in the original Chinese drugs Gou-teng, Kampo, and Choto-san. In Chinese pharmacopoeias, "Gou-teng" is composed of *U. rhynchophylla*, *U. macrophylla*, *U. hirsute*, *U. sinensis*, and *U. sessilifructus. Uncaria rhynchophylla* is also an ingredient in the original Chinese drug "Diao-teng-gou" (Sakakibara et al., 1999; Guo et al., 2018). *Uncaria rhynchophylla* (UR) and other *Uncaria* species also form part of other systems of traditional medicines such as the traditional Japanese and Native American medicine systems (Heitzman et al., 2005; Erowele and Kalejaiye, 2009). Native to China and Japan, UR is a woody climber. The stem is quadrangular, hollowed, and bears strongly curved hooked spines. The leaves

Naturally Occurring Chemicals against Alzheimer's Disease. https://doi.org/10.1016/B978-0-12-819212-2.00048-7

are simple, opposite, and bear interpetiolar stipules. The inflorescences are globose with long pedicels on the leaf axis and the species is hermaphrodite (Wiart, 2007). The hooks of UR, which are widely used in clinical practice, usually takes a couple of years to grow (Li et al., 2017).

The dried stem and hook of UR are commonly recognized as traditional treatments for epilepsy, numbness, preeclampsia, hypertension, and associated symptoms such as headaches and dizziness (Li et al., 2017; Guo et al., 2018). In Chinese traditional medicine, sun-dried pieces of hook-bearing stems of UR have been found to be effective at reducing fever in children and treating dizziness, motes in vision, and bilious disorders in adults (Wiart, 2007). The hooks and leaves of UR are also used as spasmolytics, analgesics, and sedatives.

Uncaria rhynchophylla is marketed as a healthy food product in the form of cat's claw tea, bark powder, a food supplement, and liquid extract (Junior and Dantas, 2016). A decoction or concoction is the preferred form used for UR as its bioactive compounds are readily available for uptake, distribution, absorption, and cellular metabolism (Ndagijimana et al., 2013). According to Shi et al. (2003), UR should be added last when other prescription herbs are decocted, in order to minimize the conversion of the active components. Among the various identified bioactive components of UR, the indole alkaloids groups of compounds are well known for their therapeutic effects.

Phytochemistry

In view that UR is medicinally important, several studies focusing on the phytochemistry of this species have been performed. One of the earliest phytochemical reports on UR, from 1928, revealed the isolation of rhynchophylline (Fig. 3.2.21.1), a tetracyclic oxindole alkaloid composed of an indole moiety and an indolizidine moiety. and is related to corynoxan. Later, isorhynchophylline (Fig. 3.2.21.1), the 7-epimer of rhynchophylline was isolated (Tang and Eisenbrand, 1992). To date, a variety of bioactive compounds such as glycosides, phenolics, flavonoids, sterols, tannins (Chang et al., 2012; Hou et al., 2005; Ma et al., 2009; Ndagijimana et al., 2013), and terpenoids (Endo et al., 1983; Zhang et al., 2014; Ndagijimana et al., 2013) have been identified or isolated from UR.

The major bioactive components in the stems and hooks of UR are alkaloids, which belong to two main groups, namely semiindole alkaloids or single-indole alkaloids and sesquiterpene indole alkaloids. Rhynchophylline and isorhynchophylline account for more than 43% of the total alkaloid content in UR and are the main biologically active constituents in the herb (Phillipson and Hemingway, 1975; Zhang et al., 2019). Two further related alkaloids present in UR are the isomers corynoxeine and isocorynoxeine. It is documented that approximately 97% of the total alkaloids detected in the hook, small stem, and leaves of UR are represented by the oxindole alkaloids;

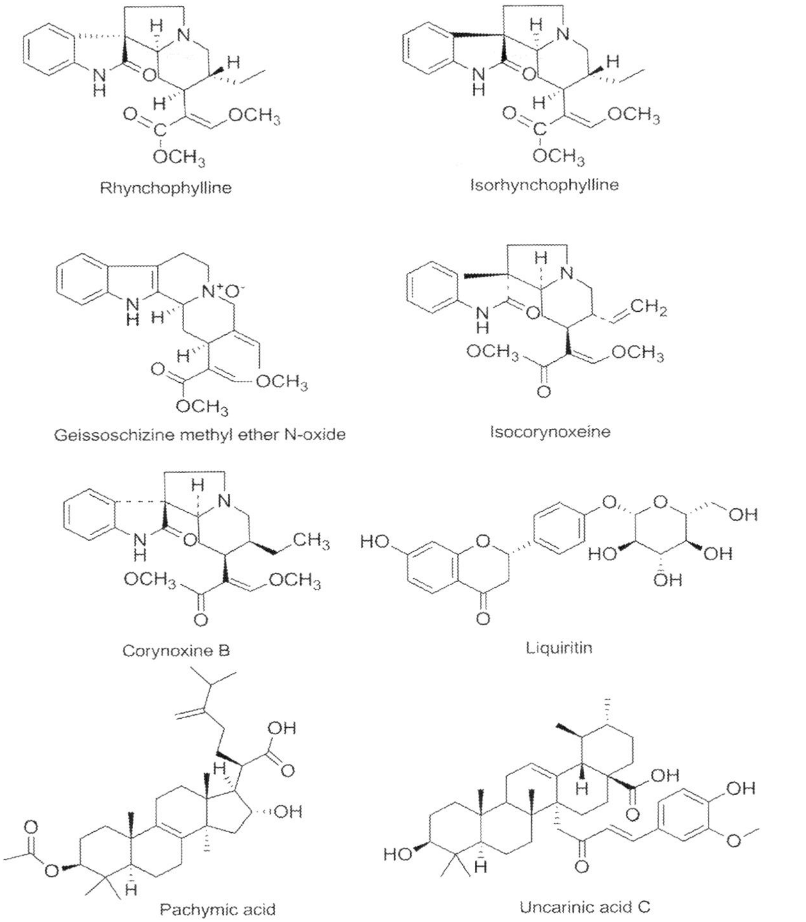

FIGURE 3.2.21.1 Chemical structure of rhynchophylline and isorhynchophylline and other compounds that showed potential against AD.

rhynchophylline, isorhynchophylline, corynoxeine, and isocorynoxe (Tang and Eisenbrand, 1992). Kong et al. (2017) reported on the occurrence of a novel tetracyclic indole alkaloid, 17-O-methyl-3,4,5,6-tetrade hydrogeissoschizine, together with seven known ones from the aerial parts of UR. Two sesquiterpene alkaloids, 5(S)-5-carboxysfrictosidine and 3,4-dehydro-5(S)-5-carboxystrictosidine have also been isolated from UR (Ndagijimana et al., 2013). Geissoschizinemethyl ether, an indole alkaloid possessing a β-alkoxy acrylic ester moiety and an α configuration of C(3)−H is also known to occur in UR (Jiang et al., 2014).

Qualitative analyses of UR have demonstrated that there are varied differences in the indole alkaloid content, depending on the cultivation methods and plant parts used. The indole alkaloids hiruteine and hirsutine have been found to be the main components in the bark of the underground part, while the alkaloids corynantheine and dihydrocorynantheine were isolated from the stem of UR (Tang and Eisenbrand, 1992). The study carried by Laus and Teppner (1996) yielded very similar results. The experimental data revealed that the leaves, lateral branches, and upper part of the stem of UR showed the presence of oxindole alkaloids such as isorhynchophylline (0.90%) and rhynchophylline (0.34%), and unsaturated oxindole alkaloids isocorynoxeine (1.19%) and corynoxeine (0.49%). Hirsutine (3.60%), hirsuteine (2.59%), corynantheine (0.68%), and dihydrocorynantheine (1.00%) were the main components in the roots, while the stem bark contained both oxindole and yohimbine indole alkaloids and the wood of the lower stem showed no detectable indole alkaloids. Kohda and coworkers (1996) illustrated that 3-α-dihydrocadambine (0.075%), hirsuteine (0.023%), and hirsutine (0.016%) were the main alkaloids in the extract of MN4K4 callus.

The major flavonoids identified from UR included trifolin (Aimi et al., 1982), kaempferol-a-L-rhamnopyranosyl-(1.6)-β-D glucopyranoside, quercetin-3-O-a-L rhamnopyranosyl-(1.6)-ß-D-galactopyranoside, rutin, hyperin, and hyperoside (Ma et al., 2009). Mei-cai et al. (2009) isolated 18 known compounds from the ethanolic extract of a twig of UR and, of the 18 compounds, noreugenin, sumreainolic, 3β,6β,19α, 24-tetrahydroxyurs-12-en-28-oic acid, 3β, 19α, 23 -trihydroxy-6-oxo-olean-12-en-28-oic acid, 3β,19α,24- trihydroxyurs-12-en-28-oic acid, cleomiscosin D, and cleomiscosin B were reported for the first time from this species. In 2016, Zhang and coresearchers isolated new ortho benzoquinones, 3-diethylamino-5-methoxy-1, 2-benzoquinone and 3-ethylamino-5-methoxy-1, 2-benzoquinone.

Compared to the intensive research into the UR hooks, limited phytochemical studies have focused on UR leaves. Considering the slow growth rate of the widely used hooks of UR, the leaves of UR represent a sustainable source of active components for medicinal uses. Li et al. (2017) performed bioassay-directed fractionation on the UR ethanolic extract, which has led to the isolation of two novel phenylpropanoid-substituted flavan-3-ols, uncariols A and B, and two pairs of new phenylpropanoid-substituted flavonol enantiomers, (±)-uncariols C and D, together with nine previously reported compounds from the leaves of UR. A comparative chemical and pharmacological study on the different sections of UR has demonstrated that the different parts have well-differentiated chemical profiles; the hooks and stems were similar and principally contain pentacyclic heteroyohimbine and cadambine-type alkaloids; tetracyclic oxindoles exclusively occurred in fruits; and d-seco pentacyclic heteroyohimbines were dominant compounds in leaves. Interestingly, the leaves displayed the highest psychiatric capacity with remarkable activity on melatonin (MT_1 and MT_2) receptors, with agonistic rates of 39.7% and

97.6%, respectively, as compared to the hooks and stems which have shown moderate to weak activity. For the 5-HT_{1A} and 5-HT_{2C} receptors, the hooks and stems were most potent, with agonistic rates of 92.6% and 83.1% (Zhang et al., 2017). Yamanaka et al. (1983) characterized 10 tertiary alkaloids in different parts of UR and the results showed that the leaves and hooks shared similar alkaloid profiles. Qu et al. (2012) reported the presence of high levels of oxindole alkaloids rhynchophylline, isorhynchophylline, corynoxeine, and isocorynoxeine in the stem with hooks, while the leaves possessed high amounts of two glycosidic indole alkaloids, namely vincoside lactam and strictosidine. Taken together, the data suggest that the leaves of UR hold immense potential as a complementary medicine.

Overview of the pharmacological activity of *U. rhynchophylla*

Antioxidant properties

The pharmacological activity of crude extracts of herbal drugs containing UR and the individual alkaloids isolated from this plant have been extensively studied. Oxidative stress is implicated in the pathology of several chronic diseases, such as cardiovascular diseases, acute and chronic kidney disease, neurodegenerative diseases (NDs), and cancer (Liguori et al., 2018). In this direction, antioxidants from natural sources are in high demand. Studies have shown that different extracts and alkaloids of UR exerted remarkable antioxidant activities (Chang et al., 2012; Kim et al., 2013; Yin et al., 2010; Zhang et al., 2013). Zhang et al. (2013) reported that polysaccharide fractions from UR, composed mainly of glucose, fructose, xylose, arabinose, and rhamnose, have showed remarkable radical-scavenging activity as well as high in vivo antioxidant potency. Li et al. (2017) reported the free radical-scavenging potential of bioactives isolated from the leaves of UR.

Antiinflammatory properties

The in vitro antiinflammatory potential of rhynchophylline and isorhynchophylline from UR in lipopolysaccharide (LPS)-activated microglial cells has been studied. The tested compounds inhibited proinflammatory mediators such as NO, TNF-α, and interleukin (IL)-1β in LPS-activated N9 cells, by downregulating iNOS protein expression and attenuating the activation of NF-κB and ERK and p38 MAPKs in N9 cells (Yuan et al., 2009). Rhynchophylline, isorhynchophylline, corynoxeine, isocorynoxeine, and vincoside lactam from leaves of UR inhibited NO release in LPS-activated primary rat cortical microglial cells (IC_{50}: 13.7–19.0 μM) (Yuan et al., 2008). Aqueous extract of UR stem with hooks (0.5–2 mg/mL) has been observed to attenuate LPS-induced NO and IL-1β secretion as well as inducible NO synthase expression in RAW 264.7 macrophages. Its antiinflammatory activity was associated with suppressed LPS-induced NF-κB, phosphorylation, degradation of IκB-α, and phosphorylation of Akt, ERK1/2, p38, and c-Jun N-terminal kinase (Kim et al., 2010). Wu and Xiao (2019) reported that UR alkaloid extract (140 mg/kg body weight/day) ameliorated preeclampsia symptoms and mitigated inflammatory responses (interleukin [IL]-6, IL-1β, tumor necrosis factor-α, and interferon-γ) in the LPS-induced preeclampsia rat model, which demonstrated that UR extract is a promising alternative therapy for preeclampsia.

Hypotensive and cardioprotective effects

One of the most common uses of UR in folk medicines is for treating hypertension and cardiovascular disease (Jung et al., 2013). These activities have been associated with the presence of certain alkaloids. Hirsutine, a major indole alkaloid isolated from UR, has been documented to exert antihypertensive and antiarrhythmic effects by reducing intracellular Ca^{2+} levels in rat aorta and the action potential in cardiac muscle. Hirsutine was also effective at protecting rat cardiomyocytes from hypoxia-induced cell death (Wu et al., 2011). The proliferation of vascular smooth muscle cells (VSMCs) induced by injury to the intima of arteries is a crucial event in the pathogenesis of hypertension, atherosclerosis, and restenosis. The suppression of VSMC growth is effective for treating cardiovascular diseases (Zhang et al., 2008). A study by Kim et al. (2008) demonstrated that corynoxeine isolated from UR significantly inhibited the platelet-derived growth factor (PDGF)-BB-induced DNA synthesis of VSMCs in a concentration-dependent manner without eliciting any toxic effect. Pretreatment of VSMCs with corynoxeine (5−50 μM) for 24 h, considerably inhibited PDGF-BB-induced extracellular signal-regulated kinase 1/2 (ERK1/2) activation and the findings suggested that corynoxeine is potentially useful for preventing and treating vascular diseases and restenosis after angioplasty. Rhynchophylline and isorhynchophylline displayed antihypertensive activity by inhibiting the vasomotor center, blocking sympathetic nerves or ganglia, and the L-type Ca^{2+} channels on the smooth muscle cells (Chunqing et al., 2000; Li et al., 2013; Zhang et al., 2004). The work carried out by Zhang et al. (2008) has demonstrated that isorhynchophylline was potent against angiotensin II-induced cell proliferation, an effect that appears to be partly ascribed by an elevated nitric oxide production, regulation of the cell cycle, and attenuation of the overexpressions of proto-oncogene c-fos, osteopontin (OPN) and proliferating cell nuclear antigen (PCNA) mRNAs related to VMSC proliferation. Guo et al. (2014) reported that isorhynchophylline showed inhibition of pulmonary arterial smooth muscle cell proliferation. Blocking of PDGF-Rb phosphorylation or regulation of G0/G1 phase cell cycle may be its potential mode of action.

Antiangiogenic properties

In vitro and in vivo studies have demonstrated that UR is a potent angiogenic agent. *Uncaria rhynchophylla* has significantly stimulated cell proliferation and migration in human umbilical vein endothelial cells and formation of capillary-like structures that play a vital role in the angiogenesis process, in a dose-dependent manner. *Uncaria rhynchophylla*-treated impregnated Matrigel led to a great increase in tube/network formation and hemoglobin content compared with basic fibroblast growth factor and hence promoted blood vessel formation (Choi et al., 2005).

Anticancer properties

Numerous studies have highlighted the anticancer effects of UR and its bioactive components. Lee et al. (2000) reported that two new pentacyclic triterpene esters, uncarinic acids A and B, isolated from the hooks of UR inhibited phospholipase Cγ1 (PLCγ1) ($IC_{50} = 35.66$, and 44.55 μM, respectively) in a dose-dependent manner; PLCγ1 is an enzyme which induces proliferation in human cancer cells and inhibits the proliferation of human cancer cells, HCT-15 (colon), MCF-7 (breast), A549 (lung), and HT-1197 (bladder) overexpressing PLC*r*1. Inhibitors of phosphatidylinositol-specific phospholipase C (PLC), especially the *Υ* isoform, is regarded as a valuable approach for the development of anticancer agents (Lee et al., 1999, 2000). The structure−activity relationships of these compounds showed that ursane moiety, *trans*-configuration and *p*-coumaroyloxy group of these pentacyclic triterpene esters are potent functional groups for inhibiting PLCγ1 activity (Lee et al., 2000, 2008).

A cDNA microarray study of UR demonstrated that hyperin isolated from dried stems upregulated 50 genes and downregulated 37 genes in SNU-668 human gastric cancer cells; the activity against many of these genes was associated with antioxidation mechanisms (Dooil et al., 2002). The work carried by Jo et al. (2008) demonstrated that the methanolic extracts of UR have significantly induced cytotoxicity and apoptosis in HT-29 human colon cancer cell via a caspase-dependent pathway. Proanthocyanidins from UR have been shown to significantly inhibit cell viability and migration ability in MDA-MB-231 breast cancer cells in a dose-dependent manner. Moreover, the bioactive proanthocyandins have induced apoptosis associated with ROS generation and inherent apoptotic signaling pathway. In the same study, a combination of proanthocyandins from UR and 5-fluorouracil (5-FU) was observed to exert a synergistic cytotoxic effect on MDA-MB-231 cells. Taken together, the data suggested that proanthocyandins from UR are promising therapeutic agents for breast cancer.

Antiviral activity

The indole alkaloid, hirsutine, of UR was recognized as an effective anti-dengue virus (DENV) compound, displaying high efficacy and low cytotoxicity. Hirsutine was potent against all DENV serotypes and was found to exert its effect by inhibiting viral particle assembly, budding, or the release step (Hishiki et al., 2017).

Uncaria rhynchophylla and central nervous system (CNS)-related activity

The therapeutic effects of UR on CNS-related ailments have been extensively studied both in vitro and in vivo. In vitro studies have revealed that UR extract significantly inhibited NMDA receptor-activated ion currents in acutely dissociated hippocampal CA1 neurons in cultured brain slices (Lee et al., 2003) and ameliorated methamphetamine-induced rat cortical neuron death (Mo et al., 2006). Suk et al. (2002) reported that UR displays neuroprotective activity by inhibiting cyclooxygenase-2 (COX-2) in rats with transient global ischemia. Based on the in vitro studies, it was also suggested that the antiinflammatory actions of UR extract possibly contribute to its neuroprotective effect. Isorhynchophylline from UR has been reported to display protective effects against ischemia- and glutamate-induced neuronal damage or death (Kang et al., 2004; Yuan et al., 2009), and suppression of 5-hydroxytryptamine (5-HT) receptor (Matsumoto et al., 2005). Neuroprotective activity displayed by hirsutine has also been associated with its ability to inhibit inflammation-mediated neurotoxicity and microglial activation (Jung et al., 2013). The anxiolytic effects of the aqueous UR extract have also been documented (Jung et al., 2006).

In vivo studies have shown that UR improved cognitive deficits induced by D-galactose and scopolamine (Suk-Chul and Dong-Ung, 2013; Xian et al., 2011). The intraperitoneal administration of the UR methanolic extract (100−1000 mg/kg at 0 and 90 min after reperfusion) showed a significant protective effect on hippocampal CA1 neurons against a 10-min induced transient forebrain ischemia. The extract has shown inhibitory activity against TNF-α and nitric oxide production in BV-2 mouse microglial cells in vitro and these antiinflammatory actions of UR extract are suggested to account for its neuroprotective property (Suk et al., 2002). To better understand the pharmacological properties of UR on the CNS, the contents are arranged as follows: anticonvulsion, anti-Parkinson's disease, and anti-Alzheimer's.

Anticonvulsion

The anticonvulsant effect of UR aqueous extract has been seen in kainic acid-induced epileptic seizures models (Hsieh et al., 1999; Hsieh et al., 2009; Lo et al., 2010).Hsieh et al., (1999) suggested that the antiepileptic effect of UR was due to its suppressive effect on lipid peroxidation in the brain. The study by Ho et al. (2014) showed that UR and rhynchophylline ameliorated kainic acid-induced epileptic seizures by regulating the immune response and neurotrophin signaling pathway, and subsequently suppressing neuron survival brain-derived neurotrophin factor (BDNF) and interleukin-1β (IL-1β) gene expression in the cortex and hippocampus. Studies have also shown that UR displayed neuroprotective effects by inhibiting microglial activation in the hippocampus (Tang et al., 2010), and reducing astrocyte proliferation and S100 calcium-binding protein B in the hippocampus (Liu et al., 2012).

Anti-Parkinson's disease

UR also has shown neuroprotective effects in in vitro and in vivo models of Parkinson's disease (PD). The study carried out by Shim et al. (2009) revealed that aqueous UR extract exerted a neuroprotective effect against 6-hydroxydopamine (6-OHDA)-induced toxicity through antioxidative and antiapoptotic activities in PC12 cells and 6-OHDA-lesioned rats. In PC12 cells, the UR extract significantly decreased cell death and ROS generation, increased total glutathione (GSH) levels, and suppressed caspase-3 activity. In the 6-OHDA-lesioned rats, post-treatment with a low dose of UR extract (5 mg/kg/day for 14 days) showed behavioral recovery and protection of dopaminergic neurons in the substantia nigra pars compacta. Inhibition of monoamine oxidase B (MAO-B) is useful for alleviating symptoms and slows down the progression of PD and Alzheimer's disease. At a dose of 100 μg/mL, methanolic extract of UR hook inhibited MAO-B activity by 85.2%. The isolated bioactive compounds, (+)-catechin and (−)-epicatechin (100 μg/mL), inhibited MAO-B activity with IC_{50} values of 88.6 and 58.9 μM, respectively (Hou et al., 2005). More recently, Lan et al. (2018) reported that UR ameliorates PD by regulating MAPK and PI3K-AKT signal pathways and suppressing heat shock protein 90 expression in the substantia nigra.

Alzheimer's disease

Accumulating evidence has shown that UR exerts beneficial effects on Alzheimer's disease (AD) patients. Notably, studies have demonstrated that UR extracts or active compounds have potently inhibited Aβ fibril formation, disassembled performed Aβ fibrils, and inhibited acetylcholinesterase (Fujiwara et al., 2006; Lin et al., 2008; Xian et al., 2011). The alkaloid, geissoschizine methyl ether N-oxide (Fig. 3.2.21.1), from UR has exhibited anti-AChE activity (Yang et al., 2012; Jiang et al., 2014). *Uncaria rhychophylla* has a neuroprotective effect on $AlCl_3$-induced AD rats by improving the brain index and on $A\beta 1-42$-induced pheochromocytoma (PC12) cells by decreasing cell damage and improving cell vitality (Huang et al., 2008, 2017). Six components (pachymic acid, liquiritin, rhynchophylline, isorhynchophylline, corynoxeine and isocorynoxeine (Fig. 3.2.21.1) from UR have been characterized and associated with the resulting therapeutic effects (Huang et al., 2017). In addition, corynoxine B (Fig. 3.2.21.1), an isomer of corynoxine, reduced the Aβ level through facilitating degradation of amyloid precursor protein (APP) by enhancing the basal level of autophagy and increasing lysosomal activity (Durairajan et al., 2013).

A study has shown that UR hinders Aβ aggregation and destabilizes the preformed Aβ aggregates (Fujiwara et al., 2006). An in vivo study using APP transgenic mice has demonstrated that UR treatment reduces $A\beta_{1\text{-}42}$ in the cerebral cortex and improved memory impairment and social interaction (Fujiwara et al., 2011). Xian et al. (2011) reported that 70% hydroethanol extract of UR hook-bearing stem and branch (200 or 400 mg/kg) notably ameliorated exploratory behavior and enhanced spatial learning and memory function in D-galactose-induced a rat cognitive deficit model. This activity was achieved due to increased acetylcholine and glutathione levels and, reduced acetylcholinesterase activity and malonodialdehyde level in the rat brains. Further research has shown that isorhynchophylline (Fig. 3.2.21.1) can improve memory deficits by counteracting oxidative stress and exhibiting an antiinflammatory effect of brain tissues in D-galactose-induced mouse (Xian et al., 2014). Previous research has shown that the neurotoxicity induced by Aβ is mediated by intracellular calcium influx in primary neurons such as hippocampal (Kelly and Ferreira, 2006) and cortical neurons (Zhu et al., 2009). Researchers have also reported that rhynchophylline and isorhynchophylline rescued PC12 cells against $A\beta_{25-35}$-induced neuronal toxicity by suppressing intracellular calcium overload and tau protein hyperphosphorylation (Xian et al., 2012), which is in accordance with inhibition of intracellular calcium influx that might be ascribed for neuroprotection against Aβ-induced cell death. Experimental data have also suggested that isorhynchophylline, at a daily dose of 20 or 40 mg/kg for 21 days, exerted a protective effect against $A\beta$25−35-induced apoptosis in PC12 cells by suppressing oxidative stress and the mitochondrial pathway of cellular apoptosis and enhancing of p-CREB expression via the PI3K/Akt/GSK-3β signaling pathway (Xian et al., 2012, 2014).

The study carried out by Shao et al. (2015) demonstrated that rhynchophylline (Fig. 3.2.21.1) has a protective effect against soluble Aβ1−42 oligomer-induced hippocampal hyperactivity in the cortex and subiculum in 5XFAD mice. In vivo electrophysiological data have shown that rhynchophylline can remold the spontaneous discharges disturbed by Aβ and neutralize the deleterious effect of $A\beta_{1-42}$ on the neural circuit. More recently, Shin et al. (2018) have reported that UR significantly inhibited Aβ aggregation and accumulation and Aβ-mediated neuropathology in the brains of 5XFAD mice.

Rhynchophylline has been suggested as an inhibitor of tyrosine kinase EphA4 receptor, leading to restoration of the synaptic impairment in the transgenic mouse models of AD (Fu et al., 2014). Based on the aforementioned facts, rhynchophylline and isorhynchophylline may be the main bioactive compounds of UR for its anti-AD activity. Uncarinic acid C (Fig. 3.2.21.1), a component of UR, has also been acknowledged as a promising compound that reduces Aβ aggregation by inhibiting the nucleation phase (Yoshioka et al., 2016).

Conclusion

Uncaria rhynchophylla is widely used in folk medicines in the treatment of various diseases. *Uncaria rhynchophylla*, on its own or as part of a polyherbal preparation, is a valuable therapeutic for vascular and neurosis diseases. Phytochemical investigations have revealed numerous structural types of chemical components. Pharmacological studies showed that UR or its isolated bioactives possess various biological attributes, especially in the areas of antioxidant, antiinflammatory, antihypertension, antiarrhythmic, anticancer, anticonvulsive, anti-PD and anti-AD properties. Numerous studies have established a broad understanding of the pharmacological value of UR and its bioactive compounds in the treatment of AD by highlighting diverse beneficial roles in neuronal survival, synaptic plasticity, and microglial activation in the CNS. In this sense, further investigation should be carried out to confirm the neuropharmacological roles of UR bioactive compounds that could be potentially exploited for designing drugs.

Acknowledgment

D. Lobine thanks the Higher Education Commission, Mauritius, for the award of the post-doctoral fellowship.

References

Aimi, N., Shito, T., Fukushima, K., ITAI, Y., Aoyama, C., Kunisawa, K., Sakai, S., Haginiwa, J., Yamsaki, K., 1982. Studies on plants containing indole alkaloids. VIII indole alkaloid glycosides and other constituents of the leaves of *Uncaria rhynchophylla* MIQ. Chem. Pharm. Bull. 30 (11), 4046–4051.

Chang, C.L., Lin, C.S., Lai, G.H., 2012. Phytochemical characteristics, free radical scavenging activities, and neuroprotection of five medicinal plant extracts. Evid. Based Complement. Alternat. Med. 2012.

Choi, D.Y., Huh, J.E., Lee, J.D., Cho, E.M., Baek, Y.H., Yang, H.R., Cho, Y.J., Kim, K.I., Kim, D.Y., Park, D.S., 2005. *Uncaria rhynchophylla* induces angiogenesis in vitro and in vivo. Biol. Pharm. Bull. 28 (12), 2248–2252.

Chunqing, S., Yi, F., Weihui, H., 2000. Different hypotensive effects of various active constituents isolated from *Uncaria rhynchophylla*. Chin. Tradit. Herb. Drugs 31 (10), 762–763.

Dooil, J., Kim, J.H., Lee, Y.H., Baek, M.I., Lee, S.E., Baek, N.I., Kim, H.Y., 2002. cDNA microarray analysis of transcriptional response to hyperin in human gastric cancer cells. J. Microbiol. Biotechnol. 12 (4), 664−668.

Durairajan, S.S.K, Huang, Y., Chen, L., Song, J., Liu, L., Li, M, 2013. Corynoxine isomers decrease levels of amyloid-β peptide and amyloid-β precursor protein by promoting autophagy and lysosome biogenesis. Mol. Neurodegener. 8 (1), 1−2.

Endo, K., Oshima, Y., Kikuchi, H., Koshihara, Y., Hikino, H., 1983. Hypotensive principles of Uncaria hooks. Planta Med. 49 (11), 188−190.

Erowele, G.I., Kalejaiye, A.O., 2009. Pharmacology and therapeutic uses of cat's claw. Am. J. Health Syst. Pharm. 66 (11), 992−995.

Fu, A.K., Hung, K.W., Huang, H., Gu, S., Shen, Y., Cheng, E.Y., Ip, F.C., Huang, X., Fu, W.Y., Ip, N.Y., 2014. Blockade of EphA4 signaling ameliorates hippocampal synaptic dysfunctions in mouse models of Alzheimer's disease. Proc. Natl. Acad. Sci. U.S.A. 111 (27), 9959−9964.

Fujiwara, H., Iwasaki, K., Furukawa, K., Seki, T., He, M., Maruyama, M., Tomita, N., Kudo, Y., Higuchi, M., Saido, T.C., Maeda, S., 2006. Uncaria rhynchophylla, a Chinese medicinal herb, has potent antiaggregation effects on Alzheimer's β-amyloid proteins. J. Neurosci. Res. 84 (2), 427−433.

Fujiwara, H., Takayama, S., Iwasaki, K., Tabuchi, M., Yamaguchi, T., Sekiguchi, K., Ikarashi, Y., Kudo, Y., Kase, Y., Arai, H., Yaegashi, N., 2011. Yokukansan, a traditional Japanese medicine, ameliorates memory disturbance and abnormal social interaction with anti-aggregation effect of cerebral amyloid β proteins in amyloid precursor protein transgenic mice. Neuroscience 180, 305−313.

Guo, Q., Yang, H., Liu, X., Si, X., Liang, H., Tu, P., Zhang, Q., 2018. New zwitterionic monoterpene indole alkaloids from *Uncaria rhynchophylla*. Fitoterapia 127, 47−55.

Guo, H., Zhang, X., Cui, Y., Deng, W., Xu, D., Han, H., Wang, H., Chen, Y., Li, Y., Wu, D., 2014. Isorhynchophylline protects against pulmonary arterial hypertension and suppresses PASMCs proliferation. Biochem. Biophys. Res. Commun. 450 (1), 729−734.

Heitzman, M.E., Neto, C.C., Winiarz, E., Vaisberg, A.J., Hammond, G.B., 2005. Ethnobotany, phytochemistry and pharmacology of Uncaria (Rubiaceae). Phytochemistry 66 (1), 5−29.

Hishiki, T., Kato, F., Tajima, S., Toume, K., Umezaki, M., Takasaki, T., Miura, T., 2017. Hirsutine, an indole alkaloid of *Uncaria rhynchophylla,* inhibits late step in dengue virus lifecycle. Front. Microbiol. 8, 1674.

Ho, T.Y., Tang, N.Y., Hsiang, C.Y., Hsieh, C.L., 2014. *Uncaria rhynchophylla* and rhynchophylline improved kainic acid-induced epileptic seizures via IL-1β and brain-derived neurotrophic factor. Phytomedicine 21 (6), 893−900.

Hou, W.C., Lin, R.D., Chen, C.T., Lee, M.H., 2005. Monoamine oxidase B (MAO-B) inhibition by active principles from *Uncaria rhynchophylla*. J. Ethnopharmacol. 100 (1−2), 216−220.

Hsieh, C.L., Tang, N.Y., Chiang, S.Y., Hsieh, C.T., Lin, J.G., 1999. Anticonvulsive and free radical scavenging actions of two herbs, *Uncaria rhynchophylla* (MIQ) Jack and *Gastrodia elata* Bl., in kainic acid-treated rats. Life Sci. 65 (20), 2071−2082.

Hsieh, C.L., Chen, M.F., Li, T.C., Li, S.C., Tang, N.Y., Hsieh, C.T., Pon, C.Z., Lin, J.G, 1999. Anticonvulsant effect of Uncaria rhynchophylla (Miq) Jack. in rats with kainic acid-induced epileptic seizure. Am. J. Chin. Med. 27 (02), 257−264.

Hsieh, C.L., Ho, T.Y., Su, S.Y., Lo, W.Y., Liu, C.H., Tang, N.Y., 2009. *Uncaria rhynchophylla* and rhynchophylline inhibit c-Jun N-terminal kinase phosphorylation and nuclear factor-κB activity in kainic acid-treated rats. Am. J. Chin. Med. 37 (02), 351−360.

Huang, H.C., Zhong, R.L., Cao, P., Wang, C.R., Yang, D.G., 2008. Effects of Goutengsan on model of Alzheimer dementia in rats by $AlCl_3$. Zhongguo Zhong Yao Za Zhi (China J. Chin. Mater. Med.) 33 (5), 553–556.

Huang, H.C., Wang, C.F., Gu, J.F., Chen, J., Hou, X.F., Zhong, R.L., Xia, Z., Zhao, D., Yang, N., Wang, J., Tan, X.B., 2017. Components of goutengsan in rat plasma by microdialysis sampling and its protection on Aβ1–42-induced PC12 cells injury. Evid. Based Complement. Alternat. Med. 2017.

Jiang, W.W., Su, J., Wu, X.D., He, J., Peng, L.Y., Cheng, X., Zhao, Q.S., 2014. Geissoschizine methyl ether N-oxide, a new alkaloid with anti-acetylcholinesterase activity from *Uncaria rhynchophylla*. Nat. Prod. Res. 29 (9), 842–847.

Jo, K.J., Cha, M.R., Lee, M.R., Yoon, M.Y., Park, H.R., 2008. Methanolic extracts of *Uncaria rhynchophylla* induce cytotoxicity and apoptosis in HT-29 human colon carcinoma cells. Plant Foods Hum. Nutr. 63 (2), 77–82.

Jung, J.W., Ahn, N.Y., Oh, H.R., Lee, B.K., Lee, K.J., Kim, S.Y., Cheong, J.H., Ryu, J.H., 2006. Anxiolytic effects of the aqueous extract of Uncaria rhynchophylla. J. Ethnopharmacol. 108 (2), 193–197.

Jung, H.Y., Nam, K.N., Woo, B.C., Kim, K.P., Kim, S.O., Lee, E.H., 2013. Hirsutine, an indole alkaloid of *Uncaria rhynchophylla*, inhibits inflammation-mediated neurotoxicity and microglial activation. Mol. Med. Rep. 7 (1), 154–158.

Junior, J.B.P, Dantas, K.G.F, 2016. Evaluation of inorganic elements in cat's claw teas using ICP OES and GF AAS. Food Chem. 196, 331–337. https://doi.org/10.1016/j.foodchem.2015.09.057.

Kang, T.H., Murakami, Y., Takayama, H., Kitajima, M., Aimi, N., Watanabe, H., Matsumoto, K., 2004. Protective effect of rhynchophylline and isorhynchophylline on in vitro ischemia-induced neuronal damage in the hippocampus: putative neurotransmitter receptors involved in their action. Life Sci. 76 (3), 331–343.

Kelly, B.L., Ferreira, A., 2006. β-amyloid-induced dynamin 1 degradation is mediated by N-methyl-D-aspartate receptors in hippocampal neurons. J. Biol. Chem. 281 (38), 28079–28089.

Kim, J.H., Bae, C.H., Park, S.Y., Lee, S.J., Kim, Y., 2010. *Uncaria rhynchophylla* inhibits the production of nitric oxide and interleukin-1β through blocking nuclear factor κB, Akt, and mitogen-activated protein kinase activation in macrophages. J. Med. Food 13 (5), 1133–1140.

Kim, Y.S., Hwang, J.W., Kim, S.E., Kim, E.H., Jeon, Y.J., Moon, S.H., Jeon, B.T., Park, P.J., 2013. Antioxidant activity and protective effects of *Uncaria rhynchophylla* extracts on t-BHP-induced oxidative stress in Chang cells. Biotechnol. Bioproc. Eng. 17 (6), 1213–1222.

Kim, T.J., Ju-Hyun, L., Jung-Jin, L., Ji-Yeon, Y., Bang-Yeon, H., Sang-Kyu, Y., Li, S., Li, G., Myoung-Yun, P., Yeo-Pyo, Y., 2008. Corynoxeine isolated from the hook of Uncaria rhynchophylla inhibits rat aortic vascular smooth muscle cell proliferation through the blocking of extracellular signal regulated kinase 1/2 phosphorylation. Biol. Pharm. Bull. 31 (11), 2073–2078.

Kong, F., Ma, Q., Huang, S., Yang, S., Fu, L., Zhou, L., Dai, H., Yu, Z., Zhao, Y., 2017. Tetracyclic indole alkaloids with antinematode activity from *Uncaria rhynchophylla*. Nat. Prod. Res. 31 (12), 1403–1408.

Lan, Y.L., Zhou, J.J., Liu, J., Huo, X.K., Wang, Y.L., Liang, J.H., Zhao, J.C., Sun, C.P., Yu, Z.L., Fang, L.L., Tian, X.G., 2018. *Uncaria rhynchophylla* ameliorates Parkinson's disease by inhibiting HSP90 expression: insights from quantitative proteomics. Cell. Physiol. Biochem. 47 (4), 1453–1464.

Laus, G., Teppner, H., 1996. Alkaloids of an *Uncaria rhynchophylla* (Rubiaceae-Coptosapelteae). Phyton 185–196.

Lee, J., Son, D., Lee, P., Kim, D.K., Shin, M.C., Jang, M.H., Kim, C.J., Kim, Y.S., Kim, S.Y., Kim, H., 2003. Protective effect of methanol extract of *Uncaria rhynchophylla* against excitotoxicity induced by N-methyl-D-aspartate in rat hippocampus. J. Pharmacol. Sci. 92 (1), 70–73.

Lee, J.S., Kim, J., Kim, B.Y., Lee, H.S., Ahn, J.S., Chang, Y.S., 2000. Inhibition of phospholipase Cγ1 and cancer cell proliferation by triterpene esters from *Uncaria rhynchophylla*. J. Nat. Prod. 63 (6), 753–756.

Lee, J.S, Yang, M.Y., Yeo, H., Kim, J., Lee, H.S., Ahn, J.S, 1999. Uncarinic acids: Phospholipase Cγ1 inhibitors from hooks of Uncaria rhynchophylla. Bioorg. Med. Chem. Lett. 9 (10), 1429–1432.

Lee, J.S., Yoo, H., Suh, Y.G., Jung, J.K., Kim, J., 2008. Structure-activity relationship of pentacylic triterpene esters from *Uncaria rhynchophylla* as inhibitors of phospholipase Cγ1. Planta Med. 74 (12), 1481–1487.

Li, P.Y., Zeng, X.R., Cheng, J., Wen, J., Inoue, I., Yang, Y., 2013. Rhynchophylline-induced vasodilation in human mesenteric artery is mainly due to blockage of L-type calcium channels in vascular smooth muscle cells. N. Schmied. Arch. Pharmacol. 386 (11), 973–982.

Li, R., Cheng, J., Jiao, M., Li, L., Guo, C., Chen, S., Liu, A., 2017. New phenylpropanoid-substituted flavan-3-ols and flavonols from the leaves of *Uncaria rhynchophylla*. Fitoterapia 116, 17–23.

Liguori, I., Russo, G., Curcio, F., Bulli, G., Aran, L., Della-Morte, D., Gargiulo, G., Testa, G., Cacciatore, F., Bonaduce, D., Abete, P., 2018. Oxidative stress, aging, and diseases. Clin. Interv. Aging 13, 757.

Lin, H.Q., Ho, M.T., Lau, L.S., Wong, K.K., Shaw, P.C., Wan, D.C., 2008. Anti-acetylcholinesterase activities of traditional Chinese medicine for treating Alzheimer's disease. Chem. Biol. Interact. 175 (1–3), 352–354.

Liu, C.H., Lin, Y.W., Tang, N.Y., Liu, H.J., Hsieh, C.L., 2012. Neuroprotective effect of *Uncaria rhynchophylla* in Kainic acid-induced epileptic seizures by modulating hippocampal mossy fiber sprouting, neuron survival, astrocyte proliferation, and S100b expression. Evid. Based Complement. Alternat. Med. 2012.

Lo, W.Y., Tsai, F.J., Liu, C.H., Tang, N.Y., Su, S.Y., Lin, S.Z., Chen, C.C., Shyu, W.C., Hsieh, C.L., 2010. *Uncaria rhynchophylla* upregulates the expression of MIF and cyclophilin A in kainic acid-induced epilepsy rats: a proteomic analysis. Am. J. Chin. Med. 38 (04), 745–759.

Ma, B., Wu, C.F, Yang, J.Y, Wang, R., Kano, Y., Yuan, D., 2009. Three new alkaloids from the leaves of Uncaria rhynchophylla. Helv. Chim. Acta 92 (8), 1575–1585.

Matsumoto, K., Morishige, R., Murakami, Y., Tohda, M., Takayama, H., Sakakibara, I., Watanabe, H., 2005. Suppressive effects of isorhynchophylline on 5-HT2A receptor function in the brain: behavioural and electrophysiological studies. Eur. J. Pharmacol. 517 (3), 191–199.

Mei-cai, D., Wei, J., Wei-wei, D., Yang, C.B, Lu, R.H, 2009. Study on Chemical Constituents of Uncaria rhynchophylla. Nat. Prod. Res. 21 (2), 242–245.

Mo, Z.X., Xu, D.D., Ken, K.L., 2006. Effects of rhynchophylline on rat cortical neurons stressed by methamphetamine. Pharmacologyonline 3, 856–861.

Ndagijimana, A., Wang, X., Pan, G., Zhang, F., Feng, H., Olaleye, O., 2013. A review on indole alkaloids isolated from *Uncaria rhynchophylla* and their pharmacological studies. Fitoterapia 86, 35–47.

Phillipson, J.D., Hemingway, S.R., 1975. Chromatographic and spectroscopic methods for the identification of alkaloids from herbarium samples of the genus Uncaria. J. Chromatogr. A 105 (1), 163–178.

Qu, J., Gong, T., Ma, B., Zhang, L., Kano, Y., Yuan, D., 2012. Comparative study of fourteen alkaloids from *Uncaria rhynchophylla* hooks and leaves using HPLC-diode array detection-atmospheric pressure chemical ionization/MS method. Chem. Pharm. Bull. 60 (1), 23–30.

Sakakibara, I., Terabayashi, S., Kubo, M., Higuchi, M., Komatsu, Y., Okada, M., Taki, K., Kamei, J., 1999. Effect on locomotion of indole alkaloids from the hooks of Uncaria plants. Phytomedicine 6 (3), 163–168.

Shao, H., Mi, Z., Ji, W.G., Zhang, C.H., Zhang, T., Ren, S.C., Zhu, Z.R., 2015. Rhynchophylline protects against the amyloid β-induced increase of spontaneous discharges in the hippocampal CA1 region of rats. Neurochem. Res. 40 (11), 2365–2373.

Shi, J.S., Yu, J.X., Chen, X.P., Xu, R.X., 2003. Pharmacological actions of Uncaria alkaloids, rhynchophylline and isorhynchophylline. Acta Pharmacol. Sin. 24 (2), 97–101.

Shim, J.S., Kim, H.G., Ju, M.S., Choi, J.G., Jeong, S.Y., Oh, M.S., 2009. Effects of the hook of *Uncaria rhynchophylla* on neurotoxicity in the 6-hydroxydopamine model of Parkinson's disease. J. Ethnopharmacol. 126 (2), 361–365.

Shin, S.J., Jeong, Y., Jeon, S.G., Kim, S., Lee, S.K., Choi, H.S., Im, C.S., Kim, S.H., Kim, S.H., Park, J.H., Kim, J.I., 2018. Shin, S.J., Jeong, Y., Jeon, S.G., Kim, S., Lee, S.K., Choi, H.S., Im, C.S., Kim, S.H., Kim, S.H., Park, J.H., Kim, J.I., Uncaria rhynchophylla ameliorates amyloid beta deposition and amyloid beta-mediated pathology in 5XFAD mice. Neurochem. Int. 121, 114–124.

Suk, K., Kim, S.Y., Leem, K., Kim, Y.O., Park, S.Y., Hur, J., Baek, J., Lee, K.J., Zheng, H.Z., Kim, H., 2002. Neuroprotection by methanol extract of *Uncaria rhynchophylla* against global cerebral ischemia in rats. Life Sci. 70 (21), 2467–2480.

Tang, N.Y., Liu, C.H., Su, S.Y., Jan, Y.M., Hsieh, C.T., Cheng, C.Y., Shyu, W.C., Hsieh, C.L., 2010. Uncaria rhynchophylla (Miq) Jack plays a role in neuronal protection in kainic acid-treated rats. Am. J. Chin. Med. 38 (02), 251–263.

Suk-Chul, S, Dong-Ung, L., 2013. Ameliorating effect of new constituents from the hooks of Uncaria rhynchophylla on scopolamine-induced memory impairment. Chin. J. Nat. Med. 11 (4), 391–395.

Tang, W., Eisenbrand, G., 1992. *Uncaria rhynchophylla* (Miq.) Jacks. In: Chinese Drugs of Plant Origin. Springer, Berlin, Heidelberg, pp. 997–1002.

Wiart, C., 2007. Plants affecting the central nervous system. In: Ethnopharmacology of Medicinal Plants: Asia and the Pacific. Humana Press, pp. 57–153.

Wu, L.X., Gu, X.F., Zhu, Y.C., Zhu, Y.Z., 2011. Protective effects of novel single compound, Hirsutine on hypoxic neonatal rat cardiomyocytes. Eur. J. Pharmacol. 650 (1), 290–297.

Wu, L.Z., Xiao, X.M., 2019. Evaluation of the effects of *Uncaria rhynchophylla* alkaloid extract on LPS-induced preeclampsia symptoms and inflammation in a pregnant rat model. Braz. J. Med. Biol. Res. 52 (6).

Xian, Y.F., Lin, Z.X., Mao, Q.Q., Hu, Z., Zhao, M., Che, C.T., Ip, S.P., 2012. Bioassay-guided isolation of neuroprotective compounds from *Uncaria rhynchophylla* against beta-amyloid-induced neurotoxicity. Evid. Based Complement. Alternat. Med. 2012.

Xian, Y.F., Lin, Z.X., Zhao, M., Mao, Q.Q., Ip, S.P., Che, C.T., 2011. *Uncaria rhynchophylla* ameliorates cognitive deficits induced by D-galactose in mice. Planta Med. 77 (18), 1977–1983.

Xian, Y.F., Mao, Q.Q., Wu, J.C., Su, Z.R., Chen, J.N., Lai, X.P., Ip, S.P., Lin, Z.X., 2014. Isorhynchophylline treatment improves the amyloid-β-induced cognitive impairment in rats via inhibition of neuronal apoptosis and tau protein hyperphosphorylation. J. Alzheimers Dis. 39 (2), 331–346.

Yang, Z.D., Duan, D.Z., Du, J., Yang, M.J., Li, S., Yao, X.J., 2012. Geissoschizine methyl ether, a corynanthean-type indole alkaloid from *Uncaria rhynchophylla* as a potential acetylcholinesterase inhibitor. Nat. Prod. Res. 26 (1), 22–28.

Yamanaka, E., Kimizuka, Y., Aimi, N., Sakai, S., Haginiwa, J., 1983. Studies of plants containing indole alkaloids. IX. Quantitative analysis of tertiary alkaloids in various parts of *Uncaria rhynchophylla* MIQ. Yakugaku Zasshi: J. Pharm. Soc. Jpn. 103 (10), 1028.

Yin, W., Duan, S., Zhang, Y., Zeng, L., XinMing, S., 2010. Antioxidant activities of different solvents extracts and alkaloids of Uncaria rhynchophylla (Miq.) Jacks. J. Guangxi Univ. Nat. Sci. 28 (1), 31–34.

Yoshioka, T., Murakami, K., Ido, K., Hanaki, M., Yamaguchi, K., Midorikawa, S., Taniwaki, S., Gunji, H., Irie, K., 2016. Semi-synthesis and structure–activity studies of uncarinic acid C isolated from *Uncaria rhynchophylla* as a specific inhibitor of the nucleation phase in amyloid β42 aggregation. J. Nat. Prod. 79 (10), 2521–2529.

Yuan, D., Ma, B., Wu, C.F., Yang, J., Zhang, L., Liu, S., Wu, L., kano, Y., 2008. Alkaloids from the Leaves of *Uncaria rhynchophylla* and Their Inhibitory Activity on NO Production in Lipopolysaccharide-Activated Microglia. J. Nat. Prod. 71 (7), 1271–1274.

Yuan, D., Ma, B., Yang, J.Y., Xie, Y.Y., Wang, L., Zhang, L.J., Kano, Y., Wu, C.F., 2009. Anti-inflammatory effects of rhynchophylline and isorhynchophylline in mouse N9 microglial cells and the molecular mechanism. Int. Immunopharm. 9 (13–14), 1549–1554.

Zhang, C., Wu, X., Xian, Y., Zhu, L., Lin, G., Lin, Z.X., 2019. Evidence on integrating pharmacokinetics to find a truly therapeutic agent for Alzheimer's disease: comparative pharmacokinetics and disposition kinetics profiles of stereoisomers isorhynchophylline and rhynchophylline in rats. Evid. Based Complement. Alternat. Med. 2019.

Zhang, F., Sun, A.S., Yu, L.M., Wu, Q., Gong, Q.H., 2008. Effects of isorhynchophylline on angiotensin II-induced proliferation in rat vascular smooth muscle cells. J. Pharm. Pharmacol. 60 (12), 1673–1678.

Zhang, J.G., Geng, C.A., Huang, X.Y., Chen, X.L., Ma, Y.B., Zhang, X.M., Chen, J.J., 2017. Chemical and biological comparison of different sections of *Uncaria rhynchophylla* (Gou-Teng). Eur. J. Mass Spectrom. 23 (1), 11–21.

Zhang, L., Koyyalamudi, S.R., Jeong, S.C., Reddy, N., Bailey, T., Longvah, T., 2013. Immunomodulatory activities of polysaccharides isolated from *Taxillus chinensis* and *Uncaria rhyncophylla*. Carbohydr. Polym. 98 (2), 1458–1465.

Zhang, W.B., Chen, C.X., Sim, S.M., Kwan, C.Y., 2004. In vitro vasodilator mechanisms of the indole alkaloids rhynchophylline and isorhynchophylline, isolated from the hook of *Uncaria rhynchophylla* (Miquel). N. Schmied. Arch. Pharmacol. 369 (2), 232–238.

Zhang, Y.B., Yang, W.Z., Yao, C.L., Feng, R.H., Yang, M., Guo, D.A., Wu, W.Y., 2014. New triterpenic acids from *Uncaria rhynchophylla*: chemistry, NO-inhibitory activity, and tandem mass spectrometric analysis. Fitoterapia 96, 39–47.

Zhang, Q., Lei, C., Le-Jian, H.U., Wen-Yuan, L.I.U., Feng, F., Wei, Q.U., 2016. Two new ortho benzoquinones from *Uncaria rhynchophylla*. Chin. J. Nat. Med. 14 (3), 232–235.

Zhu, J.T., Choi, R.C., Xie, H.Q., Zheng, K.Y., Guo, A.J., Bi, C.W., Lau, D.T., Li, J., Dong, T.T., Lau, B.W., Chen, J.J., 2009. Hibifolin, a flavonol glycoside, prevents β-amyloid-induced neurotoxicity in cultured cortical neurons. Neurosci. Lett. 461 (2), 172–176.

Chapter 3.2.22

Alpinia officinarum

Arpan Mukherjee[1], Gowardhan Kumar Chouhan[1], Saurabh Singh[1], Koustav Chatterjee[2], Akhilesh Kumar[1], Anand Kumar Gaurav[1], Durgesh Kumar Jaiswal[1], Jay Prakash Verma[1]

[1]*Institute of Environment and Sustainable Development, Banaras Hindu University, Varanasi, Uttar Pradesh, India;* [2]*Department of Biotechnology, Visva-Bharati, Santiniketan, West Bengal, India*

Introduction

Alzheimer disease (AD) is one of the most progressive diseases causing amnesia and movement dysfunction in the elderly population around the world (Chu et al., 2012). The etiology of this disease is highly complex and unrevealed. Deposition of the amyloid beta protein amyloid-β (Aβ) is believed to be a key pathologic event that is mainly responsible for development of AD (Gilbert, 2014; Sun et al., 2015; De Strooper and Karran, 2016). Amyloid-β is mainly made up of 40−42 amino acids formed by aberrant proteolytic cleavage of amyloid precursor protein (APP). Amyloid-β is the component that is frequently found in senile plaques, and it occurs in the brains of humans with AD. However, monomeric forms of amyloid-β42 are most pathogenic in the early stage of AD (Huang et al., 2016). Some major malfunctioning effects of amyloid-β42 are induction of synaptic dysfunction, neurite dystrophy, and dendritic simplification (Larson and Lesné, 2012; Benilova et al., 2012). These are associated with cholinergic dysfunction, oxidative stress, and selective loss of cholinergic neurons (Craig et al., 2011; Agostinho et al., 2010). Another characteristic associated with the AD patient is increased level of acetylcholinesterase (AChE), the key enzyme that hydrolyzes the neurotransmitter acetylcholine (Orhan et al., 2006). Increased oxidative stress due to lacking balance between antioxidative agent and reactive oxygen species production causes neuronal loss and ultimately leads to neurodegenerative diseases (Javed et al., 2012). Thus, current efforts against AD have been started that aim to develop drugs against destructive effects of Aβ oligomers. A number of natural products have been identified that show effective results against Aβ-induced neurotoxicity (Yamada et al., 2015; Awasthi et al., 2016). Moreover, the

Naturally Occurring Chemicals against Alzheimer's Disease. https://doi.org/10.1016/B978-0-12-819212-2.00038-4

signalling pathway induced due to Aβ can be downregulated by, for instance, curcumin and its related products found in the ginger family (*Zingiberaceae*). Ginger is an annual plant and cultivated in many countries including India, China, Vietnam, and Thailand due to its abilities of Aβ downregulation and memory enhancement (Garcia-Alloza et al., 2007; Xiao et al., 2010; Ahmed and Gilani, 2011; Tiwari et al., 2014; Zhao et al., 2015). Resveratrol is a stilbenoid, a type of natural phenolic product found in many plants, which has reported multineuroprotective effects against Aβ (Rege et al., 2014, 2015). Moreover, galangin is a major flavonoid derived from the rhizomes of *Alpinia officinarum*. Several studies have shown galangin to be an antagonist against Aβ due to antioxidative and free radical scavenging activity (Guo et al., 2010). Two naturally occurring compounds, diarylheptanoids (diphenylheptanoids) 7-(4-hydroxyphenyl)-1-phenyl-4E-hepten-3-one and 7-(4-hydroxy-3-methoxyphenyl)-1-phenyl-4E-hepten-3-one, were extracted from the rhizomes of *A. officinarum* and found to promote new neuronal cells in vivo (Tang et al., 2015). In this chapter, we will discuss the different properties of *A. officinarum* and its application as traditional medicine in different diseases, focusing on the role of *A. officinarum* in AD.

A. officinarum used for medicinal purpose

Varieties around the world of *Alpinia*

Alpinia, a flowering plant of the ginger family, *Zingiberaceae*, was named after Prospero Alpine who was an Italian botanist specializing in exotic plants of the 17th century (Simonetti, 1990). The species variation for the plant is huge with its nativity ranging from Asia, Australia, and the Pacific Islands, basically occupying tropical and subtropical climate. This is the largest genus of *Zingiberaceae* family, with about 230 species, with the important ones being discussed further.

***Alpinia rafflesiana*:** It is native to Peninsular Malaysia and has two varieties namely, var. *rafflesiana* and var. *hirtior* (Ridl.) Holttum. The difference between the two can be spotted in their indumentum and the length of primary bracts. *A. rafflesiana* var. *rafflesiana* can grow from 0.5 to 2 m in length with leaves glabrous on the upper surface (except the midrib) and velvety on the lower surface and margins. It can grow from the lowland to the hill forest up to 938 m altitude, mainly in primary forest (Holttum, 1950). Cardamomin, which is used for its antiinflammatory activity in cellular models of inflammation, is isolated from the fruits of *A. rafflesiana* (Sung et al., 2012).

***Alpinia galanga*:** This species is prevalent in the Asiatic region, concentrated mainly in the Southeast Asia region, where it is used for cooking purposes. Its vernacular name in the region varies from lengkuas and greater galangal to blue ginger, kulanjan in Hindi, and kholinjan in Urdu (Peter and Babu, 2012). The related species are *A. officinarum* Hance and *A. calcarata* Rosc. They are also known as lesser galangal. *A. officinarum* is known by several local names: in Japan (as *ryokyo*); in Turkey and Iran (as *havlıcan* and *khoulanjan*); in India, (*chitrarathai chooranam* and *aichhia* or *dum aidu*) and in china (as *gao linang jiang* or *heha*) (Abubakar et al., 2018). Ancient reports reveal that the extractive juices from these plants were used as an Ayurvedic drug, with India, Thailand, and Indonesia being its major suppliers. It is reported to have been originated from Southern China and Java and also along the sub-Himalayan region of Bihar, West Bengal, and Assam. The current cultivation of *A. galangal* is reported in southeast Asian countries such as India, Bangladesh, China, and Surinam.

***Alpinia zerumbet*:** This is prevalent in the Australian region and is native to East Asia. It also commonly called shell ginger, and often called by names such as pink porcelain lily, variegated ginger, or butterfly ginger. It looks similar to culinary ginger, i.e., *Zingiber officinale*. Rhizomes produce stout, slightly curved stems with evergreen leaves. It produces an inflorescence on old growth and produces drooping racemes at the end of leafy stems rather directly from rhizomes. Flowers appear to be pearlescent seashells, when in the bud form, which is how it got the name shell ginger. Flowers produce a slight fragrance during the blooming season and are often followed by striated fruits (Holttum, 1950).

***Alpinia purpurata*:** It is also called red ginger and is prevalent in the Pacific Islands. It also grows well in places such as Hawaii, the Caribbean islands, Asiatic region, and many Central American countries. It is also found in Samoa where its flower is the national flower called "teuilla." It has long, brightly colored red bracts, which look like the bloom, but the true flower is the small white flower on the top of it (Peter and Babu, 2012). They are often marked with two varieties, where the red is called the jungle king, and the pink is called the jungle queen.

A. officinarum application in other disease

A. officinarum has been used for very long time in the Asian cultures to prevent different types of diseases (Zhang et al., 2013). All part (aerial parts, leaves, roots, and rhizomes) of the *A. officinarum* plants were used for medicinal purpose either directly or by preparing an extract with the help of various polar and nonpolar solvents. The solvents that are used to prepare the medicinal extract of *A. officinarum* are methanol, ethanol, water, ethyl-acetate, petroleum ether, hexane, dichloromethane, and chloroform. Preparation of *A. officinarum* plant extract has some specific techniques including ultrasonication, soaking, Soxhlet extraction, and maceration, and the fractionization process of the *A. officinarum* extracts is conducted by solvent–solvent partition method (Basri et al., 2017).

Different solvent extracts of the plant parts have different disease control properties. For example, methanol extract of the areal parts and rhizome works as antiinflammatory, antioxidant, and anticancer agent, and methanol extract of leaf and root work as antiproliferative and antioxidant agent. The ethanol acetate extracts of rhizomes work as antiinflammatory, antioxidant, anticancer, and inhibitory agent of some enzymes. Ethanol extract of rhizomes and root of *A. officinarum* play a role as antibacterial, antioxidant, and anticancer agent. Hexane extract of leaf and rhizomes of *A. officinarum* show high antiproliferative components. Di-chloromethane extract of rhizome showed anticancer activity, whereas leaf extract showed antiproliferative activity. Chloroform extract of leaf and rhizome extract showed antiproliferative, antioxidant, and anticancer activity, respectively, and petroleum extract of rhizome showed antiinflammation property (Basri et al., 2017). Aqueous extract of rhizomes of *A. officinarum* showed antioxidant and antiproliferative properties, and aqueous extract of leaves were used in enteritis and arthritis (Usha et al., 2016; Basri et al., 2017). Diethyl ether and hydroalcoholic extracts of *A. officinarum* suppress the inflammatory cytokines, work as antibacterial components, inhibit osteoclast genesis, and reduce stomach ache (Tao et al., 2006).

A. officinarum application in Alzheimer disease

AD is the most common neurodegenerative disease, causing dementia and death in aging peoples. The major symptoms of AD include cognitive dysfunction and loss of primary memory. At present, the treatment of AD is mostly focused on inhibitors of acetylcholinesterase. *A. officinarum* (galangal) is a medicinal plant, and its dried rhizomes are a source of acetyl cholinesterase inhibitors. It is widely used for remedy of multiple diseases in addition to AD (Ding et al., 2019). The rhizome of this plant is traditionally used as a flavoring in food and in medicine. Galangal has many biologically active substances, such as flavonoids, diarylheptanoids, essential oils, phenylpropanoids, glycosides, and many others (Tang et al., 2018). Galangin showed potential effect to inhibit the production of β-amyloid and acetylcholinesterase, used for treatment of AD. *A. officinarum* has several acetylcholinesterase inhibitors such as galantamine, huperzine, and rivastigmine. Among most known flavonoids, galangin, isolated from *A. officinarum*, showed highest inhibitory effect on acetylcholinesterase. It is a naturally occurring flavonoid that inhibits β-amyloid production (Guo et al., 2010). In addition, galangin has antioxidative and free radical scavenging activity. Galangin contains a 3-hydroxyflavone backbone, and it is one of the major flavonoids found in *A. officinarum*. The mechanism may be related to modulation of signalling pathways. The major mechanism involved in inhibition is gene modification at the transcription level, mainly methylation and acetylation (Zeng et al., 2015). So, the aformentioned suggest that flavonoid galangal is useful for further development of drugs to inhibit acetylcholinesterase and cure

AD. Thus, *A. officinarum* has increasing and important applications in the medicinal herb industry.

Working mechanism of *A. officinarum* on Alzheimer disease

Among neurodegenerative agents or compounds known so far, amyloid-β (Aβ) peptide is one of the major components that causes neurodegenerative disorder like AD in a direct mechanism (Gilbert, 2014). Abnormal cleavage of APP releases 40−42 amino acids containing Aβ peptides that deposit at the synapses and exert pathogenicity (Sun et al., 2015; Larson and Lesné, 2012). 7-(4-Hydroxyphenyl)-1-phen-yl-4E-hepten-3-one (AO-1), a diarylheptanoid like 7-(4-Hydroxyphenyl)-1-phenyl-4E-hepten-3-one extracted from *A. officinarum*, has a profound effect on neural differentiation and neurite outgrowth (Benilova et al., 2012). AO-1 also showed to protect neurons by inhibiting apoptosis and dendrite atrophy induced by Aβ. AO-1 inhibits caspase mediated pathway of apoptosis and dendrite protection through PI3K-mTOR mediated signalling pathway. The study has further suggested that it protects neurons from pathogenicity, arising during AD (Tang et al., 2015; Huang et al., 2016). Acetylcholinesterase (AChE) and butyryl cholinesterase (BChE) commonly function as the regulators of acetylcholine during neurotransmission. Though AChE is found in the neuromuscular junction and BChE originates in the white matter and glia of the brain, both have established an association with the progression of AD through the accumulation of amyloid-β, but in a distinct way. A stable complex that is formed between AChE and senile plaque has shown to have increased the neurotoxicity of amyloid components, whereas BChE is observed to be associated with the Aβ plaques in the cerebral cortex of an AD brain (Talesa, 2001; Sultan, 2016). Thus, the regulations of these cholinesterase have created a significant target area in AD research. For decades, the effects of several plant extracts have been tested to reduce the neurotoxicity. Galanga (*A. officinarum* Hance), known as a traditional medicinal plant found in southeast Asian countries, contains flavonoids, glycosides, and diarylheptanoids as important chemical compounds and has multiple biologic roles such as antiinflammatory, anticancerous, antioxidant, etc. Recent evidence has demonstrated that the water-ethanolic extract of galangal has a profound effect on the inhibition of AChE (Köse et al., 2015). Galangin is one of the crucial flavonoids of galanga that contain a 3-hydroxyflavone backbone. Ellman assay has revealed that galangin has the highest inhibitory effect on AChE over the activity of other flavonoids evaluated (Guo et al., 2010). On the other hand, the results of the enzymatic assay of the various flavonoids for their BChE inhibitory role showed that the galangin reversibly inhibits the BChE variants more than other flavonoids (Katalini et al., 2014). The mechanism of inhibition of galangin on AChE and BChE is still a matter of research. The overall mechanistic approach of *A. officinarum* isolated compounds is illustrated in Fig. 3.2.22.1.

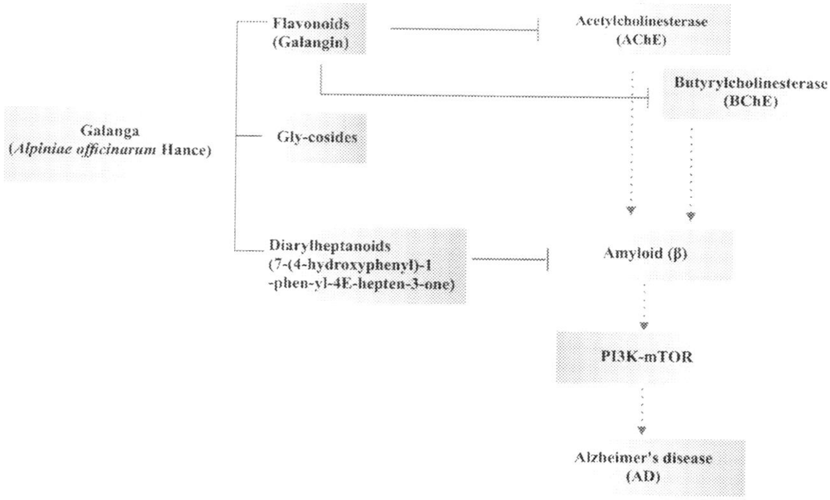

FIGURE 3.2.22.1 The mechanistic approach of *Alpinia officinarum* extracted compounds on the inhibition of neurotoxicity and AD.

Clinical studies and safety issues

The ancient prescription Liangfu Pill, consisting of *A. officinarum* and *Cyperus rotundus* L., is now widely used for treatment of stomach pain, dysmenorrhea caused by cold, and qi stagnation. On the basis of pharmacologic and clinical practices, A. *officinarum* has the effect of inhibiting *Helicobacter pylori*, *Staphylococcus aureus*, *Bacillus cereus*, *Pseudomonas aeruginosa*, and *Escherichia coli* and antioxidation (Srividya et al., 2010; Zhang et al., 2010). There are few clinical reports of *A. officinarumon* as its side effects for various organ dietary-related to the use of galangal during their application. Liu et al. (2015) reported no significant adverse effects on weight, skin, respiratory system, or limb activity of rabbits. As safety evaluation, Karunarathne et al. (2018) observed that the undiluted crude galangal extract showed negligible irritation on nonabraded skin of New Zealand white rabbits with 0.25 primary irritation index, whereas the abraded skin of the rabbits showed irritation for all tested dilutions of galangal extracts: 0.75, 0.5, 0.25, and 0.125 g/mL. Also, the single oral dose of galangal extract at 2000 mg/kg did not produce mortality or significant changes in the general behavior, body weight, feed intake, or biochemical analysis (ALT, AST, BUN, and creatinine levels) of Wistar rats compared to the control. However, 2000 and 50 mg/kg body weight of galangal extract were highly toxic to Wistar rats when administered intraperitoneally.

Conclusion

Plants are the sole sources of a lot of medicinal components. *A. officinarum* plant parts like aerial, leaves, roots, and rhizomes and its extract work to suppress infection, pain, fever, inflammation, oxidation, cancer, arthritis, and AD. Galangin is one of the crucial flavonoids of galanga that contain a 3-hydroxyflavone backbone. Ellman assays have revealed that galangin has the highest inhibitory effect on AChE over the activity of other flavonoids evaluated. On the other hand, results of the enzyme assay of different flavonoids for their BChE inhibitory role demonstrated that galangin reversibly inhibits BChE variants more than other flavonoids. The mechanism of inhibition of galangin on AChE and BChE is still a matter of research. Promising and scientific data from epidemiologic areas as well as in vitro and animal model studies suggest that flavonoid galangin may affect several diseases, but we need more clinical trial data to support this mechanism more scientifically. From the preceding report, we can conclude that *A. officinarum* and its flavonoid galangin has a potential role to suppress some fatal disease, but we need further detailed scientific study and clinical trial to understand the other beneficial roles on health of humans, and its study is required to minimize the gap of hypothetical mechanisms on AD control.

Acknowledgement

The authors are thankful to Head and Director of Institute of Environment and Sustainable Development, Banaras Hindu University, for providing a lab facility for research and development.

References

Abubakar, I.B., Malami, I., Yahaya, Y., Sule, S.M., 2018. A review on the ethnomedicinal uses, phytochemistry and pharmacology of *Alpinia officinarum* Hance. J. Ethnopharmacol. 224, 45–62.

Agostinho, P., Cunha, A.R., Oliveira, C., 2010. Neuroinflammation, oxidative stress and the pathogenesis of Alzheimer's disease. Curr. Pharm. Des. 16 (25), 2766–2778.

Ahmed, T., Gilani, A.H., 2011. A comparative study of curcuminoids tomeasure their effect on inflammatory and apoptotic gene expressionin an Abeta plus ibotenic acid-infused rat model of Alzheimer'sdisease. Brain Res. 1400, 1–18.

Awasthi, M., Singh, S., Pandey, V.P., Dwivedi, U.N., 2016. Alzheimer's disease: an overview of amyloid beta dependent pathogenesis and its therapeutic implications along with *in silico* approaches emphasizing the role of natural products. J. Neurol. Sci. 361, 256–271.

Basri, A.M., Taha, H., Ahmad, N., 2017. A review on the pharmacological activities and phytochemicals of *Alpinia officinarum* (Galangal) extracts derived from bioassay-guided fractionation and isolation. Pharmacogn. Rev. 11 (21), 43.

Benilova, I., Karran, E., De Strooper, B., 2012. The toxic Aβ oligomer and Alzheimer's disease: an emperor in need of clothes. Nat. Neurosci. 15 (3), 349.

Chu, Y.F., Chang, W.H., Black, R.M., Liu, J.R., Sompol, P., Chen, Y., Cheng, I.H., 2012. Crude caffeine reduces memory impairment and amyloid β1−42 levels in an Alzheimer's mouse model. Food Chem. 135 (3), 2095−2102.

Craig, L.A., Hong, N.S., McDonald, R.J., 2011. Revisiting the cholinergic hypothesis in the development of Alzheimer's disease. Neurosci. Biobehav. Rev. 35 (6), 1397−1409.

De Strooper, B., Karran, E., 2016. The cellular phase of Alzheimer's disease. Cell 164 (4), 603−615.

Ding, P., Yang, L., Feng, C., Xian, J.C.,A., 2019. Research and application of *Alpinia officinarum* in medicinal field. Chin. Herb. Med.

Garcia-Alloza, M., Borrelli, L.A., Rozkalne, A., Hyman, B.T., Bacskai, B.J., 2007. Curcumin labels amyloid pathology *in vivo*, disrupts existing plaques, and partially restores distorted neurites in an Alzheimer's mouse model. J. Neurochem. 102, 1095−1104.

Gilbert, B.J., 2014. Republished: the role of amyloid β in the pathogenesis of Alzheimer's disease. Postgrad. Med. J. 90 (1060), 113−117.

Guo, A.J., Xie, H.Q., Choi, R.C., Zheng, K.Y., Bi, C.W., Xu, S.L., Tsim, K.W., 2010. Galangin, a flavanol derived from Rhizoma *Alpiniae Officinarum*, inhibits acetylcholinesterase activity in vitro. Chem. Biol. Interact. 187 (1-3), 246−248.

Holttum, R.E., 1950. The zingiberaceae of the Malay Peninsula. Gard. Bull. Singapore 13 (1), 1−249.

Huang, X., Tang, G., Liao, Y., Zhuang, X., Dong, X., Liu, H., Shi, L., 2016. 7-(4-Hydroxyphenyl)-1-phenyl-4E-hepten-3-one, a Diarylheptanoid from *Alpinia officinarum*, protects neurons against amyloid-β induced toxicity. Biol. Pharm. Bull. b16-00411.

Javed, H., Khan, M.M., Ahmad, A., Vaibhav, K., Ahmad, M.E., Khan, A., et al., 2012. Rutin prevents cognitive impairments by ameliorating oxidative stress and neuroinflammation in rat model of sporadic dementia of Alzheimer type. Neuroscience 210, 340−352.

Katalini, M., Bosak, A., Kovarik, Z., 2014. Flavonoids as inhibitors of human butyryl cholinesterase variants. Food Technol. Biotechnol. 52 (1), 64.

Köse, L.P., Gülcin, I., Gören, A.C., Namiesnik, J., Martinez-Ayala, A.L., Gorinstein, S., 2015. LC−MS/MS analysis, antioxidant and anticholinergic properties of galanga (*Alpinia officinarum* Hance) rhizomes. Ind. Crops Prod. 74, 712−721.

Karunarathne, P.U.H.S., Thammitiyagodage, M., Weerakkody, N., 2018. Safety evaluation of galangal (*Alpinia galanga*) extract for therapeutic use as an antimicrobial agent. Int. J. Pharm. Sci. Res. 9 (11), 4582−4590.

Larson, M.E., Lesné, S.E., 2012. Soluble Aβ oligomer production and toxicity. J. Neurochem. 120, 125−139.

Liu, J.X., Sun, Y.H., Li, C.P., 2015. Volatile oils of Chinese crude medicines exhibit antiparasitic activity against human Demodex with no adverse effects in vivo. Exp. Ther. Med. 9 (4), 1304−1308.

Orhan, G., Orhan, I., Sener, B., 2006. Recent developments in natural and synthetic drug research for Alzheimer's disease. Lett. Drug Des. Discov. 3 (4), 268−274.

Peter, K.V., Babu, K.N., 2012. Introduction to herbs and spices: medicinal uses and sustainable production. In: Handbook of herbs and spices. Woodhead Publishing, pp. 1−16.

Rege, S.D., Geetha, T., Broderick, T.L., Babu, J.R., 2015. Resveratrol protects *β*amyloid-induced oxidative damage and memory associated proteins in H19-7 hippocampal neuronal cells. Curr. Alzheimer Res. 12, 147−156.

Rege, S.D., Geetha, T., Griffin, G.D., Broderick, T.L., Babu, J.R., 2014. Neuroprotective effects of resveratrol in Alzheimer's disease pathology. Front. Aging Neurosci. 6, 218.

Simonetti, G., 1990. Simon & Schuster's Guide to Herbs and Spices. Simon & Schuster.

Srividya, A.R., Dhanabal, S.P., Misra, V.K., Suja, G., 2010. Antioxidant and antimicrobial activity of Alpinia officinarum. Indian J. Pharm. Sci. 72 (1), 145.

Sultan, D., 2016. Butyryl cholinesterase as a diagnostic and therapeutic target for Alzheimer's disease. Curr. Alzheimer Res. 13 (10), 1173−1177.

Sun, X., Chen, W.D., Wang, Y.D., 2015. β-Amyloid: the key peptide in the pathogenesis of Alzheimer's disease. Front. Pharmacol. 6, 221.

Sung, B., Prasad, S., Gupta, S.C., Patchva, S., Aggarwal, B.B., 2012. Regulation of inflammation-mediated chronic diseases by botanicals. Adv. Bot. Res. 62, 57−132. Academic Press.

Talesa, V.N., 2001. Acetylcholinesterase in Alzheimer's disease. Mech. Ageing Dev. 122 (16), 1961−1969.

Tang, G., Dong, X., Huang, X., Huang, X.J., Liu, H., Wang, Y., Shi, L., 2015. A natural diarylheptanoid promotes neuronal differentiation via activating ERK and PI3K-Akt dependent pathways. Neuroscience 303, 389−401.

Tang, X., Xu, C., Yagiz, Y., Simonne, A., Marshall, M.R., 2018. Phytochemical profiles, and antimicrobial and antioxidant activities of greater galangal [*Alpiniagalanga* (Linn.) Swartz.] flowers. Food Chem. 255, 300−308.

Tao, L., Wang, Z.T., Zhu, E.Y., Lu, Y.H., Wei, D.Z., 2006. HPLC analysis of bioactive flavonoids from the rhizome of *Alpinia officinarum*. South Afr. J. Botany 72 (1), 163−166.

Tiwari, S.K., Agarwal, S., Seth, B., Yadav, A., Nair, S., Bhatnagar, P., Karmakar, M., Kumari, M., Chauhan, L.K.S., Patel, D.K., Srivastava, V., Singh, D., Gupta, S.K., Tripathi, A., Chaturvedi, R.K., Gupta, K.C., 2014. Curcumin loaded nanoparticles potently induce adult neurogenesis and reverse cognitive deficits in Alzheimer's disease model *via* canonical Wnt/β-catenin pathway. ACS Nano 8, 76−103.

Usha, S., Rajasekaran, C., Siva, R., 2016. Ethnoveterinary medicine of the Shervaroy Hills of Eastern Ghats, India as alternative medicine for animals. J. Tradit. Complement. Med. 6 (1), 118−125.

Xiao, Z., Lin, L., Liu, Z., Ji, F., Shao, W., Wang, M., Liu, L., Li, S., Li, F., Bu, X., 2010. Potential therapeutic effects of curcumin: relationship to microtubule-associated proteins 2 in Aβ1−42 insult. Brain Res. 1361, 115−123.

Yamada, M., Ono, K., Hamaguchi, T., Noguchi-Shinohara, M., 2015. Naturalphenolic compounds as therapeutic and preventive agents for cerebral amyloidosis. Adv. Exp. Med. Biol. 863, 79−94.

Zeng, H., Huang, P., Wang, X., Wu, J., Wu, M., Huang, J., 2015. Galangin-induced down-regulation of BACE1 by epigenetic mechanisms in SH-SY5Y cells. Neuroscience 294, 172−181.

Zhang, B.B., Dai, Y., Liao, Z.X., Ding, L.S., 2010. Three new antibacterial active diarylheptanoids from Alpinia officinarum. Fitoterapia 81 (7), 948−952.

Zhang, W., Tang, B., Huang, Q., Hua, Z., 2013. Galangin inhibits tumor growth and metastasis of B16F10 melanoma. J. Cell. Biochem. 114 (1), 152−161.

Zhao, H., Li, N., Wang, Q., Cheng, X., Li, X., Liu, T., 2015. Resveratrol decreases theinsoluble Aβ1-42 level in hippocampus and protects the integrity of the blood-brainbarrier in AD rats. Neuroscience 310, 641−649.

Chapter 3.2.23

Himatanthus lancifolius (Müll.Arg.) Woodson

Devina Lobine, Mohamad Fawzi Mahomoodally
Department of Health Sciences, Faculty of Medicine and Health Sciences, University of Mauritius, Réduit, Mauritius

Abbreviations

AChE acetylcholinesterase
AD Alzheimer's disease
AlkF alkaloid-rich fraction
BChE butylcholinesterase
HL *Himatanthus lancifolius* (Müll.Arg.)

Introduction

Himatanthus lancifolius (Müll.Arg.) (HL) Woodson (Apocynaceae), previously known as *Plumeria lancifolia*, is a shrub that has a long history in South American folk medicine. It is a Brazilian native plant and is popularly called agoniada. Some *Himatanthus* species have been reported to exert antimolluscicidal, antispasmodic, and antiinflammatory activities, as revealed by pharmacological investigations (Baggio et al., 2005).

The dried stem bark of HL is largely used to treat ulcers, skin diseases, asthma, syphilis, stomach disorder, headache, and fatigue, and is used as a febrifuge and emenagogue, and mainly to stimulate uterine contractions (Souza et al., 2004; Lima et al., 2010; Barros et al., 2013). HL is officially registered in Brazilian Pharmacopeia I (1929) and is a component of commercial products meant to treat dysmenorrhea and improve menopausal symptoms (Rattmann et al., 2005).

Naturally Occurring Chemicals against Alzheimer's Disease. https://doi.org/10.1016/B978-0-12-819212-2.00039-6

Phytochemistry

Limited chemical investigations have been carried on HL. It is reported that HL is rich in indole alkaloids, with uleine as its main compound (Nardin et al., 2010) (Fig. 3.2.23.1). The presence of iridoids such as glucosylplumeride and the alkaloid demethoxyaspidospermine has also been reported in HL (França et al., 2000; Rattmann et al., 2005).

Overview of pharmacological properties

Like other *Himatanthus* species, HL has barely been documented in the scientific literature for its biological activities. Baggio et al. (2005) investigated the gastroprotective property of the indole alkaloid–rich fraction (AlkF) obtained from the bark of HL using both in vitro and in vivo methods. Administration of AlkF protected rats from induced gastric lesions by ethanol (ED_{50} = 30 mg/kg, orally) and decreased the hypersecretion of gastric acid (ED_{50} = 82 mg/kg, intradermally) induced by pylorus ligature. AlkF exerts its gastroprotective effects through activation of a number of cytoprotective mechanisms (glutathione-dependent) including enzymatic and nonenzymatic antioxidants by inhibiting gastric acid secretion through the blockade of H^+/K^+-ATPase activity. Three main indole alkaloids—uleine (53%), its isomer (13%), and demethoxyaspidormine (23.8%)—and traces of other five alkaloids were characterized from the AlkF. The findings support the popular use of this plant species as an antiulcer agent.

The alkaloidal fraction from HL bark has shown broad-spectrum in vitro antimicrobial activity against pathogenic microorganisms such as *Staphylococcus aureus* including methicillin-resistant *S. aureus* strains, *Staphylococcus epidermidis*, *Enterococcus faecalis*, *Escherichia coli*, *Pantoea agglomerans*, and *Acinetobacter baumannii*, and the animal pathogen *S. aureus* canine (Souza et al., 2004).

HL has also shown the potential to exert antiinflammatory properties through various mechanisms. Nardin et al. (2008) investigated the potential effects of the uleine-rich fraction (URF) of HL bark on human leukocytes in vitro, the main promoters of the inflammatory response. The URF was

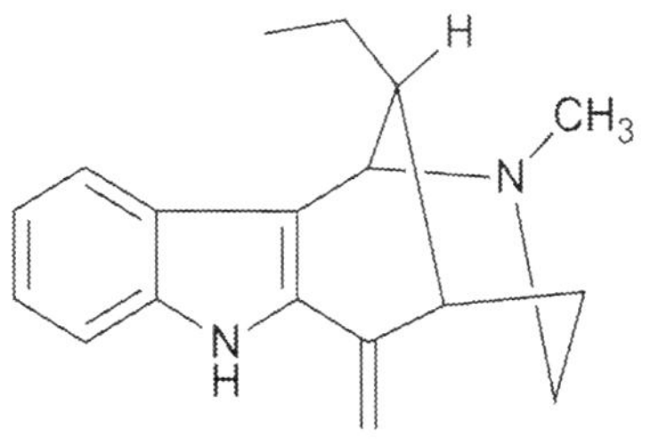

FIGURE 3.2.23.1 Structure of uleine.

observed to hinder the migration of casein-induced granulocytes and their adhesion to extracellular protein matrix (fibronectin and vitronectin) significantly, along with mononuclear cells, by downregulating the expression of integrins receptors ($\alpha4\beta1$ and $\alpha5\beta1$). Further investigation has shown that URF (10^{-5}−1 μg/mL) can potentially modulate the immune system by suppressing the proliferation of phytohemagglutinin (PHA)-induced lymphocytes by blocking their transformation into blast-dividing cells. In addition, URF was observed to show protection to human lymphocytes from PHA-induced toxicity (Nardin et al., 2010).

The AlkF of HL has significantly inhibited the cell growth of leukemic cell lines, Daudi (0.1−10 μg/mL), K562 (1−10 μg/mL), and REH cells (10−100 μg/mL), while displaying a modest effect on normal marrow cell proliferation (IC_{50} = 584.48 μg/mL). The AlkF was not toxic to any cells up to 10 μg/mL. The flow cytometric expression of Annexin-V and 7-AAD in K562 and Daudi cells demonstrated that increasing doses of AlkF for 48 h did not induce cells to undergo apoptosis, which suggests cytostatic activity for tumor (Lima et al., 2010). Another study showed that URF from HL suppressed the proliferation of Daudi and REH cells in a dose-dependent manner, in particular for Daudi cells, for which immunosuppressive effects of URF were observed at doses up to 10 μg/mL (Nardin et al., 2010).

Seidl et al. (2010) investigated the acetylcholinesterase (AChE)-inhibiting properties of HL extracts and its bioactive compound, uleine, in vitro. The dichloromethane and ethyl acetate fractions (5 mg/mL) were significantly potent against AChE by 54.73% ± 0.6% and 74.20% ± 2.3%, respectively. Uleine was the main active compound present in both fractions, and it displayed inhibitory activity against AChE (IC_{50}: 0.45 μM). More recently, a study was carried to investigate the ability of uleine isolated from the stem bark of HL, to interact with the cholinergic and amyloidogenic pathways involved in the pathogenesis of Alzheimer's disease (AD). Uleine has showed strong inhibitory capacity against AChE and butylcholinesterase (IC_{50} 279.0 ± 4.5 and 24.0 ± 1.5 μM, respectively) and β-secretase (IC_{50} 180 ± 22 nM) and significantly hindered the self-aggregation of amyloid beta peptide. In vitro cytotoxicity analysis using an MTT cell proliferation assay indicated that uleine elicited no toxic effect on PC12 or SH-SY5Y neuronal cells (Seidl et al., 2017). However, so far, human studies specific to cognition have not been performed using HL. Preclinical and clinical safety and toxicity data also have not reported.

Conclusion

The pharmaceutical industry has continuously searched for new lead compounds with better therapeutic action and minimal or no side effects. Although limited, the studies showed that HL has good pharmacological action. The data amassed here show that HL and its bioactive uleine hold huge potential for managing AD.

Acknowledgment

D. Lobine wishes to thank the Higher Education Commission, Mauritius, for the award of the postdoctoral fellowship.

References

Baggio, C.H., Otofuji, G.D.M., de Souza, W.M., de Moraes Santos, C.A., Torres, L.M.B., Rieck, L., de Andrade Marques, M.C., Mesia-Vela, S., 2005. Gastroprotective mechanisms of indole alkaloids from *Himatanthus lancifolius*. Planta Med. 71 (08), 733–738.

Barros, P.M., de Couto, N.M., Silva, A.S., Barbosa, W.L., 2013. Development and validation of a method for the quantification of an alkaloid fraction of *Himatanthus lancifolius* (Muell. Arg.) Woodson by ultraviolet spectroscopy. J. Chem. 2013.

França, O.O., Brown, R.T., Santos, C.A.M., 2000. Uleine and demethoxyaspidospermine from the bark of *Plumeria lancifolia*. Fitoterapia 71 (2), 208–210.

Lima, M.P.D., Hilst, L.F., Mattana, F.V.R., Santos, C.A.D.M., Weffort-Santos, A.M., 2010. Alkaloid-rich fraction of *Himatanthus lancifolius* contains anti-tumor agents against leukemic cells. Braz. J. Pharm. Sci. 46 (2), 273–280.

Nardin, J.M., de Souza, W.M., Lopes, J.F., Florao, A., de Moraes Santos, C.A., Weffort-Santos, A.M., 2008. Effects of *Himatanthus lancifolius* on human leukocyte chemotaxis and their adhesion to integrins. Planta Med. 74 (10), 1253–1258.

Nardin, J.M., Lima, M.P., Machado, J.C., Hilst, L.F., de Moraes Santos, C.A., Weffort-Santos, A.M., 2010. The uleine-rich fraction of *Himatanthus lancifolius* blocks proliferative responses of human lymphoid cells. Planta Med. 76 (07), 697–700.

Rattmann, Y.D., Terluk, M.R., Souza, W.M., Santos, C.A., Biavatti, M.W., Torres, L.B., Mesia-Vela, S., Rieck, L., da Silva-Santos, J.E., Maria, C.D.A., 2005. Effects of alkaloids of *Himatanthus lancifolius* (Muell. Arg.) Woodson, Apocynaceae, on smooth muscle responsiveness. J. Ethnopharmacol. 100 (3), 268–275.

Seidl, C., Aimbire de Moraes Santos, C., De simone, A., Bartolini, M., Maria Weffort Santos, A., Andrisano, V., 2017. Uleine disrupts key enzymatic and non-enzymatic biomarkers that leads to Alzheimer's disease. Curr. Alzheimer Res 14 (3), 317–326.

Seidl, C., Correia, B.L., Stinghen, A.E., Santos, C.A., 2010. Acetylcholinesterase inhibitory activity of uleine from *Himatanthus lancifolius*. Z. Naturforsch. C Biosci. 65 (7–8), 440–444.

Souza, W.M., Stinghen, A.E.M., Santos, C.A.M., 2004. Antimicrobial activity of alkaloidal fraction from barks of *Himatanthus lancifolius*. Fitoterapia 75 (7–8), 750–753.

Chapter 3.2.24

Nelumbo nucifera

Firoz Akhter[1], Asma Akhter[1], Victor W. Day[2], Erika D. Nolte[3], Suman Bhattacharya[1], Mohd Saeed[4]
[1]*Department of Surgery, Columbia University Irving Medical Center, New York, NY, United States;* [2]*X-ray Crystallography Laboratory, University of Kansas, Lawrence, KS, United States;* [3]*Department of Pharmacology and Toxicology, Higuchi Bioscience Center, University of Kansas, Lawrence, KS, United States;* [4]*Department of Biology, College of Sciences, University of Hail, Hail, Saudi Arabia*

Nutrition and cognitive function

Nutritional deficiency is one of the crucial reasons for cognitive decline in the elderly. Some groups of vitamin B (folic acid) and dietary antioxidants such as vitamin C (ascorbic acid) and vitamin E (tocopherol) are potentially important nutrients to control or reduce the impairment of brain function by scavenging the negative effects of reactive oxygen species (ROS) and free radicals (Zhao et al., 2016). It has also been reported that excessive consumption of cholesterol-rich foods and a high-fat diet impair cognitive function (Fu et al., 2017) [17]. *Nelumbo nucifera* (Family: Nymphaeaceae) has been generally used as a vegetable or traditional herbal medicine, as well as an antiinflammatory, antipyretic, hemostatic, and sedative agent (Du et al., 2012), signifying that it could possess CNS effects. Although each part, including the fruit seeds, contains potent therapeutic activity, a high level of ascorbic acid is a commonly known therapeutic agent in multiple ways. A growing body of evidence suggests that the neuroprotective property of ascorbic acid is one of the important constituents of *N. nucifera* and has a profound effect on cognitive decline and memory impairment or dementia. Impairment of cognitive function is characterized by reduced blood flow in the hypothalamus of the brain and generation of ROS, which induce excess production of free radicals in the brain and brain mitochondrial dysfunction due to free radical-induced inflammation (Phaniendra et al., 2015). All these are generally considered to be natural processes in aging and in the pathogenesis of AD (Guzik and Touyz, 2017). The possible mechanism by which *N. nucifera*

Naturally Occurring Chemicals against Alzheimer's Disease. https://doi.org/10.1016/B978-0-12-819212-2.00040-2

exerts its beneficial neuroprotective effects is by reducing the level of low-density lipoprotein (LDL) cholesterol, which is a risk factor for cardiovascular disorders, increasing high-density lipoprotein (HDL) cholesterol (Dinu et al., 2018; Akhter, 2019), and lowering oxidative stress induced by lipid peroxidation by increasing antioxidant activity, which improves mitochondrial function and cognitive activity (Casas et al., 2014) and lowers the production of inflammatory markers (C-reactive protein) in the brain (Grosso et al., 2014).

Glycative and oxidative stress and Alzheimer's disease

Advanced glycation end products (AGEs) and ROS-induced oxidative stress have been hypothesized in the development of various diseases and disease complications, and identified as a key factor in AD pathogenesis (Ahmad et al., 2014; Akhter et al., 2020; Khan, 2020). The glycoxidative stress of AGEs in AD pathogenesis was originally ascribed to its physicochemical properties, including protein–protein crosslinking, but current studies increasingly emphasize the role of AGEs in cellular perturbation and signaling events, especially in oxidative stress response (Ahmad et al., 2013; Akhter et al., 2013). In AD, glyceraldehyde-derived AGEs and receptor for AGE (RAGE) are the main sources of neurotoxicity through cerebral β-amyloidosis (Emanuele et al., 2005; Akhter et al., 2017b; Yan et al., 2018) (Fig. 3.2.24.1). Long-lived proteins, including amyloid-β (Aβ), have been found to be modified by AGE adducts, and a recent study has revealed that the formation of AGE-modified exacerbates the Aβ toxicity of Aβ (Li et al., 2013). Studies revealed that in both the neurofibrillary tangles and senile plaques of patients with AD, AGEs have been accumulating (Singh et al., 2016), and RAGE appears to be tangled in the transport of Aβ through the blood–brain barrier (BBB) (Deane et al., 2009). In addition, diabetes-induced AD pathology seems to have more severe and highly accumulated AGEs in the brain compared with AD alone (Biessels and Despa, 2018).

Maillard reaction produces hydroxyl and superoxide-free radicals, which interact in an acute phase reaction with AGE-modified proteins, inducing oxidative stress (Akhter et al., 2014, 2015, 2016; Shahab et al., 2014; Akhter et al., 2017a; Ahmad et al., 2018; Li et al., 2019). In addition, induced free radicals have been shown to cause oxidative stress of neighboring neurons by triggering peripheral macrophages and chemotaxis of mononuclear phagocytes to induce neuronal inflammation by cytokines (interleukin [IL]-1 and IL-6, and tumor necrosis factor [TNF]-α) (Mosley et al., 2006). Remarkably, the immunoreactivity of IL-6 was earlier confirmed in plaques in AD pathology, and elevated IL-6 concentrations have been measured in brains of AD patients (Wu et al., 2015).

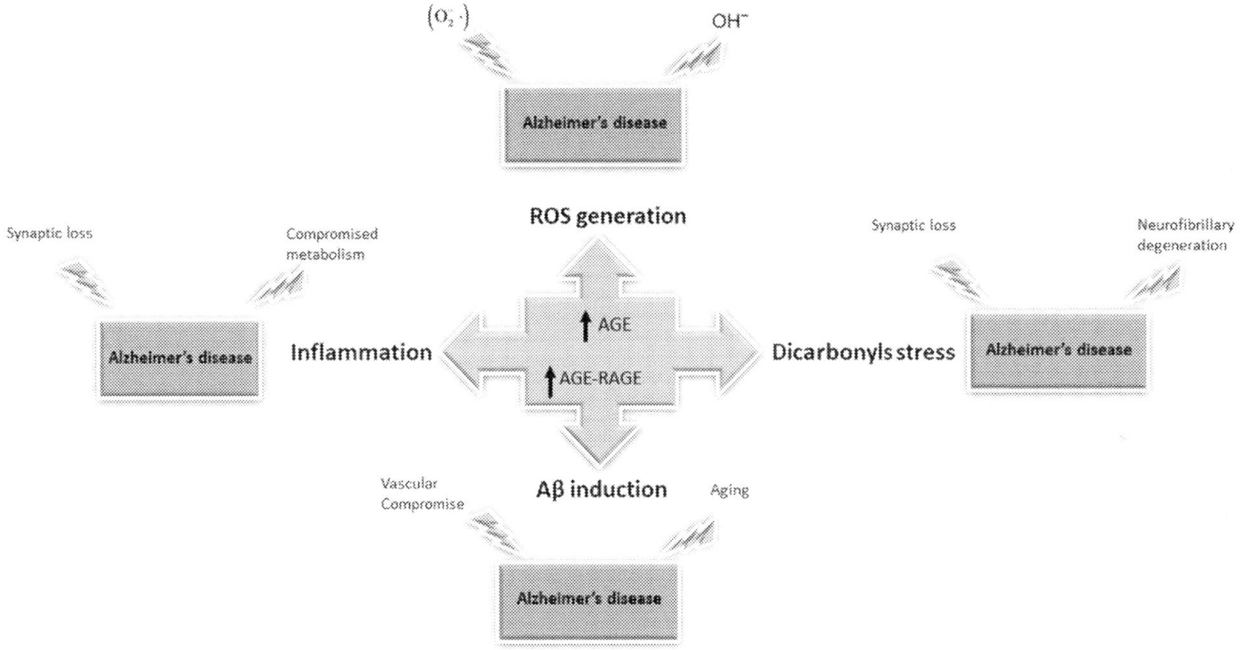

FIGURE 3.2.24.1 A simplified overview of the involvement of advanced glycation end product and advanced glycation end product–receptor for advanced glycation end product interaction in Alzheimer's disease pathology.

Constituents of *Nelumbo nucifera*

N. nucifera has various chemical constituents in different parts of plants (Fig. 3.2.24.2). Here are the major biologically active constituents with respect to plant parts:

Seed: Ascorbic acid, benzoic acid, sodium hydrogen sulfite.

Embryo: Nuciferine, pronuciferine, lotusine, rutin, hyperin, and dimethyl coclaurine.

Stamen: Linalool, luteolin glucoside, dehydroanonaine, anonaine, armepavine, β-sitosterol, kaempferol-3-*O*-β-D glucoronide, asimilobine, dimethyl coclaurine, lirinidine, dehydronuciferine, quercetin, liriodenine, dehydroemetine, isoquercitrin, nornuciferine, *N*-methylasimilobine, *N*-methylcoclaurine, *N*-methylisococlaurine, *N*-norarmepavine, roemerine.

Flower: Quercetin, luteolin, luteolin glucoside, kaempferol, kaempferol-3-*O*-glucoside, isoquercitrin.

Leaf: Roemerine, nuciferine, nornuciferine, armepavine, pronuciferine, *N*-nornuciferine, anonaine, liriodenine, quercetin, tartaric acid, gluconic acid, acetic acid, malic acid, ginnol, nonadecane, succinic acid.

Though these constituents are biologically active, ascorbic acid has been identified as a potentially strong antioxidant in AD pathology. Mukherjee et al. investigated that the rhizome extract of *N. nucifera* inhibited acetylcholinesterase activity and increased acetylcholine concentration in the brain of AD patients and thereby increased cholinergic function ((Mukherjee et al., 2007). It was also reported that, in the dentate gyrus of the hippocampus, the rhizome extract of *N. nucifera* improves memory and learning by enhancing neurogenesis (Yang et al., 2008).

Role of *Nelumbo nucifera* as a potent antiglycation treatment and antioxidant

In the qualitative analysis of *N. nucifera* plants, a total of 30 compounds, including flavonoids, alkaloids, and proanthocyanidins, were identified. Ascorbic acid is one of the main components of *N. nucifera* that exerts a neuroprotective effect in the AD brain (Fig. 3.2.24.3). In the western United States more than 10% of the population commonly consume ascorbic acid as a nutritional supplement after multivitamins (Krone and Ely, 2004). Ascorbic acid is absorbed by the epithelial cells of the intestine, while around 90% is absorbed by the distal part of the small intestine. Blood plasma and organ tissues contain around 60 and 900 μmol/L of ascorbate, respectively, while concentrations of ascorbic acid in the brain are more than 10-fold. Ascorbic acid has been exposed to free radicals and cellular perturbation (Dickinson et al., 2002), has been defined as one of the most powerful reductants and

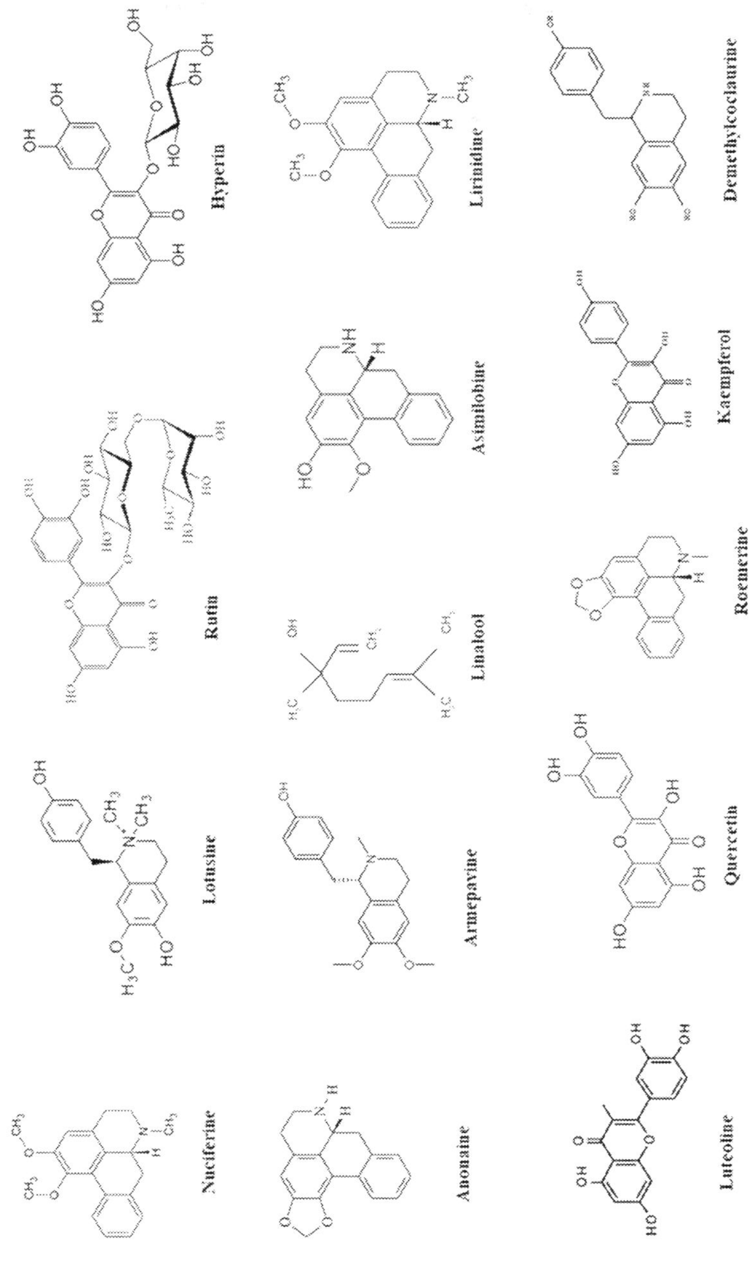

FIGURE 3.2.24.2 Various therapeutic active chemical constituents of *Nelumbo nucifera*.

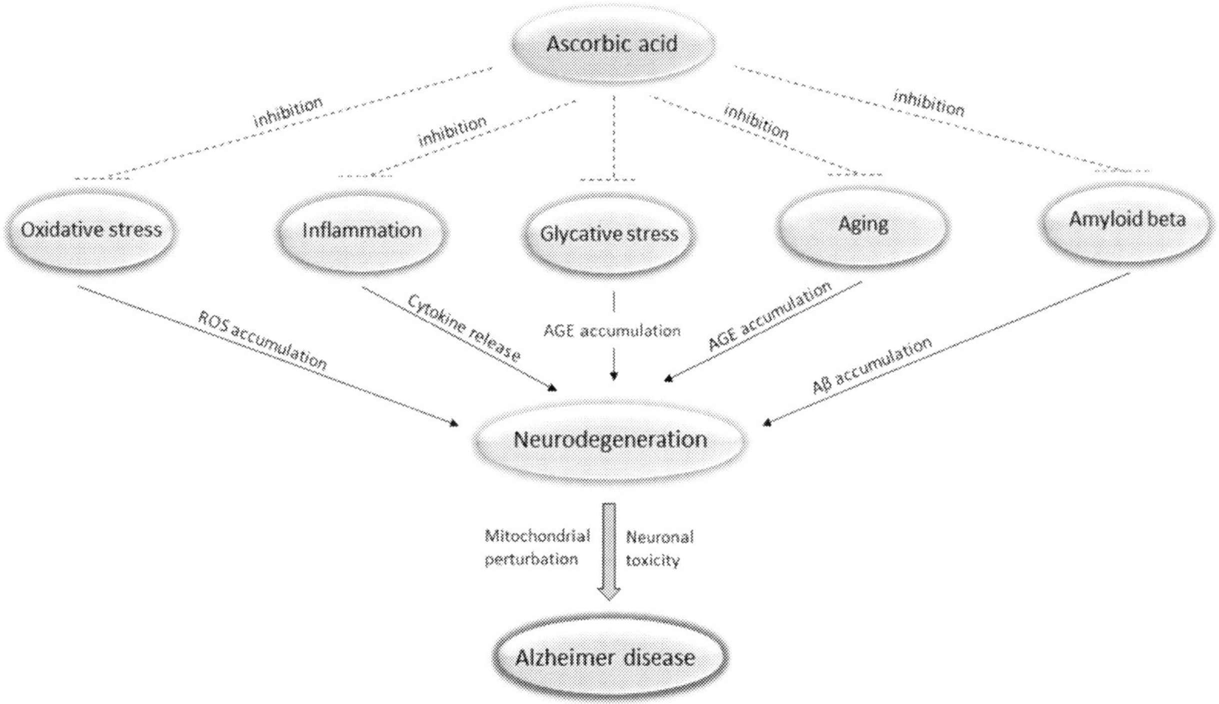

FIGURE 3.2.24.3 Ascorbic acid is at the crossroads of biological aging, amyloid accumulation, glycative stress, inflammation, and oxidative stress, with a potential role in the onset of neurodegeneration or Alzheimer's disease.

radical scavengers, and is the first defense mechanism against radicals in blood circulation (Yimcharoen et al., 2019). In patients with neurodegeneration, it is suggested that ascorbic acid reduces complications mediated through free radical damage from bimolecular glycation and glucose oxidation (Linetsky et al., 2008). Ascorbic acid and tocopherol work synergistically; therefore ascorbic acid has been revealed to be proficient at restoring α-tocopherol from tocopheroxyl free radicals, which are formed upon inhibition of lipid peroxidation by tocopherol (Bursac-Mitrovic et al., 2016).

Preventive effect of *Nelumbo nucifera* in amyloid accumulation and cognitive impairment

Glycoxidative stress is closely associated with abundant age-related diseases, including AD (Inagi, 2014). The most important component of *N. nucifera* is ascorbic acid, which has been found to have great beneficial effects on AD pathology (Akhter et al., 2017c). The mechanisms allied with ascorbic acid neuroprotection comprise scavenging activity against free radical generation, AGE production, metal ion chelation, neuroinflammation, and the conquest of Aβ (Akhter et al., 2019). Aβ contains copper, zinc, and iron due to the presence of metal-binding sites. Induced neurotoxicity is considered to initiate from the generation of Aβ-mediated ROS through trace metals (Monacelli et al., 2017). Aβ is considered to be the primary factor of AD pathogenesis (Sun et al., 2015), which is produced through the cleavage of amyloid precursor protein (APP) by beta-secretase-1 with neurotoxic oligomer accretion. Aβ (1−42) accumulation in the brain results in induced neuronal toxicity to glycoxidative stress (O'Brien and Wong, 2011), impairment in synaptic plasticity, neuroinflammation (Martella et al., 2018), and neuroapoptosis. Induced Aβ interferes with mitochondrial dynamics through hydroxyl radical generation, biomolecular glycoxidation, and lipid peroxidation. AGE−RAGE interaction induces proinflammatory cytokines through the activation of various pathways (Akhter et al., 2017c). Furthermore, lower concentrations of pentosidine were observed in the CSF of Alzheimer's victims, compared to non-Alzheimer's subjects, in support of a role for perturbed AGE metabolism in AD pathology (Monacelli et al., 2014). In addition, it has been reported that increased neuronal death and apoptosis induced by Aβ peptide could be prevented by orally administered vitamin C, which reduces glycoxidative stress and proinflammatory cytokines (Hartel et al., 2004). Additionally, cells pretreated with ascorbic acid also exhibited a reduction in basal rates of endogenous Aβ production (Monacelli et al., 2017). Antioxidant, antiglycation, or metal chelator treatment might be promising to delay AD progression (Gilgun-Sherki et al., 2003).

Nelumbo nucifera and cognitive impairment: in vitro evidence

AD is a major public health problem in the United States and the rest of the world. It has been reported that isoflurane, an anesthetic agent, showed a significant effect on memory impairment and cognitive decline by increasing the levels of ROS, which induce mitochondrial dysfunction and decrease ATP levels that ultimately stimulate caspase-3 activation (Thomsen et al., 2017). Ascorbic acid treatment showed important effects in the reduction of isoflurane-induced caspase-3 activation by decreasing the accumulation of ROS in the H4-APP cell line, attenuating the reduction of ATP levels, and inhibiting the opening of mPTP in B104 cell lines (Cheng et al., 2015). Ascorbic acid concentration is higher in the brain than in all most all other organs. There is new evidence that the concentration of ascorbic acid is very strong in CSF and brain parenchyma (Hashimoto et al., 2017). In CSF or in brain parenchyma, during DNA methylation, 5-hydroxymethylcytosine reduced to 5-methylcytosine, and ascorbic acid plays an important role as a cofactor in this reaction. It is also vital in neuronal repair and new cell generation and it should also be mentioned here that ascorbic acid is involved in the posttranscriptional modification of many genes (Blaschke et al., 2013). In the brain, ascorbic acid is transported into the choroid plexus and in neurons by a sodium-dependent vitamin C transporter. There are two sodium-dependent vitamin C transporters: SVCT-1 and SVCT-2. SVCT-2 is the only transporter that expresses and transports ascorbic acid in the brain (Gess et al., 2010). The expression of these two transporters is regulated by disease condition and age.

Nelumbo nucifera and cognitive impairment: in vivo evidence

The biological effect of *N. nucifera* in health and disease has been actively studied for more than 80 years ago. As an electron donor, it executes many important biological functions (Du et al., 2012). With the loss of two electrons, ascorbic acid reversibly oxidizes to form dehydroascorbic acid. Ascorbate also protects in vivo biomolecules from oxidation because it is a highly effective water-soluble antioxidant (Mescic Macan et al., 2019). Humans cannot synthesize ascorbic acid in vivo, so to carry out the biological function, the human brain relies on a dietary source of ascorbic acid. Various animal studies have shown that vitamin C plays an important role in neurodevelopment, more specifically the general development of neurons and myelin formation by influencing neuronal differentiation (Hansen et al., 2014). It has also been reported that ascorbic acid modulates the cholinergic, catecholinergic, and glutaminergic systems of the brain and affects synaptic neurotransmission by inhibiting the interaction between specific neurotransmitters and their receptors (da Encarnacao et al., 2018). Sinha and coworkers found that in the presence of ascorbic acid and iron, amyloid-β_{1-42} showed antioxidant effects and inhibited the synthesis of H_2O_2 through metal chelation in the mitochondria of rat brain (Sinha et al., 2013). Induction of diabetes in animals by injecting streptozotocin stimulates oxidative stress, which reduces the expression of SVCT-2 in brain parenchyma, which is a phenomenon identical to AD (Weber et al., 1996). These mice have much lower concentrations of ascorbic acid in the brain, which develop rapid seizures and increase mortality.

The synthesis of blood vessel formation in the brain is called angiogenesis, and ascorbic acid is very much essential for the formation of procollagen, which acts as an adhesive that gives support and shape to blood vessels in the brain (Thomsen et al., 2017). Impairment of ascorbic acid function or reduction of vitamin C level in CSF is linked with vascular dysfunction (May, 2016). Studies have revealed that pretreatment with ascorbic acid in a mouse model prior to lipopolysaccharide induction improved lipopolysaccharide-induced cognitive decline or memory deficits by inhibiting the activation of microglia and reducing the production of proinflammatory cytokines TNF-α and IL-1β. It also decreased malondialdehyde and superoxide dismutase activity in the hippocampus region (Zhang et al., 2018). It was also reported that ascorbic acid supplement alleviated or reduced all causes of dementia in (AD) pathogenesis. In a clinical study, it has been reported that the use of vitamin C supplement significantly reduced all causes of dementia, AD, and cognitive impairment (Basambombo et al., 2017).

Nelumbo nucifera, behavioral studies, and neuropathology

AD is a common and important form of neurodegenerative disorder. There are many mechanisms in the genesis of AD but the accumulation of Aβ oligomer in the brain and abnormal phosphorylation of tau are the main concerns (Rajmohan and Reddy, 2017). Studies showed that continuous treatment of ascorbic acid for 6 months in mice regenerates behavioral dysfunction and decreases the synthesis of Aβ oligomers in the brain by inhibiting brain oxidative damage through reduced synthesis of oxidative stress markers and proinflammatory cytokines in the hippocampus (Moretti et al., 2017). Chronic administration of aluminum chloride is one of the pathological reasons for AD. Olajide and coworkers found that in the rat, daily treatment of ascorbic acid (100 mg/kg) significantly lowered behavioral deficits through the reduction of membrane lipid peroxidation, and modulated neuronal bioenergetics (Olajide et al., 2017). It has been reported that, in terms of lifespan, ascorbic acid is essential for the formation and stability of collagen and carnitine, two of the most important ingredients for the enzyme dopamine-β-hydroxylase (Mane and Kamatham, 2018). It is also important in T-cell proliferation through the inhibition of T-cell apoptosis, an important phenomenon for life-span (Saffarpour and Nasirinezhad, 2017). Ascorbic acid could be beneficial for treating depression and behavior. It is important to mention that the antidepressant effect and behavior of ascorbic acid is mediated by the inhibition of *N*-methyl-D-aspartate receptors and modulates neuronal excitability through inhibition of the potassium (K^+) channel (Holtzheimer and Nemeroff, 2008).

Future direction

From the foregoing explanations with both clinical and biological aspects, it can be mentioned that *N. nucifera* and its constituents, mainly ascorbic acid, could be of beneficial effect for the treatment of neuropathological disorders. It has been reported that various cellular and biochemical approaches indicate that ascorbic acid exerts its beneficial effect on the peripheral nervous system via the reduction of oxidative stress and Aβ inhibition, and increases the oxidative stress protective factor (TMX1), which itself works as a therapeutic agent (Phan et al., 2018). It is evident that ascorbic acid has shown defensive and therapeutic efficacy against glycoxidative stress damage. Additionally, the potentiality of ascorbic acid on the reduction of BBB dysfunction has been identified as a new likely remedial mechanism in AD pathology. The clinical impact of ascorbic acid on AD pathology is yet to be decisively recognized. Furthermore, additional inquiries regarding the amount and duration of ascorbic acid ingestion are positively advisable. Conjoining other anti-glycation products, antioxidants, and Aβ inhibitors with ascorbic acid may demonstrate value for the prevention of AD by reducing oxidative and glycoxidative stress.

References

Ahmad, S., et al., 2013. Studies on glycation of human low density lipoprotein: a functional insight into physico-chemical analysis. Int. J. Biol. Macromol. 62, 167–171.

Ahmad, S., et al., 2018. Do all roads lead to the Rome? The glycation perspective! Semin. Canc. Biol. 49, 9–19.

Ahmad, S., et al., 2014. Glycoxidation of biological macromolecules: a critical approach to halt the menace of glycation. Glycobiology 24 (11), 979–990.

Akhter, F., et al., 2017a. Detection of circulating auto-antibodies against ribosylated-LDL in diabetes patients. J. Clin. Lab. Anal. 31 (2).

Akhter, F., et al., 2019. AGE exacerbate amyloid beta (Aβ) induced Alzheimer pathology: a systemic overview. In: Networking of Mutagens in Environmental Toxicology, pp. 159–170.

Akhter, F., et al., 2017b. Mitochondrial perturbation in Alzheimer's disease and diabetes. Prog. Mol. Biol. Transl. Sci. 146, 341–361.

Akhter, F., et al., 2015. Acquired immunogenicity of calf thymus DNA and LDL modified by D-ribose: a comparative study. Int. J. Biol. Macromol. 72, 1222–1227.

Akhter, F., et al., 2016. Antigenic role of the adaptive immune response to D-ribose glycated LDL in diabetes, atherosclerosis and diabetes atherosclerotic patients. Life Sci. 151, 139–146.

Akhter, F., et al., 2014. An immunohistochemical analysis to validate the rationale behind the enhanced immunogenicity of D-ribosylated low density lipo-protein. PLoS One 9 (11), e113144.

Akhter, F., et al., 2017c. Toxicity of protein and DNA-AGEs in neurodegenerative diseases (NDDs) with decisive approaches to stop the deadly consequences. Perspect. Environ. Toxicol. 99–124.

Akhter, F., et al., 2013. Bio-physical characterization of ribose induced glycation: a mechanistic study on DNA perturbations. Int. J. Biol. Macromol. 58, 206–210.

Akhter, F., 2019. Therapeutic efficacy of Boerhaavia diffusa (Linn.) root methanolic extract in attenuating streptozotocin-induced diabetes, diabetes-linked hyperlipidemia and oxidative-stress in rats. Biomed. Res. Ther. 6 (7), 3293–3306.

Akhter, F., et al., 2020. High Dietary Advanced Glycation End Products Impair Mitochondrial and Cognitive. J. Alzheimers Dis. 165, 178.

Basambombo, L.L., et al., 2017. Use of vitamin E and C supplements for the prevention of cognitive decline. Ann. Pharmacother. 51 (2), 118–124.

Biessels, G.J., Despa, F., 2018. Cognitive decline and dementia in diabetes mellitus: mechanisms and clinical implications. Nat. Rev. Endocrinol. 14 (10), 591–604.

Blaschke, K., et al., 2013. Vitamin C induces tet-dependent DNA demethylation and a blastocyst-like state in ES cells. Nature 500 (7461), 222–226.

Bursac-Mitrovic, M., et al., 2016. Effects of L-ascorbic acid and alpha-tocopherol on biochemical parameters of swimming-induced oxidative stress in serum of Guinea pigs. Afr. J. Tradit. Complementary Altern. Med. 13 (4), 29–33.

Casas, R., et al., 2014. The immune protective effect of the Mediterranean diet against chronic low-grade inflammatory diseases. Endocr. Metab. Immune Disord. Drug Targets 14 (4), 245–254.

Cheng, B., et al., 2015. Vitamin C attenuates isoflurane-induced caspase-3 activation and cognitive impairment. Mol. Neurobiol. 52 (3), 1580–1589.

da Encarnacao, T.G., et al., 2018. Dopamine promotes ascorbate release from retinal neurons: role of D1 receptors and the exchange protein directly activated by cAMP type 2 (EPAC2). Mol. Neurobiol. 55 (10), 7858–7871.

Deane, R., et al., 2009. Clearance of amyloid-beta peptide across the blood-brain barrier: implication for therapies in Alzheimer's disease. CNS Neurol. Disord. Drug Targets 8 (1), 16–30.

Dickinson, P.J., et al., 2002. Neurovascular disease, antioxidants and glycation in diabetes. Diabetes Metab. Res. Rev. 18 (4), 260–272.

Dinu, M., et al., 2018. Mediterranean diet and multiple health outcomes: an umbrella review of meta-analyses of observational studies and randomised trials. Eur. J. Clin. Nutr. 72 (1), 30–43.

Du, J., et al., 2012. Ascorbic acid: chemistry, biology and the treatment of cancer. Biochim. Biophys. Acta 1826 (2), 443–457.

Emanuele, E., et al., 2005. Circulating levels of soluble receptor for advanced glycation end products in Alzheimer disease and vascular dementia. Arch. Neurol. 62 (11), 1734–1736.

Fu, Z., et al., 2017. Long-term high-fat diet induces hippocampal microvascular insulin resistance and cognitive dysfunction. Am. J. Physiol. Endocrinol. Metab. 312 (2), E89–e97.

Gess, B., et al., 2010. Sodium-dependent vitamin C transporter 2 (SVCT2) is necessary for the uptake of L-ascorbic acid into Schwann cells. Glia 58 (3), 287–299.

Gilgun-Sherki, Y., et al., 2003. Antioxidant treatment in Alzheimer's disease: current state. J. Mol. Neurosci. 21 (1), 1–11.

Grosso, G., et al., 2014. Mediterranean diet and cardiovascular risk factors: a systematic review. Crit. Rev. Food Sci. Nutr. 54 (5), 593–610.

Guzik, T.J., Touyz, R.M., 2017. Oxidative stress, inflammation, and vascular aging in hypertension. Hypertension 70 (4), 660–667.

Hansen, S.N., et al., 2014. Does vitamin C deficiency affect cognitive development and function? Nutrients 6 (9), 3818–3846.

Hartel, C., et al., 2004. Effects of vitamin C on intracytoplasmic cytokine production in human whole blood monocytes and lymphocytes. Cytokine 27 (4–5), 101–106.

Hashimoto, K., et al., 2017. Increased levels of ascorbic acid in the cerebrospinal fluid of cognitively intact elderly patients with major depression: a preliminary study. Sci. Rep. 7 (1), 3485.

Holtzheimer, P.E., Nemeroff, C.B., 2008. Novel targets for antidepressant therapies. Curr. Psychiatr. Rep. 10 (6), 465–473.

Inagi, R., 2014. Glycative stress and glyoxalase in kidney disease and aging. Biochem. Soc. Trans. 42 (2), 457–460.

Khan, MY., 2020. The neoepitopes on methylglyoxal (MG) glycated LDL create autoimmune response; autoimmunity detection in T2DM patients with varying disease duration. Cell Immunol 351, 104062.

Krone, C.A., Ely, J.T., 2004. Ascorbic acid, glycation, glycohemoglobin and aging. Med. Hypotheses 62 (2), 275–279.

Li, X.H., et al., 2013. Glycation exacerbates the neuronal toxicity of beta-amyloid. Cell Death Dis. 4, e673.

Li, Y., et al., 2019. The non-enzymatic glycation of LDL proteins results in biochemical alterations - a correlation study of Apo B100-AGE with obesity and rheumatoid arthritis. Int. J. Biol. Macromol. 122, 195–200.

Linetsky, M., et al., 2008. Glycation by ascorbic acid oxidation products leads to the aggregation of lens proteins. Biochim. Biophys. Acta 1782 (1), 22–34.

Mane, S.D., Kamatham, A.N., 2018. Ascorbyl stearate and ionizing radiation potentiate apoptosis through intracellular thiols and oxidative stress in murine T lymphoma cells. Chem. Biol. Interact. 281, 37–50.

Martella, G., et al., 2018. Synaptic plasticity changes: hallmark for neurological and psychiatric disorders. Neural Plast. 2018, 9230704.

May, J.M., 2016. Ascorbic acid repletion: a possible therapy for diabetic macular edema? Free Radic. Biol. Med. 94, 47–54.

Mescic Macan, A., et al., 2019. Therapeutic perspective of vitamin C and its derivatives. Antioxidants 8 (8).

Monacelli, F., et al., 2017. Vitamin C, aging and Alzheimer's disease. Nutrients 9 (7).

Monacelli, F., et al., 2014. Pentosidine determination in CSF: a potential biomarker of Alzheimer's disease? Clin. Chem. Lab. Med. 52 (1), 117–120.

Moretti, M., et al., 2017. Preventive and therapeutic potential of ascorbic acid in neurodegenerative diseases. CNS Neurosci. Ther. 23 (12), 921–929.

Mosley, R.L., et al., 2006. Neuroinflammation, Oxidative Stress and the Pathogenesis of Parkinson's Disease. Clin. Neurosci. Res. 6 (5), 261–281.

Mukherjee, P.K., et al., 2007. Screening of Indian medicinal plants for acetylcholinesterase inhibitory activity. Phytother Res. 21 (12), 1142–1145.

O'Brien, R.J., Wong, P.C., 2011. Amyloid precursor protein processing and Alzheimer's disease. Annu. Rev. Neurosci. 34, 185–204.

Olajide, O.J., et al., 2017. Ascorbic acid ameliorates behavioural deficits and neuropathological alterations in rat model of Alzheimer's disease. Environ. Toxicol. Pharmacol. 50, 200–211.
Phan, V., et al., 2018. Characterization of naive and vitamin C-treated mouse schwann cell line MSC80: induction of the antioxidative thioredoxin related transmembrane protein 1. J. Proteome Res. 17 (9), 2925–2936.
Phaniendra, A., et al., 2015. Free radicals: properties, sources, targets, and their implication in various diseases. Indian J. Clin. Biochem. 30 (1), 11–26.
Rajmohan, R., Reddy, P.H., 2017. Amyloid-beta and phosphorylated tau accumulations cause abnormalities at synapses of Alzheimer's disease neurons. J. Alzheimers Dis. 57 (4), 975–999.
Saffarpour, S., Nasirinezhad, F., 2017. Functional interaction between N-methyl-D-aspartate receptor and ascorbic acid during neuropathic pain induced by chronic constriction injury of the sciatic nerve. J. Basic Clin. Physiol. Pharmacol. 28 (6), 601–608.
Shahab, U., et al., 2014. Immunogenicity of DNA-advanced glycation end product fashioned through glyoxal and arginine in the presence of Fe^{3+}: its potential role in prompt recognition of diabetes mellitus auto-antibodies. Chem. Biol. Interact. 219, 229–240.
Singh, S.K., et al., 2016. Overview of Alzheimer's disease and some therapeutic approaches targeting abeta by using several synthetic and herbal compounds. Oxid. Med. Cell Longev. 2016, 7361613.
Sinha, M., et al., 2013. Antioxidant role of amyloid beta protein in cell-free and biological systems: implication for the pathogenesis of Alzheimer disease. Free Radic. Biol. Med. 56, 184–192.
Sun, X., et al., 2015. Beta-amyloid: the key peptide in the pathogenesis of Alzheimer's disease. Front. Pharmacol. 6, 221.
Thomsen, M.S., et al., 2017. The vascular basement membrane in the healthy and pathological brain. J. Cerebr. Blood Flow Metabol. 37 (10), 3300–3317.
Weber, P., et al., 1996. Vitamin C and human health–a review of recent data relevant to human requirements. Int. J. Vitam. Nutr. Res. 66 (1), 19–30.
Wu, Y.Y., et al., 2015. Alterations of the neuroinflammatory markers IL-6 and TRAIL in Alzheimer's disease. Dement Geriatr. Cogn. Dis. Extra 5 (3), 424–434.
Yan, S.F., et al., 2018. Identification and characterization of amyloid-beta accumulation in synaptic mitochondria. Methods Mol. Biol. 1779, 415–433.
Yang, W.M., et al., 2008. Novel effects of *Nelumbo nucifera* rhizome extract on memory and neurogenesis in the dentate gyrus of the rat hippocampus. Neurosci. Lett. 443 (2), 104–107.
Yimcharoen, M., et al., 2019. Effects of ascorbic acid supplementation on oxidative stress markers in healthy women following a single bout of exercise. J. Int. Soc. Sports Nutr. 16 (1), 2.
Zhang, X.Y., et al., 2018. Vitamin C alleviates LPS-induced cognitive impairment in mice by suppressing neuroinflammation and oxidative stress. Int. Immunopharm. 65, 438–447.
Zhao, Z.L., et al., 2016. Antioxidant activities of crude phlorotannins from Sargassum hemiphyllum. J. Huazhong Univ. Sci. Technol. Med. Sci. 36 (3), 449–455.

Chapter 3.2.25

Zingiber officinale

Tanuj Joshi[1], Laxman Singh[2], Arvind Jantwal[1], Sumit Durgapal[1], Jyoti Upadhyay[3], Aadesh Kumar[1], Mahendra Rana[4]

[1]*Department of Pharmaceutical Sciences, Kumaun University, Nainital, Uttarakhand, India;* [2]*G.B. Pant National Institute of Himalayan Environment & Sustainable Development, Kosi-katarmal, Almora, Uttarakhand, India;* [3]*School of Health Sciences, University of Petroleum and Energy Studies, Dehradun, Uttarakhand, India;* [4]*Department of Pharmaceutical Sciences, Center for Excellence in Medicinal Plants and Nanotechnology, Kumaun University, Nainital, Uttarakhand, India*

Introduction

Ginger rhizomes have been used as spice for more than 2000 years (Vasconcelos et al., 2018). The plant consists of dried rhizomes of *Zingiber officinale* Roscoe belonging to the family Zingiberaceae.

Sanskrit: Ausadha, Muhausadha, Nagara, Visva, Adrak
English: Ginger, ginger root.
Hindi: Adrak, Ada, Saunth
Bengali: Ada, Suntha (API Vol. 1, The Wealth of India, 2009)

It is a herbaceous perennial plant about 1 m in height. Rhizomes of the plant are thick-lobed, pale yellowish in color; the rhizomes are laterally compressed, bearing oblique branches and a bud at the apex (Kokate et al., 2007).

Distribution and cultivation

Ginger is said to be a native plant of South-East Asia. It is cultivated in Africa, Australia, Taiwan, and many other countries. More than 35% of the production of ginger is from India. The majority of its cultivation is carried out in tropical and subtropical regions, and requires a warm and humid climate, with well-distributed rainfall (150−300 cm). The plant can grow easily from sea level up to an altitude of 1500 m and requires heavy manuring Fig. 3.2.25.1 shows ginger crop. (The Wealth of India, 2009).

Naturally Occurring Chemicals against Alzheimer's Disease. https://doi.org/10.1016/B978-0-12-819212-2.00041-4

FIGURE 3.2.25.1 Ginger plant in a field. *Picture credit: Laxman Singh.*

Chemical composition

The large number of chemicals present in ginger can easily be associated with its pharmacological effect. Ginger has been characterized as a functional food due to its composition. The drug contains calcium, carbohydrates, carotene, dietary fiber, fat, iron, protein, and vitamin C (Ali et al., 2008; Kumari and Gupta, 2016). Other than vitamin C, the vitamins present are thiamine, niacin, and riboflavin. Starch is the principal carbohydrate present. The characteristic odor of ginger is due to the presence of essential oil (oil of ginger). It is greenish yellow in color, soluble in alcohol, and is aromatic, but the oil lacks the pungent taste of ginger. Constitution of the oil is zingiberol, zingiberene, ar-curcumene, farnesene, β-bisabolene, β-elemene, β-sesquiphellandrene, and γ-selinene (The Wealth of India, 2009).

Role of *Zingiber officinale* in Alzheimer's disease

A characteristic feature of Alzheimer's disease (AD) is the generation of senile plaques. This is due to the amyloid beta (Aβ) protein. Phytochemicals like gingerol are crucial for *Z. officinale*'s effects. In an experiment, [6]-gingerol's actions were evaluated against Aβ (25−35) proteins, which cause SHSY5Y cell death. Reactive nitrogen and oxygen species generated by Aβ (25−35) were suppressed by gingerol. Also it restored the levels of glutathione (an important endogenous antioxidant agent) decreased by Aβ (25−35). Enzymes like γ-glutamylcysteine ligase and heme oxygenase-1 are involved with glutathione and heme synthesis, respectively. Their protein and messenger ribonucleic acid expression levels were upregulated by gingerol. Activation of nuclear factor erythroid 2-related factor 2 by gingerol was responsible for the aforementioned effect (Lee et al., 2011). Thus by enhancing the antioxidant capacity of the body, gingerol leads to therapeutic benefits in AD. Lipid peroxidation induced by Fe^{2+} ions is also inhibited by extract of ginger. Lipid peroxidation-inhibiting ability can be due to the antioxidant activity of ginger extract. Various varieties of ginger provide neuroprotection due to their antioxidant activity (Bellé et al., 2004; Oboh et al., 2012; Mirmosayyeb et al., 2017). The antiinflammatory activity of *Z. officinale* can also be responsible for its actions. The beneficial effect of *Z. officinale* on neurodegenerative disorders was based on the antiinflammatory potential of *Z. officinale*. The rhizomes of *Z. officinale* were extracted with hexane for evaluation of its antiinflammatory potential. In the study, *Z. officinale* inhibited various factors associated with inflammation. Suppression of nuclear factor kappa-light-chain-enhancer of activated B cells (NF-κB) in an experiment on aged rats supported the antiinflammatory actions of *Z. officinale*. Thus besides antioxidant activity, antiinflammatory activity might also be responsible for the efficacy of *Z. officinale* in AD (Jung et al., 2009; Mirmosayyeb et al., 2017). Monocytic THP-1 cells from the human body are inhibited due to the actions of ginger by a variety of proinflammatory stimuli and the decrease in gene expressions, which are related to inflammation in cells that are like microglia (Ghayur et al., 2008). Gingerol studies were also done on another important phytochemical of *Z. officinale* called 6-shogaol. Researchers evaluated the activation of microglia by 6-shogaol. An inhibitory effect on nitric oxide and nitric oxide synthase, which were induced by lipopolysaccharide (LPS), was produced by 6-shogaol. 6-Shogaol was thus beneficial in inhibiting the activation of microglia by LPS (Ha et al., 2012). Phytochemicals like terpenoids, flavanoids, tannins, and alkaloids present in ginger extract are responsible for inhibiting acetylcholinesterase activity in rat's brain. Acetylcholinesterase (AChE)-inhibiting activity is responsible for augmenting cholinergic transmission, which can be beneficial in AD (Oboh et al., 2012; Mathew and Subramanian, 2014; Mirmosayyeb et al., 2017).

Phytochemicals of ginger can target AD by various mechanisms. However, mechanisms of many of these components have not been elaborated to a great extent. Thus a computer-aided study was performed to study the detailed action of these phytoconstituents. The four components of ginger extract were reported to possess inhibitory activities on human acetylcholinesterase (*Hss*AChE). *Hss*AChE activity of the four components of ginger extract was compared with inhibitors (with known *Hss*ACHE activity) and the standard drug donepezil. It was shown by the results of the study that two compounds showed good *Hss*AChE-inhibiting activity and the effects of these compounds can be as promising as donepezil (Cuya et al., 2017, Fig. 3.2.25.2).

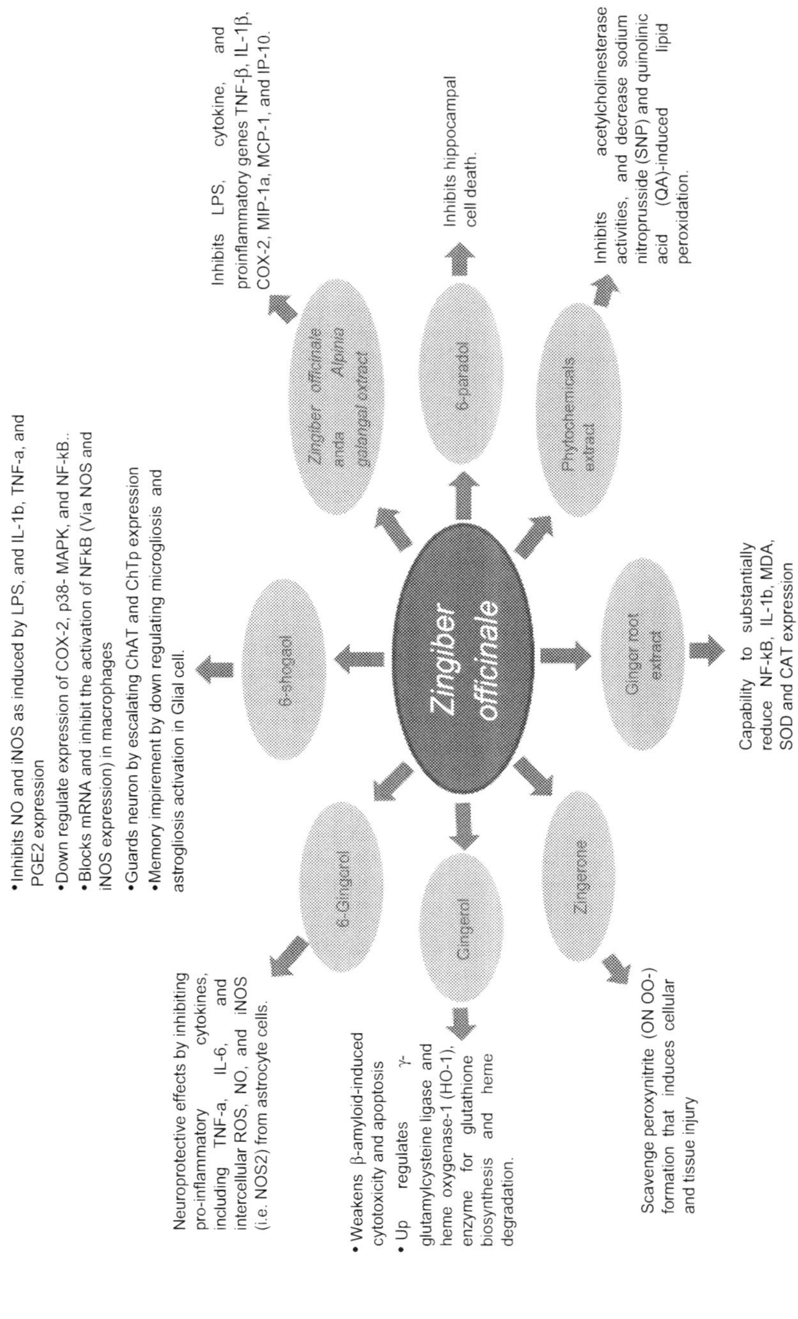

FIGURE 3.2.25.2 Mode of action of *Zingiber officinale*.

In vitro and *in vivo* studies of *Zingiber officinale* with respect to Alzheimer's disease

Various researchers have performed both *in vitro* and *in vivo* studies on *Z. officinale* targeting AD. In Wistar rats the effectiveness of ginger rhizome oil (GO) and ginger rhizome extract (GE) in reducing the effects of AD was determined. GE, GO, and memantine (standard drug) were administered to rats according to their groups, respectively. AD was induced in all groups by aluminum chloride treatment except in the negative control group (normal saline). Positive control received vehicle plus aluminum chloride. A colorimeter was used to determine AChE activity in the brain. The Elisa method was used for evaluation of caspase-3, p53, NF-κB, and brain acetylcholine (ACh). Acetylcholinetransferase (AChT) activity was determined in the brain using immunohistochemical procedures. Also histopathological studies of brain tissue sections were carried out. When positive control was compared with negative control it was found that NF-κB, caspase-3, AChE, and p53 levels were elevated and ACh levels were decreased in rats belonging to positive control. Also negative reaction was associated with AChT activity in the positive control group. Amyloid plaques were observed in the positive control group on histopathological examination. Groups in which AD was induced and given treatment with GE and GO showed significant improvements in the disturbed biochemical parameters associated with AD. The increased parameters were decreased and the decreased parameters were increased by GE and GO treatment. Also treatment with GE and GO showed association of positive reaction with AChT activity. Histopathological examination revealed that there was the disappearance of most amyloid plaques and improvement of brain morphological structure in rats treated with GE and GO. The reason for GE and GO in reducing the damaging effects of AD in the study can be attributed to their anticholinesterase, antiinflammatory, and antiapoptotic potential (Ahmed et al., 2014). Extract of ginger in methanol was evaluated for antiamyloidogenic, cholinesterase inhibition, and antioxidant and neuroprotectant abilities in a study. Quercetin and the content of phenol in extract were also evaluated. Ginger showed high antioxidant activity and butylcholinesterase, acetylcholinesterase inhibitory activity. Also in rat hippocampus cell culture, ginger extract protected the cells against toxicity produced by Aβ. The survival of cells was enhanced by ginger extract dose dependently. Ginger extract also prevented Aβ oligomer formation and led to dissociation of already formed oligomers. Since prevention of oligomer is important in the treatment of AD, ginger extract by its antiamyloidogenic activity can show effectiveness in AD (Mathew and Subramanian, 2014). In a study, both *in vitro* and *in vivo* analyses were carried out on ginger for its benefits in AD. Phytochemical studies were also performed. The *in vitro* efficacy of compounds isolated from ginger extract, ginger oil (GEO), and ginger methanolic extract (GME) against AD were evaluated by determining their AChE inhibitory potential using Ellman's assay. The antioxidant activity of GEO and GME and six isolated compounds was tested by diphenyl picryl hydrazyl radical (DPPH) scavenging assay. Also the action of GEO and GME against AD was evaluated by inducing AD in rats through the administration of aluminum chloride. The compounds isolated from the ginger extract were 8-shogaol, methoxy-[10]-gingerol, methoxy-[6]-gingerol, 6-gingerol, 6-shogaol, and diacetoxy-[6]-gingerdiol. The results indicated that GEO and GME showed moderate inhibitory effect against AChE. The isolated compounds showed *in vitro antioxidant* activity in DPPH radical scavenging assay. GME showed a better effect than GMO in scavenging DPPH radicals. The abilities of learning as well as memory were improved by GME and GMO rats with AD. The morphological structure of the brain tissues also showed improvement and there was disappearance of the majority of amyloid plaques due to treatment with GME and GMO (Fathy et al., 2015). A preparation called WS-5 containing extracts of *Chaenomeles sinensis*, *Curcuma longa*, and *Z. officinale* was evaluated in a study by *in vitro* and *in vivo* methods against AD. In the *in vivo* model, AD was induced by Aβ. WS-5 showed inhibitory actions against AChE. Aβ oligomerization was prevented by inhibition of aggregation of $A\beta_{1-42.}$ Antioxidant activity of WS-5 was found to be high as evaluated by DPPH radical scavenging assay and confirmed by total antioxidant assay. WS-5 also showed antiinflammatory activity. Results showed that beneficial effects on learning and memory were produced by WS-5. The accumulation of Aβ plaques in mice was also decreased. Also active constituents of the extract responsible for AChE activity were identified using high-performance liquid chromatography (Kim et al., 2019). A different study showed that dementia produced in mice with Aβ (1−42) and scopolamine administration was treated by 6-shogaol (a phytoconstituent of *Z. officinale*) by improving cognition. This action of 6-shogaol was attributed to the elevation in hippocampal post-synaptic proteins, nerve growth factor, and inhibition of inflammatory mediators (Moon et al., 2014; Shal et al., 2018). Expression of brain-derived neurotrophic factor choline acetyltransferase was increased and reactive oxygen species (ROS) release was decreased through signaling mediated by tropomyosin receptor kinase B in HT22 hippocampal neuronal cells treated with H_2O_2. This study demonstrates the neuroprotective effect of 6-shogaol (Shim and Kwon, 2012; Shal et al., 2018). Also 6-shogaol showed beneficial neuroprotective effect in a model of LPS-induced inflammation. NF-κB, inducible nitric oxide synthase, and cyclooxygenase-2 were reduced and the release of proinflammatory cytokines was suppressed in astrocytes given LPS treatment (Shim et al., 2011; Shal et al., 2018). Thus these studies on shogaol highlight the important role that this phenolic phytochemical of *Z. officinale* can play in AD. In a study, spatial recognizing behavior and acquisition recalling were evaluated in rats by using the Y-maze test and shuttle box test. Improvements in acquisition, recall, and retention were observed in rats after administration of *Z. officinale* (Gharibi et al., 2013). Researchers have found that *Z. officinale* has a neuroprotective effect against monosodium glutamate toxicity (Singh et al., 2014). An experiment evaluated 6-shogaol and 6-gingerol, which are the main constituents of *Z. officinale*, on glycogen synthase kinase-3 beta (GSK3β). GSK3β helps in the cleavage of amyloid precursor protein to produce increased amounts of Aβ. Thus inhibition of GSK3β can lower the levels of Aβ and hence protect against neurotoxicity. The study showed that GSK3β enzyme was inhibited by 6-shogaol and 6-gingerol. SHSY-5Y cells were also protected against toxicity by 6-gingerol and 6-shogaol (Yerer et al., 2017). In an experiment, the beneficial effects of *Z. officinale* in rats treated with aluminum chloride for induction of AD were studied. The animals were divided into five groups. Ginger extract was used in doses of 108 and 216 mg/day. Rivastigmine was used as the standard drug. The results indicated that rivastigmine and ginger ameliorated symptoms of neurodegeneration in rats with AD (Mahdy et al., 2014). A study was also conducted in which 27 herbs were selected and their extraction was performed with the help of chloroform and aqueous methanol (90%). Results showed that *Z. officinale* was beneficial in providing protection to cells against damage produced by β-amyloid protein. Similar results were shown by extracts of other plants as well. In comparison with other extracts, ginger extract and *Curcuma aromatica* extract were found to be very effective (Kim et al., 2007).

Clinical studies on *Zingiber officinale* with respect to Alzheimer's disease

Clinical studies on *Z. officinale* have been performed by researchers for the evaluation of its effect in AD. *Z. officinale* was evaluated for its effect with respect to cognitive functions in middle-aged healthy women. The study was randomized and the total number of women used in the study was 60. After grouping of the participants, *Z. officinale* was given to two groups and placebo was given to one group. The treatment was given to the participants once a day for a total duration of 2 months. Women used in the study were evaluated for cognitive function and working memory at the start of the study before treatments were started, 1 month later, and 2 months later. Groups that were treated with ginger extract showed an increase in cognition and working memory (Saenghong et al., 2012). Davaie Loban (a polyherbal preparation) containing *Z. officinale* was evaluated for its beneficial effects in AD in a study. Davaie Loban, which has been mentioned in traditional texts as an herbal medicine for the treatment of forgetfulness, was evaluated in clinical trials for its effectiveness in mild to moderate AD. Davaie Loban capsules were used in the study. The patients used in the study were greater than 50 years of age and suffered from mild to moderate AD. Twenty patients belonging to the placebo group and 24 patients belonging to the Davaie Loban-treated group completed the study. The results of the study showed that Davaie Loban may be an effective medicine in patients suffering from mild to

TABLE 3.2.25.1 Toxicological studies of *Zingiber officinale* and understanding the basic mechanism against Alzheimer disease.

S. No.	Plant extract/compounds	Experimental model animal	Mechanism of drug effect	References
1.	Kami-untan-to (KUT), Kampo medicine (13 herbs)	Male ddY mice	KUT extract considerably reduces the mortality rate and impairs memory-allied behavior in thiamine-deficient mice, possibly via triggering of cholinergic neurons in the hippocampus.	Nakagawasai et al. (2004)
2.	Zhokumei-to, a Kampo formula (9 herbs)	Male ddY mice	Impairs amyloid β (25–35)-induced memory loss and aids in synaptic reconstruction by increasing the synaptophysin level in the hippocampus and cerebral cortex.	Tohda et al. (2004)
3.	6-shogaol	RAW 264.7 cells	Blocks mRNA and protein expression of inducible NOS (iNOS) and cyclooxygenase-2 (COX-2) in lipopolysaccharide (LPS)-induced macrophages via triggering of PI3K/Aκt/IκB kinases IKK and mitogen-activated protein kinase (MAPK), which in turn inhibit the activation of nuclear factor-κB (NF-κB).	Pan et al. (2008)
4.	6-shogaol	HT22 cells	Guards neurons by escalating the expression of choline acetyltransferase and choline transporter via increase in brain-derived neurotrophic factor.	Shim and Kwon (2012)

5.	6-gingerol	C57Bl and Balb/C mice	Inhibits the production of proinflammatory cytokines (interleukin [IL]-1β, IL-12, tumor necrosis factor [TNF]-α) via LPS-induced macrophage stimulation.	Tripathi et al. (2007)
6.	Gingerol	SH-SY5Y and HT22 cells	Attenuates β-amyloid-induced (Aβ (25−35)) cytotoxicity and apoptosis. Besides this, upregulates the antioxidant expressions of γ-glutamylcysteine ligase and heme oxygenase-1, prominent enzyme systems for glutathione biosynthesis and heme degradation.	Lee et al. (2011)
7.	Zingerone	YPEN-1 cells	Scavenges peroxynitrite (ONOO−) formation that induces cellular and tissue injury.	Shin et al. (2005)
8.	Extract of *Z. officinale*	BV-2 cells	Inhibits transcription of inflammatory mediator genes through the MAPK and NF-κB signaling pathways and exhibits antiinflammatory activity.	Jung et al. (2009)
9.	Extract of *Z. officinale*	Male Wistar rats	Impairs focal cerebral memory and diminishes brain damage.	Wattanathorn et al. (2011)
10.	6-gingerol	C6 rat astroglioma cells and Sprague−Dawley rats	Exhibits neuroprotective effects in LPS-induced disorder and suppresses the release of proinflammatory cytokines, including TNF-α, IL-6, and increases intercellular reactive oxygen species, nitric oxide (NO), and iNOS (i.e., NOS_2) from astrocyte overactivation and improves cognitive ability.	Zhang et al. (2018)

Continued

TABLE 3.2.25.1 Toxicological studies of *Zingiber officinale* and understanding the basic mechanism against Alzheimer disease.—cont'd

S. No.	Plant extract/compounds	Experimental model animal	Mechanism of drug effect	References
11.	Ginger root extract	Female Sprague–Dawley rats	High dose of root extract shows lower level of NF-κB, IL-1β, malondialdehyde, superoxide dismutase, and catalase expression than the rats that received moderate- or low-dose concentration.	Zeng et al. (2013)
12.	6-gingerol and 6-shogaol	SHSY-5Y (ATCC; CRL-2266) human neuroblastoma cells	Inhibits the activity of glycogen synthase kinase-3 beta resulting in reduction of Aβ-induced neurotoxicity.	Yerer et al., 2017
13.	Water-extractable phytochemicals of red ginger (*Z. officinale* var. Rubra) and white ginger (*Z. officinale* var. Roscoe)	Adult male Wistar strain albino rats	Inhibits acetylcholinesterase activities, and considerably decreases sodium nitroprusside and quinolinic acid-induced lipid peroxidation.	Oboh et al. (2012)
14.	6-shogaol	Mice	Considerably trims down glial cell (microgliosis and astrogliosis) activation in intrahippocampal and scopolamine-induced memory impairment.	Moon et al. (2014)

15.	6-shogaol	Sprague–Dawley rats	Inhibits the release of NO• and iNOS as induced by LPS, and exerts its antiinflammatory activity by inhibiting the production of IL-1β, TNF-α, and prostaglandin E2. Concurrently downregulates the expression of COX-2, p38, MAPK, and NF-κB.	Ha et al. (2012)
16.	Ginger extract (*Z. officinale* and *Alpinia galangal*)	Human monocytic THP-1 cells	Inhibits LPS, cytokine, and Aβ peptide- induced expression of the proinflammatory genes TNF-β, IL-1β, COX-2, macrophage inflammatory protein 1α, monocyte chemoattractant protein-1, and interferon-gamma inducible protein 10.	Grzanna et al. (2004)
17.	Ginger extract	Adult rat hippocampal cells	Improves cell survival against Aβ-induced toxicity and increases high antioxidant activity.	Mathew and Subramanian (2014)
18.	6-paradol	Primary hippocampal cells	Inhibits hippocampal cell death.	Choi et al. (2017)
19.	Ginger extract	Adult male Sprague–Dawley rats	Taints 3,4-methylenedioxymethamphetamine-induced neurotoxicity, and downregulates Bcl-2 with upregulation of Bax gene expression.	Mehdizadeh et al. (2012)

moderate AD (Tajadinia et al., 2015). *Z. officinale* has been evaluated with respect to the treatment of AD in very few human studies, so there is tremendous opportunity for researchers to investigate the effectiveness of *Z. officinale* in AD.

Toxicological studies

Toxicological studies of *Zingiber officinale* are mentioned in Table 3.2.25.1.

Conclusion

Z. officinale is a plant that carries huge potential to cure a number of physiological conditions. It is a drug that has been used in traditional and folk medicine for over two millennia and is mentioned in the Ayurveda. Despite all these factors, ginger basically finds its use as a spice and hence its potential has yet to be realized. *Z. officinale* is a great functional food with low toxicity. It is a good source for various phytochemicals, which offer their antioxidant and antiinflammatory properties. Because of these antioxidant and antiinflammatory potentials, it is used in the effective management and treatment of AD. Oxidative stress along with the inflammatory cascades are the main underlying causes for the pathogenesis of AD. Food supplementation with ginger may be an approach to reduce AD.

References

Ahmed, H.H., Zaazaa, A.M., Bosy, A., Abd El-Motelp, July—August 2014. *Zingiber officinale* and Alzheimer's disease: evidences and mechanisms. Int. J. Pharmaceut. Sci. Rev. Res. 27 (2), 142—152. Article No. 21.

Ali, H., Blunden, G., Tanira, M.O., Nemmar, A., 2008. Some phytochemical, pharmacological and toxicological properties of ginger (*Zingiber officinale* Roscoe): a review of recent research. Food Chem. Toxicol. 46, 409—420.

Bellé, N.A., Dalmolin, G.D., Fonini, G., Rubin, M.A., Rocha, J.B., 2004. Polyamines reduces lipid peroxidation induced by different pro-oxidant agents. Brain Res. 1008, 245—251.

Choi, J.G., Kim, Y.S., Jeong, M., Oh, M.S., 2017. Pharmacotherapeutic potential of ginger and its compounds in age-related neurological disorders. Pharmacol. Therapeut. 182, 56—59.

Cuya, T., Baptista, L., Costa França, T.C., 2017. A molecular dynamics study of components of the ginger (*Zingiber officinale*) extract inside human acetylcholinesterase: implications for Alzheimer disease. J. Biomol. Struct. Dyn. 3843—3855. https://doi.org/10.1080/07391102.2017.1401004.

Fathy, M.M., Eid, H.H., Hussein, M.A., Ahmed, H.H., Hussein, A.A., 2015. The role of *Zingiber officinale* in the treatment of Alzheimer's disease: in-vitro and in-vivo evidences. Res. J. Pharmaceut. Biol. Chem. Sci. 6, 735—749.

Gharibi, A., Khalili, M., Kiasalari, Z., Hoseinirad, M., 2013. The effect of *Zingiber officinalis* L. on learning and memory in rats. J. Bas. Clin. Pathophysiol. 2, 2013—2014.

Ghayur, M.N., Gilani, A.H., Ahmed, T., Khalid, A., Nawaz, S.A., Agbedahunsi, J.M., Choudhary, M.I., Houghton, P.J., 2008. Muscarinic, Ca^{++} antagonist and specific butyrylcholinesterase inhibitory activity of dried ginger extract might explain its use in dementia. J. Pharm. Pharmacol. 60, 1375–1383.

Grzanna, R., Phan, P., Polotsky, A., Lindmark, L., Frondoza, C.G., 2004. Ginger extract inhibits β-amyloid peptide–induced cytokine and chemokine expression in cultured THP-1 monocytes. J. Alternative Compl. Med. 10 (6), 1009–1013.

Ha, S.K., Eunjung Moon, E., Ju, M.S., Kim, D.H., Ryu, J.H., Oh, M.S., Kim, S.Y., 2012. 6-shogaol, a ginger product, modulates neuroinflammation: a new approach to neuroprotection. Neuropharmacology 63, 211–223.

Jung, H.W., Yoon, C.H., Park, K.M., Han, H.S., Park, Y.K., 2009. Hexane fraction of *Zingiberis* Rhizoma Crudus extract inhibits the production of nitric oxide proinflammatory cytokines in LPS-stimulated BV2 microglial cells via the NF kappaB pathway. Food Chem. Toxicol. 47 (6), 1190–1197.

Kim, D.S.H.L., Kim, J.-Y., Han, Y.S., 2007. Alzheimer's disease drug discovery from herbs: neuroprotectivity from β-amyloid (1–42) insult. J. Alternative Compl. Med. 13, 333–340.

Kim, J.E., Shrestha, A.C., Kim, H.S., Ham, H.N., Hyeong Kim, J., Jee Kim, Y., Jeong Noh, Y., Kim, S.J., Keun Kim, D., Kwon Jo, H., Sung Kim, D., Hyun Moon, K., Ho Lee, J., Ok Jeong, K., Yoon Leem, J., 2019. WS-5 extract of *Curcuma longa, Chaenomeles sinensis,* and *Zingiber officinale* contains anti-AChE compounds and improves β-amyloid-induced memory impairment in mice. Evid. Base Compl. Alternative Med. 2019, 1–16.

Kokate, C.K., purohit, A.P., Gokhale, S.B., 2007. In: Prakashan, N. (Ed.), Pharmacognosy, pp. 408–410.

Kumari, S., Gupta, A., 2016. Nutritional composition of dehydrated Ashwagandha, Shatavari, and Ginger Root powder. Int. J. Home Sci. 2 (3), 68–70.

Lee, C., Park, G.H., Kim, C.Y., Jang, J.H., 2011. [6]-gingerol attenuates β-amyloid-induced oxidative cell death via fortifying cellular antioxidant defense system. Food Chem. Toxicol. 49 (6), 1261–1269.

Mahdy, K.A., Nadia, A.M.G., Marrie, A.E.-F.H., Yassin, N.A.Z., El-Shenawy, S.M.A., Farrag, A.R.H., Bassant, M.M.I., 2014. Protective effect of ginger (*Zingiber officinale*) on Alzheimer's disease induced in rats. J. Neuroinf. Dis. 5, 1–10.

Mathew, M., Subramanian, S., 2014. In vitro evaluation of anti-Alzheimer effects of drying ginger (*Zingiber officinale* Roscoe) extract. Indian J. Exp. Biol. 52, 606–612.

Mehdizadeh, M., Dabaghian, F., Nejhadi, A., Fallah-huseini, H., Choopani, S., Shekarriz, N., Molavi, N., Basirat, A., Kazorgah, F.M., Samzadeh-Kermani, A., Asl, S.S., 2012. *Zingiber officinale* alters 3,4-methylenedioxymethamphetamine-induced neurotoxicity in rat brain. Cell J. (Yakhteh) 14 (No 3), 177–185. Autumn.

Mirmosayyeb, O., Tanhaei, A., Sohrabi, H.R., Martins, R.N., Tanhaei, M., Amin Najafi, M., Ali, S., Meamar, R., 2017. Possible role of common spices as a preventive and therapeutic agent for Alzheimer's disease. Int. J. Prev. Med. 8, 5.

Moon, M., Kim, H.G., Choi, J.G., Oh, H., Lee, P.K., Ha, S.K., et al., 2014. 6-shogaol, an active constituent of ginger, attenuates neuroinflammation and cognitive deficits in animal models of dementia. Biochem. Biophys. Res. Commun. 449, 8–13. https://doi.org/10.1016/j.bbrc.2014.04.121.

Nakagawasai, O., Yamadera, F., Iwasaki, K., Arai, H., Taniguchi, R., Tan-No, K., Sasaki, H., Tadano, T., 2004. Effect of kami-untan-to on the impairment of learning and memory induced by thiamine-deficient feeding in mice. Neuroscience 125, 233–241.

Oboh, G., Ademiluyi, A.O., Akinyemi, A.J., 2012. Inhibition of acetylcholinesterase activities and some pro-oxidant induced lipid peroxidation in rat brain by two varieties of ginger (*Zingiber officinale*). Exp. Toxicol. Pathol. 64, 315–319.

Pan, M.H., Hsieh, M.C., Hsu, P.C., et al., 2008. 6-shogaol suppressed lipopolysaccharide-induced up-expression of iNOS and COX-2 in murine macrophages. Mol. Nutr. Food Res. 52 (12), 1467–1477.

Saenghong, N., Wattanathorn, J., Muchimapura, S., Tongun, T., Piyavhatkul, N., Banchonglikitkul, C., Kajsongkram, T., 2012. *Zingiber officinale* improves cognitive function of the middle-aged healthy women. Evid. base Compl. Alternative Med. 1–9.

Shal, B., Ding, W., Ali, H., Kim, Y.S., Khan, S., 2018. Anti-neuroinflammatory potential of natural products in attenuation of Alzheimer's disease. Front. Pharmacol. 9, 1–17.

Shim, S., Kwon, J., 2012. Effects of [6]-shogaol on cholinergic signaling in HT22 cells following neuronal damage induced by hydrogen peroxide. Food Chem. Toxicol. 50, 1454–1459. https://doi.org/10.1016/j.fct.2012.02.014.

Shim, S., Kim, S., Choi, D.S., Kwon, Y.B., Kwon, J., 2011. Anti-inflammatory effects of [6]-shogaol: potential roles of HDAC inhibition and HSP70 induction. Food Chem. Toxicol. 49, 2734–2740. https://doi.org/10.1016/j.fct.2011.08.012.

Shin, S.G., Kim, J.Y., Chung, H.Y., Jeong, J.C., 2005. Zingerone as an antioxidant against peroxynitrite. J. Agric. Food Chem. 53 (19), 7617–7622.

Singh, S.K., Patel, J.R., Bachle, D., 2014. A review on *Zingiber officinale*: a natural gift. Int. J. Pharm. Biol. Sci. 5, 508–525.

Tajadinia, H., Saifadinic, R., Choopania, R., Mitra, M., Kamalinejadd, M., Haghdoost, A.A., 2015. Herbal medicine Davaie Loban in mild to moderate Alzheimer's disease: a 12-week randomized double-blind placebo-controlled clinical trial. Compl. Ther. Med. 23, 767–772.

The Wealth of India, 2009. A Dictionary of Indian Raw Materials, vol. III. C.S.I.R., New Delhi, India, pp. 89–102.

Tohda, C., Matsumoto, N., Zou, K., Meselhy, M.R., Komatsu, K., 2004. Abeta(25–35)-induced memory impairment, axonal atrophy, and synaptic loss are ameliorated by M1, a metabolite of protopanaxadiol- type saponins. Neuropsychopharmacology 29, 860–868.

Tripathi, S., Maier, K.G., Bruch, D., Kittur, D.S., 2007. Effect of 6-gingerol on proinflammatory cytokine production and costimulatory molecule expression in murine peritoneal macrophages. J. Surg. Res. 138 (2), 209–213.

Vasconcelos, M.S., Mota, E.F., Rochette, N.F.G., et al., 2018. Ginger (*Zingiber officinale* Roscoe). In: "Nabavi, S.M., Silva, A.S. (Eds.), Nonvitamin and Nonmineral Nutritional Supplements. Academic Press, pp. 235–239.

Wattanathorn, J., Jittiwat, J., Tongun, T., Muchimapura, S., Ingkaninan, K., 2011. *Zingiber officinale* mitigates brain damage and improves memory impairment in focal cerebral ischemic rat. Evid. Based Complement. Alternat. Med. 2011, 429–505.

Yerer, M.B., Tiryaki, M.K., Demirpolat, E., 2017. GSK-3beta inhibitory effects of 6-gingerol and 6-shogaol help to the recovery of SHSY-5Y cells after amyloid beta1–42 oligomer or aggregate toxicity. J. Cell. Biotechnol. 2, 145–157.

Zeng, G.F., Zhang, Z.Y., Lu, L., Xiao, D.Q., Zong, S.H., He, J.M., 2013. Protective effects of ginger root extract on Alzheimer disease-induced behavioral dysfunction in rats. Rejuvenation Res. 16 (2), 124–133.

Zhang, F., Zhang, J.G., Yang, W., Xu, P., Xiao, Y.L., Zhang, H.T., 2018. 6-Gingerol attenuates LPS-induced neuroinflammation and cognitive impairment partially via suppressing astrocyte overactivation. Biomed. Pharmacother. 107, 1523–1529.

Index

Note: 'Page numbers followed by "*f*" indicate figures and "*t*" indicate tables.'

A

AA. *See* Arachidonic acid (AA)
Acetoxyvalerenic acid (AVA), 117–118. *See also* Valerenic acid (VA)
  anesthetic action of, 120
  brain-derived neurotrophic factor (BDNF), 119
  chemical structure of, 118, 118f
  clinical studies, 122, 123t
  dementia, 120
  occurrence of, 118
  sedative and anxiolytic potential, 121
Acetylcholine (ACh), 16–17, 94, 373
Acetylcholinesterase (AChE)
  Alpinia officinarum, 457
  ashwagandha, 402
  Bacopa monnieri, 247
  Celastrus paniculatus, 425
  cognition, 94
  huperzine A, 127–128
  hyperactivation of, 109
  hypericin (HP), 159–160
  inhibition of, 94
  Lepidium meyenii, 331–332
  Magnolia officinalis (MO), 365
  Melissa officinalis (MO), 231–233
  nobiletin, 189
  Panax ginseng (PG), 219–220
  protopine, 170–171
Acetylcholinesterase inhibitors (AChEIs), 2, 94–95, 95f, 228–229
Acorus calamus L.
  antirheumatic and pain-relieving activity, 339
  cholinergic system, 343–344
  clinical role of, 345
  clinical trials, 344–345
  cognitive impairment, 343–344
  cytotypes, 338–339
  parts of, 337, 338f
  pharmacology of, 342–343
  phytochemical investigation of, 339–341, 340t–342t
  safety issues, 344–345
Adenosine monophosphate (AMP), 171
ADHD. *See* Attention deficit hyperactivity disorder (ADHD)
Advanced gl ycation end products (AGEs), 468, 469f
Adverse drug reactions, 41, 88–89
Alpinia galanga, 454–455
Alpinia officinarum, 453–455
  application of, 455–457
  clinical studies, 458
  mechanism of, 457, 458f
  safety issues, 458
Alpinia purpurata, 455
Alpinia rafflesiana, 454
Alpinia zerumbet, 455
Aluminum chloride, 375–376
Alzheimer's Disease International (ADI), 409–410
Amaryllidaceae, 83
Aminophylline, 273
Amyloid-β (Aβ) protein, 94, 293–294, 453–454
  Collinsonia canadensis L., 375
  Convolvulus prostratus, 412
  Ginkgo biloba, 207–208
  Melissa officinalis (MO), 233–234
  nobiletin, 190–191
  Zingiber officinale, 482–483
Amyloidogenesis, 207–208
Amyloid precursor protein (APP), 176, 318, 359, 453–454
  curcumin, 51–52, 54–56
  metabolism of, 11
  Nelumbo nucifera, 473
  omega 3 polyunsaturated fatty acids (PUFAs), 71
  rivastigmine, 94
Analgesics, 273

Anti-Alzheimer's disease (AD) activity, 170–171, 431
Convolvulus prostratus
acetylcholinesterase (AChE), 412
amyloid β (Aβ) inhibition, 412
cognitive function, 414–416
lipoxygenase (LOX), 412
neuroprotection, 412–414
Antianxiety drugs, 274
Antiapoptotic effect, 205–207
Antibiotics, 274
Anticoagulants, 274
Antidiabetic agents, 274
Antihypertensive drugs, 274
Antiinflammatory drugs, 53, 274
Antineoplastic agents, 274–275
Antiobesity agents, 275
Antiplatelet drugs, 274
Antispasmodic agents, 275
Anxiety, 286, 417–418
Apigenin, 295–296, 295f
Apolipoprotein E (ApoE), 110–111
Apoptosis, 205–206, 304, 465
APP. *See* Amyloid precursor protein (APP)
Arachidonic acid (AA), 69
Ascorbic acid, 470–473, 472f
Ashwagandha, 11–12, 20. *See also Withania somnifera*
aqueous extract of, 402
cognitive and psychomotor performance, 402
roots of, 401
water extract of, 403
Asiaticoside, 261–262
Attention deficit hyperactivity disorder (ADHD), 159
Axona, 141
Ayurveda, 11, 16–17, 245, 411. *See also specific plants*

B

Bach, 337
Bacopa monnieri, 20–21, 167–168, 243–244
adverse effects of, 252
chemical constituents, 245, 246t
clinical studies, 250–252
mechanism of, 246–247
taxonomical classification of, 244–246, 245f
in vitro methods, 247–250
in vivo methods, 247–250
Balm mint, 17–18
BDNF. *See* Brain-derived neurotrophic factor (BDNF)
Berberine (BBR), 147, 150f
pharmacological properties
antiamyloidogenic activity, 150
antiapoptotic activity, 149
anticholinesterase activity, 149–150
antiinflammatory activity, 149
antioxidant activity, 148–149
chemical structure, 148, 148f
side effects of, 151
toxicological profile of, 151
Bertholletia excelsa, 379–380
adverse effect, 384
anticarcinogenic properties, 380
antioxidant properties, 380
chemical constituents, 380, 381f
clinical studies, 382–383
distribution of, 380
mode of action, 380–382, 381f
Biomarkers, 375–376, 416–417
stress, 382
resveratrol, 40
Tinospora cordifolia, 355
Blood-brain barrier (BBB), 41
Collinsonia canadensis L., 375
Nelumbo nucifera, 468
nobiletin, 187–188
quercetin (Que), 111
rivastigmine, 101
Brahmi. *See Bacopa monnieri*
Brain-derived neurotrophic factor (BDNF), 119, 294–296
Brazil nuts, 379. *See also Bertholletia excelsa*
Butyryl cholinesterase (BChE), 95
Alpinia officinarum, 457
Celastrus paniculatus, 425
Panax ginseng (PG), 219–220

C

CA. *See Centella asiatica* (CA)
Calamus oil, 338–339
Capric triglyceride
chemistry of, 139–140, 140f
ketone bodies (KBs), 145
medical food, 141
natural sources of, 139
properties of, 140
synthesis of, 139–140
treatment, 143–145
Caprylic acid, 139
antioxidants, 140–141
chemistry of, 139–140, 140f
dietary supplement, 140
ketogenic diet, 140

property of, 140
stability and resistance, 140
β-Carotene, 373
Carvacrol, 375–376
Catalase, 204–205
Catharanthus roseus, 200–201
Cat's claw, 437–438
Celastrus paniculatus, 425–426, 426f, 431
botanical description, 426
chemical constituents, 428–429, 428t–429t
dosage and administration, 432, 432t
medicinal parts, 427
pharmacological activities, 430–431
physical properties, 427–428
plant taxonomy, 426–427
safety issue, 432
therapeutic effects, 429–430, 430t
Centella asiatica (CA), 18–19, 167–168, 257–258, 258f, 261t, 355
chemical constituents, 259, 259f
clinical studies, 263–264
distribution of, 259
pharmacological activity, 260, 260t
in vitro studies, 261–262
in vivo studies, 261–262
Central nervous system (CNS), 188, 231
Convolvulus prostratus, 420–421
alcohol addiction, 419–420
anxiety, 417–418
depression, 416–417
epilepsy (seizure disorder), 418
Huntington's disease, 418
stimulation, 418–419
stress, 417
rosemary, 276
Uncaria rhynchophylla (UR), 444
anticonvulsion, 444–445
anti-Parkinson's disease (PD), 445
Valeriana officinalis, 286
Cerebrospinal fluid (CSF), 98, 474
Chamomile, 294
Choline acetyltransferase (ChAT), 366
Cholinergic dysfunction, 318
Cholinergic neurons, 53
Cholinesterase inhibitors (ChEIs), 87–88, 167, 178, 201–202
Chotosan, 190–191
Chronic fatigue syndrome (CFS), 159
Citrus aurantium, 187
Citrus reticulata, 187
Citrus unshiu, 187
Collinsonia canadensis L., 374–376
common names, 373–374
traditional uses of
fragrant deodorants, 374
kidney function, 374
Commiphora wightii, 321. *See also* Guggulu
Common balm, 17–18
Computed tomography, 177–178
Congophilic amyloid angiopathy, 68
Convolvulus pluricaulis, 19, 410–411, 421
Convolvulus prostratus, 409–410
anti-Alzheimer's disease (AD) activity
acetylcholinesterase (AChE), 412
amyloid β (Aβ) inhibition, 412
cognitive function, 414–416
lipoxygenase (LOX), 412
neuroprotection, 412–414
bioactive compounds of, 411, 411f
central nervous system (CNS), 420–421
alcohol addiction, 419–420
anxiety, 417–418
depression, 416–417
epilepsy (seizure disorder), 418
Huntington's disease, 418
stimulation, 418–419
stress, 417
oxidation and oxidative stress, 420
toxicity studies, 421
Corydalis yanhusuo, 171–172
Curcuma longa, 11–12, 14–16
Curcumin, 14, 39
bicactivities of, 49–50, 51f
chemical structure of, 49–50, 50f
clinical studies/trials, 53–54, 55t
impacts of, 54–57
phytomedicinal properties of, 52–53
treatment of, 57–58
Curcuminoids, 14–16
Cyclin-dependent kinase 5 (CDK5), 413–414
Cyclooxygenase-2 (COX-2), 170, 365–366
Cyclosporine, 275
CYP3A4, 111
Cytochrome P-450 isozyme, 86, 99

D

Dasatinib, 111
Dementia, 1, 11, 50, 67–68, 109, 147, 176
acetoxyvalerenic acid (AVA), 120
capric triglyceride, 141
Celastrus paniculatus, 431
galantamine, 83
Galanthus nivalis, 304–305
Ginkgo biloba, 199–200, 204
rivastigmine, 93
valerenic acid (VA), 120
Demethoxyaspidospermine, 464
Demethoxycurcumin, 39

Depression, 416–417
Diabetes, 1
Diazepam, 417
Diuretics, 275
Docosahexaenoic acid (DHA), 58, 66, 66f, 69–70
Donepezil, 53, 87–88, 94–95, 260
Dopamine receptor (DR-2), 70
Drosophila melanogaster, 413–414

E

Eicosapentaenoic acid (EPA), 66, 66f
Elevated plus-maze (EPM), 415–416
Endoplasmic reticulum (ER), 54–56, 149
Environmental risk factors, 1, 2f
Epilepsy (seizure disorder), 418

F

Fatty acids, 139, 392–393
  essential, 65–66
  nonessential, 65–66
  saturated, 65
  unsaturated, 65
Flavonoids, 206, 360–363, 428–429, 453–454, 456–457
  Ginkgo biloba, 12–14
  guggulu, 321–322
  health properties, 185
  Valeriana officinalis, 283
5-Fluorouracil (5-FU), 443
Forced swimming test (FST), 322–323
Fractionated coconut oil, 139
Fumaria indica, 171
Fumaria schleicheri, 172

G

Galangin, 456–457
Galantamine, 53, 94–95, 308
  adverse drug reactions, 88–89
  biological sources, 84
  chemistry of, 84, 85f
  cholinesterase inhibitors, 87–88
  neurodegenerative diseases, 83
  pharmacology of, 84
    absorption, 85
    acetylcholinesterase (AChE) inhibition, 85
    distribution, 85–86
    metabolism, 86
    nicotine acetylcholine receptors (nACHR) modulation, 85
    renal excretion, 86
  plant sources, 83
  toxicology, 88–89
  in vitro production of, 84
Galantamine hydrobromide, 83
Galanthus nivalis, 301, 303f, 304–309, 309f
  botany of, 302–303
  clinical studies, 309–311
  distribution and habitat, 301–302
  phytochemistry, 303–304
  structure–activity relationship, 303–304
Galanthus woronowii, 307
Gamma-aminobutyric acid (GABA), 247, 285–286, 419–420
Genetic risk factors, 1, 2f
Ginkgo biloba, 12–14, 167–168
  amyloid-β (Aβ) protein, 207–208
  amyloidogenesis, 207–208
  antiapoptotic effect, 205–207
  antiinflammatory effect, 207
  antioxidant activity, 204–205
  broad-spectrum analysis, 202, 203f
  chemical structures, 200–201, 201f–202f
  classification of, 200–201
  constituents of, 204
  mechanism of, 208–209
  mitochondrial function, 205
  twig of, 201–202, 203f
Ginsenosides, 217–218
Globalization, 22–23
Glucosides, 352–353
Glucosylplumeride, 464
Glutamatergic antagonist, 2
Glutathione peroxidase (GPx), 381–382
Glutathione reductase (GR), 413–414
Glutathion reductase (GSH), 204–205
Glycative stress, 468
Glycoxidative stress, 473
Glycyrrhiza glabra, 18
Gou-teng, 437–438
Guduchi, 16–17, 351. *See also Tinospora cordifolia*
Guggulu, 317–318, 320f
  antioxidants, 319f, 320
  chemistry, 321–322
  clinical studies, 324
  drug interactions, 325–326
  hypercholesterolemia, 321
  mechanism of, 324–325
  preclinical studies, 322–324
  safety issues, 325

H

Haldi, 14
Hemorrhoids, 374
Hepatoprotective activity, 171

High-density lipoprotein (HDL) cholesterol, 467–468
Himatanthus lancifolius (Müll.Arg.) (HL)
dried stem bark of, 463
pharmacological properties, 464–465
phytochemistry, 464, 464f
Homeopathy, 18
Homeostasis, 69–70
Honokiol, 360–363
Human umbilical cord derived mesenchymal stem cells (hUC-MSCs), 40
Huntington's disease, 418
Huperzia serrata, 130, 132f
Huperzine A, 127–128, 128f
bacterial biotransformation products, 132–133, 134f
Ceriporia lacerate HS-ZJUT-C13A biotransformation products, 132–133, 134f
chemistry of, 128
12-*epi*-lycopodine, 132–133, 133f
Huperzia serrata, 130, 132f
synthesis, 130, 131f
ZT-1, 129, 130f
5-Hydroxytryptamine 1A (5-HT_{1A}) receptors, 179
Hypercholesterolemia, 1, 321
Hypericin (HP), 157
pharmacological activities of, 158–161
acetylcholinesterase (AChE), 159–160
β-amyloid peptides, 160–161
antiinflammatory, 161
antioxidant activity, 161
safety issues, 161–162
structures of, 157, 157f
Hypericum perforatum L., 156, 156f
Hyperphosphorylation, 51–52
Hypertension, 1
Hypothalamic–pituitary–adrenal axis activity, 304

I

Indian ginseng, 401
Indian pennywort, 18–19
Indole alkaloids, 464
Inducible nitric oxide synthase (iNOS), 317–318, 365–366
Inflammatory cytokines, 51–52
Insomnia, 121, 221, 286, 431
Intractable epilepsy, 140
Irritable bowel syndrome, 159
2-Isopropyl-5-methylphenol, 375
Isorhynchophylline, 438–439

K

Ketogenesis, 142, 142f
Ketolysis, 142, 142f
Ketone bodies (KBs), 141, 143
Kijitsu, 187
Korean ginseng, 217. *See also Panax ginseng* (PG)

L

Lactic acid dehydrogenase (LDH), 331–332
Lemon balm, 17–18. *See also Melissa officinalis* (MO)
Lepidium meyenii
chemical composition of, 330–331, 331f
clinical studies of, 333
mechanism of, 331–332
pharmacological actions of, 331–332
toxicity profile of, 333
Leucojum aestivum L., 84
α-Linolenic acid, 66, 66f
Lipid peroxidation (LPO), 288
Lipopolysaccharide (LPS), 374, 482–483
Lipoxygenase (LOX), 52–53, 412
Liquorice, 18
Low-density lipoprotein (LDL) cholesterol, 467–468
Lycoris radiata, 84

M

Maca, 330–331. *See also Lepidium meyenii*
Magnolia officinalis (MO), 359–360
bioactive compounds, memory enhancement, 359–360, 361t–363t
compounds isolation, 365, 365f
extract and components, 367
neuroprotective effects, 363–367
root barks and stems, 360
secondary metabolites (SMs), 363–367
Magnolol, 360–363
Malkangni, 425–426
Mandragora officinarum, 117
Mandukaparni, 18–19, 257–258. *See also Centella asiatica* (CA)
Matricaria recutita L., 293–296, 297f
clinical studies, 296
preclinical studies, 295–296, 295f
side effects of, 296
Matrix metalloproteinase (MMP), 331–332
MCI. *See* Moderate cognitive impairment (MCI)

Medicinal plants, 3, 3t–6t, 11, 15t, 49–50, 109, 155, 225. *See also specific plants*
  Bacopa monnieri, 20–21
  Centella asiatica (CA), 18–19
  chamomile, 294
  Convolvulus pluricaulis, 19
  Curcuma longa, 14–16
  dose administration, 22
  drug discovery, traditional knowledge, 21, 22f
  Ginkgo biloba, 12–14
  globalization, 22–23
  Glycyrrhiza glabra, 18
  mass cultivation, 23
  Melissa officinalis (MO), 17–18
  rosemary, 271
  Salvia officinalis, 16
  Tinospora cordifolia, 16–17
  traditional medicine system, 12, 13t–14t
  Withania somnifera, 20
Mediterranean-type diet, 57–58
Medium chain saturated fatty acid (MCFA), 139
Medium chain triglyceride (MCT), 139, 143
Melissa officinalis (MO), 17–18, 225–226
  cultivation of, 227
  mechanism of, 232f
    acetylcholinesterase (AChE), 231–233
    amyloid-β (Aβ) protein, 233–234
    antiinflammatory activity, 235
    antioxidant activity of, 234–235
  phytochemical profile, 227–228
  safety issues, 235–236
  taxonomy of, 226, 226f
  traditional uses, 227–228
  treatment of
    acetylcholinesterase inhibitors (AChEI), 228–229
    N-methyl-D-aspartate (NMDA) receptors, 229–230, 230t
Memantine, 178, 306, 430–431
4-*O*-Methylhonokiol, 360–363, 365
Mitochondria stress, 149
Mitogen-activated protein kinases (MAPKs), 188–189
Moderate cognitive impairment (MCI), 199–200
Monounsaturated fatty acids (MUFAs), 65, 380
Mucuna pruriens, 355

N

Naphthodianthrones, 157
Narcissus aestivum, 84
Narcissus confusus, 84
Nardostachys chinensis, 286
Nelumbo nucifera, 467–468
  amyloid accumulation, 473
  antiglycation treatment, 470–473
  antioxidants, 470–473
  cognitive impairment, 473
    in vitro, 474
    in vivo, 474–475
  constituents of, 470, 471f
  neuropathology, 475–476
Neurofibrillary tangles (NFTs), 318
  cholinergic neurons, 94
  curcumin, 51–52, 57
Nitric oxide (NO), 170, 317–318, 442
  curcumin, 58
  inhibitory effect, 482–483
Nivalin, 303
N-methyl-D-aspartate (NMDA) receptors, 229–230, 230t
Nobiletin, 186f
  chemistry of, 187
  human studies, 191–192
  metabolism and distribution, 187–188
  pharmacological effects of
    amyloid-β (Aβ) protein, 190–191, 192f
    antineuroinflammatory activities, 188–189
    cholinesterase inhibitory activities, 189
    cognitive impairment, 190
    neuroprotective activities, 188–189
    neuroprotective effect, 189
  sources of, 187
  structure of, 185, 186f
Non-steroidal anti-inflammatory drugs (NSAIDs), 39–40
Norepinephrine (NE), 121
Nutritional deficiency, 467–468

O

Obovatol, 360–363
Obsessive–compulsive disorder (OCD), 159
Ocimum sanctum, 354
Oleanolic acid types, 217–218
Omega 6 fatty acids, 66–68
Omega 3 polyunsaturated fatty acids (PUFAs), 66
  biological properties of, 69–70, 70f
  clinical trials, 71, 72t–77t

cognition, 67–68
dietary sources, 66–67, 67t
dopaminergic neurodevelopmental functioning, 70
importance of, 71
neuronal functions, 67–68
Oxidative stress (OS), 11, 148, 170, 380, 468
nobiletin, 189
Urtica diocia, 393–395

P

Palmatine, 352–353
Panax ginseng (PG), 167–168, 217
bioactive constituents, 217–218
distribution of, 217
dosage of, 220
mechanism of, 218–219
pharmacological studies, 219–220
side effects of, 221
species, 217
systematic classification, 217, 218t
taxonomy of, 217
Parkinson's disease (PD), 95, 243
Pediomelum cuspidatum, 34
Pentylenetetrazole (PTZ), 122
Perkin's method, 169
Phenolics, 2
Phlegmariurus squarrosus, 130
Phyllanthus emblica, 354
Physostigma venenosum, 95
Physostigmine, 306
Phytohemagglutinin (PHA), 464–465
Platelet-activating factor (PAF), 209
Poliomyelitis, 83
Polygonum cuspidatum, 33
Polymethoxyflavonoids (PMFs), 185. *See also* Nobiletin
Polyphenols, 185
Polyunsaturated fatty acids (PUFAs), 65. *See also* Omega 3 polyunsaturated fatty acids (PUFAs)
Brazil nuts, 380
types of, 67
Positron emission tomography (PET), 187–188
Premenstrual syndrome (PMS), 159
Presenilin-1, 94
Presenilin-2, 94
Proanthocyanidins, 443
Prostaglandin E2, 170
Protopanaxadiol (PPD), 217–218
Protopanaxatriol (PPT), 217–218
Protopine
anticancer activity, 171–172
antioxidant activity, 171–172
biological activities of, 170–171
hepatoprotective activity, 171
plant species, 168, 168t
synthesis of, 169, 169f
Protopine hydrochloride, 169
Pseudohypericin, 157, 157f
PUFAs. *See* Polyunsaturated fatty acids (PUFAs)
Purnella vulgaris, 200–201

Q

Quercetin (Que), 112f
anti-Alzheimer's disease (AD) activity of
clinical studies, 111
preclinical studies, 110–111
chemical structure of, 110, 110f
side effects, 111
toxicological profile, 111

R

Razadine, 83
Reactive nitrogen species (RNS), 317–318
Reactive oxygen species (ROS), 51–52, 148–149
Ginkgo biloba, 204–205
Lepidium meyenii, 331–332
Matricaria recutita L., 295
Nelumbo nucifera, 467–468, 473
nobiletin, 189
Panax ginseng (PG), 219–220
Urtica diocia, 393–395
Zingiber officinale, 485–487
Receptor for advanced glycation end product (RAGE), 468
Reminyl, 83
Renal excretion
galantamine, 86
resveratrol, 36
Resveratrol
adverse drug reactions, 41
antiaging effects, 33
cis and *trans* forms of, 36, 36f
clinical trials, 41–42
human umbilical cord derived mesenchymal stem cells (hUC-MSCs), 40
memory performance, 38
molecular structure of, 36, 36f
neuroprotective effect, 38

Resveratrol (*Continued*)
non-steroidal anti-inflammatory drugs (NSAIDs), 39–40
pharmacokinetics of
absorption, 34
distribution, 34
excretion, 36
metabolism, 34–36
semisynthetic prenylated derivatives, 39
sirtuins, 37
sources of, 33
natural, 34, 35t
stilbene synthase, 33
toll-like receptors (TLRs), 37
toxicology of, 41
Reversible acetylcholinesterase (ACh) inhibitor, 83
Reversible dementias, 304–305
Rhynchophylline, 438
Richweed, 373–374
Rivastigmine, 53, 87–88, 94–95
chemistry of, 96
drug interactions, 99–100
hybrids and actions, 100, 101f
inhalation, 103
injections, 102
intranasal route, 101–102
mechanism of, 98–99
oral bioavailability, 103
pharmacodynamics of, 99
pharmacokinetics of, 99
pharmacophore of, 95
sources of, 96, 96f–97f
synthesis of, 97, 97f
tolerability, 103
transdermal, 103
ROS. *See* Reactive oxygen species (ROS)
Rosemary, 271
bioactive compounds, 271, 272f
central nervous system (CNS), 276
classification of, 271, 272t
drug interactions, 273–275
side effects, 276, 277f

S

Salicylates, 275
Salmonella typhimurium, 355–356
Salvia officinalis, 16
Saponification, 139–140
Saturated fatty acids (SFAs), 65
Brazil nuts, 380
Scopolamine, 331–332
Seasonal affective disorder (SAD), 159
Secondary metabolites, 2–3, 147, 271
Secondary metabolites (SMs), 2–3, 147, 271, 293–294, 363–367
β-Secretase, 219–220
β-Secretase cleaving enzyme I *(BACE1)*, 359
Selenium, 380–382
Serotonin (5-HT), 70
Shankhpushpi, 19, 411, 421–422
Short-term memory loss, 50
Siddha, 11, 18
Simultaneous distillation-extraction (SDE) process, 339–341
Sirtuin1 (Sirt1), 33
activation of, 38
expression of, 37
Snowdrops. *See Galanthus nivalis*
Somatization disorder, 159
Spinosin, 178–180
Sporadic Alzheimer's disease (AD), 94
Sporadic dementia of the Alzheimer's type (SDAT), 56–57
St. John's Wort (SJW), 157
Steam hydrolysis, 139–140
Steroids, 352–353
Stilbene synthase, 33
Stinging nettle, 392–393
Stone root, 373–374
Superoxide dismutase (SOD), 204–205, 413–414
Sweet flag, 337
Synthetic drugs, 49

T

Tacrine, 260
Tail suspension test (TST), 322–323
Taxus baccata, 200–201
Terpenoids, 2
Thiobarbituric acid-reactive substances (TBARS), 235
Thymol, 375
Tinospora cordifolia, 16–17, 352f
adverse effects, 355–356
biologic activities, 353
chemical constituents, 352–353
classification of, 351
distribution of, 351
medicinal uses, 353
memory disorders, 353–355
side effects of, 355–356
Toll-like receptors (TLRs), 37
Touhi, 187

Traditional Chinese medicine (TCM), 11–12
Traditional medicinal systems, 12, 13t–14t, 21–22, 49
Transfer latency (TL), 414–415
Triglycerides, 65
3,4',5-Trihydroxystilbene, 33. *See also* Resveratrol
Trisphaeridine, 304
Turmeric, 14, 49–50. *See also* Curcumin

U

Uleine-rich fraction (URF), 464–465, 464f
Unani, 11, 18
Uncaria rhynchophylla (UR)
central nervous system (CNS), 444
anticonvulsion, 444–445
anti-Parkinson's disease (PD), 445
dried stem and hook of, 438
pharmacological activity of
antiangiogenic properties, 443
anticancer properties, 443
antiinflammatory properties, 441–442
antioxidant properties, 441
antiviral activity, 444
hypotensive and cardioprotective effects, 442
phytochemistry, 438–441, 439f
Ungernia victoris, 84
Unsaturated fatty acids, 65
Urinary tract infections, 374
Urtica diocia, 392
chemical constituents, 392–393
inflammatory cascades, 395–396
mechanism of, 392–393
in natural habitat, 390–391, 391f
oxidative stress (OS), 393–395
toxicologic studies, 396–397

V

Valerenic acid (VA), 117–118. *See also* Acetoxyvalerenic acid (AVA)
anesthetic action, 120
anticonvulsant, 122
antiinflammatory, 121
brain-derived neurotrophic factor (BDNF), 119
chemical structure of, 118, 118f
clinical studies, 122, 123t
CYP3A4-mediated metabolism, 122
dementia, 120
gastrointestinal (GI) activity, 122
insomnia, 121
occurrence of, 118
physical and psychological stress, 121
Valerian, 283
Valeriana officinalis, 284f
clinical studies, 288–289
mechanism of, 286, 287f
medicinal uses of, 285–286
natural chemicals, 283–284, 284t, 285f, 287f
safety issues, 289
Vascular smooth muscle cells (VSMCs), 442
Veratrum grandiflorum, 33–34
Vitamin B, 467–468
Vitamin C, 373, 467–468, 474–475
Vitamin E, 260, 373, 467–468
Vitis vinifera, 34
Volatile organic compounds (VOCs), 339–341

W

Winter cherry, 401
Withaferin A, 401–402, 404
Withania somnifera, 11–12, 20, 167–168, 200–201, 355
pharmacokinetics, 401–402
toxicological studies, 404
Withanolide A, 403
Withanoside IV, 403

Z

Zingiber officinale, 482f
chemical composition, 482
clinical studies, 487–492
cultivation of, 481
distribution of, 481
mode of action, 483, 484f
rhizomes of, 481
toxicological studies, 488t–491t, 492
in vitro and in vivo studies of, 485–487